Lecture Notes in Physics

Springer
Berlin
Heidelberg
New York
Barcelona
Hong Kong
London
Milan
Paris
Singapore
Tokyo

Physics and Astronomy

ONLINE LIBRARY

http://www.springer.de/phys/

The Editorial Policy for Proceedings

The series Lecture Notes in Physics reports new developments in physical research and teaching – quickly, informally, and at a high level. The proceedings to be considered for publication in this series should be limited to only a few areas of research, and these should be closely related to each other. The contributions should be of a high standard and should avoid lengthy redraftings of papers already published or about to be published elsewhere. As a whole, the proceedings should aim for a balanced presentation of the theme of the conference including a description of the techniques used and enough motivation for a broad readership. It should not be assumed that the published proceedings must reflect the conference in its entirety. (A listing or abstracts of papers presented at the meeting but not included in the proceedings could be added as an appendix.)

When applying for publication in the series Lecture Notes in Physics the volume's editor(s) should submit sufficient material to enable the series editors and their referees to make a fairly accurate evaluation (e.g. a complete list of speakers and titles of papers to be presented and abstracts). If, based on this information, the proceedings are (tentatively) accepted, the volume's editor(s), whose name(s) will appear on the title pages, should select the papers suitable for publication and have them refereed (as for a journal) when appropriate. As a rule discussions will not be accepted. The series editors and Springer-Verlag will normally not interfere with the detailed editing except in fairly obvious cases or on technical matters.

Final acceptance is expressed by the series editor in charge, in consultation with Springer-Verlag only after receiving the complete manuscript. It might help to send a copy of the authors' manuscripts in advance to the editor in charge to discuss possible revisions with him. As a general rule, the series editor will confirm his tentative acceptance if the final manuscript corresponds to the original concept discussed, if the quality of the contribution meets the requirements of the series, and if the final size of the manuscript does not greatly exceed the number of pages originally agreed upon. The manuscript should be forwarded to Springer-Verlag shortly after the meeting. In cases of extreme delay (more than six months after the conference) the series editors will check once more the timeliness of the papers. Therefore, the volume's editor(s) should establish strict deadlines, or collect the articles during the conference and have them revised on the spot. If a delay is unavoidable, one should encourage the authors to update their contributions if appropriate. The editors of proceedings are strongly advised to inform contributors about these points at an early stage.

The final manuscript should contain a table of contents and an informative introduction accessible also to readers not particularly familiar with the topic of the conference. The contributions should be in English. The volume's editor(s) should check the contributions for the correct use of language. At Springer-Verlag only the prefaces will be checked by a copy-editor for language and style. Grave linguistic or technical shortcomings may lead to the rejection of contributions by the series editors. A conference report should not exceed a total of 500 pages. Keeping the size within this bound should be achieved by a stricter selection of articles and not by imposing an upper limit to the length of the individual papers. Editors receive jointly 30 complimentary copies of their book. They are entitled to purchase further copies of their book at a reduced rate. As a rule no reprints of individual contributions can be supplied. No royalty is paid on Lecture Notes in Physics volumes. Commitment to publish is made by letter of interest rather than by signing a formal contract. Springer-Verlag secures the copyright for each volume.

The Production Process

The books are hardbound, and the publisher will select quality paper appropriate to the needs of the author(s). Publication time is about ten weeks. More than twenty years of experience guarantee authors the best possible service. To reach the goal of rapid publication at a low price the technique of photographic reproduction from a camera-ready manuscript was chosen. This process shifts the main responsibility for the technical quality considerably from the publisher to the authors. We therefore urge all authors and editors of proceedings to observe very carefully the essentials for the preparation of camera-ready manuscripts, which we will supply on request. This applies especially to the quality of figures and halftones submitted for publication. In addition, it might be useful to look at some of the volumes already published. As a special service, we offer free of charge LATEX and TEX macro packages to format the text according to Springer-Verlag's quality requirements. We strongly recommend that you make use of this offer, since the result will be a book of considerably improved technical quality. To avoid mistakes and time-consuming correspondence during the production period the conference editors should request special instructions from the publisher well before the beginning of the conference. Manuscripts not meeting the technical standard of the series will have to be returned for improvement.

For further information please contact Springer-Verlag, Physics Editorial Department II, Tiergartenstrasse 17, D-69121 Heidelberg, Germany

Series homepage – http://www.springer.de/phys/books/lnpp

Zhangxin Chen Richard E. Ewing
Zhong-Ci Shi (Eds.)

Numerical Treatment of Multiphase Flows in Porous Media

Proceedings of the International Workshop
Held at Beijing, China, 2-6 August 1999

 Springer

Editors

Zhangxin Chen
Department of Mathematics
Southern Methodist University
Box 750156
Dallas, TX 75275-0156, USA

Richard E. Ewing
Institute for Scientific Computation
Texas A&M University
College Station, TX 77843-3404, USA

Zhong-Ci Shi
Institute of Computational Mathematics
Chinese Academy of Sciences
P.O. Box 2719
Beijing 100080, P.R. China

Library of Congress Cataloging-in-Publication Data.

Die Deutsche Bibliothek - CIP-Einheitsaufnahme
Numerical treatment of multiphase flows in porous media : proceedings
of the international workshop, held at Beijing, China, 2 - 6 August
1999 / Zhangxin Chen ... (ed.). - Berlin ; Heidelberg ; New York ;
Barcelona ; Hong Kong ; London ; Milan ; Paris ; Singapore ; Tokyo :Springer, 2000
  (Lecture notes in physics ; Vol. 552)
  (Physics and astronomy online library)
  ISBN 3-540-67566-3

ISSN 0075-8450
ISBN 3-540-67566-3 Springer-Verlag Berlin Heidelberg New York

Springer-Verlag Berlin Heidelberg New York
a member of BertelsmannSpringer Science+Business Media GmbH

© Springer-Verlag Berlin Heidelberg 2000
Printed in Germany

The use of general descriptive names, registered names, trademarks, etc. in this publication does not imply, even in the absence of a specific statement, that such names are exempt from the relevant protective laws and regulations and therefore free for general use.

Typesetting: Camera-ready by the authors/editors
Cover design: *design & production*, Heidelberg

Printed on acid-free paper
SPIN: 10770796   55/3144/du - 5 4 3 2 1 0

Preface

The need to predict, understand, and optimize complex physical and chemical processes occurring in and around the earth, such as groundwater contamination, oil reservoir production, discovering new oil reserves, and ocean hydrodynamics, has been increasingly recognized. Despite their seemingly disparate natures, these geoscience problems have many common mathematical and computational characteristics. The techniques used to describe and study them are applicable across a broad range of areas.

The study of the above problems through physical experiments, mathematical theory, and computational techniques requires interdisciplinary collaboration between engineers, mathematicians, computational scientists, and other researchers working in industry, government laboratories, and universities. By bringing together such researchers, meaningful progress can be made in predicting, understanding, and optimizing physical and chemical processes.

The International Workshop on Fluid Flow and Transport in Porous Media was successfully held in Beijing, China, August 2–6, 1999. The aim of this workshop was to bring together applied mathematicians, computational scientists, and engineers working actively in the mathematical and numerical treatment of fluid flow and transport in porous media. A broad range of researchers presented papers and discussed both problems and current, state-of-the-art techniques.

Over seventy people from Australia, Bulgaria, China, France, Ghana, Germany, Norway, Russia, South Korea, Spain, Taiwan, and the United States of America attended this workshop and more than fifty papers were presented on a variety of subjects in mathematical theory, numerical methods, parallel computation, optimization, surface water and ocean modeling, chemically reactive phenomena, atmospheric modeling, multiscale phenomena, and media characterization.

This book contains thirty-eight selected papers presented at the workshop. They cover recent advances and developments of a wide range of numerical issues of multiphase fluid flow and transport in porous media. The porous media considered range from ordinary media to fractured and deformable ones. The physical and mathematical models treated involve a variety of flows from single phase compressible flow to multiphase, multicomponent flow with mass interchange between phases. The numerical methods studied range from standard finite difference and finite element methods to nonstandard mixed finite element and characteristics-based techniques. The computational algorithms developed utilize classical fast iterative solvers and modern multigrid and domain decomposition approaches combined with local grid refinement techniques. The characteristics-based techniques for advection-dominated flow and transport processes are emphasized in this book; the classical modified methods of characteristics to newly developed locally conservative Eulerian–Lagrangian methods are addressed.

Financial support for the workshop was generously provided by the National Natural Science Foundation of China, the Beijing Institute of Applied Physics and Computational Mathematics, the Chinese State Key Basic Research Project, the US Army Research Office-Far East, the Office of Naval Research International Field Office Asia, and the US Air Force Asian Office of Aerospace Research and Development. We also thank the local organizers at the International Center of Computational Physics, Beijing, China, for their hard work and hospitality.

Zhangxin Chen, Richard E. Ewing, and Zhong-Ci Shi

May 29, 2000

Contents

List of Contributors

Clarisse Alboin
INRIA-Rocquencourt, BP 105
78153 Le Chesnay Cedex, France
Clarisse.Alboin@inria.fr

Mohamed Al-Lawatia
Department of Mathematics and Statistics
Sultan Qaboos University, P.O. Box 36
Al-Khod Postal Code 123, Muscat, Sultanate of Oman
allawati@squ.edu.om

Todd Arbogast
Department of Mathematics; C1200
The University of Texas at Austin
Austin, TX 78712, U.S.A.
arbogast@math.utexas.edu

Peter Bastian
IWR, Universität Heidelberg, Im Neuenheimer Feld 368
69120 Heidelberg, Germany
Peter.Bastian@iwr.uni-heidelberg.de

Hongsen Chen
Department of Mathematics
University of Wyoming
Laramie, Wyoming, U.S.A.
hchen@uwyo.edu

Zhangxin Chen
Department of Mathematics
Southern Methodist University, Box 750156
Dallas, TX 75275-0156, U.S.A.
zchen@mail.smu.edu

Aijie Cheng
Department of Mathematics

Shandong University
Jinan 250100, P. R. China
ajcheng@math.sdu.edu.cn

Tatiana Chernogorova
Faculty of Mathematics and Informatics
The Sofia University St. Kl. Ohridski
5 J.Bouchier bul., BG-1163 Sofia, Bulgaria
chernogorova@fmi.uni-sofia.bg

Xia Cui
Laboratory of Computational Physics
Institute of Applied Physics and Computational Mathematics
P.O. Box 8009-26, Beijing 100088, P. R. China
lucycui@yahoo.com

Ronald A. DeVore
Department of Mathematics
University of South Carolina, Columbia
South Carolina 29208, U.S.A.
devore@math.sc.edu

Craig C. Douglas
Department of Computer Science
University of Kentucky, 325 McVey Hall - CCS
Lexington, KY 40506-0045, U.S.A.
douglas@ccs.uky.edu

Jim Douglas, Jr.
Department of Mathematics
Purdue University, West Lafayette
IN 47907-1395, U.S.A.
douglas@math.math.purdue.edu

Magne S. Espedal
Department of Mathematics
University of Bergen
Bergen, Norway
magne.espedal@mi.uib.no

Richard E. Ewing
Institute for Scientific Computation
Texas A&M University, College Station
Texas 77843-3404, U.S.A.
ewing@isc.tamu.edu

Tigui Fan
Department of Physics
Chifeng Teachers' College
Chifeng, 024000, P. R. China

Gunnar E. Fladmark
Department of Mathematics
University of Bergen
Bergen, Norway
gunnar.fladmark@nho.hydro.com

Vladimir Garanzha
Computing Center, Russian Academy of Sciences
40, Vavilov Str., 117967, Moscow, RUSSIA
garan@ccas.ru

Jesús García-Lafuente
E.T.S. Ingenieros Informáticos
Departmento de Física Aplicada II
Universidad de Málaga
Campus del Teatinos, E-29081, Málaga, Spain
glafuente@ctima.uma.es

Ian Gladwell
Department of Mathematics
Southern Methodist University
Dallas, Texas 75275, U.S.A.
gladwell@seas.smu.edu

Rainer Helmig
CAB, Universität Braunschweig, Pockelsstrasse 3
38106 Braunschweig, Germany
r.helmig@tu-bs.de

Norbert Herrmann
Inst. f. Angew. Mathematik
Univ. Hannover, Welfengarten 1
D-30167 Hannover, Germany
herrmann@ifam.uni-hannover.de

Jonathan Hu
Department of Mathematics
University of Kentucky, 715 Patterson Office Tower
Lexington, KY 40506-0027, U.S.A.
jhu@ms.uky.edu

Chieh-Sen Huang
Department of Applied Mathematics
National Sun Yat-sen University
Kaohsiung 80424, Taiwan, R.O.C.
huangcs@math.nsysu.edu.tw

Oleg Iliev
Institute for Industrial Mathematics (ITWM)
Erwin-Schroedinger str., bld. 49
D-67663 Kaiserslautern, Germany
iliev@itwm.uni-kl.de

Mohamed Iskandarani
Institute of Marine and Coastal Sciences
Rutgers University, P.O. Box 231
New Brunswick, NJ 08903-0231, U.S.A.
mohamed@ahab.rutgers.edu

Jérôme Jaffré
INRIA-Rocquencourt, BP 105
78153 Le Chesnay Cedex, France
Jerome.Jaffre@inria.fr

Hartmut Jakobs
CAB, Universität Braunschweig, Pockelsstrasse 3
38106 Braunschweig, Germany
h.jakobs@tu-bs.de

Vladimir Konshin
Computing Center, Russian Academy of Sciences
40, Vavilov Str., 117967, Moscow, RUSSIA
konshin@ccas.ru

Markus Kowarschik
Lehrstuhl für Systemsimulation (IMMD 10)
Institut für Informatik
Universität Erlangen-Nürnberg
Martensstrasse 3, D-91058 Erlangen, Germany
kowarschik@informatik.uni-erlangen.de

Raytcho Lazarov
Department of Mathematics
Texas A&M University, College Station
TX 77843-3368, U.S.A.
lazarov@math.tamu.edu

Xiuren Lei
Department of Applied mathematics
South China University of Technology
510640, Guangzhou, P. R. China
mahpeng@scut.edu.cn

Baoyan Li
Department of Mathematics
Southern Methodist University, Box 750156
Dallas, TX 75275-0156, U.S.A.
bli@mail.smu.edu

Xianggui Li
Department of Mathematics
University of Petroleum
Dongying, P. R. China
maths@hdpu.edu.cn

Yuanxiang Li
Laboratory of Software Engineering
Wuhan University, 430072
Wuhan, P. R. China
yxli@whu.edu.cn

Dong Liang
School of Mathematics and System Sciences
Shandong University, Jinan
Shandong 250100, P. R. China
dliang@math.sdu.edu.cn

Jiangguo Liu
Department of Mathematics
University of South Carolina, Columbia
South Carolina 29208, U.S.A.
jliu0@math.sc.edu

Stephen L. Lyons
Upstream Strategic Research Center
Mobil Technology Company
Dallas, Texas, U.S.A.
steve_l_lyons@email.mobil.com

Yuanle Ma
Institute of Nuclear Energy Technology
Tsinghua University, Beijing 100084, P. R. China

Zhibo Ma
Laboratory of Computational Physics
Institute of Applied Physics and Computational Mathematics
P.O. Box 8009-26, Beijing 100088, P. R. China
mazhibo@hotmail.com

Pingbing Ming
Institute of Computational Mathematics
Chinese Academy of Sciences
PO Box 2719, Beijing, P. R. China
mpb@lsec.cc.ac.cn

Jan-Erik Nordtvedt
Department of Physics
University of Bergen
Allegt.55, N-5008 Bergen, Norway
jan-erik.nordtvedt@rf.no

Geir Åge Øye
Department of Mathematics
University of Bergen
Bergen, Norway
geir.oye@mi.uib.no

Dimitrios Papavassiliou
School of Chemical Engineering and Materials Science
The University of Oklahoma
Norman, Oklahoma 73019
dvpapava@ou.edu

Hong Peng
Department of Computer Science
South China University of Technology
510640, Guangzhou, P. R. China
mathpeng@scut.edu.cn

Felipe Pereira
Instituto Politécnico da
Universidade do Estado do Rio de Janeiro
Nova Friburgo, RJ, Brazil 28601-970
pereira@brie.iprj.uerj.br

Guan Qin
Upstream Strategic Research Center
Mobil Technology Company

Dallas, Texas, U.S.A.
guan_qin@email.mobil.com

Volker Reichenberger
IWR, Universität Heidelberg, Im Neuenheimer Feld 368
69120 Heidelberg, Germany
Volker.Reichenberger@iwr.uni-heidelberg.de

Hilde Reme
Department of Mathematics
University of Bergen
Bergen, Norway
hilde.reme@mi.uib.no

Jean E. Roberts
INRIA-Rocquencourt, BP 105
78153 Le Chesnay Cedex, France
Jean.Roberts@inria.fr

Ulrich Rüde
Lehrstuhl für Systemsimulation (IMMD 10)
Institut für Informatik
Universität Erlangen-Nürnberg
Martensstrasse 3, D-91058 Erlangen, Germany
ruede@informatik.uni-erlangen.de

Thomas F. Russell
Department of Mathematics
University of Colorado at Denver
P.O. Box 173364, Campus Box 170
Denver, CO 80217-3364, U.S.A.
thomas.russell@cudenver.edu

Christophe Serres
IPSN/DES/SESID, BP 6
92265 Fontenay Aux Roses Cedex, France
Christophe.Serres@ipsn.cea.fr

Dongwoo Sheen
Department of Mathematics
Seoul National University
Seoul 151-742, Korea
sheen@math.snu.ac.kr

Zhong-Ci Shi
Institute of Computational Mathematics

Chinese Academy of Sciences
PO Box 2719, Beijing, P. R. China
shi@lsec.cc.ac.cn

Anna M. Spagnuolo
Department of Mathematics and
Institute for Scientific Computation
Texas A&M University, College Station
Texas 77843-3368, U.S.A.
annas@math.tamu.edu

Sam Subbey
Department of Physics
University of Bergen
Allegt.55, N-5008 Bergen, Norway
sam.subbey@rf.no

Jiachang Sun
Institute of Software
Chinese Academy of Sciences
Beijing, P. R. China
zhy@mail.rdcps.ac.cn

Tong Sun
Institute for Scientific Computation
Texas A&M University
College Station, Texas 77843, U.S.A.
tsun@math.tamu.edu

Francisco R. Villatoro
E.T.S. Ingenieros Industriales
Departmento de Lenguajes y Ciencias de la Computación
Universidad de Málaga, Campus del Ejido, E-29013
Málaga, Spain
villa@lcc.uma.es

Gaohong Wang
Department of Mathematics
Shandong University
Jinan 250100, P. R. China
ajcheng@math.sdu.edu.cn

Hong Wang
Department of Mathematics
University of South Carolina, Columbia

South Carolina 29208, U.S.A.
hwang@math.sc.edu

Xuewen Wang
INRIA-Rocquencourt, BP 105
78153 Le Chesnay Cedex, France
Xuewen.Wang.Alboin@inria.fr

Yan-Fei Wang
Department of Applied Math. & Physics
Hebei University of Technology
Tianjin 300130, P. R. China
maph@hebut.edu.cn

Ziting Wang
Department of Mathematics
University of Petroleum
Dongying, P. R. China
Wangzt@sunctr.hdpu.edu.cn

Christian Weiss
Lehrstuhl für Rechnertechnik und Rechnerorganisation (LRR-TUM)
Institut für Informatik
Technische Universität München
D-80290 München, Germany
weissc@in.tum.de

Yonghong Wu
School of Mathematics and Statistics
Curtin University of Technology
Perth, WA 6845, Australia
yhwu@cs.curtin.edu.au

Ting-Yan Xiao
Department of Applied Math. & Physics
Hebei University of Tecnology
Tianjin 300130, P. R. China
maph@hebut.edu.cn

Shengwu Xiong
Laboratory of Software Engineering
Wuhan University, 430072
Wuhan, P. R. China
yxli@whu.edu.cn

Hong Xu
INRS-Telecommunications
16 Place du Commerce le-des-Soeurs (Verdun)
Quebec, Canada H3E 1H6
xuhong@inrs-telecom.uquebec.ca

Xuejun Xu
Institute of Computational Mathematics
Chinese Academy of Sciences
P.O.Box 2719, Beijing 100080, P. R. China
xxj@lsec.cc.ac.cn

David P. Yale
Upstream Strategic Research Center
Mobil Technology Company
Dallas, Texas, U.S.A.
david_p_yale@email.mobil.com

Danping Yang
Department of Mathematics
University of Shandong
Jinan, P. R. China
dpang@math.sdu.edu.cn

Dequan Yang
Department of Physics
Inner Mongolia National Teachers' College
Tong Liao, 028043 P. R. China

Xinyu Yang
Department of Physics
Inner Mongolia National Teachers' College
Tong Liao, 028043 P. R. China

Li-Ming Yeh
Department of Applied Mathematics
National Chiao-Tung University
Hsinchu, Taiwan, R.O.C.
liming@math.nctu.edu.tw

Xijun Yu
Laboratory of Computational Physics
Institute of Applied Physics and Computational Mathematics
P.O.Box 8009-26, Beijing 100088, P. R. China
y_xijun@hotmail.com

Yirang Yuan
Institute of Mathematics
Shandong University
Jinan, Shandong, 250100, P. R. China
yryuan@math.sdu.edu.cn

Guoyou Zhang
Department of Physics
Inner Mongolia National Teachers' College
Tong Liao, 028043, P. R. China

Huaiyu Zhang
Institute of Software
Chinese Academy of Sciences
Beijing, P. R. China
zhy@mail.rdcps.ac.cn

Wen Zhang
Department of Mathematics and Statistics
Oakland University
Rochester, Michigan 48309, U.S.A.
wzhang@na-net.ornl.gov

Weidong Zhao
Department of Mathematics
Shandong University
Jinan, P. R. China
wdzhao@math.sdu.edu.cn

Zhongsheng Zhao
Department of Physics
Liao Ning University
Shen Yang, 110000 P. R. China

Jianshi Zhu
Laboratory of Computational Physics
Institute of Applied Physics and Computational Mathematics
P.O. Box 8009-26, Beijing 100088, P. R. China
mazhibo@hotmail.com

Xiufen Zou
Wuhan University of Hydraulic and Electric Engineering
Wuhan 430072, P. R. China

Mathematical and Numerical Techniques in Energy and Environmental Modeling

ZHANGXIN CHEN   RICHARD E. EWING

Abstract

Mathematical models have been widely used to predict, understand, and optimize many complex physical processes, from semiconductor or pharmaceutical design to large-scale applications such as global weather models to astrophysics. In particular, simulation of environmental effects of air pollution is extensive. Here we address the need for using similar models to understand the fate and transport of groundwater contaminants and to design in situ remediation strategies.

Three basic problem areas need to be addressed in the modeling and simulation of the flow of groundwater contamination. First, one obtains an effective model to describe the complex fluid/fluid and fluid/rock interactions that control the transport of contaminants in groundwater. This includes the problem of obtaining accurate reservoir descriptions at various length scales and modeling the effects of this heterogeneity in the reservoir simulators. Next, one develops accurate discretization techniques that retain the important physical properties of the continuous models. Finally, one develops efficient numerical solution algorithms that utilize the potential of the emerging computing architectures. We will discuss recent advances and describe the contribution of each of the papers in this book in these three areas.

KEYWORDS: reservoir simulation, mathematical models, partial differential equations, numerical algorithms

1   Introduction

The objective of reservoir simulation is to understand the complex chemical and physical fluid flow processes occurring in an underground reservoir sufficiently well so as to be able to predict the fate and optimize remediation of groundwater contaminants. Toward that end, one must be able to predict the performance of the reservoir under various remediation schemes. To do this, a model of the reservoir and its flow processes must be constructed to yield information about the complex phenomena accompanying different remediation strategies.

There are four major stages to the modeling process. First, a physical model of the flow processes is developed incorporating as much geology, chemistry, and physics as is deemed necessary to describe the essential phenomena. Second, a mathematical formulation of the physical model is obtained, usually involving coupled systems of nonlinear, time-dependent partial differential equations. Third, once the properties of the mathematical model, such as existence, uniqueness, and regularity of the solution, are sufficiently well understood, a discretized numerical model of the mathematical equations is produced. A numerical model is determined that has the required properties of accuracy and stability and that produces solutions representing the basic physical features as well as possible without introducing spurious phenomena associated with the specific numerical scheme. Finally, a computer code capable of efficiently performing the necessary computations for the numerical model is developed. The total modeling process encompasses aspects of each of these four intermediate steps.

The modeling process is not completed with one pass through these four steps. Once a computer code has been developed which gives concrete quantitative results for the total model, this output is compared with corresponding measured observations of the physical process. If the results do not match well, one iterates back through the complete modeling process, changing the various intermediate models in ways to obtain a better correlation between the physical measurements and the computational results. Often, many iterations of this modeling loop are necessary to obtain reasonable models for the highly complex physical phenomena describing contaminant remediation processes.

The trends in reservoir modeling are contained in three broad topics: 1) obtaining better reservoir descriptions and incorporating these descriptions in reservoir simulators, 2) modeling the complex multiphase flow processes and developing accurate discretization schemes for these models, and 3) developing algorithms that can exploit the potential of the emerging computing architectures (particularly the potential parallelism of the parallel/vector architectures). We will briefly discuss these major trends in this paper and describe the contribution of each of the papers in this book in these areas.

2   Reservoir Characterization

The processes of both single and multiphase flow involve convection, or physical transport, of fluids through a heterogeneous reservoir. The equations used to simulate this flow at a macroscopic level are variations of Darcy's law. Darcy's law has been derived via a volume averaging of the Navier-Stokes equations [85], which govern flow through the reservoir at a microscopic or pore-volume level. Reservoirs themselves have scales of heterogeneity ranging from pore-level to field scales. In the standard averaging process for Darcy's law, many important physical phenomena which may eventually govern the macroscopic flow are lost. The further averaging of the reservoirs and fluid properties necessary to use grid blocks of the size of 10–10^2 meters in field-scale simulators further complicates the modeling process. Certain techniques have been developed to address these scaling problems. One of them is discussed in Arbogast's paper in this book on scaling up fine grid information to coarse scales in an approximation to a nonlinear two-phase flow problem in porous media.

3   Model Equations for Reservoir Flow

Although the techniques that we will discuss apply equally well to the recovery of hydrocarbon and the transport of contaminants through the saturated or unsaturated soil zones, we will describe the multiphase flow processes in the terminology of transport of contaminants in groundwater.

The simplest and the most popular is the model of fully saturated incompressible reservoirs. In this case the water (or the liquid) phase occupies the whole pore space and the flow is due to the nonuniform pressure distribution. The mathematical formulation is based on the mass balance equation and Darcy's law (see, e.g., [7, 69])

$$\nabla \cdot (\rho \mathbf{u}) = F \quad \text{and} \quad \mathbf{u} = -\frac{\mathbf{K}}{\mu}(\nabla p - \rho \mathbf{g}) \qquad \text{in } \Omega,\, t > 0, \qquad (3.1)$$

where $\Omega \subset \mathbf{R}^3$ represents a reservoir, $\mathbf{u}$ is the volumetric flux of water, F is the source or sink of fluid, ρ is the fluid density, $\mathbf{K}$ is the absolute permeability tensor, μ is the dynamic fluid viscosity, p is the fluid pressure, and $\mathbf{g}$ is the acceleration vector due to gravity.

Darcy's law provides a relation between the volumetric flux in the mass conservation equation and the pressure in the fluid. This relation is valid for viscous dominated flows which occur at relatively low velocities.

The transport of a contaminant that is dissolved in the water is described by the following equation:

$$\frac{\partial(\theta c)}{\partial t} + \nabla \cdot (\rho \mathbf{u} c) - \nabla \cdot (\theta \mathbf{D} \nabla c) + \beta \theta c = G(c) \qquad \text{in } \Omega, \, t > 0, \qquad (3.2)$$

where c is the concentration of the contaminant, $\mathbf{D}$ is the dispersion tensor, β is the reaction rate, $\theta = \phi \rho$, ϕ is the porosity, and G is the source/sink term. The form of the diffusion/dispersion tensor $\mathbf{D}$ is given by

$$\mathbf{D} = d_m I + |\mathbf{u}| \left[d_l E(\mathbf{u}) + d_t (I - E(\mathbf{u})) \right],$$

where $E_{ij}(\mathbf{u}) = u_i u_j / |\mathbf{u}|^2$, d_m is the molecular diffusion coefficient, and d_l and d_t are the longitudinal and transverse dispersion coefficients, respectively. In general, $d_l \approx 10 d_t$, but this may vary greatly with different soils, fractured reservoir, etc. Also, the viscosity μ in equation (3.1) is assumed to be determined by some mixing rule. In addition to equations (3.1) and (3.2), initial and boundary conditions are specified. The flow at injection and production wells is modeled in equations (3.1) and (3.2) via point or line sources and sinks.

When either an air or vapor phase or a nonaqueous phase liquid contaminant (NAPL) is present, the equations describing two phase, immiscible flow in a horizontal reservoir are given by

$$\begin{aligned} \frac{\partial(\phi \rho_w S_w)}{\partial t} - \nabla \cdot \left(\mathbf{K} \frac{\rho_w k_{rw}}{\mu_w} \nabla(p_w - \rho_w \mathbf{g}) \right) &= q_w \rho_w, \\ \frac{\partial(\phi \rho_a S_a)}{\partial t} - \nabla \cdot \left(\mathbf{K} \frac{\rho_a k_{ra}}{\mu_a} \nabla(p_a - \rho_a \mathbf{g}) \right) &= q_a \rho_a, \end{aligned} \qquad (3.3)$$

where the subscripts w and a refer to the water and air phases, respectively, S_i is the saturation, p_i is the pressure, ρ_i is the density, k_{ri} is the relative permeability, μ_i is the viscosity, and q_i is the external flow rate, each with respect to the ith phase. The saturations sum to unity and thus one of them can be eliminated. The pressure between the two phases is described by the capillary pressure

$$p_c(S_w) = p_a - p_w.$$

Although formally the system in equations (3.1) and (3.2) seems quite different from system (3.3), the latter system may be rearranged in a form which very closely resembles the former system. To use the same basic simulation techniques in our sample computations to treat both miscible and immiscible displacement, we will follow the ideas of Chavent [17].

The global pressure p and total velocity $\mathbf{v}$ formulation of a two-phase water (w) and air (a) flow model is given by the following equations [30]:

$$S_a c_a \frac{dp}{dt} + \nabla \cdot \mathbf{v} = -\frac{\partial \phi(p)}{\partial t} + q(x, S_w, p),$$

$$\mathbf{v} = -\mathbf{K}\lambda(\nabla p - \mathbf{G}_\lambda), \qquad (3.4)$$

$$\phi \frac{\partial S_w}{\partial t} + \nabla \cdot (f_w \mathbf{v} - \mathbf{G}_1 - \mathbf{D}(S_w) \cdot \nabla S_w) = -S_w \frac{\partial \phi(p)}{\partial t} + q_w,$$

where $\mathbf{G}_1 = \mathbf{K}\lambda_a q_w (\rho_a - \rho_w)\mathbf{g}$. The global pressure and total velocity are defined by

$$p = \frac{1}{2}(p_w + p_a) + \frac{1}{2}\int_{S_c}^{S_w} \frac{\lambda_a - \lambda_w}{\lambda} \frac{dp_c}{d\xi} d\xi \qquad \text{and} \qquad \mathbf{v} = \mathbf{v}_w + \mathbf{v}_a,$$

where $p_c(S_c) = 0$. Further, $d/dt \equiv \phi(\partial/\partial t) + \mathbf{v}_a/S_a \cdot \nabla$, $\lambda = \lambda_w + \lambda_a$ is the total mobility, and $\lambda_i = k_{ri}/\mu_i$, $i = w, a$, is the mobility for water and air. The gravity force G_λ and capillary diffusion term $D(S)$ are expressed as

$$\mathbf{G}_\lambda = \frac{\lambda_w \rho_w + \lambda_a \rho_a}{\lambda} \mathbf{g} \qquad \text{and} \qquad \mathbf{D}(S_w) = -\mathbf{K}\lambda_a f_w \frac{dp_c}{dS_w}, \qquad (3.5)$$

and the compressibility c_a and fractional flow of water f_w are defined by

$$c_a = \frac{1}{\rho_a} \frac{d\rho_a}{dp_a} \qquad \text{and} \qquad f_w = \frac{\lambda_w}{\lambda}.$$

We note that in this formulation the only diffusion/dispersion term is capillary mixing described by equation (3.5).

The equations presented above describe multiphase and multicomponent flow in reservoirs. They can be used to simulate various production strategies in an attempt to understand and optimize hydrocarbon recovery or remediation strategies for contaminant removal. However, to use these equations effectively, parameters that describe the soil, rock, and fluid properties for the particular reservoir application must be input into the model. The relative permeabilities, which are nonlinear functions of the water saturation, can be estimated via laboratory experiments using reservoir cores and resident fluids. In the groundwater literature, often both the specific storativity $\theta = \rho_w S_w$ and the relative permeabilities are estimated using parameter fitting of certain function forms (see, e.g., van Genuchten [80]). The popularity of the van Genuchten fits comes from the fact that they produce smooth, differentiable functions that are easy to handle numerically. Similarly, fluid viscosities are relatively easy to obtain. However, the permeability tensor

K, the porosity ϕ, the capillary pressure curve $p_c(S_w)$, and the diffusion and dispersion coefficients are effective values that must be obtained from local properties via scaling techniques. In addition, the inaccessibility of the reservoir to measurement of even the local properties increases the difficulties. See [49, 52, 53, 73, 84] and the references contained therein for a survey of parameter estimation and history-matching techniques which have been applied.

Even if complete information is known about the reservoir properties in a highly heterogeneous reservoir, the problem of how to represent this reservoir on coarse-grid blocks of different length scales still remains. The power of supercomputers must be brought to bear for simulation studies using homogenization and statistical averaging to represent fine-scale phenomena on coarser grids.

Most of the papers in this book deal with the model equations presented in this section; other papers handle slightly different flow equations such as the elastic porous medium model by Chen-Ewing-Lyons-Qin-Sun-Yale, the non-Darcy well models by Garanzha-Konshin-Lyons-Papavassiliou-Qin, the lattice Boltzmann models for energy equations by Y. Li, the non-Newtonian flow by Ming-Shi, the omega equations by Villatoro-Garcia-Lafuente, and the Volterra integral equations of the first kind for the determination of capillary pressure functions by Subbey-Nordtvedt and Wang-Xiao. The mathematical and numerical techniques developed here can be possibly extended to more complicated model equations [24, 25, 33]. Error analysis for the numerical methods described in this paper for equation (3.4) has been carried out in [23, 26, 29].

4   Mathematical Results

The differential equations presented in the previous section have been studied in the past few decades. Existence of weak solutions to these equations under the assumption that the fluids are incompressible has been established [1, 2, 3, 18, 20, 28, 66, 70, 71]. Recently, uniqueness and regularity of the weak solutions has been studied [20, 21]. The degeneracy and strong coupling of these differential equations makes it very hard to study them. In particular, these mathematical properties for compressible fluids have not been obtained yet. Also, due to the degeneracy and strong coupling, the solutions do not have much regularity [20, 21]. Any attempt to obtain error estimates for the numerical methods described in this paper and other papers in this book for these differential equations has to respect the minimum regularity [29].

5 Mixed Methods for Accurate Velocity Approximations

There are two major sources of error in the methods currently being utilized for finite difference discretizations of equation (3.4). The first occurs in the approximation of the fluid pressure and velocity. The second comes from the techniques for upstream weighting to stabilize the saturation equation in (3.4). In this section, we describe the mixed finite element method for the accurate approximation of the total velocity $\mathbf{v}$. Some alternate upstream-weighting techniques developed from a finite element context were presented in Ewing *et al.* [50].

Among the disadvantages of the conforming discretizations are the lack of local mass conservation of the numerical model and some difficulties in computing the phase velocities needed in the transport and saturation equations. The straightforward numerical differentiation is far from being justifiable in problems formulated in a highly heterogeneous reservoir with complex geometry. On the other hand, the mixed finite element method [14] offers an attractive alternative. In fact, this method conserves mass cell by cell and produces a direct approximation of the two variables of interest–pressure and velocity. Below we explain briefly the mixed finite element method for the pressure equation.

To describe the mixed method we introduce two Hilbert spaces

$$W = L^2(\Omega), \qquad \boldsymbol{V} = \left\{ \varphi \in L^2(\Omega)^3, \ \nabla \cdot \varphi \in L^2(\Omega) \right\}.$$

The pressure equation is written in the following mixed weak form: Find $(p, \mathbf{v}) \in W \times \boldsymbol{V}$ such that

$$
\begin{aligned}
(A\mathbf{v}, \varphi) - (p, \nabla \cdot \varphi) &= (\mathbf{G}_\lambda, \varphi) & &\forall \, \varphi \in \boldsymbol{V}, \ t > 0, \\
\left(C(p, S_a) \frac{\partial p}{\partial t}, \psi \right) + (\nabla \cdot \mathbf{v}, \psi) &= (f(p, S_w), \psi) & &\forall \, \psi \in W, \ t > 0,
\end{aligned}
$$

(5.1)

with $p(0) \in L^2(\Omega)$ being the given initial pressure, where $C(p, S_a) = S_a c_a$ and $A = (\mathbf{K}\lambda)^{-1}$. We note that A is always symmetric and positive definite which leads to a well defined problem. This is in contrast to system (3.3) where the relative permeability $k_{r\alpha}$ vanishes when the phase α is absent in some subregion of Ω. We note that if there were nonhomogeneous boundary conditions on $\partial\Omega$ they should be added to the right-hand side $(f(p, S_w), \psi)$. Corresponding changes in the bilinear forms in the left-hand side should be introduced in the case of Robin boundary conditions. Obviously, equation (5.1) forms a nonlinear problem. To solve it, one can use the Picard linearization (see, e.g., [18]) or any other feasible approach.

We triangulate the domain Ω into elements, say, simplexes, rectangular parallelepipeds, and/or prisms, with characteristic diameter h. Let $W_h \subset W$ and $\boldsymbol{V}_h \subset \boldsymbol{V}$ be the Raviart-Thomas-Nedelec [76, 75], the Brezzi-Douglas-Fortin-Marini [12], the Brezzi-Douglas-Marini [13], the Brezzi-Douglas-Durán-Fortin [11], or the Chen-Douglas [22] mixed finite element space associated with the triangulation and time discretization $t_n = n\Delta t$, $n = 0, 1, \dots$. The mixed finite element solution $(P^n, \boldsymbol{V}^n) \in W_h \times \boldsymbol{V}_h$ satisfies

$$\begin{aligned}
(A^n \boldsymbol{V}^n, \boldsymbol{\varphi}_h) - (\nabla \cdot \boldsymbol{\varphi}_h, P^n) &= (\mathbf{G}_\lambda^n, \boldsymbol{\varphi}_h) && \forall\, \boldsymbol{\varphi}_h \in \mathbf{V_h}, \\
\frac{1}{\Delta t}(C^n(P^n - P^{n-1}), \psi_h) + (\nabla \cdot \boldsymbol{V}^n, \psi_h) &= (f^n, \psi) && \forall\, \psi_h \in W_h,
\end{aligned} \tag{5.2}$$

with $P^0 \in W_h$ expressed through given initial datum.

This is an implicit in time Euler approximation of a nonlinear problem which can be solved by Picard or Newton iterations. Obviously, one can easily formulate the Crank-Nicolson scheme.

The resulting system of linear equations has the form of a saddle point problem defined on a pair of finite dimensional spaces W_h and $\boldsymbol{V}_h$:

$$\begin{pmatrix} \mathbf{A} & \mathbf{B}^T \\ \mathbf{B} & -\mathbf{D} \end{pmatrix} \begin{pmatrix} \boldsymbol{V}^n \\ P^n \end{pmatrix} = \begin{pmatrix} F \\ G \end{pmatrix},$$

where $F \in \boldsymbol{V}_h$ and $G \in W_h$ are given and $P^n \in W_h$ and $\boldsymbol{V}^n \in \boldsymbol{V}_h$ represent the unknown approximate solution on the time level t_n. Here $\mathbf{A} : \boldsymbol{V}_h \mapsto \boldsymbol{V}_h$ is a linear symmetric and positive definite operator, the linear map $\mathbf{B}^T : W_h \mapsto \boldsymbol{V}_h$ is the adjoint of $\mathbf{B} : \boldsymbol{V}_h \mapsto W_h$, and $\mathbf{D} : W_h \mapsto W_h$ is either $(1/\Delta t)\mathbf{M}$ with $\mathbf{M}$ similar to the mass matrix in W_h for time dependent problems or $\mathbf{0}$ for steady state problems. The existence and uniqueness of a solution is guaranteed by the fact that the pair of spaces $(W_h, \boldsymbol{V}_h)$ satisfies the *inf-sup* condition of Babuška-Brezzi [14].

This is an indefinite system with a large number of unknowns. Such type of systems is more difficult to solve compared with the definite systems. However, the popularity of the mixed method has increased considerably as a consequence of the progress made in recent years in developing efficient algorithms for solving this indefinite system (see, e.g., [4, 9, 10, 19, 31, 32, 34, 56, 64, 65, 77]). The mixed method for approximating accurately the total velocity **v** is used in this paper. This method is further examined and exploited in the papers by Arbogast, Douglas-Pereira-Yeh, Garanzha-Konshin-Lyons-Papavassiliou-Qin, Huang-Spagnuolo, Ming-Shi, Qin-Wang-Ewing-Espedal, Wang-Li, Russell, D. Yang, and Yuan in this book.

6 Characteristics-Based Techniques

In multiphase or multicomponent flow models, the convective, hyperbolic part is a linear function of the velocity. An operator-splitting technique has been developed to solve the purely hyperbolic part by time stepping along the associated characteristics [40, 54, 55, 78]. We first obtain the non-divergence form of equation (3.2) with $\theta = 1$ by using the product rule for differentiation on the $\nabla \cdot (\mathbf{u}c)$ term and applying equation (3.1) to obtain

$$\phi \frac{\partial c}{\partial t} + \mathbf{u} \cdot \nabla c - \nabla \cdot (\mathbf{D}\nabla c) = q(\tilde{c} - c) . \tag{6.1}$$

Next, the first and second terms in equation (6.1) are combined to form a directional derivative along what would be the characteristics for the equation if the tensor **D** were zero. The resulting equation is

$$\nabla \cdot (\mathbf{D}\nabla c) + q(\tilde{c} - c) = \phi \frac{\partial c}{\partial t} + \mathbf{u} \cdot \nabla c \equiv \phi \frac{\partial c}{\partial \tau} . \tag{6.2}$$

The system obtained by modifying equations (3.1) and (3.2) in this way is solved sequentially. An approximation for **u** is first obtained at time level $t = t^n$ from a solution of equation (3.1) with the fluid viscosity μ evaluated via some mixing rule at time level t^{n-1}. Equation (3.1) can be solved by the mixed finite element method for a more accurate fluid velocity as in the last section. Let $C^n(x)$ and $\mathbf{U}^n(x)$ denote the approximations of $c(x, t)$ and $\mathbf{u}(x, t)$, respectively, at time level $t = t^n$. The directional derivative is then discretized along the "characteristic" mentioned above as

$$\phi \frac{\partial c}{\partial \tau}(x, t^n) \approx \phi \frac{C^n(c) - C^{n-1}(\bar{x}^{n-1})}{\Delta t} , \tag{6.3}$$

where $\bar{x}^{n-1}$ is defined for an x as

$$\bar{x}^{n-1} = x - \frac{\mathbf{U}^n(x)\Delta t}{\phi}. \tag{6.4}$$

This technique is a discretization back along the "characteristic" generated by the first-order derivatives from equation (6.2). Equations (6.3) and (6.4) are useful if the characteristics do not change much in each time step. In general, several "micro" steps may be necessary to trace accurately the characteristic back through a full time step. Although the advection-dominance in the original equation (6.2) makes it non-self-adjoint, the form with the directional derivative is self-adjoint and discretization techniques for self-adjoint equations can be utilized. This modified method of characteristics (MMOC) can be combined with either finite difference or finite element spatial discretizations.

In multiphase flow, the convective part is nonlinear. A similar operator-splitting technique to solve the saturation equation in (3.4) needs reduced time steps because the pure hyperbolic part may develop shocks. An operator-splitting technique has been developed for multiphase flow [35, 36, 37, 43, 44, 45] which retains the long time steps in the characteristic solution without introducing serious discretization errors.

When the gravity term is ignored and the porosity of the reservoirs does not change with time in equation (3.4), the operator splitting gives the following set of equations (with $S = S_w$):

$$\phi\frac{\partial \bar{S}}{\partial t} + \frac{d}{dS}\mathbf{f}^m(\bar{S}) \cdot \nabla \bar{S} \equiv \phi\frac{d\bar{S}}{d\tau} = 0, \quad t_m \leq t \leq t_{m+1}, \tag{6.5}$$

and

$$\phi\frac{\partial S}{\partial \tau} + \nabla \cdot (\mathbf{b}^m(S)S) - \nabla \cdot (D(S)\nabla S) = \mathbf{q}(\mathbf{x}, t), \quad t_m \leq t \leq t_{m+1}, \tag{6.6}$$

with proper initial and boundary conditions. As noted earlier, the saturation S is coupled to the pressure/velocity equations, which are solved by the mixed finite element method described in the last section.

The splitting of the fraction flow function into the two parts $\mathbf{f}^m(S) + \mathbf{b}(S)S$ is constructed [44] such that $\mathbf{f}^m(S)$ is linear in the shock region, $0 \leq S \leq S_1 < 1$, and $\mathbf{b}(S) \equiv 0$ for $S_1 \leq S \leq 1$. Further, equation (6.5) produces the same unique physical solution after a shock has been completely formed as

$$\frac{\partial S}{\partial t} + \nabla \cdot (\mathbf{f}^m(S) + \mathbf{b}(S)S) = 0, \tag{6.7}$$

with an entropy condition imposed. This means that, for a fully developed shock, the characteristic solution of equation (6.5) always produces a unique solution and, as in the single-phase case, we may use long time steps Δt without loss of accuracy.

The solution of equation (6.6) via a variational method leads to the following Petrov-Galerkin equations:

$$
\begin{aligned}
B(S_h^m, \varphi_i) &\equiv (\phi S_h^{m+1}, \varphi_i) - (\Delta t \mathbf{b}(\mathbf{x}, t^m) S_h^{m+1}, \nabla \varphi_i) \\
&\quad + (\Delta t D(\mathbf{x}, t^m) \nabla S_h^{m+1}, \nabla \varphi_i) \qquad (6.8) \\
&= (g_h^m(\mathbf{x}, t^m), \varphi_i), \quad i = 1, 2, \dots, N, \ S_h^m \in M_h, \ \varphi_i \in N_h,
\end{aligned}
$$

where M_h and N_h are the trial and test spaces spanned by $\{\theta_i\}$ and $\{\varphi_i\}$, $i = 1, 2, \dots, N$, respectively. $B(\cdot, \cdot)$ given by equation (6.8) is a nonsymmetrical bilinear form with spatially-dependent coefficients.

To obtain equation (6.8), we have used the characteristic solution from equation (6.5) to approximate $(\partial/\partial\tau)S$ and the nonlinear coefficients in equation (6.6). The nonsymmetry in the bilinear form $B(\cdot, \cdot)$ is caused by the nonlinearity of the convective part of the equation, represented by the term $\mathbf{b}(S)S$. This term balances the diffusion forces in the shock region after a traveling front has been established.

Unfortunately, the MMOC techniques described above generally do not conserve mass. Also, the proper method for treating boundary conditions in a conservative and accurate manner using these techniques is not obvious. Recently, Celia, Russell, Herrera, and Ewing have devised Eulerian-Lagrangian localized adjoint methods (ELLAM) [16, 68], a set of schemes that are defined expressly for conservation of mass properties.

The ELLAM formulation was motivated by localized adjoint methods [15, 67], which are one form of the optimal test function methods [6, 37, 39, 44]. We briefly describe these methods. Let

$$
Lu = f, \qquad x \in \Sigma = \Omega \quad \text{or} \quad (x, t) \in \Sigma = \Omega \times J
$$

denote a partial differential equation in space or space-time, where J is the time interval of interest. Integrating against a test function ϕ, we obtain the weak form

$$
\int_\Sigma Lu\phi d\omega = \int_\Sigma f\phi d\omega.
$$

If we choose test functions ϕ to satisfy the formal adjoint equation $L^*\phi = 0$ and $\phi = 0$ on the boundary, except at certain nodes or edges denoted by ℓ_i, then integration by parts (the divergence theorem in higher dimensions) yields

$$
\sum_i \int_{\ell_i} uL^*\phi d\omega = \int_\Sigma f\phi d\omega.
$$

Various different test functions can be used to focus upon different types of information. Herrera has built an extensive theory around this concept; see [67] for references. The theory is quite general and can deal with situations where distributions do not apply, such as when both u and ϕ are discontinuous.

As in the work of Demkowitz and Oden [39], we want to localize these test functions to maintain sparse matrices. Certain choices of space-time test functions which are useful for linear equations of the form (3.2) have been described in [39, 79].

The ELLAM techniques have been extended to a wide variety of applications [33, 38, 57, 58, 59, 60, 61, 62, 63, 81, 82, 83]. Optimal order error estimates have been developed for advection [59], advection-diffusion [62, 82], advection-reaction [57, 58, 59, 61, 62, 83], and advection-diffusion-reaction [60, 81] systems. These techniques are further studied and applied in the papers by DeVore-Wang-Liu-Xu, Qin-Wang-Ewing-Espedal, Wang-Al-Lawatia, and Wang-Liang-Ewing-Lyons-Qin in this book.

Recently, in the study of computational geosciences, a new locally conservative Euler-Lagrangian method (LCELM) was introduced by Douglas, Pereira, and Yeh [42]. This technique is an extension of the characteristic-mixed method for transport-dominated diffusion processes introduced by Arbogast and Wheeler [5]. The extension properly treats nonlinear problems. It was shown [42] that the LCELM technique is superior to the MMOC and the modified method of characteristics with adjust advection (MMOCAA) [41] techniques. The LCELM conserves mass locally, the MMOCAA does it globally, and the MMOC does not at all. The LCELM technique is further considered in the papers by Douglas-Pereira-Yeh and Huang-Spagnuolo in this book. Both the LCELM and characteristic-mixed techniques in [5] can be formulated in the ELLAM framework.

Because of its simplicity, the MMOC technique is still popular. It is being applied for compressible flow problems in the papers by Cheng-Wang, D. Yang, Yu-Wu, Yuan, and Zhao in this book.

7 Local Grid-Refinement and Domain Decomposition Techniques

Many time-dependent fluid flow problems involve both large-scale processes and highly localized phenomena that are often critical to the overall chemical and physical behaviors of the flows. For large-scale applications, it is frequently impossible to use a uniform grid which is sufficiently fine to resolve the local phenomena without yielding numbers of unknowns that will overburden even the largest of today's supercomputers. Since these local processes are often dynamic, efficient numerical simulation requires the ability to perform dynamic self-adaptive local grid refinement. The need for adaptive techniques has provided the impetus for the development of local grid-refinement software tools, some of which are used in day-to-day applications for small- to mid-size problems. Software and engineering tools capable of dynamic local grid refinement need to be developed for large-scale, fluid flow applications. The adaptive grid-refinement algorithms must also be closely matched with the architecture features of the new advanced computers to take advantage of possible vector and parallel capabilities.

Normally, local refinement must be performed if a fluid interface is located within the coarse-grid block to resolve the solution there. A slightly different strategy is to make the region of local refinement big enough so that we can use the same refinements for several of the large time steps. This local patch refinement technique [8, 27, 51, 74] has proved to be very effective for obtaining local resolution around fixed singular points such as wells in a reservoir. Adaptive grid-refinement techniques utilizing the patch technique have been presented in several surveys (see, e.g., [46, 47]). The patch technique has been incorporated efficiently in existing multiphase industrial reservoir simulation codes. Results for the SPE Comparison projects number 1 and 2 were presented in [48]. The local refinement was both efficient and effective since excellent results were obtained without destroying the efficiency of the original codes.

The difficult problem with these techniques is the communication of the solution between the fine and coarse grids. The domain decomposition technique described in [72] gives accurate and efficient treatment of the communication problem. General domain decomposition techniques are applied and analyzed for flow problems in the papers by Alboin-Jaffré-Roberts-Wang-Serres, Bastian-Chen-Ewing-Helmig-Jakobs-Reichenberger, Li-Ma, Ma-Zhu, Reme-Oye-Espedal-Fladmark, Sheen, Shi-Xu, D. Yang, and H. Zhang in this book.

Acknowledgments. This work is supported in part by National Science Foundation grants DMS-9626179, DMS-9972147, and INT-9901498, by EAP grant R825207-01, and by a gift grant from the Mobil Technology Company.

References

[1] Alt, H. W. and di Benedetto, E., Nonsteady flow of water and oil through inhomogeneous porous media, *Ann. Scuola Norm. Sup. Pisa Cl. Sci.* **12** (1985), 335–392.

[2] Antontsev, S. N., Kazhikhov, A. V., and Monakhov, V. N., Boundary-Value Problems in the Mechanics of Nonuniform Fluids, Studies in Mathematics and its Applications, Amsterdam, 1990.

[3] Arbogast, T. J., The existence of weak solutions to single porosity and simple dual-porosity models of two-phase incompressible flow, *Nonlin. Analysis: Theory, Methods, and Appl.* **19** (1992), 1009–1031.

[4] Arbogast, T. J. and Chen, Z., On the implementation of mixed methods as nonconforming methods for second order elliptic problems, *Math. Comp.* **64** (1995), 943–972.

[5] Arbogast, T. J. and Wheeler, M. F., A characteristic-mixed finite element method for advection-dominated transport problems, *SIAM J. Numer. Anal.* **32** (1995) 404–424.

[6] Barrett, J. W. and Morton, K. W., Approximate symmetrization and Petrov-Galerkin methods for diffusion-convection problems, *Comp. Meth. Appl. Mech. and Eng.* **45** (1984), 97–122.

[7] Bear, J, *Dynamics of Fluids in Porous Media*, Dover Publications, 1988.

[8] Bramble, J. H., Ewing, R. E., Pasciak, J. E., and Schatz, A. H., A preconditioning technique for the efficient solution of problems with local grid refinement, *Comp. Meth. Appl. Mech. and Eng.* **67** (1988), 149–159.

[9] Bramble, J. H. and Pasciak, J., A preconditioning technique for indefinite system resulting from mixed approximations of elliptic problems, *Math. Comp.* **50** (1988), 1–18.

[10] Bramble, J. H., Pasciak, J. E., and Vassilev, A., Analysis of the inexact Uzawa algorithm for saddle point problems, *SIAM J. Numer. Anal.* **34** (1997), 1072–1092.

[11] Brezzi, F., Douglas, J., Jr., Durán, R., and Fortin, M., Mixed finite elements for second order elliptic problems in three variables, *Numer. Math.* **51** (1987), 237–250.

[12] Brezzi, F., Douglas, J., Jr., Fortin, M., and Marini, L., Efficient rectangular mixed finite elements in two and three space variables, *RAIRO Modèl. Math. Anal. Numér* **21** (1987), 581–604.

[13] Brezzi, F., Douglas, J., Jr., and Marini, L., Two families of mixed finite elements for second order elliptic problems, *Numer. Math.* **47** (1985), 217–235.

[14] Brezzi, F. and Fortin, M., *Mixed and Hybrid Finite Methods*, Springer-Verlag, New York, 1991.

[15] Celia, M. A., Herrera, I., Bouloutas, E., and Kindred, J. S., A new numerical approach for the advection-diffusive transport equation, *Numerical Methods for PDEs* **5** (1989), 203–226.

[16] Celia, M. A., Russell, T. F., Herrera, I., Ewing, R. E., An Eulerian-Lagrangian localized adjoint method for the advection-diffusion equation, *Advances in Water Resources* **13** (1990), 187–206.

[17] Chavent, G., *A new formulation of diphasic incompressible flows in porous media*, in Lecture Notes in Mathematics, Vol. 503, Springer-Verlag, 1976.

[18] Chavent, G. and Jaffre, J., *Mathematical Models and Finite Elements for Reservoir Simulation: Single Phase, Multiphase and Multicomponent Flows Through Porous Media*, North-Holland, Amsterdam, 1986.

[19] Chen, Z., Equivalence between and multigrid algorithms for nonconforming and mixed methods for second order elliptic problems, *East-West J. Numer. Math.* **4** (1996), 1–33.

[20] Chen, Z., Degenerate two-phase incompressible flow I: existence, uniqueness and regularity of a weak solution, *J. Diff. Equations*, to appear.

[21] Chen, Z., Degenerate two-phase incompressible flow IV: Regularity, stability and stabilization, submitted.

[22] Chen, Z. and Douglas, J., Jr., Prismatic mixed finite elements for second order elliptic problems, *Calcolo* **26** (1989), 135–148.

[23] Chen, Z., Espedal, M. S., and Ewing, R. E., Continuous-time finite element analysis of multiphase flow in groundwater hydrology, *Applications of Mathematics* **40** (1995), 203–226.

[24] Chen, Z. and Ewing, R. E., From single-phase to compositional flow: applicability of mixed finite elements, *Transport in Porous Media* **27** (1997), 225–242.

[25] Chen, Z. and Ewing, R. E., Comparison of various formulations of three-phase flow in porous media, *J. Comp. Physics* **132** (1997), 362–373.

[26] Chen, Z. and Ewing, R. E., Fully-discrete finite element analysis of multiphase flow in groundwater hydrology, *SIAM J. Numer. Anal.* **34** (1997), 2228–2253.

[27] Chen, Z. and Ewing, R. E., Local mesh refinement for degenerate two-phase incompressible flow problems, The Proceedings of the Ninth International Colloquium on Differential Equations, D. Bainov, ed., Plovdiv, Bulgaria, 1999, pp. 85–90.

[28] Chen, Z. and Ewing, R. E., Mathematical analysis for reservoir models, *SIAM J. Math. Anal.* **30** (1999), 431–453.

[29] Chen, Z. and Ewing, R. E., Degenerate two-phase incompressible flow III: Optimal error estimates, *Numer. Math.*, to appear.

[30] Chen, Z., Ewing, R. E., and Espedal, M. S., Multiphase flow simulation with various boundary conditions, in Computational Methods in Water Resources, A. Peters, G. Wittum, B. Herrling, U. Meissner, C. A. Brebbia, W. G. Gray, and G. F. Pinder, eds., Kluwer Academic Publishers, Netherlands, 1994, pp. 925–932.

[31] Chen, Z., Ewing, R. E., Kuznetsov, Y., Lazarov, R., and Maliassov, S., Multilevel preconditioners for mixed methods for second order elliptic problems, *Numer. Linear Alg. and Appl.* **3** (1996), 427–453.

[32] Chen, Z., Ewing, R. E., and Lazarov, R., Domain decomposition algorithms for mixed methods for second order elliptic problems, *Math. Comp.* **65** (1996), 467–490.

[33] Chen, Z., Qin, G., and Ewing, R. E., Analysis of a compositional model for fluid flow in porous media, *SIAM J. Appl. Math.*, to appear.

[34] Cowsar, L., Mandel, J., and Wheeler, M., Balancing domain decomposition for mixed finite elements, *Math. Comp.* **64** (1995), 989–1015.

[35] Dahle, H. K., Adaptive characteristic operator splitting techniques for convection-dominated diffusion problems in one and two space dimensions, in IMA Volumes in Mathematics and Its Applications, volume II, Springer Verlag, 1988, pp 77–88.

[36] Dahle, H. K., Espedal, M. S., and Ewing, R. E., Characteristic Petrov-Galerkin subdomain methods for convection diffusion problems, in IMA Volume 11, Numerical Simulation in Oil Recovery, M.F. Wheeler, ed., Springer-Verlag, Berlin, 1988, pp. 77–88.

[37] Dahle, H. K., Espedal, M. S., Ewing, R. E., and Sævareid, O., Characteristic adaptive sub-domain methods for reservoir flow problems, *Numerical Methods for PDEs*, **6** (1990), 279–309.

[38] Dahle, H. K., Ewing, R. E., and Russell, T., Eulerian-Lagrangian localized adjoint methods for a nonlinear advection-diffusion equation, *Comput. Meth. Appl. Mech. Eng.* **122** (1995), 223–250.

[39] Demkowitz, L. and Oden, J. T., An adpative characteristic Petrov-Galerkin finite element method for convection-dominated linear and nonlinear parabolic problems in two space variables, *Comp. Meth. Appl. Mech. and Eng.* **55** (1986), 63–87.

[40] Douglas, J., Jr. and Russell, T., Numerical methods for convection dominated diffusion problems based on combining the modified method of characteristics with finite element or finite difference procedures, *SIAM J. Numer. Anal.* **19** (1982), 871–885.

[41] Douglas, J., Jr., Furtado, F., and Pereira, F., On the numerical simulation of waterflooding of heterogeneous petroleum reservoirs, *Computational Geosciences* **1** (1997) 155–190.

[42] Douglas, J., Jr., Pereira, F., and Yeh, L., A locally conservative Eulerian-Lagrangian numerical method and its application to nonlinear transport in porous media, to appear.

[43] Espedal, M. S. and Ewing, R. E., Petrov-Galerkin subdomain methods for two-phase immiscible flow, *Comp. Meth. Appl. Mech. and Eng.* **64** (1987), 113–135.

[44] Espedal, M. S., Ewing, R. E., and Russell, T., Mixed methods, operator splitting, and local refinement techniques for simulation on irregular grids, in Proceedings 2nd European Conference on the Mathematics of Oil Recovery, D. Guerillot and O. Guillon, eds., Editors Technip, Paris, 1990, pp. 237–245.

[45] Espedal, M. S., Ewing, R. E., Russell, T., and Sævareid, O., Reservoir simulation using mixed methods, a modified method of characteristics, and local grid refinement, in Proceedings of Joint IMA/SPE European Conference on the Mathematics of Oil Recovery, Cambridge University, July 25–27, 1989.

[46] Espedal, M. S., Hansen, R., Langlo, P., Sævareid, O., and Ewing, R. E., Heterogeneous porous media and domain decomposition methods, Proceedings 2nd European Conference on the Mathematics of Oil Recovery, D. Guerillot and O. Guillon, eds., Paris, Editors Technip, 1990, pp. 157–163.

[47] Espedal, M. S., Hansen, R., Langlo, P., Sævareid, O., and Ewing, R. E., Efficient adaptaive procedures for fluid flow, *Comp. Meth. Appl. Mech. Eng.* **55** (1986), 89–103.

[48] Ewing, R. E., Boyett, B. A., Babu, D. K., and Heinemann, R. F., Efficient use of locally refined grids for multiphase reservoir simulation,

in Proceedings of Tenth Society of Petroleum Engineers Symposium on Reservoir Simulation, SPE 18413, Houston, Texas, February 6–8 1989, pp. 55–70.

[49] Ewing, R. E. and George, J. H., Identification and control of distributed parameters in porous media flow, Distributed Parameter Systems, F. Kappel, K. Kunisch, and W. Schappacher, eds., Lecture Notes in Control and Information Sciences, volume 75, Springer-Verlag, Berlin, 1985, pp. 145–161.

[50] Ewing, R. E., Heinemann, R. T., Koebbe, J. V., and Prasad, U. S., Velocity weighting techniques for fluid displacement, *Comp. Meth. Appl. Mech. Eng.* **64** (1987), 137–151.

[51] Ewing, R. E., Lazarov, R. D., and Vassilevski, P. S., Local refinement techniques for elliptic problems on cell-centered girds, II: Optimal order two-grid iterative methods, *Numer. Linear Algebra with Appl.* **1** (1994), 337–368.

[52] Ewing, R. E., Pilant, M. S., Wade, J. G., Watson, A. T., Estimating parameters in scientific computation: A survey of experience from oil and groundwater modeling, *IEEE Computational Science & Engineering* **1** (1994), 19–31.

[53] Ewing, R. E., Pilant, M. S., Wade, J. G., Watson, A. T., Identification and control problems in petroleum and groundwater modeling, Control Problems in Industry (I. Lasciecka and B. Morton, eds.), Progress in Systems and Control Theory, 21, Birkhauser, Basel, 119–149.

[54] Ewing, R. E., Russell, T. F., and Wheeler, M. F., Simulation of miscible displacement using mixed methods and a modified method of characteristics, in Proceedings Seventh SPE Symposium on Reservoir Simulation, SPE 12241, San Francisco, November 15–18 1983, pp. 71–82.

[55] Ewing, R. E., Russell, T. F., and Wheeler, M. F., Convergence analysis of an approximation of miscible displacement in porous media by mixed finite elements and a modified method of characteristics, *Comp. Meth. Appl. Mech. Eng.* **47** (1984), 73–92.

[56] Ewing, R. E., Shen, J., and Vassilevski, P. S., Vectorizable preconditioners for mixed finite element solution of second-order elliptic problems, *International Journal of Computer Mathematics* **44** (1992), pp. 313–327.

[57] Ewing, R. E. amd Wang, H., An optimal-order estimate for Eulerian-Lagrangian localized adjoint methods for variable-coefficient advection-reaction problems, *SIAM J. Numer. Anal.* **33** (1996), 318–348.

[58] Ewing, R. E. amd Wang, H., An Eulerian-Lagrangian localized adjoint method for variable-coefficient advection-reaction problems, in Advances in Hydro-Science and Engineering, S. Wang, ed., volume 1, Part B, University of Mississippi Press, 1993, pp. 2010–2015.

[59] Ewing, R. E. amd Wang, H., Eulerian-Lagrangian localized adjoint methods for linear advection or advection-reaction equations and their convergence analysis, *Computational Mechanics* **12** (1993), 97–121.

[60] Ewing, R. E. amd Wang, H., Eulerian-Lagrangian localized adjoint methods for variable coefficient advection-diffusive-reactive equations in groundwater contaminant transport, in Advances in Optimization and Numerical Analysis, S. Goméz and J.P. Hennart, eds., volume 275, Kluwer Academic Publishers, Netherlands, 1994, pp. 185–205.

[61] Ewing, R. E. amd Wang, H., Eulerian-Lagrangian localized adjoint methods for reactive transport in groundwater, in Environmental Studies: Mathematical Computational, and Statistical Analysis, IMA Volume in Mathematics and its Application, M.F. Wheeler, ed., volume 79, Springer-Verlag, Berlin, 1995, pp. 149–170.

[62] Ewing, R. E. amd Wang, H., Optimal-order convergence rate for Eulerian-Lagrangian localized adjoint method for reactive transport and contamination in groundwater, *Numer. Meth. in PDE's* **11** (1995), 1–31.

[63] Ewing, R. E., Wang, H., and Russell, T., Eulerian-Lagrangian localized adjoint methods for convection-diffusion equations and their convergence analysis, *IMA J. Numer. Anal.* **15** (1995), 405–459.

[64] Ewing, R. E. and Wang, J., Analysis of mixed finite element methods on locally refined grids, *Numer. Math.* **63** (1992), 183–194.

[65] Ewing, R. E. and Wang, J., Analysis of multilevel decomposition iterative methods for mixed finite element methods, R.A.I.R.O. 28(4) (1994), pp. 377–398.

[66] Feng, X., On existence and uniqueness for a coupled system modeling miscible displacement in porous media, *J. Math. Anal. Appl.*, **194** (1995), 441–469.

[67] Herrera, I., Unified formulation of numerical methods I. Green's formula for operators in discontinuous fields, *Numer. Meth. for PDEs* **1** (1985), 25–44.

[68] Herrera, I., Ewing, R. E., Celia, M. A., and Russell, T. F., Eulerian-Lagrangian localized adjoint method: The theoretical framework, *Numer. Meth. for PDE's* **9** (1993), 431–457.

[69] Hittel, D., *Fundamentals of Soil Physics*, Academic Press, 1980.

[70] Kroener, D. and Luckhaus, S., Flow of oil and water in a porous medium, *J. Diff. Equations* **55** (1984), 276–288.

[71] Kružkov, S. N. and Sukorjanskiĭ, S. M., Boundary problems for systems of equations of two-phase porous flow type; statement of the problems, questions of solvability, justification of approximate methods, *Math. USSR Sbornik* **33** (1977), 62–80.

[72] Langlo, D. and Espedal, M., Heterogeneous reservoir models, two-phase immiscible flow in 2d, in Mathematical Modeling in Water Resources, Computational Methods in Water Resources, T. F. Russell, R. E. Ewing, C. A. Brebbia, W. G. Gray, and G. F. Pinder, eds., IX, volume 2, Elsevier Applied Science, London, 1992, pp. 71–80.

[73] Lin, T. and Ewing, R. E., Parameter estimation for distributed systems arising in in fluid flow problems via time series methods, in Proceedings of Conference on "Inverse Problems", Oberwolfach, West Germany, 1986. Birkhauser, Berlin, pp. 117–126.

[74] McCormick, S. and Thomas, J., The fast adaptive composite grid methods for elliptic boundary value problems, *Math. Comp.* **46** (1986), 439–456.

[75] Nedelec, J., Mixed finite elements in $\Re^3$, *Numer. Math.* **35** (1980), 315–341.

[76] Raviart, R. and Thomas, J., A mixed finite element method for second order elliptic problems, Lecture Notes in Mathematics, vol. 606, Springer, Berlin, 1977, pp. 292–315.

[77] Rusten, T. and Winther, R., A preconditioned iterative method for saddle point problems, *SIAM J. Matrix Anal. Appl.* **13** (1992), 887–904.

[78] Russell, T., The time-stepping along characteristics with incomplete iteration for Galerkin approximation of miscible displacement in porous media, *SIAM J. Numer. Anal.* **22** (1985), 970–1013.

[79] Russell, T., Eulerian-Lagrangian localized adjoint methods for advection-dominated problems, In Proceedings of 13th Biennial Conference on Numerical Analysis, Dundee, Scotland, June 27–30 1989. Pitmann Publishing Company.

[80] van Genuchten, M., A closed form equation for predicting the hydraulic conductivity in soils, *Soil Sci. Soc. Am. J.* **44** (1980), 892–898.

[81] Wang, H., Ewing, R. E., and Celia, M. A., Eulerian-Lagrangian localized adjoint methods for reactive transport with biodegradation, *Numer. Meth. for PDEs* **11** (1995), 229–254.

[82] Wang, H., Ewing, R. E., and Russell, T. F., Eulerian-Lagrangian localized adjoint methods for variable-coefficient convection-diffusion problems arising in groundwater applications, in Computational Methods in Water Resources, IX, Numerical Methods in Water Resources, volume 1, T. F. Russell, R. E. Ewing, C. A. Brebbia, W. G. Gray, and G. F. Pinder, eds., Elsevier Applied Science, London, 1992, pp. 25–32.

[83] Wang, H., Lin, T., and Ewing, R. E., Eulerian-Lagrangian localized adjoint methods with domain decomposition and local refinement techniques for advection-reaction problems with discontinuous coefficients, in Computational Methods in Water Resources, IX, Numerical Methods in Water Resources, volume 1, T. F. Russell, R. E. Ewing, C. A. Brebbia, W. G. Gray, and G. F. Pinder, eds., Elsevier Applied Science, London, 1992, pp. 17–24.

[84] Watson, A. T., Wade, J. G., and Ewing, R. E., Parameter and system identification for fluid flow in underground reservoirs, in Proceedings of the Conference, Inverse Problems and Optimal Design in Industry, Philadelphia, PA, July 8–10 1994.

[85] Whitaker, S., Flow in porous media II: The governing equations for immiscible two-phase flow, *Transport in Porous Media* **1** (1986), 102–125.

Domain Decomposition for Some Transmission Problems in Flow in Porous Media

CLARISSE ALBOIN   JÉRÔME JAFFRÉ
JEAN E. ROBERTS   XUEWEN WANG
CHRISTOPHE SERRES

Abstract

A variety of models are considered: one-phase flow in a porous medium, two-phase flow in a porous medium with two rock types, and one-phase flow in a porous medium with fractures. For each of these models the domain of calculation is divided into subdomains corresponding to the physics of the problem. Then it is shown how to rewrite the problems as interface problems to use nonoverlapping domain decomposition.

KEYWORDS: porous media flow, domain decomposition

1   Introduction

Domain decomposition methods have been studied for the most part as algebraic tools for solving problems on parallel machines: see [10] for a review of these methods. However, in many models of flow in porous media arising from environmental problems in the subsurface as well as from reservoir simulation, the domain of calculation is naturally divided into subdomains corresponding to the physics of the problem. Therefore, it is reasonable to construct nonoverlapping domain decomposition methods which can take into account the coupling of the physical phenomena taking place in the subdomains. In these methods one rewrites the global problem as a problem with unknowns on the subdomain interfaces.

After presenting the single phase flow in §2, we show in §3 and §4 how it applies to more complex problems: two-phase flow in a porous medium with two rock types and one-phase flow in a porous medium with fractures.

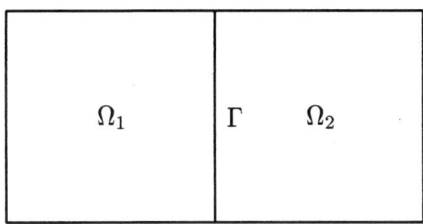

Figure 1: The domain Ω divided into two subdomains Ω_1 and Ω_2.

2   Single Phase Darcy Flow

We consider first the simple case of an incompressible single phase flow in a porous medium. The flow is governed by the equations

$$\text{div}\,\vec{\varphi} = 0, \quad \vec{\varphi} = -K\,\vec{\nabla}p, \quad \text{in } \Omega, \tag{2.1}$$

where Ω is a polygonal domain. The unknowns $\vec{\varphi}$ and p are the Darcy velocity and the fluid pressure and the coefficient K is the absolute permeability which may depend on $x \in \Omega$. To equations (2.1) we add the boundary conditions

$$p = p_d \quad \text{on } \partial\Omega_D, \quad \vec{\varphi} \cdot \vec{n} = q_d \quad \text{on } \partial\Omega_N, \tag{2.2}$$

where $\vec{n}$ denotes the outward normal to Ω. $\partial\Omega_D$ is the part of the boundary of Ω supporting Dirichlet boundary conditions and $\partial\Omega_N$ the part supporting Neumann boundary conditions, with $\partial\Omega = \partial\Omega_D \cup \partial\Omega_N$.

Now we divide the domain Ω into two polygonal subdomains Ω_1 and Ω_2 and we denote by Γ the interface between the subdomains: $\Gamma = \overline{\Omega}_1 \cap \overline{\Omega}_2$ (see Fig. 1).

If we denote by $(\vec{\varphi}_i, p_i), i = 1, 2$, the restriction of the solution of the system of equations (2.1) and (2.2) to the subdomain Ω_i, then we have

$$\begin{aligned} \text{div}\,\vec{\varphi}_i &= 0, \quad \vec{\varphi}_i = -K_i\,\vec{\nabla}p_i \quad \text{in } \Omega_i, \\ p_i &= p_d \quad \text{on } \partial\Omega_i \cap \partial\Omega_D, \quad \vec{\varphi} \cdot \vec{n}_i = q_d \quad \text{on } \partial\Omega_i \cap \partial\Omega_N, \end{aligned} \tag{2.3}$$

with the transmission conditions

$$p_1 = p_2, \quad \vec{\varphi}_1 \cdot \vec{n}_1 = \vec{\varphi}_2 \cdot \vec{n}_2 \text{ on } \Gamma. \tag{2.4}$$

Here $\vec{n}_i$ is the outward normal to Ω_i. These conditions express continuity of the pressure and mass conservation.

Let us now discretize the subdomains Ω_i with a mesh $\mathcal{T}_i$ of triangles or parallelograms. To simplify we will assume that the two subdomain meshes are conforming in the sense that their union forms a regular discretization of the whole domain Ω. We denote by $\mathcal{E}_i$ the set of edges associated to $\mathcal{T}_i$.

To approximate the problem we use the Raviart-Thomas mixed finite elements of lowest order [9, 1]. For this purpose we introduce the approximation spaces

$$\vec{X}_i(g) = \{\vec{v} \in \vec{RT}_0(\Omega_i) \mid \vec{v} \cdot \vec{n}_i = g \text{ on } \partial\Omega_i \cap \partial\Omega_N\},$$

$$M_i = \{q \in L^2(\Omega_i) \mid q_{|C} \in P_0(C), C \in \mathcal{T}_i\}, \quad N = \prod_{E \in \Gamma} P_0(E).$$

Here $\vec{RT}_0(\Omega_i)$ denotes the Raviart-Thomas space of lowest order [8]; functions in this space are uniquely defined by their flux through the edges of $\mathcal{E}_i$. The Darcy velocity $\vec{\varphi}_i$ is calculated in this space. P_0 denotes the space of constants and M_i and N are spaces of piecewise constant functions defined on Ω_i and on Γ respectively. The pressure p_i inside Ω_i is approximated in M_i and the pressure on Γ denoted by λ is approximated in N. We will use the same notation for the approximating functions as for the solution of the continuous problem and we assume now that the boundary data functions p_d and q_d are constant on each interval of the discretized boundary $\partial\Omega$.

To solve the problem by nonoverlaping domain decomposition techniques we follow ideas from [5] to reduce the problem in Ω to an interface problem on Γ. We introduce Dirichlet-to-Neumann operators $\mathcal{S}_i$ associated to each subdomain Ω_i as follows. Given λ_i in N we solve the problem

Find $\vec{\varphi}_i \in \vec{X}_i(q_d)$, $p_i \in M_i$ such that

$$\int_{\Omega_i} \mathrm{div}\vec{\varphi}_i = 0,$$

$$\int_{\Omega_i} K^{-1}\vec{\varphi}_i \cdot \vec{v} - \int_{\Omega_i} p_i \, \mathrm{div}\vec{v} + \int_{\partial\Omega_i \cap \partial\Omega_D} p_d \, \vec{v} \cdot \vec{n}_i \tag{2.5}$$

$$+ \int_\Gamma \lambda_i \, \vec{v} \cdot \vec{n}_i = 0, \quad \forall \vec{v} \in \vec{X}_i(0).$$

Then we define $\mathcal{S}_i$ by

$$\mathcal{S}_i(\lambda_i) = \vec{\varphi}_i \cdot \vec{n}_i \mid_\Gamma . \tag{2.6}$$

Our problem can now be rewritten as the interface problem

Find $\lambda_i \in N, i = 1, 2$, such that $\lambda_1 = \lambda_2$, $\mathcal{S}_1(\lambda_1) + \mathcal{S}_2(\lambda_2) = 0$. (2.7)

The first equality corresponds to pressure continuity while the second corresponds to mass conservation (see equation (2.4)). The Dirichlet-to-Neumann operators $\mathcal{S}_i$ are affine. Denote by $\overline{\mathcal{S}}_i$ the linear part of $\mathcal{S}_i$ and by $\widehat{\mathcal{S}}_i$ the constant part: $\mathcal{S}_i(\lambda_i) = \overline{\mathcal{S}}_i(\lambda_i) + \widehat{\mathcal{S}}_i$. With $\lambda = \lambda_1 = \lambda_2$, the domain decomposition method is reduced to the linear problem

$$\text{Find } \lambda \in N \text{ such that } (\mathcal{S}_1 + \mathcal{S}_2)\lambda = F \tag{2.8}$$

where $F = \widehat{\mathcal{S}}_1 + \widehat{\mathcal{S}}_2$. One can now apply a conjugate gradient method to calculate λ. Preconditionners have been studied. For instance a Neumann-Neumann preconditionner has been presented and analyzed in [6]. For the case of a decomposition with many subdomains where some do not touch the boundary of Ω, the balancing domain preconditionner is robust with respect to strong variations of the permeability K [7, 3].

In the following sections we show how these domain decomposition techniques apply to a variety of situations.

3 Two-Phase Flow with Two Rock Types

3.1 Formulation of the problem

We consider two-phase incompressible flow and we assume that the domain Ω is divided into two subdomains Ω_i, each subdomain corresponding to a rock type. This means that not only are the porosity and the absolute permeability different in Ω_1 and in Ω_2 but the relative permeability and capillary pressure curves are also.

Two-phase flow is formulated in terms of a saturation equation and a pressure equation using the global pressure [2]. We assume the flow to be incompressible and we neglect gravity.

The saturation equation expresses volume conservation for the wetting phase (which is equivalent to mass conservation since the flow is assumed to be incompressible), so inside each subdomain Ω_i we have

$$\begin{aligned} \Phi_i \frac{\partial S_i}{\partial t} + \operatorname{div} \vec{\varphi}_{wi} &= 0, \\ \vec{\varphi}_{wi} = \vec{r}_i + \vec{f}_i, \quad \vec{r}_i = -K_i a_i(S_i)\vec{\nabla} S_i, \quad \vec{f}_i &= K_i b_i(S_i)\vec{\varphi}_i, \end{aligned} \tag{3.1}$$

where $S_i = S_{wi}$ is the saturation of the wetting phase ($0 < S_i < 1$). Here Φ_i and K_i denote the porosity and the absolute permeability, and $\vec{\varphi}_i$ is the total Darcy velocity, the sum of the Darcy velocities of the wetting and the nonwetting phases:

$$\vec{\varphi}_i = \vec{\varphi}_{wi} + \vec{\varphi}_{nwi}.$$

The coefficients a_i and b_i depend on the mobilities k_{wi} and k_{nwi} and the capillary pressure p_{ci} which are functions of the saturation:

$$a_i = \frac{k_{wi} k_{nwi}}{k_{wi} + k_{nwi}} \frac{dp_{ci}}{dS}, \quad b_i = \frac{k_{wi}}{k_{wi} + k_{nwi}}.$$

The capillary pressure is $p_{ci} = p_{nwi} - p_{wi}$ where p_{nwi} and p_{wi} denote the pressures in the nonwetting and wetting phases.

Plugging the first equation of (3.1) into the second, one obtains for saturation equation a nonlinear parabolic equation of diffusion-advection type. The vector $\vec{r}_i$, the diffusive contribution to $\vec{\varphi}_{wi}$, is due to capillary effects and $\vec{f}_i$, the advective contribution to $\vec{\varphi}_{wi}$, depends on the total Darcy velocity $\vec{\varphi}_i$ which is given by the pressure equation that we now describe.

The pressure equation expresses the conservation of the total volume of the two phases. Since the flow is assumed to be incompressible this takes the form

$$\operatorname{div} \vec{\varphi}_i = 0, \quad \vec{\varphi}_i = -K_i d_i(S_i) \vec{\nabla} p_i, \tag{3.2}$$

where the global pressure p_i is given by

$$p_i = \frac{1}{2}(p_{wi} + p_{nwi}) + \gamma_i(S). \tag{3.3}$$

The coefficients γ_i and d_i are functions of the saturation S:

$$\gamma_i = \int_0^S \left(b_i(S) - \frac{1}{2}\right) \frac{dp_{ci}}{dS}, \quad d_i = k_{wi} + k_{nwi}.$$

Continuity of the phase pressures p_{wi} and p_{nwi} implies that the capillary pressure p_{ci}, and consequently the saturation S_i, is continuous, and that the global pressure p_i is also continuous (see definition (3.3)). Also, because of phase conservation, the normal components of the phase Darcy velocities $\vec{\varphi}_{wi}$ and $\vec{\varphi}_{nwi}$, and consequently the normal components of the total Darcy velocity $\vec{\varphi}_i$, are continuous across any hypersurface.

Now we come to the transmission condition across Γ and we assume here for sake of simplicity that the two capillary pressure curves have the same endpoints.

Across the interface Γ we still have phase conservation and continuity of the phase pressures. This latter condition implies that the capillary pressure is continuous since it is the difference of the phase pressures, and that the quantity $p - \gamma$ is continuous (see equation (3.3). Thus for the pressure equation we have the transmission conditions

$$p_1 - \gamma_1(S_1) = p_2 - \gamma_2(S_2), \quad \vec{\varphi}_1 \cdot \vec{n}_1 = \vec{\varphi}_2 \cdot \vec{n}_2. \tag{3.4}$$

This implies that in general the global pressure p is discontinuous across Γ. The second equation of (3.4) enforces conservation of the global mass of the two phases.

For the saturation equation the transmission conditions are

$$p_{c1}(S_1) = p_{c2}(S_2), \quad \vec{\varphi}_{w1} \cdot \vec{n}_1 = \vec{\varphi}_{w2} \cdot \vec{n}_2 \quad \text{on } \Gamma. \tag{3.5}$$

Thus the saturation is discontinuous in general; the second equation enforces conservation of the mass of the wetting phase. One should note that the first equation is a nonlinear transmission condition for the saturation.

3.2 The pressure equation

To equations (3.2) we add boundary conditions (2.2). Equation (3.4) shows that the situation differs now from that in §2 in that the pressure p is discontinuous across Γ with a given jump.

We discretize with the same ideas as in §2, and we introduce the Dirichlet-to-Neumann operator $\mathcal{S}_i$ which associates to $\lambda_i \in N$ the flow rate $\vec{\varphi}_i \cdot n_i$ where $\vec{\varphi}_i$ is the solution of

Find $\vec{\varphi}_i \in \vec{X}_i(q_d)$, $p_i \in M_i$ such that

$$\begin{aligned}
&\int_{\Omega_i} \text{div}\vec{\varphi}_i = 0, \\
&\int_{\Omega_i} (K_i d_i(S_i))^{-1} \vec{\varphi}_i \cdot \vec{v} - \int_{\Omega_i} p_i \, \text{div}\vec{v} + \int_{\partial\Omega_i \cap \Gamma_D} p_d \, \vec{v} \cdot \vec{n}_i \\
&\qquad\qquad + \int_\Gamma (\gamma_i(S_i) + \lambda_i) \, \vec{v} \cdot \vec{n}_i = 0, \quad \vec{v} \in \vec{X}_i(0).
\end{aligned} \tag{3.6}$$

With this definition of the Dirichlet-to-Neumann operators the problem reduces again to problems (2.7) and (2.8). In this case λ represents $p - \gamma(S)$.

3.3 The saturation equation

To equations (3.1) we add the boundary conditions

$$S = S_d \quad \text{on } \partial\Omega_{SD}, \quad \vec{\varphi}_w \cdot \vec{n} = q_{wd} \quad \text{on } \partial\Omega_{SN}. \tag{3.7}$$

Discretizing the saturation equation is more complex than discretizing the pressure equation. Indeed it is a nonlinear parabolic equation, often advection dominated, with a diffusion term which degenerates when the saturation is minimum or maximum. We propose the use of a semi-implicit Euler discretization in time [4]. When calculating the saturation at the $n + 1$ time level the advection term is lagging in time at the n time level and calculated with upstream values of the saturation, while the diffusion term is calculated at the $n + 1$ time level with the nonlinear coefficient a lagging also at the previous time level. Thus at each time step one has to solve only a linear system to calculate the saturation.

We assume that the data S_d and q_{wd} are constant on each interval of the boundary $\partial \Omega$ and the approximation spaces for S_i and $\vec{\varphi}_{wi}$ are the same as in the previous sections. Actually since the advective part of the flow $\vec{f}_i$ is calculated at the previous time level, the main flow unknowm is the diffusive part of the flow $\vec{r}_i$.

To simplify the presentation, we assume that the domain Ω is rectangular, discretized with rectangles with sides parallel to the x_1 and x_2 coordinates axes so we can use a cell-centered finite volume method. Also the interface Γ is supposed to be parallel to the x_2 axis as in Fig. 1. We denote by $\partial \Omega_{iB}, \partial \Omega_{iT}, \partial \Omega_{iL}$ and $\partial \Omega_{iR}$ the bottom, top, left and right parts of the boundary $\partial \Omega_i$.

With these assumptions the discretized saturation equation is

Find $\vec{r}_i \in \vec{X}_i(r_d^n)$, $S_i \in M_i$ such that

$$
\begin{aligned}
&\int_C \Phi_i \frac{S_i^{n+1} - S_i^n}{\Delta t} + \int_{\partial C} F_{wi}^{n*} = 0 \quad \forall C \in \mathcal{T}_i, \\
&F_{wi}^{n*} = \vec{r}_i^{n+1} \cdot \vec{n_C} + F_i^n, \\
&F_i^n = \vec{\varphi}_i^n \cdot \vec{n_C}\, b_i(S_{i-}^n), \\
&\vec{r}_i^{n+1} \cdot \vec{n}_E \mid_E = -\overline{K}_{iE}^H a(\overline{S}_{iE}^n) \frac{S_{iC_{E1}}^{n+1} - S_{iC_{E2}}^{n+1}}{h_E} \quad \forall E \in \mathcal{E}_i, E \not\subset \partial \Omega_{SN}, \\
&S_{iC_{E1}}^{n+1} = S_d \quad \text{when } E \subset \partial \Omega_{SD} \cap (\partial \Omega_{iB} \cup \partial \Omega_{iL}), \\
&S_{iC_{E2}}^{n+1} = S_d \quad \text{when } E \subset \partial \Omega_{SD} \cap (\partial \Omega_{iT} \cup \partial \Omega_{iR}), \\
&S_{iC_{E2}}^{n+1} = \lambda_1 \quad \text{when } E \subset \Gamma,\ i = 1, \\
&S_{iC_{E1}}^{n+1} = \lambda_2 \quad \text{when } E \subset \Gamma,\ i = 2,
\end{aligned}
\tag{3.8}
$$

with the transmission conditions

$$
p_{c1}(\lambda_1) = p_{c2}(\lambda_2), \quad F_{w1}^{n*} + F_{w2}^{n*} = 0 \text{ on } \Gamma. \tag{3.9}
$$

We used the notation: $\vec{n}_C$ is the outward normal to ∂C, $\vec{n}_E$ is the normal to the edge E pointing in the positive x_1 direction if E is vertical or pointing in the positive x_2 direction if E is horizontal, S_{i-}^n is the saturation value which is upstream with respect to $\vec{\varphi}_i^n$, and $r_d^n = q_{wd}^n - F_i^n$. When E is an interior edge we denote by C_{E1} and C_{E2} the two cells adjacent to E, $\overline{K}_{iE}^{H}$ denotes the harmonic average of K in these two cells, $\overline{S}_{iE}^{n}$ denotes the standard average of $S_{iC_{E1}}^{n+1}$ and $S_{iC_{E2}}^{n+1}$, and h_E is equal to the space discretization step. When E is a boundary edge $\overline{K}_{iE}^{H}$ is just the value of K in the neighbouring edge and h_E is equal to half the space discretization step.

To rewrite problem (3.8) and (3.9) as an interface problem we proceed as before and we introduce for each subdomain Ω_i the linear Dirichlet-to-Neumann operator $\mathcal{S}_i$ defined as

$$\mathcal{S}_i(\lambda_i) = (\vec{r}_i^{n+1} \cdot \vec{n}_i + F_i^n)\,|_\Gamma,$$

where $\vec{r}_i^{n+1}$ is calculated by solving equations (3.8) inside each subdomain Ω_i with a given λ_i on Γ.

Our problem can now be rewritten as the interface problem

$$\begin{aligned}
&\text{Find } \lambda_i \in N, i = 1, 2, \text{ such that} \\
&p_{c1}(\lambda_1) = p_{c2}(\lambda_2), \quad \mathcal{S}_1(\lambda_1) + \mathcal{S}_2(\lambda_2) = 0.
\end{aligned} \tag{3.10}$$

One observes that the first equation is nonlinear and that the second equation implies that $\vec{r} \cdot n$ is discontinuous across Γ. One way to solve problem (3.10) is to use incomplete Newton iterations with a preconditionned GMRES.

Remark. It is possible to make the diffusion term fully implicit. This would lead to use a nonlinear Dirichlet-to-Neumann operator.

3.4   A numerical example

As an example we consider the displacement of a nonwetting fluid by a wetting fluid with a mobility 10 times larger. The domain of calculation has two regions with different rock types, that on the left having an absolute permeability 5 times larger than that on the right. The injection is parallel to the interface Γ between the two rock types (see Fig. 2).

The capillary functions are given by the standard formula

$$p_c(S) = J(S)\sqrt{\frac{\Phi}{K}}$$

and are shown in Fig. 3. Since K is larger in Ω_1 than in Ω_2, the capillary pressure is smaller in Ω_1 than in Ω_2, so at the interface we can expect a discontinuous saturation smaller in Ω_1. Figure 4 shows numerical results at a certain time. The picture on the right shows the total Darcy velocity field $\vec{\varphi}$ at a certain time. Since the injection rate is constant along the bottom

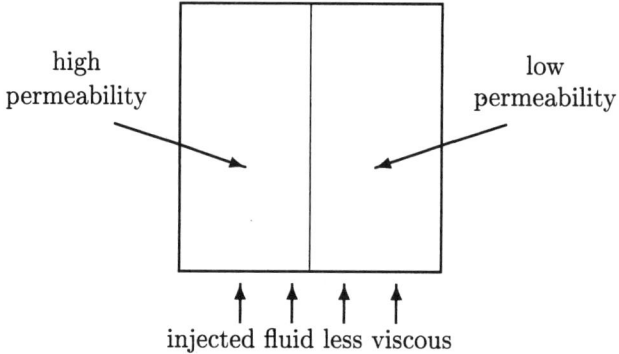

Figure 2: An example of a displacement in a medium with two rock types.

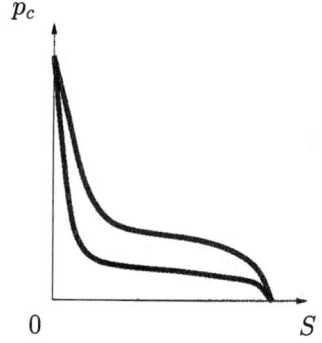

Figure 3: Capillary pressure curves, p_{c1} lower and p_{c2} upper.

boundary and the permeability is low on the right the Darcy velocity in Ω_2 turns to the left in the vicinity of the injection boundary. The picture on the left shows the saturation of the injected fluid at the same time. One can observe the discontinuity at the interface. Note that along this interface, where the saturation is not equal to zero, the saturation is smaller on the left than on the right because the capillary pressure curve is smaller on the left than on the right. However further away from this interface it is the opposite: the saturation, as the absolute permeability, is larger on the left than on the right.

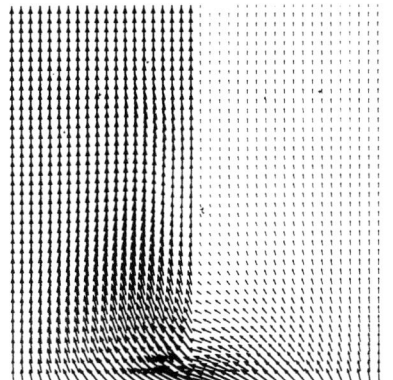 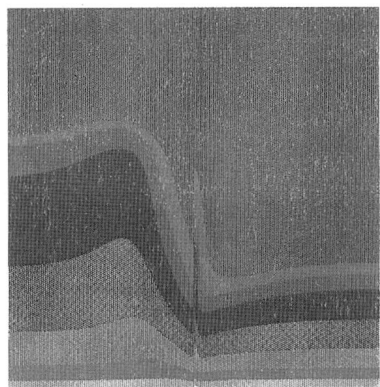

Figure 4: Calculated total Darcy velocity (left) and saturation (right).

4 Domain Decomposition with Fractures

4.1 Formulation of the problem

In this section the domain Ω is divided into two subdomains $\Omega_i, i = 1, 2$, by a fracture Γ which is also a porous medium but with higher permeability. This fracture is assumed to have a width small compared to the size of the whole domain, so in the numerical model it is modelled as the interface Γ between the subdomains.

Here we are interested in understanding the interaction between the flow in the subdomains and the flow in the fracture. We assume that the flow in the subdomains as well as in the fracture is governed by Darcy's law, is incompressible and we neglect gravity.

Interaction between the fracture and the subdomains is assumed to satisfy mass conservation and pressure continuity. Thus we consider the set of equations: in the subdomains:

$$\text{div}(\vec{\varphi}_i) = 0, \quad \vec{\varphi}_i = -K_i \vec{\nabla} p_i \quad \text{in } \Omega_i,$$
$$p_i = p_f \text{ on } \partial\Omega_i \cap \Gamma, \ p_i = p_d \text{ on } \partial\Omega_i \cap \partial\Omega_D, \ \vec{\varphi}_i \cdot \vec{n}_i = q_d \text{ on } \partial\Omega_i \cap \partial\Omega_N,$$

and in the fracture:

$$\frac{\partial \varphi_f}{\partial x_f} = \vec{\varphi}_1 \cdot \vec{n}_1 + \vec{\varphi}_2 \cdot \vec{n}_2, \ \varphi_f = -\sigma_f \, K_f \frac{\partial p_f}{\partial x_f} \quad \text{on } \Gamma, \tag{4.1}$$
$$p_f = p_{f,d} \text{ on } \partial\Gamma_D, \ \varphi_f = q_{fd} \text{ on } \partial\Gamma_N.$$

Here K_f, σ_f, p_f and φ_f denote the permeability, the width, the pressure and the flow rate in the fracture, and $\partial/\partial x_f$ denotes the derivative along the fracture.

At the extremities of the fracture there are Dirichlet (on $\partial\Gamma_D$) or Neumann (on $\partial\Gamma_N$) boundary conditions. If $\partial\Gamma_D$ touches $\partial\Omega_i \cap \partial\Omega_D$ then pressure continuity implies that the pressure data must satisfy $p_d = p_{fd}$.

The first equation in (4.1) expresses mass conservation for the flow in the fracture. The righthand side in this equation is the contribution of the subdomain flow to the fracture flow.

We proceed now as in §2, defining the same Dirichlet-to-Neumann operators (2.5) and (2.6) and using the same notation. Moreover on Γ the pressure p_f is approximated in a space M_f of functions constant on each interval, while the flow rate φ_f is approximated in $X_f(q_{fd})$ a space of continuous piecewise linear functions which are equal to q_{fd} on $\partial\Gamma_N$.

This results in the interface problem

Find $\varphi_f \in X_f(q_{fd})$, $p_f \in N$ such that
$$\int_{\Gamma} S_1(p_f) + S_2(p_f) + \frac{\partial\varphi_f}{\partial x_f} = 0,$$
$$\int_{\Gamma} (\sigma_f K_f)^{-1} \varphi_f v - \int_{\Gamma} p_f \frac{\partial v}{\partial x_f} + (p_{fd} v)|_{\partial\Gamma_D} = 0 \quad \forall v \in X_f(q_0).$$

Comparing with problem (2.8) we observe that the problem to be solved here is a global equation on the interface Γ.

4.2   A numerical experiment

To illustrate the model we consider an ideal dimensionless problem. The domain is an horizontal rectangular slice of porous medium, of dimensions 2×1, with a given pressure on the left and right boundaries and no flow conditions on the top and bottom boundaries. In the domain the permeability is equal to one. The domain is divided into two equally large sub-domains by a linear fracture parallel to the x_2 axis. In the fracture we chose $\sigma_f K_f = 2$. For example the fracture could be of width 0.1 and could have a permeability equal to 20.

Two cases are considered. A symmetric case where pressures on the left and on the right boundaries of the domain are equal. So the flow is driven only by the fracture and is symmetric. In the other case there is a pressure drop from the right boundary to the left one. Then the flow is a combination of the flow in the fracture and that going from left to right in the rest of the porous medium. Flow in the fracture is driven by a pressure drop of 10

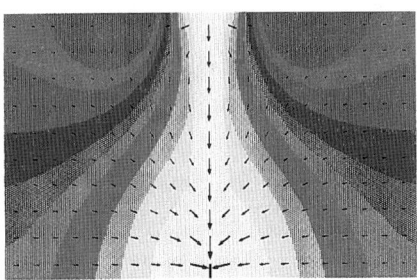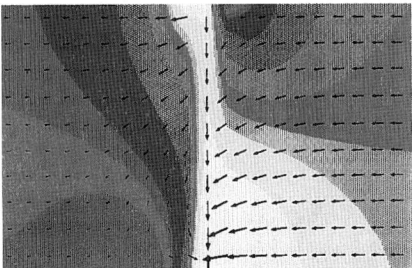

Figure 5: Calculated Darcy's velocity for a symmetric and a nonsymmetric flow pattern

between the two extremities of the fracture for the first case and a pressure drop of 5 for the second case.

Numerical results are shown on Fig. 5. Arrows represent the flow field with length proportional to the magnitude of the velocity. The gray scale represents the magnitude of the velocity with the lightest color corresponding to the largest velocity. We see that there is actual flow interaction between the fracture and the rest of the porous medium. In particular one can observe that some fluid is coming out of the fracture and then is coming back into it. In the nonsymmetric case we notice also that even though most of the flow is attracted into the fracture, there is still some flow on the left part of the domain pointing toward the left.

5 Conclusions

Studying a few examples of flow in porous media, we met a variety of transmission conditions which are nonstandard: discontinuity of the scalar variable (pressure or saturation), discontinuity of the flow rate variable (capillary flow), nonlinear transmission condition (saturation), nonlocal transmission condition (porous medium with fractures). In spite of this variety of situations we showed that domain decomposition techniques based on Dirichlet-to-Neumann operators can be used to set these problems as interface problems to be solved using domain decomposition algorithms.

References

[1] Brezzi, F. and Fortin, M., *Mixed and Hybrid Finite Element Methods*, Springer Verlag, Berlin, 1991.

[2] Chavent, G. and Jaffré, J., *Mathematical Models and Finite Elements for Reservoir Simulation*, volume 17 of *Studies in Mathematics and its Applications*. North Holland, Amsterdam, Amsterdam, 1986.

[3] Cowsar, L., Mandel, J., and Wheeler, M., Balancing domain decomposition for mixed finite elements, *Math. of Comp.* **64** (1993), 989–1015.

[4] Douglas, J., Jr. and Dupont, T., Galerkin methods for parabolic equations, *SINUM* **7** (1970), 575–626.

[5] Glowinski, R. and Wheeler, M., Domain decomposition and mixed finite element methods for elliptic problems, in Glowinski, R. et al., editor, *Proceedings of the First Symposium on Domain Decomposition Methods for PDEs*, SIAM, Philadelphia, 1987, 144–172.

[6] Le Tallec, P., De Roeck, Y.-H., and Vidrascu, M., Domain decomposition methods for large linearly elliptic three dimensional problems, *J. Comp. Appl. Math.* **34** (1991). 341–362.

[7] Mandel, J., Balancing domain decomposition, *Comm. in Numerical Methods in Engineering* **9** (1993), 233–241.

[8] Raviart, P.-A. and Thomas, J.-M., A mixed finite element method method for second order elliptic problems, in I. Galligani and E. Magenes, editors, Mathematical Aspects of Finite Element Methods; Lecture Notes in Mathematics 606, Springer, Berlin, 1977, 292–315.

[9] Roberts, J. E. and Thomas, J.-M., Mixed and hybrid methods, in P.G. Ciarlet and J.L. Lions, editors, Handbook of Numerical Analysis Vol.II, North Holland, Amsterdam, 1991, 523–639.

[10] Smith, B., Bjorstadt, P., and Gropp, W., *Domain Decomposition: Parallel Multilevel Methods for Elliptic Partial Differential Equations*, Cambridge University Press, 1996.

Numerical Subgrid Upscaling
of Two-Phase Flow in Porous Media

TODD ARBOGAST

Abstract

We present an approach and numerical results for scaling up fine grid information to coarse scales in an approximation to a nonlinear parabolic system governing two-phase flow in porous media. The technique allows upscaling of the usual parameters porosity and relative and absolute permeabilities, and also the location of wells and capillary pressure. Some of these are critical nonlinear terms that need to be resolved on the fine scale, or serious errors will result. Upscaling is achieved by explicitly decomposing the differential system into a coarse-grid-scale operator coupled to a subgrid-scale operator, which we localize by imposing a closure assumption. We approximate the coarse-grid-scale operator with a mixed finite element method that has a second order accurate velocity coupled implicitly to the subgrid scale. The subgrid-scale operator is approximated locally by a first order accurate mixed method. A numerical Greens influence function technique allows us to solve these subgrid problems independently of the coarse-grid approximation. No explicit macroscopic coefficients nor pseudo-functions result. The method is easily seen to be optimally convergent in the case of a single linear parabolic equation.

KEYWORDS: upscaling, subgrid, numerical Greens functions, porous media

1 Introduction

In many physical problems, there are scales that are too fine to resolve on any reasonable computational mesh. The objective of upscaling or homogenization is to replace the governing equations by a simpler set of equations for which the solution can be resolved on a reasonable coarse-scale mesh and approximates the average behavior of the solution of the governing equations. In its simplest form, one replaces the coefficients of the governing equations with an effective or macroscopic coefficient [5]. This works well in certain situations [9, 1], but not so well in others. Often it is necessary to change the form of the governing equations to obtain a suitable coarse-scale model [3];

there is no *a priori* reason to expect otherwise. Such is especially the case when nonlinearities are present, since it is well known that a function of an average is *not* the average of the function. Various techniques are used in this case, including homogenization [1, 10], the definition of pseudo-functions (altered forms of the nonlinear functions that appear in the governing equations), the use of renormalization methods [8], and many other techniques.

In terms of the simulation of flow and transport in a porous medium, our goal for nonlinear upscaling in this paper will be to resolve some of the finer scales in the solution directly, so that no loss of accuracy due solely to averaging will result. We will then be able to incorporate directly into the simulation relative and absolute permeability, porosity, capillary pressure, and well location information on scales smaller than the computational grid. That is, our nonlinear functions such as relative permeability and capillary pressure need not be modified, since the fine scales have been sufficiently resolved. Our technique is based on numerics. We assume that a fine grid fully represents the important physical scales, and that our computational grid is somewhat coarser. That is, perhaps some other homogenization technique has elevated the problem to a reasonable fine scale, but this fine scale is still too fine to compute over.

We present our ideas by considering first a problem representing incompressible, single-phase flow in a porous medium in the absence of gravity:

$$\nabla \cdot \mathbf{u} = f, \quad \mathbf{u} = -K \nabla p, \tag{1.1}$$

where p is the pressure, K is the permeability divided by the fluid viscosity, $\mathbf{u}$ is the Darcy velocity, and f represents the wells. For simplicity, set $\mathbf{u} \cdot \nu = 0$ on the external boundary.

As an outline of the paper, we present in §2 a derivation of the upscaled equations for single phase flow, including a definition of our closure assumption. In §3 we give a mixed finite element approximation of the equations that is compatible with the closure assumption. A solution technique based on the computation of numerical Greens functions is given in §4. The accuracy of the method is discussed in §5. Finally in §6, we present briefly the two-phase problem, followed in §7 by some numerical results.

2 Derivation of the Upscaled Equations

We rewrite (1.1) in variational form as: Find $p \in W = L^2$ and $\mathbf{u} \in \mathbf{V} = H(\text{div})$ such that

$$
\begin{aligned}
\int \nabla \cdot \mathbf{u}\, w\, dx &= \int f\, w\, dx & \forall w \in W, \\
\int K^{-1}\, \mathbf{u} \cdot \mathbf{v}\, dx = -\int \nabla p \cdot \mathbf{v}\, dx &= \int p\, \nabla \cdot \mathbf{v}\, dx & \forall v \in \mathbf{V},
\end{aligned}
\tag{2.1}
$$

where $H(\text{div}) = \{ \mathbf{v} \in (L^2)^3 : \nabla \cdot \mathbf{v} \in L^2, \mathbf{v} \cdot \nu = 0 \text{ on the external boundary} \}$.

Let W_c and $\mathbf{V}_c$ be the computationally resolvable parts of W and $\mathbf{V}$, and δW and $\delta \mathbf{V}$ the remainders. That is,

$$
\begin{aligned}
W &= W_c \oplus \delta W, & \mathbf{V} &= \mathbf{V}_c \oplus \delta \mathbf{V}, \\
p &= p_c + \delta p \in W_c \oplus \delta W, & \mathbf{u} &= \mathbf{u}_c + \delta \mathbf{u} \in \mathbf{V}_c \oplus \delta \mathbf{V}.
\end{aligned}
$$

2.1 Separation of scales

Separate the fine and δ-scales by restricting to appropriate test functions in the variational formulation. For the coarse scale, we have

$$
\begin{aligned}
\int \nabla \cdot (\mathbf{u}_c + \delta \mathbf{u})\, w_c\, dx &= \int f\, w_c\, dx & \forall w_c \in W_c, \\
\int K^{-1}\, (\mathbf{u}_c + \delta \mathbf{u}) \cdot \mathbf{v}_c\, dx &= \int (p_c + \delta p)\, \nabla \cdot \mathbf{v}_c\, dx & \forall \mathbf{v}_c \in \mathbf{V}_c,
\end{aligned}
\tag{2.2}
$$

and for the δ-scale,

$$
\begin{aligned}
\int \nabla \cdot (\mathbf{u}_c + \delta \mathbf{u})\, \delta w\, dx &= \int f\, \delta w\, dx & \forall \delta w \in \delta W, \\
\int K^{-1}\, (\mathbf{u}_c + \delta \mathbf{u}) \cdot \delta \mathbf{v}\, dx &= \int (p_c + \delta p)\, \nabla \cdot \delta \mathbf{v}\, dx & \forall \delta \mathbf{v} \in \delta \mathbf{V}.
\end{aligned}
\tag{2.3}
$$

If we were to ignore the δ-scales (i.e., perform no upscaling), then we would simply set $\delta \mathbf{u} = 0$, $\delta p = 0$, and use only the coarse equation (2.2). Upscaling concerns the treatment of these other terms and (2.3).

Given $(\mathbf{u}_c, p_c, f)$, we can solve for

$$
\delta \mathbf{u} = \Phi_u(\mathbf{u}_c, p_c, f) \quad \text{and} \quad \delta p = \Phi_p(\mathbf{u}_c, p_c, f),
\tag{2.4}
$$

where Φ is a multi-linear operator. Thus (2.2) becomes

$$
\begin{aligned}
\int \nabla \cdot (\mathbf{u}_c + \Phi_u(\mathbf{u}_c, p_c, f))\, w_c\, dx &= \int f\, w_c\, dx & \forall w_c \in W_c, \\
\int K^{-1}\, (\mathbf{u}_c + \Phi_u(\mathbf{u}_c, p_c, f)) \cdot \mathbf{v}_c\, dx & \\
= \int (p_c + \Phi_p(\mathbf{u}_c, p_c, f))\, \nabla \cdot \mathbf{v}_c\, dx & & \forall \mathbf{v}_c \in \mathbf{V}_c,
\end{aligned}
\tag{2.5}
$$

posed only on the coarse scale. We remark that no approximation has been made yet; all scales are fully resolved by (2.3)–(2.5). However, these two equations are intrinsically coupled, since Φ is a nonlocal operator.

2.2   Closure assumption (localization approximation)

Define a coarse computational grid and assume that

$$\delta \mathbf{V} \cdot \nu = 0 \qquad \text{on } \partial E_c \tag{2.6}$$

for each coarse element E_c. Then Φ (i.e., (2.3)) becomes a local operator:

$$\int_{E_c} \nabla \cdot (\mathbf{u}_c + \delta\mathbf{u})\, \delta w\, dx = \int_{E_c} f\, \delta w\, dx \qquad \forall\, \delta w \in \delta W|_{E_c},$$
$$\int_{E_c} K^{-1} (\mathbf{u}_c + \delta\mathbf{u}) \cdot \delta\mathbf{v}\, dx = \int_{E_c} (p_c + \delta p)\, \nabla \cdot \delta\mathbf{v}\, dx \qquad \forall\, \delta\mathbf{v} \in \delta V|_{E_c}.$$
$$\tag{2.7}$$

Condition (2.6) is our closure assumption. We have assumed that all net flux between coarse elements occurs only on the coarse scale.

3   Mixed Finite Element Approximation

We assume that nested fine and coarse computational grids are used, and let h and H be the grid spacings, respectively. The fine grid is assumed to be essentially what is needed to fully resolve the physical scales. Generally speaking, we envision H/h as a moderate integer (4 to 10, say).

We approximate (2.4)–(2.7) by a mixed finite element method. Other discretizations could be employed; however, the local conservation of these methods make them attractive for porous media simulation [12]. They approximate both the pressure and Darcy velocity directly.

Coarse BDDF$_1$ velocities δ-scale RT$_0$ velocities

FIGURE 1: The degrees of freedom of the approximating spaces.

As depicted in Fig. 1, we approximate the coarse equation (2.5) on the coarse grid with BDDF$_1$ spaces (BDM$_1$ in 2-D) [6, 7]. These spaces have 1 degree of freedom per coarse element for the pressure approximation, W_c, and 3 degrees of freedom per coarse element face (2 per edge in 2-D) for the velocity, $\mathbf{V}_c$. They are second order accurate in H for the velocity and first order for the pressure. The fine grid is used for the δ-equation (2.7). We use within each coarse element RT$_0$ spaces [11]. These have one degree

of freedom per fine element for the pressure approximation, δW, subject
to the requirement of orthogonality to W_c that the average over the coarse
element vanishes, and RT_0 has one degree of freedom per fine element face
for the velocity, δV, subject to the closure assumption (2.6). They are first
order accurate in h for both the pressure and velocity for pressures with zero
average over coarse elements and velocities with zero normal components on
coarse element boundaries.

We remark that pressure is resolved fully on the fine scale, and formally
approximated to first order in h. The fact that the BDDF$_1$ space approximates velocity to second order compensates for the closure assumption, which
assumes all net flow between coarse elements is on the coarse scale. Without
this choice, the results degrade significantly [4].

4   Solution by Numerical Greens Functions

We present now a technique to solve the system of equations efficiently. Before
we elaborate, the outline of the technique is as follows.

1. Pre-solve for the influence of the coarse scale on the δ-scale. These are
   small disjoint problems, one for each coarse element, by (2.6). These
   pre-solutions are numerical Greens functions for the δ-problems (2.7).
2. Solve the coarse scale problem (2.5), accounting for the pre-solution
   response of the δ-scale to the coarse scale in (2.4).
3. Post-solve to combine results to form the fine-scale solution.

Integrals over coarse elements of the form $W_c * \delta W$ vanish by orthogonality.
Since $\nabla \cdot V_c = W_c$ and $\nabla \cdot \delta V = \delta W$, integrals of $\nabla \cdot \delta V * W_c$ and $\nabla \cdot V_c * \delta W$
also vanish; thus, several terms in the equations below vanish.

4.1   Pre-solution

Locally on each coarse element E_c, let $v_{c,i} \in V_c$ have flux only at a single
degree of freedom ($i = 1, ..., 18$ in 3-D and $i = 1, ..., 8$ in 2-D). Then

$$\mathbf{u}_c = \sum_i \alpha_i \mathbf{v}_{c,i}. \tag{4.1}$$

Solve the following problems for the numerical Greens functions.
Nonhomogeneous terms. Find $\delta \mathbf{u}_0 \in \delta V$ and $\delta p_0 \in \delta W$ such that

$$\begin{aligned}
\int \nabla \cdot \delta \mathbf{u}_0 \, \delta w \, dx &= \int f \, \delta w \, dx & \forall \, \delta w \in \delta W|_{E_c}, \\
\int K^{-1} \delta \mathbf{u}_0 \cdot \delta \mathbf{v} \, dx &= \int \delta p_0 \, \nabla \cdot \delta \mathbf{v} \, dx & \forall \, \delta \mathbf{v} \in \delta V|_{E_c}.
\end{aligned} \tag{4.2}$$

Influence of $\mathbf{v}_{c,i}$. For each i, find $\delta\mathbf{u}_i \in \delta\mathbf{V}$ and $\delta p_i \in \delta W$ such that

$$\int \nabla \cdot \delta\mathbf{u}_i \, \delta w \, dx = 0 \qquad\qquad \forall \, \delta w \in \delta W|_{E_c},$$
$$\int K^{-1}(\mathbf{v}_{c,i} + \delta\mathbf{u}_i) \cdot \delta\mathbf{v} \, dx = \int \delta p_i \, \nabla \cdot \delta\mathbf{v} \, dx \qquad \forall \, \delta\mathbf{v} \in \delta\mathbf{V}|_{E_c}. \tag{4.3}$$

Note that the combinations

$$\delta\mathbf{u}_0 + \textstyle\sum_i \alpha_i \, \delta\mathbf{u}_i \equiv \Phi_u(\mathbf{u}_c, f) = \delta\mathbf{u},$$
$$\delta p_0 + \textstyle\sum_i \alpha_i \, \delta p_i \equiv \Phi_p(\mathbf{u}_c, f) = \delta p \tag{4.4}$$

depend linearly on the (as yet unknown) nodal values of $\mathbf{u}_c$ and on the numerical Greens functions and give $\delta\mathbf{u}$ and δp solving (2.7).

4.2   Coarse solution

Given the numerical Greens functions and the implicit representation of the upscaling operator (4.4), we can now reformulate (2.5) as a problem for the course unknowns only. We find $\mathbf{u}_c \in \mathbf{V}_c$ and $p_c \in W_c$ such that

$$\int \nabla \cdot \mathbf{u}_c \, w_c \, dx = \int f \, w_c \, dx \qquad\qquad \forall \, w_c \in W_c,$$
$$\int K^{-1}(\mathbf{u}_c + \Phi_u(\mathbf{u}_c, f)) \cdot \mathbf{v}_c \, dx = \int p_c \, \nabla \cdot \mathbf{v}_c \, dx \qquad \forall \, \mathbf{v}_c \in \mathbf{V}_c. \tag{4.5}$$

We rewrite this system with $\mathbf{v}_c = \mathbf{v}_j$ using that

$$\mathbf{u} = \mathbf{u}_c + \delta\mathbf{u} = \delta\mathbf{u}_0 + \textstyle\sum_i \alpha_i \, (\mathbf{v}_{c,i} + \delta\mathbf{u}_i),$$
$$p = p_c + \delta p = p_c + \delta p_0 + \textstyle\sum_i \alpha_i \, \delta p_i, \tag{4.6}$$

and using (4.2) and (4.3) with $\delta\mathbf{v} = \delta\mathbf{u}_j$ and orthogonality as

$$\textstyle\sum_i \alpha_i \int \nabla \cdot (\mathbf{v}_{c,i} + \delta\mathbf{u}_i) \, w_c \, dx = \int f \, w_c \, dx,$$
$$\textstyle\sum_i \alpha_i \int K^{-1}(\mathbf{v}_{c,i} + \delta\mathbf{u}_i) \cdot (\mathbf{v}_{c,j} + \delta\mathbf{u}_j) \, dx$$
$$= \int (p_c + \delta p_0) \, \nabla \cdot (\mathbf{v}_{c,j} + \delta\mathbf{u}_j) \, dx - \int K^{-1} \delta\mathbf{u}_0 \cdot (\mathbf{v}_{c,j} + \delta\mathbf{u}_j) \, dx.$$

Thus the method is similar to an "optimal test function" method where we replace $\mathbf{v}_c$ by $\mathbf{v}_{c,j} + \delta\mathbf{u}_j$; however, we also add some nonhomogeneous terms that improve the accuracy over such "optimal" methods.

4.3   Post-solution

Given $\mathbf{u}_c$, p_c, and the numerical Greens functions, compute (4.6) on the fine scale to obtain a "fully resolved" approximation of the true solution.

5   Accuracy

Denote by $\| \cdot \|$ the L^2-norm, and by P_{W_c}, P_{W_f}, and $P_{\delta W}$ the L^2-projections into W_c, the full fine grid space $W_f = W_c \oplus \delta W$, and δW, respectively.

If we solve the entire problem (2.1) for $(\mathbf{u}_f, p_f) \in \mathrm{RT}_0$ over the entire fine mesh or for $(\mathbf{u}_c, p_c) \in \mathrm{BDDF}_1$ (or BDM_1 in 2-D) on the coarse mesh, we see the following error estimates [11, 6, 7], where $(\mathbf{u}, p)$ is the true solution.

Theorem 5.1 *For RT_0 with no upscaling on the fine mesh, for $\mathbf{v} \in \mathbf{V}_f$ such that $\nabla \cdot \mathbf{v} = P_{W_f} f$,*

$$\|K^{-1/2}(\mathbf{u} - \mathbf{u}_f)\| \le \inf_{\mathbf{v}} \|K^{-1/2}(\mathbf{u} - \mathbf{v})\| \le Ch, \quad \|p - p_f\| \le Ch.$$

For $BDDF_1$ (or BDM_1) with no upscaling on the coarse mesh, for $\mathbf{v} \in \mathbf{V}_c$ such that $\nabla \cdot \mathbf{v} = P_{W_c} f$,

$$\|K^{-1/2}(\mathbf{u} - \mathbf{u}_c)\| \le \inf_{\mathbf{v}} \|K^{-1/2}(\mathbf{u} - \mathbf{v})\| \le CH^2, \quad \|p - p_c\| \le CH.$$

The upscaling technique displays elements of both estimates above. It is easy to prove the following error estimate.

Theorem 5.2 *For $BDDF_1$ (or BDM_1) upscaled with the RT_0 subgrid approximation, for $\mathbf{v} \in \mathbf{V}_c + \delta\mathbf{V}$ such that $\nabla \cdot \mathbf{v} = P_{W_f} f$, $\mathbf{v}_c \in \mathbf{V}_c$ such that $\nabla \cdot \mathbf{v}_c = P_{W_c} f$, and $\delta\mathbf{v} \in \delta\mathbf{V}$ such that $\nabla \cdot \delta\mathbf{v} = P_{\delta W} f$,*

$$\|K^{-1/2}(\mathbf{u} - (\mathbf{u}_c + \delta\mathbf{u}))\| \le \inf_{\mathbf{v}} \|K^{-1/2}(\mathbf{u} - \mathbf{v})\|$$
$$\le \inf_{\bar{\mathbf{u}},\mathbf{v}_c,\delta\mathbf{v}} \left\{ \|K^{-1/2}(\bar{\mathbf{u}} - \mathbf{v}_c)\| + \|K^{-1/2}(\mathbf{u} - \bar{\mathbf{u}} - \delta\mathbf{v})\| \right\},$$
$$\|p - p_c\| \le CH.$$

While it is difficult to interpret the velocity error, we should expect better bounds than in Theorem 5.1; that is, we should expect the velocity error to be second order. Our numerical results suggest that this is indeed the case, and that the error has in fact no simple form (as indicated in the theorem). We consider two test cases in which $K = 1$ on a unit square domain and we use Dirichlet boundary conditions and f defined from the imposed true solution $p(x, y) = xy^3 + x^2y \cos(xy)$ or $1/(1 + \exp(10x + 10y^2 - 3y - 5))$, respectively.

It is readily apparent from the data in Table 1, that if H/h is fixed, the error in pressure is $O(H) = O(h)$, and the error in velocity is $O(H^2) = O(h^2)$, as we would expect. However, if H/h is *not* fixed, the results are much less predictable. For the tests reported and a few more conducted, the best fit of the L^2-error in $\mathbf{u}$ is $E_u^1 = 150H^2 + 360H^{1/2}h^{3/2}$ and $E_u^2 = 1000H^2 + 8000H^{1/4}h^{7/4}$, respectively for the two cases, and for the pressure p, the error is $E_p^1 = 0.36h + 0.0003H$ and $E_p^2 = 0.4h + 0.0006H^{1.7}h^{-0.7}$, respectively. Thus the error depends in a complicated way on the solution and probably on H/h, but p is first order and $\mathbf{u}$ is second order accurate.

It is interesting to note from Table 1 that the error in the pressure is dominated by the fine mesh size, nearly independently of the coarse mesh.

		Case 1				Case 2	
$1/h$	$1/H$	Pressure	Velocity	$1/h$	$1/H$	Pressure	Velocity
10	10	0.0359	5.63	10	10	0.0438	173.64
20	20	0.0180	1.42	20	20	0.0219	47.08
40	40	0.0090	0.36	40	40	0.0109	12.03
80	80	0.0045	0.16	80	80	0.0055	3.03
10	2	0.0359	46.79	10	2	0.0440	413.35
20	4	0.0180	11.14	20	4	0.0220	185.70
40	8	0.0090	2.76	40	8	0.0109	43.25
80	16	0.0045	0.69	80	16	0.0055	10.82
160	4	0.0023	8.97	80	2	0.0095	273.92
160	8	0.0022	2.26	80	4	0.0060	159.92
160	16	0.0022	0.60	80	8	0.0055	36.50
160	32	0.0022	0.18	80	16	0.0055	10.82
10	2	0.0359	46.79	10	2	0.0440	413.35
20	2	0.0180	41.09	20	2	0.0230	310.56
40	2	0.0091	39.39	40	2	0.0134	281.47
80	2	0.0046	38.94	80	2	0.0095	273.92

Table 1. Some L^2-errors for Cases 1 and 2.

6   Two-phase Immiscible, Incompressible Flow

We now turn to a nonlinear problem describing the flow of two immiscible, incompressible fluids in a porous medium, such as oil and water. For phase $j = w, o$ (i.e., water and oil), let s_j, $\mathbf{u}_j$, and p_j be the phase saturations, Darcy velocities, and pressures. Let $s = s_w = 1 - s_o$, ϕ be the porosity, K the absolute permeability, g the gravitational constant, and z the depth. The mobilities are related to the relative permeabilities and fluid viscosities as $\lambda_j(s) = k_{r,j}(s)/\mu_j$ and $\lambda(s) = \lambda_w(s) + \lambda_o(s)$, and $P_c(s) = p_o - p_w$ is the capillary pressure. Conservation of mass of each phase gives the governing equations. After reformulation, we obtain the following (see, e.g., [2]).

Pressure equation:

$$\nabla \cdot \mathbf{u} = f \equiv f_w + f_o, \quad \mathbf{u} = -K\lambda(s)(\nabla p - \rho(s)\nabla z),$$

where the global pressure and density are

$$p = p_o - \int_0^s \frac{\lambda_w(\sigma)}{\lambda(\sigma)} P_c'(\sigma) \, d\sigma \quad \text{and} \quad \rho(s) = \frac{\lambda_w(s)}{\lambda(s)} \rho_w + \frac{\lambda_o(s)}{\lambda(s)} \rho_o.$$

Saturation equation:

$$\phi \frac{\partial s}{\partial t} + \nabla \cdot \mathbf{u}_w = f_w(s), \quad \mathbf{u}_w = -K \nabla q(s) + \gamma(\mathbf{u}, s),$$

where the "complementary" potential and γ are

$$q(s) = - \int_0^s \frac{\lambda_w(\sigma)\lambda_o(\sigma)}{\lambda(\sigma)} P_c'(\sigma) \, d\sigma,$$

$$\gamma(\mathbf{u}, s) = \frac{\lambda_w(s)}{\lambda(s)} \mathbf{u} - K \frac{\lambda_w(s)\lambda_o(s)}{\lambda(s)} (\rho_o - \rho_w) g \nabla z.$$

We use sequential time splitting, a backward Euler time discretization, and integration-by-parts (3 times) to obtain the following variational form and time approximation for $\Delta t > 0$ and time levels $t^n = n\Delta t$.

Pressure equation:

$$\int \nabla \cdot \mathbf{u}^n \, w \, dx = \int f^n \, w \, dx,$$

$$\int (\lambda(s^{n-1})K)^{-1} \mathbf{u}^n \cdot \mathbf{v} \, dx = \int p^n \, \nabla \cdot \mathbf{v} \, dx + \int \rho(s^{n-1}) \nabla z \cdot \mathbf{v} \, dx.$$

Saturation equation: (wherein $w \in W_f = W_c \oplus \delta W$)

$$\int_{E_f} \phi \frac{s^n - s^{n-1}}{\Delta t} w \, dx + \int_{E_f} \nabla \cdot \psi^n \, w \, dx + \int_{\partial E_f} \gamma(\mathbf{u}^n, s_{\text{up}}^n) \cdot \nu \, w \, ds$$
$$= \int_{E_f} f_w^n(s^n) \, w \, dx,$$

$$\int K^{-1} \psi^n \cdot \mathbf{v} \, dx = \int q(s^n) \nabla \cdot \mathbf{v} \, dx,$$

where $\psi^n = -K \nabla q(s^n)$, $\mathbf{u}_w^n = \psi^n + \gamma(\mathbf{u}^n, s^n)$, and we use one-point upstream weighting on the term involving γ.

As in the single-phase case, we separate the solution into coarse and fine scales $\mathbf{V}_c \oplus \delta \mathbf{V}$ or $W_c \oplus \delta W$:

$$\mathbf{v} = \mathbf{v}_c + \delta \mathbf{v}, \quad \mathbf{u} = \mathbf{u}_c + \delta \mathbf{u}, \quad \psi = \psi_c + \delta \psi,$$
$$w = w_c + \delta w, \quad p = p_c + \delta p.$$

Because the saturation equation is parabolic, it turns out that we do *not* need to decompose $s \in W_f$.

The pressure equation is linear and independent of the saturation equation, given s^{n-1}. We can solve for the upscaled $\mathbf{u}^n$ and p^n as above.

We linearize the saturation equation with Newton-Raphson, and solve for changes in s^n and ψ^n, given $\mathbf{u}^n$, using numerical Greens functions as in the linear case above. Upstream weighting on the fine scale destroys our localization assumption. To circumvent this, we simply use the old Newton result for the upstream value when it traces out of a coarse element.

7   Some Numerical Examples

We present two 2-D examples to illustrate our upscaling technique. In both, we have a square domain with uniform rectangular grids. The initial water saturation is 0.2. An injection well is placed in the lower left corner injecting water at a rate of $0.2\,\mathrm{m}^2/\mathrm{day}$, and a production well is in the adjacent corner. Time steps vary from 1 day initially to 25 days near the end of the simulations. The porosity is 0.25, but the permeability is heterogeneous.

7.1   Example 1

In this example, we have a 40 meter by 40 meter domain with a 40×40 fine grid. The base 10 logarithm of the permeability field is shown in Fig. 2.

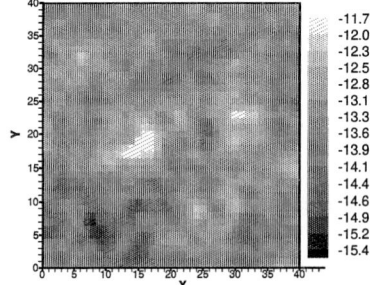

FIGURE 2: The log of the permeability field for Example 1.

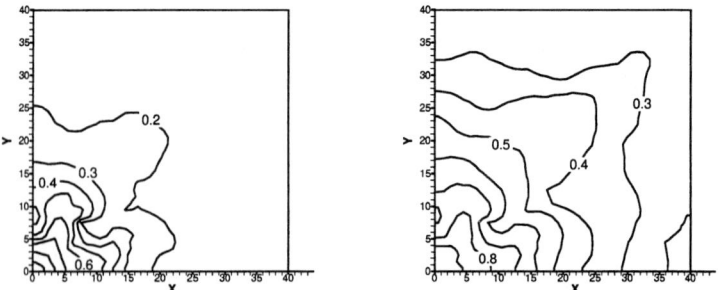

FIGURE 3:Fine-scale saturation at 100 and 500 days for Example 1.

As can be seen in Figs. 3–4, the upscaling procedure approximates the saturation quite well. Here we use a 5×5 coarse grid, so on each coarse block, we have an 8×8 subgrid for the δ-problems. The coarse solution (Fig. 5) completely fails to resolve the flow and location of the wells.

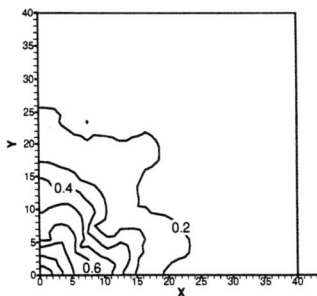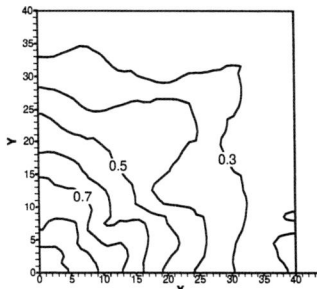

FIGURE 4:5 × 5 Upscaled saturation at 100 and 500 days for Example 1.

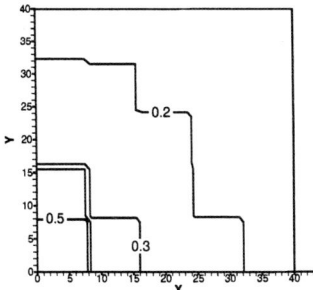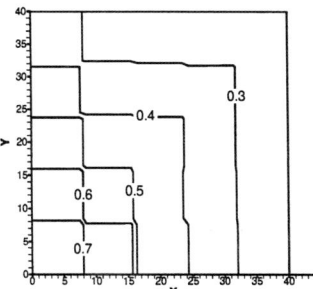

FIGURE 5:Coarse-scale saturation at 100 and 500 days for Example 1.

In Figs. 6–7, we show the results of upscaling with 2 × 2 and 8 × 8 coarse grids (20 × 20 and 5 × 5 subgrids). Both perform quite well. The coarsest example does a very good job near the well, but the performance degrades later in time a bit as the flow reaches the middle of the domain.

The number of degrees of freedom used in these examples is given in Table 2. The coarse solution is woefully inadequate; however, for the cost of a global problem of the same size, we can upscale to a very reasonable level of resolution.

	Coarse grid		
	2 × 2	5 × 5	8 × 8
Coarse Velocity	8	80	224
Upscale Velocity	3048	2880	2784
Fine Velocity	6240	6240	6240
Coarse Pressure	4	25	64
Upscale Pressure	1600	1600	1600
Fine Pressure	1600	1600	1600

Table 2. Number of degrees of freedom for Example 1.

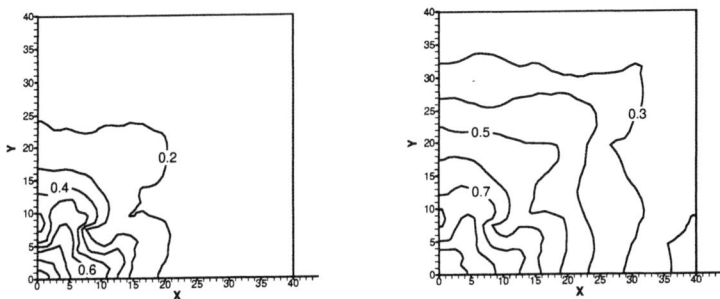

FIGURE 6:2 × 2 Upscaled saturation at 100 and 500 days for Example 1.

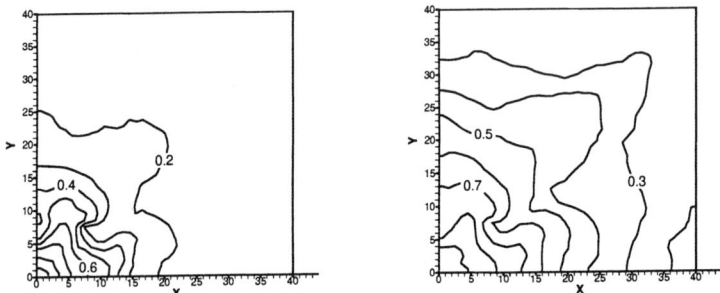

FIGURE 7:8 × 8 Upscaled saturation at 100 and 500 days for Example 1.

7.2   Example 2

In the second example, we have a 24 × 24 meter domain with a 24 × 24 fine grid and a 4 × 4 coarse grid. The base 10 logarithm of the permeability is depicted in Fig. 8. It has two high permeability streaks, akin to fractures.

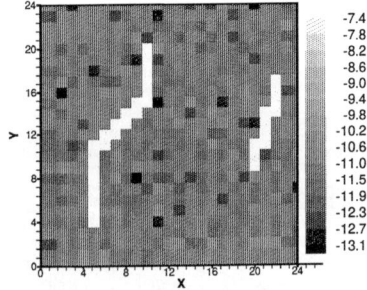

FIGURE 8:The log of the permeability field for Example 2.

Fig. 9 shows that the saturation is very difficult to resolve. The upscaling technique does a reasonable job following the flow into the first high permeability streak. The coarse solution in Fig 10, however, completely misses the high permeability streaks. It shows an overall tendency to flow right to left rather than the proper direction down to up. The number of degrees of freedom used in this example is given in Table 3.

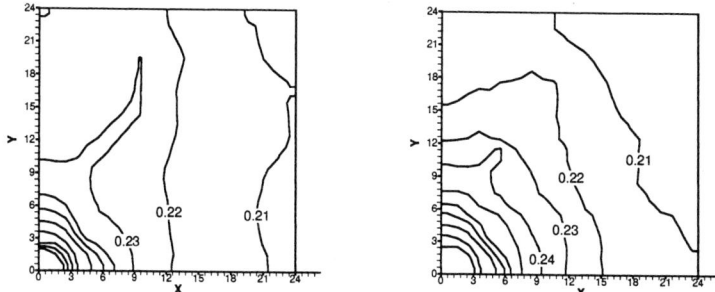

FIGURE 9:Fine-scale and 4 × 4 Upscaled saturation at 20 days for Example 2.

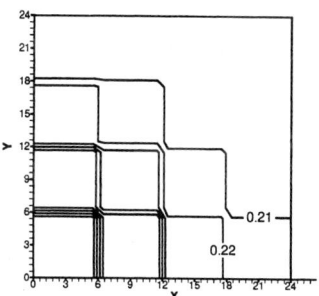

FIGURE 10:Coarse-scale saturation at 20 days for Example 2.

	Velocity	Pressure
Coarse	48	16
Upscale	1008	576
Fine	2208	576

Table 3. Number of degrees of freedom for Example 2.

Some timing results are given in Table 4. The pressure equation is solved with Jacobi preconditioned conjugate gradients, the saturation equation by Jacobi preconditioned orthomin, and the upscaling numerical Greens functions by a direct solver. The time to compute the 24 × 24 fine scale solution

is quite high, since the problem is very poorly conditioned. In contrast, the
4 × 4 coarse problem is solved easily. In this example, the upscaled problem
takes about as long to solve as the coarse problem; computing the numerical
Greens functions takes little extra time and gives a much improved solution.

Steps	Fine	Upscale	Coarse
1	2:03	0:10	0:08
2	1:48	0:09	0:08
3–10	11:32	1:10	1:05
11–20	13:29	1:28	1:21
21–36		2:21	2:13
37–48	35:47	2:02	1:52
49–58		1:42	1:33
59–65	30:26	1:21	1:14
66–72		1:21	1:14
73–76	19:22	0:46	0:42
77–84	38:48	1:34	1:25

Table 4. Some timing results for Example 2.

8   Conclusions

Our upscaling approach improves the resolution of the computed solution. It
allows recovery of fine-scale pressure, velocity, and saturation, so it incorporates fine-scale information and nonlinearities directly, thereby circumventing
the need to define pseudo-functions. The technique resolves positions of wells
within grid blocks, is efficient to compute, has good convergence properties,
can be applied at each time step of a time dependent problem, and can be
applied to a nonlinear problem during a Newton linearization step.

Acknowledgments. This work was supported by the U.S. National Science
Foundation under grants DMS-9707015 and SBR-9873326.

References

[1] Amaziane, B., Bourgeat, A., and Koebbe, J., Numerical simulation and
    homogenization of two-phase flow in heterogeneous porous media, *Trans-
    port in Porous Media* **9** (1991), 519–547.
[2] Arbogast, T., The existence of weak solutions to single-porosity and sim-
    ple dual-porosity models of two-phase incompressible flow, *J. Nonlinear
    Analysis: Theory, Methods, and Applic.* **19** (1992), 1009–1031.

[3] Arbogast, T., Gravitational forces in dual-porosity systems. I. Model derivation by homogenization and II. Computational validation of the homogenized model, *Transport in Porous Media* **13** (1993), 179–220.

[4] Arbogast, T., Minkoff, S. E., and Keenan, P. T. An operator-based approach to upscaling the pressure equation, in *Computational Methods in Water Resources XII*, v. 1, V. N. Burganos et al., eds., Computational Mechanics Publications, Southampton, U.K., 1998, 405–412.

[5] Bensoussan, A., Lions, J. L., and Papanicolaou, G., *Asymptotic Analysis for Periodic Structure*, North Holland, Amsterdam, 1978.

[6] Brezzi, F., Douglas, J., Jr., Duràn, R., and Fortin, M., Mixed finite elements for second order elliptic problems in three variables, *Numer. Math.* **51** (1987), 237–250.

[7] Brezzi, F., Douglas, J., Jr., and D. Marini, L., Two families of mixed elements for second order elliptic problems, *Numer. Math.* **88** (1985), 217–235.

[8] Christie, M. A., Mansfield, M., King, P. R., Barker, J. W., and Culverwell, I. D., A renormalization-based upscaling technique for WAG floods in heterogeneous reservoirs, SPE 29127, in *Thirteenth Symp. on Reservoir Simulation*, Society of Petroleum Engineers, Feb. 1995, 353–361.

[9] Durlofsky, L. J., Numerical calculation of equivalent grid block permeability tensors for heterogeneous porous media, *Water Resources Research* **27** (1991), pp. 699–708.

[10] Hou, T. Y. and Wu, X. H., A multiscale finite element method for elliptic problems in composite materials and porous media, *J. Comput. Phys.* **134** (1997), 169–189.

[11] Raviart, R. A. and Thomas, J. M., A mixed finite element method for 2nd order elliptic problems, in *Mathematical Aspects of the Finite Element Method*, Springer-Verlag, New York, 1977, 292–315.

[12] Russell, T. F. and Wheeler, M. F., Finite element and finite difference methods for continuous flows in porous media, in *The Mathematics of Reservoir Simulation*, R. E. Ewing, ed., Society for Industrial and Applied Mathematics, Philadelphia, 1983, Chapter II.

Numerical Simulation of Multiphase Flow in Fractured Porous Media

Peter Bastian   Zhangxin Chen
Richard E. Ewing   Rainer Helmig
Hartmut Jakobs   Volker Reichenberger

Abstract

The simulation of realistic multiphase flow problems in porous media requires efficient solution methods and means for handling the complicated structure of the media. The numerical toolbox UG has been developed to be able to use the approaches–multigrid, unstructured grids, adaptivity, and parallel computing–for a wide range of applications. An extensive library of discretization schemes has been implemented with UG in the MUFTE project. In this paper we describe a finite volume method for two-phase flow in fractured porous media and present results for the parallel multigrid solution of a gas-water air sparging problem.

KEYWORDS: multiphase flow, fractures, numerical toolbox, unstructured grids, multigrid, parallel computing

1   Introduction

The simulation of realistic problems from environmental engineering, oil reservoirs, and waste disposal sites requires the combination of different numerical methods. To overcome the difficulties associated with the complicated structure of porous media and the highly nonlinear behavior of multiphase systems, a combination of unstructured grids, adaptive local grid refinement, multigrid methods, and parallel computing is necessary. However, the complexity of a program with such a combination increases with each component by several orders of magnitude when compared to a structured sequential program. UG [2] provides a toolbox that combines these approaches for workstations and parallel computers and paves the way to use them for multiphase flow problems. Based on the program MUFTE, several applications for the simulation of multiphase, multicomponent processes have been implemented.

After an overview of a two-phase flow problem, we present a finite volume method that is part of MUFTE/UG and its extension for discontinuous and fractured porous media. We focus on the simulation of gas-water systems in the fractured media and present results for a gas-water simulation on parallel computers, gas infiltration into fractured porous media, and gas flow in single fractures. Future work will involve applications of MUFTE/UG to multiphase, multicomponent flows with mass interchange between phases.

2   The Numerical Toolbox UG

The main objective of the UG project is to provide a fundamental framework on which complex applications can be built. The basic building blocks of its software include geometry representation, mesh generation and its modification, discretization, solvers for linear and nonlinear systems, mesh refinement and error estimation, and tools for postprocessing and visualization. Since full support in all generality for each of these items is hardly possible within the framework of a research project, the UG project focuses on unstructured grids, local grid refinement, robust multigrid methods, and parallel computing on MIMD-type supercomputers. These concepts are present throughout the whole design of UG. The modules of UG are shown in Fig. 0.

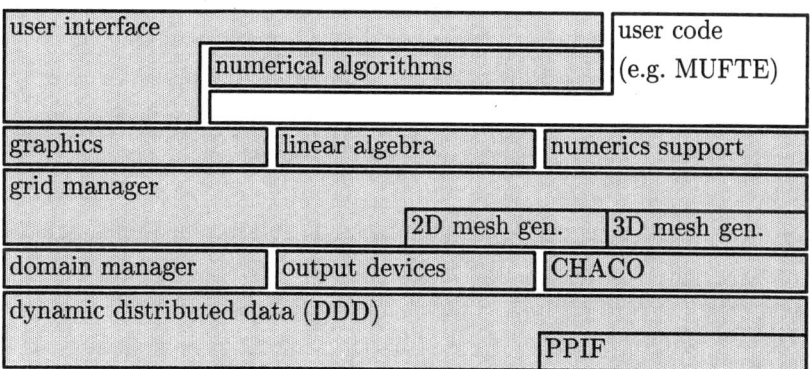

FIGURE 0: The modular structure of UG.

The dynamical distributed data (DDD) layer [4] is responsible for creating and maintaining the distributed unstructured mesh data structure and the associated vectors and sparse matrices. DDD can migrate object copies of a node to other processors, for example, while automatically maintaining its references to neighboring objects and the corresponding interface lists. To ensure portability and efficiency, DDD uses the parallel processor interface (PPIF), which provides the low-level message passing function based on MPI, PVM, and several other vendor specific message passing systems.

The domain manager handles geometry representation, based on two different implementations. The first implementation, the "standard domain", is based on a piecewise description of the boundary. Each boundary segment is defined by a mapping from a parameter space $\Lambda \subset \mathcal{R}^{d-1}$ to $\mathcal{R}^d$, $d = 2, 3$. In the second implementation, the "linear geometry model" (LGM), each boundary segment is an unstructured triangular mesh in 3D space. It is useful for complicated domains that do not allow for a parameterization. The output devices implement support for graphical user interfaces on X11 and Macintosh and deliver graphic outputs to files in the postscript and PPM format. The "low" module provides the basic function, like file I/O and timing. CHACO [8] is a graph partitioning package that is included for load balancing of the grid.

The grid manager is at the heart of UG. It is responsible for the creation and modification of the unstructured mesh data structure. For the mesh creation process, two grid generators are part of UG; the 3D mesh generator was developed by Schöberl [15]. The notion of hierarchical grids is an important part of the UG design. It is assumed that the domain geometry is simple enough to be stored on each processor and that an initial mesh can be generated that is much coarser than the final mesh used to compute the solution. The grid hierarchy is also used for the construction of finite element spaces used in the multigrid method. To ensure flexibility in the construction of the finite element spaces, degrees of freedom can be associated with nodes, edges, and element faces (in 3D). Grids can consist of elements of several types, triangles and quadrilaterals in 2D and tetrahedrons, pyramids, prisms, and hexahedrons in 3D. Meshes can be locally refined and derefined, based on geometrical criteria or through error estimators.

The graphics module on the top of the grid manager is responsible for the visualization of meshes and solutions in 2D and 3D. The visualization is parallelized and can be sent to any output device (screen or file). The linear algebra implements sparse matrix-vector operations and iterative solvers. The numerical support includes functions commonly needed in finite element and finite volume discretization schemes.

Linear and nonlinear solvers and time-stepping schemes are located in the numerical method module. Multigrid can be used as a preconditioner for Krylow subspace methods (Bi-CGstab, GMRES); several smoothers are available for the multigrid. Numerical algorithms are implemented as a set of classes which can be used directly or from which the application programmer can inherit to add new components or replace existing ones. This ensures flexibility and expandability together with the graphical *user interface* and the scripting language which drive UG applications.

UG has been used as a framework for applications from several fields: linear elasticity, elastoplacticity, Navier-Stokes equations, and density driven flow. In this paper we focus on its application to multiphase flow, which is explained in the next section.

The multiphase, multicomponent module MUFTE includes several applications. The processes which can be implemented via MUFTE may have up to three phases and up to three components. The phase transitions have been taken into account via Henry's, Raoult's, and Dalton's Law, and miscible flow can be thus treated. Nonisothermal processes can also be described, taking the energy equation into account. Future work will involve the addition of a more general phase behavior package to MUFTE.

3 Governing Equations for Two-Phase Flow

We consider the flow of two immiscible fluids in a porous medium $\Omega \subset \mathcal{R}^d$, $d = 2, 3$. Let $\mathcal{T} = (0, T)$ be the time interval of interest. We focus on the phases water and gas, but the consideration below is also valid for a general *wetting* phase and a *non-wetting* phase, each consisting of a component.

Conservation of mass for both phases $\alpha = w, g$ is expressed by

$$\frac{\partial(\Phi \varrho_\alpha S_\alpha)}{\partial t} + \nabla \cdot (\varrho_\alpha \mathbf{u}_\alpha) = \varrho_\alpha q_\alpha \qquad \text{in } \Omega \times \mathcal{T},$$

where $\Phi(\mathbf{x})$ is the porosity, $S_\alpha(\mathbf{x}, t)$ the saturation, $\varrho_\alpha(p_\alpha)$ the density, $p_\alpha(\mathbf{x}, t)$ the pressure, $\mathbf{u}_\alpha(\mathbf{x}, t)$ the macroscopic velocity, and $q_\alpha(\mathbf{x}, t)$ the source/sink term of phase α. The two phases completely fill the void space of the porous medium:

$$S_w + S_g = 1 .$$

Darcy's law is used to describe the fluid flow at a macroscopic level

$$\mathbf{u}_\alpha = -\frac{k_{r\alpha}}{\mu_\alpha} \mathbf{K}(\nabla p_\alpha - \varrho_\alpha \mathbf{g}),$$

where $K(\mathbf{x})$ is the absolute permeability tensor, $k_{r\alpha}(\mathbf{x}, S_\alpha)$ the relative permeability, $\mu_\alpha(p_\alpha)$ the dynamic viscosity of phase α, and $\mathbf{g}$ the gravitational vector. The relative permeability obeys the constraint $0 \leq k_{r\alpha}(S_\alpha) \leq 1$.

The pressure difference between two phases is expressed by the macroscopic capillary pressure

$$p_c(\mathbf{x}, t) = p_g(\mathbf{x}, t) - p_w(\mathbf{x}, t) .$$

Although $0 \leq S_w, S_g \leq 1$, the wetting fluid can not be removed from a porous medium by pure displacement and the non-wetting fluid can not be replaced completely once it has entered the medium. If we do not take phase transition effects into account, a certain amount of each phase remains in the pore space, the residual saturation $S_{\alpha r}$.

The *effective saturations* of the two phases are given by

$$\bar{S}_w = \frac{S_w - S_{wr}}{1 - S_{wr} - S_{gr}}, \qquad \bar{S}_g = \frac{S_g - S_{gr}}{1 - S_{wr} - S_{gr}} ,$$

where $0 \leq \bar{S}_\alpha \leq 1$ and $\bar{S}_w + \bar{S}_g = 1$.

In multiphase systems, p_c is a function of the wetting phase saturation: $p_c = p_c(S_w)$. One often used capillary pressure function is the Brooks-Corey capillary pressure function:

$$p_c(S_w) = p_d \bar{S}_w^{-\frac{1}{\lambda}} .$$

In the Brooks-Corey function the parameter p_d refers to the entry pressure of the porous medium (the pressure required to displace the wetting phase from the largest pores) and λ is related to the pore size distribution of the material. Typical values for λ range from 0.2 to 3. Small λ values describe single grain size materials while large λ values indicate highly non-uniform materials.

The relative permeabilities $k_{r\alpha}$ depend on the saturation of phase α. They can be evaluated by the Brooks-Corey relative permeability functions

$$k_{rw}(S_w) = \bar{S}_w^{\frac{2+3\lambda}{\lambda}} , \qquad k_{rg}(S_g) = \bar{S}_g^2 \left(1 - \left(1 - \bar{S}_g\right)^{\frac{2+\lambda}{\lambda}} \right) .$$

4 Discretization

4.1 Phase pressure-saturation formulation

For the equations in the previous section only two out of the four unknowns S_w, S_g, p_w, and p_g can be chosen as independent unknowns. Substituting

$$S_w = 1 - S_g, \qquad p_g = p_w + p_c(1 - S_g),$$

we obtain the (p_w, S_g) formulation in $\mathcal{T} \times \Omega$:

$$
\begin{aligned}
\frac{\partial(\Phi \varrho_w(1 - S_g))}{\partial t} + \nabla \cdot (\varrho_w \mathbf{u}_w) - \varrho_w q_w &= 0, \\
\frac{\partial(\Phi \varrho_g(S_g))}{\partial t} + \nabla \cdot (\varrho_g \mathbf{u}_g) - \varrho_g q_g &= 0,
\end{aligned}
\tag{4.1}
$$

with Darcy's velocities $\mathbf{u}_w$ and $\mathbf{u}_g$ given by

$$
\begin{aligned}
\mathbf{u}_w &= -\frac{k_{rw}}{\mu_w} \mathbf{K}(\nabla p_w - \varrho_w \mathbf{g}), \\
\mathbf{u}_g &= -\frac{k_{rg}}{\mu_g} \mathbf{K}(\nabla p_g + \nabla p_c(1 - S_g) - \varrho_g \mathbf{g}).
\end{aligned}
\tag{4.2}
$$

Initial and boundary conditions of Neumann or Dirichlet type are

$$
\begin{aligned}
p_w(\mathbf{x}, 0) &= p_{w0}(\mathbf{x}), & S_g(\mathbf{x}, 0) &= S_{g0}(\mathbf{x}) \text{ in } \Omega, \\
p_w(\mathbf{x}, t) &= p_{wd}(\mathbf{x}, t) \text{ on } \Gamma_{wd}, & S_g(\mathbf{x}, t) &= S_{gd}(\mathbf{x}, t) \text{ on } \Gamma_{gd}, \\
\varrho_w \mathbf{u}_w \cdot \mathbf{n} &= \phi_w(\mathbf{x}, t) \text{ on } \Gamma_{wn}, & \varrho_g \mathbf{u}_n \cdot \mathbf{n} &= \phi_g(\mathbf{x}, t) \text{ on } \Gamma_{gn}.
\end{aligned}
$$

In a similar way a (p_n, S_g) formulation can be derived. While flux type boundary conditions can be specified for both phases in both formulations, Dirichlet type boundary conditions can only be applied for variables present in these equations. A further comparison of the equations reveals that the (p_w, S_g) formulation should be used if $\bar{S}_g$ is bounded away from 1 while the (p_g, S_w) formulation should be used if $\bar{S}_w$ is bounded away from 1.

4.2 Finite volumes for two-phase flow

To ensure the applicability of the discretization method for equations (4.1) and (4.2) to a wide range of applications, a fully implicit and fully coupled finite volume method is chosen. The method is described in detail in [3], so we restrict ourselves to the core ideas.

The finite volume method requires the construction of a secondary mesh. In the present case of vertex centered finite volumes, the secondary mesh

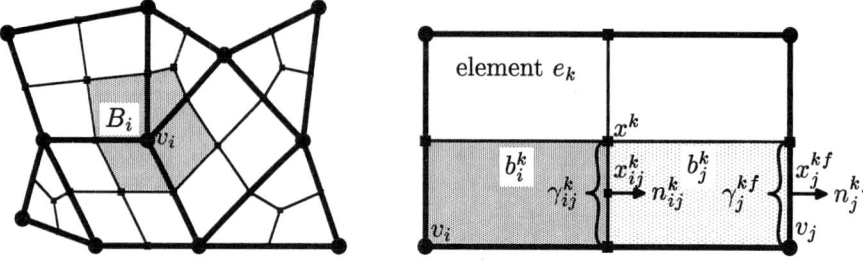

Figure 1: Secondary mesh and notation for control volumes.

$\mathcal{B}$ is constructed by connecting element barycenters with edge midpoints as shown in Fig. 1 in 2D. In 3D, the element barycenters are first connected to element face barycenters and these are then connected with edge midpoints. Each *control volume* B_i belongs to a grid vertex v_i. The intersection of a control volume B_i with element k is denoted by b_i^k (*sub-control volume*).

We use the index set $I = \{1, \ldots, N\}$, with N being the number of vertices of the grid and the index sets $I_{\alpha d} = \{i \in I \mid \mathbf{x}_i \notin \Gamma_{\alpha d}\}$ for vertices that are not part of a Dirichlet boundary. Along with these index sets, we define appropriate subsets of the standard conforming finite element space $V_h \subset H^1(\Omega)$

$$V_h = \{v \in C^0(\overline{\Omega}) \mid v \text{ is (multi-) linear on } t \in E_h\}$$

and the non-conforming test space W_h

$$W_h = \{w \in L^2(\Omega) \mid w \text{ is constant on each } B_j \in \mathcal{B}\}.$$

For V_h and W_h, we have

$$\forall i, j \in I \; \forall \varphi_i \in V_h : \; \varphi_i(\mathbf{x}_j) = \delta_{ij} \quad \text{and} \quad \forall i, j \in I \; \forall \psi_i \in W_h : \; \psi_i(\mathbf{x}_j) = \delta_{ij} \,.$$

The spaces $V_{\alpha h d}$ and $W_{\alpha h d}$ used for the discretization depend on time t:

$$V_{\alpha h d}(t) = \{v \in V_h \mid v(\mathbf{x}_i) = S_{\alpha d}(\mathbf{x}_i, t), \; i \in I \backslash I_{\alpha d}\},$$
$$W_{\alpha h d} = \{w \in W_h \mid w(\mathbf{x}_i) = 0, \; i \in I \backslash I_{\alpha d}\}.$$

The weak formulation of the present problem is: Find $p_{wh}(t) \in V_{whd}(t)$ and $S_{gh}(t) \in V_{ghd}(t)$ such that, for $w_{\alpha h} \in W_{\alpha h d}$ ($\alpha = w, g$) and $t \in \mathcal{T}$,

$$\frac{\partial}{\partial t} M_{\alpha h}(p_{wh}(t), S_{gh}(t), w_{\alpha h}) + A_{\alpha h}(p_{wh}(t), S_{gh}(t), w_{\alpha h}) \tag{4.3}$$
$$+ Q_{\alpha h}(t, p_{wh}(t), S_{gh}(t), w_{\alpha h}) = 0,$$

with the accumulation terms $M_{\alpha h}$, the internal flux terms $A_{\alpha h}$, and the source/sink and boundary flux terms $Q_{\alpha h}$ being given by

$$M_{\alpha h}(p_{wh}(t), S_{gh}(t), w_{\alpha h}) = \sum_{i \in I} w_{\alpha h}(\mathbf{x}_i) \int_{B_i} \Phi \varrho_\alpha S_{\alpha h}\, d\mathbf{x},$$

$$A_{\alpha h}(p_{wh}(t), S_{gh}(t), w_{\alpha h}) = \sum_{i \in I} w_{\alpha h}(\mathbf{x}_i) \int_{\partial B_i \cap \Omega} \varrho_\alpha \mathbf{u}_\alpha \cdot \mathbf{n}\, ds,$$

$$Q_{\alpha h}(t, p_{wh}(t), S_{gh}(t), w_{\alpha h}) = \sum_{i \in I} w_{\alpha h}(\mathbf{x}_i) \left(\int_{\partial B_i \cap \Gamma_{\alpha n}} \phi_\alpha\, ds - \int_{B_i} \varrho_\alpha q_\alpha\, d\mathbf{x} \right),$$

where in the first equation S_{wh} is replaced by $1 - S_{gh}$.

The precise evaluation of the quantities on the discretized domain is explained in detail in [1, 3]. The interior flux term is evaluated using an upwinding scheme. As an example, we show the accumulation term for the gas phase:

$$M_{nh}(p_{wh}, S_{gh}, \psi_i) = \int_{b_i} \Phi_h \varrho_{nh} S_{gh}\, d\mathbf{x} \approx \sum_{k \in E(i)} \Phi_i \varrho_{n,i} \mathbf{S}_{g,i} \mathrm{meas}(\mathrm{b}_\mathrm{i}^\mathrm{k}),$$

where Φ_i, $\varrho_{n,i}$, and $\mathbf{S}_{g,i}$ are the discrete values of the corresponding quantities at grid vertex i.

5 Interface Conditions

Media with discontinuous properties, such as fractured media, require a special treatment of the solution at these discontinuities. Consider a medium Ω that consists of two subregions Ω^I and Ω^{II}, with different absolute permeabilities. For an isotropic medium, we have $\mathbf{K}(\mathbf{x}) = k(\mathbf{x})\mathbf{I}$, with $k = k^I$ in Ω^I and $k = k^{II}$ in Ω^{II}. If $k^I > k^{II}$, then Ω^I represents a coarse material compared to Ω^{II}. Other properties, such as the porosity, may be different in the subregions as well.

We assume that Ω is initially fully saturated with water and that Ω^{II} lies above Ω^I. Fig. 2 shows the sketch of typical Brooks-Corey capillary pressure functions. If a gas phase is entering the medium from below, it will eventually reach the interface between Ω^I and Ω^{II}, but will only enter Ω^{II} if the capillary pressure is large enough. This pressure is called the *threshold pressure* or *non-wetting phase entry pressure*; it corresponds to p_d in the Brooks-Corey capillary pressure functions. In Fig. 2, the saturation that has to be reached in order for the gas phase to enter Ω^{II} is indicated by S_w^*.

If both fluids are present in both regions and we consider a point on the interface, we can look at this point from either side of the interface. If we

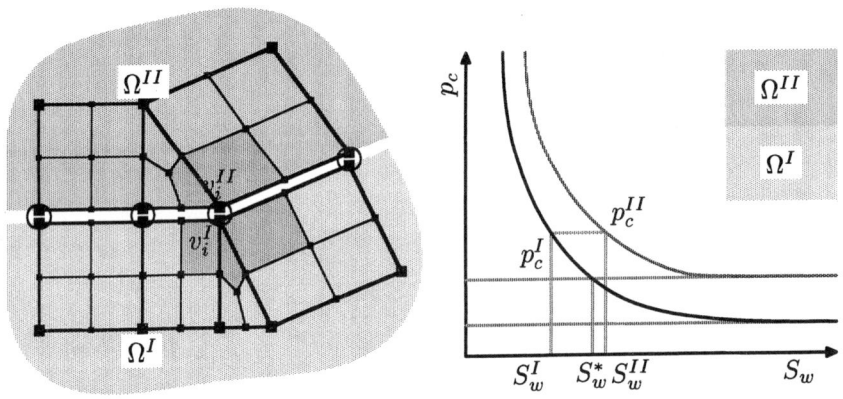

Figure 2: Grid at an interface and capillary pressure for discontinuous media.

choose to look from Ω^I, the wetting phase saturation is S_w^I (with $S_w^I < S_w^*$ since the fluid is already present in both regions). Since the capillary pressure is an intensive variable, it is a continuous function on the whole medium Ω, so $p_c^I(S_w^I) = p_c^{II}(S_w^{II})$. Consequently, the extensive variable saturation is discontinuous at the interface.

To incorporate this discontinuity into the (p_w, S_g) formulation, note that no mass is lost or produced at the interface: $\varrho_w \mathbf{u}_w \cdot \mathbf{n}$ and $\varrho_g \mathbf{u}_g \cdot \mathbf{n}$ are continuous across the interface. Furthermore, we use the *extended capillary pressure condition* [16] for points on the interface. It uses the threshold saturation S_w^* that can be found from $p_c^I(S_w^*) = p_c^{II}(1)$ and defines S_n^{II} in terms of the inverted capillary pressure function for Ω^{II} if S_w^* has been reached from Ω^I:

$$S_n^{II} = \begin{cases} 0, & S_g^I < S_g^* = 1 - S_w^*, \\ 1 - (p_c^{II})^{-1}(p_c^I(1 - S_g^I)), & S_g^I \geq S_g^* = 1 - S_w^*. \end{cases}$$

This ensures that if $S_g^I < 1 - S_w^*$ we have $S_g = 0$ in Ω^{II}; i.e., if the threshold pressure has not been reached, the non-wetting phase will not enter. Note that the capillary pressure is only defined if both phases exist and will only be defined in Ω^{II} if the non-wetting phase has entered. See [5] for a theoretical and numerical examination.

The interface conditions are incorporated into the discretization by imagining duplicated vertices on the interface as shown on the left in Fig. 2. The grid has to resolve interfaces with element edges or faces in order for the interface conditions to work.

The pressure p_w is continuous and its value is identical for both vertices.

To represent discontinuous values for the saturation in v_i^I and v_I^{II}, we define the vector $\mathbf{p}_{cmin}$, with $\mathbf{p}_{cmin,i} = \min_{k \in E(i)} p_c(\mathbf{x}^k, 1 - \mathbf{S}_{n,i})$, where $E(i)$ is the set of indices of all elements having vertex v_i as a corner. Using $\mathbf{p}_{cmin}$, we can compute the saturation S_n at vertex v_i with respect to element e_k:

$$\hat{S}_{n,i,k} = \begin{cases} \mathbf{S}_{n,i}, & \text{if } p_c(\mathbf{x}^k, 1 - \mathbf{S}_{n,i}) = \mathbf{p}_{cmin,i}, \\ 0, & \mathbf{p}_{cmin,i} < p_c(\mathbf{x}^k, 1), \\ 1 - S, & \text{where } S \text{ solves } p_c(\mathbf{x}^k, S) = \mathbf{p}_{cmin,i}. \end{cases}$$

The evaluation of saturation with respect to e_k for any $\mathbf{x} \in \bar{e}_k$ is done by

$$S_{nh}|_{e_k}(\mathbf{x}) = \sum_{m \in V(k)} \hat{S}_{n,m,k} \varphi_m(\mathbf{x}),$$

where $V(k)$ is the set of all vertices of element k. Also, all quantities that depend on the saturation are now evaluated element-wise (dependent on $\hat{S}_{n,i,k}$).

6   3D Air Sparging

The first example simulates the bubbling of air in a 3D heterogeneous porous medium. The domain is 5 meters high, about 4 by 5 meters wide and contains three lenses with different sand properties. The finite volume method with interface conditions in the (p_w, S_g) formulation is used.

The boundary conditions are $p_w = 10^5$ [Pa], $\phi_n = 0$ on the top boundary, $\phi_n = \phi_w = 0$ on the sides and bottom, and $\phi_n = -3 \cdot 10^{-3}$ [kg/(sm^2)], $\phi_w = 0$ at the inlet boundaries on the bottom. See Fig. 3 for the location of the inlet boundaries.

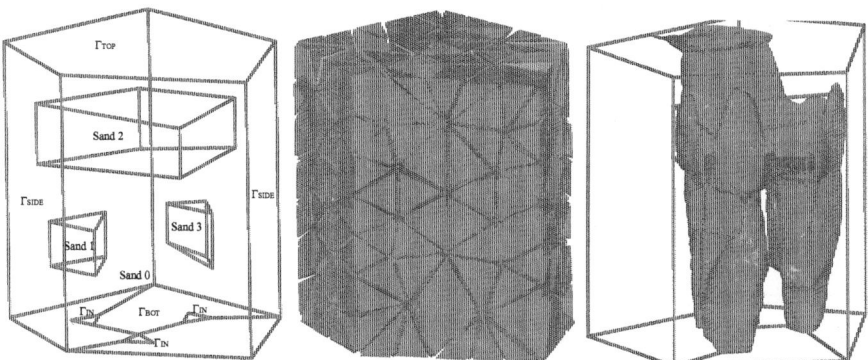

Figure 3: Geometry, coarse grid and isosurface for $S_g = 0.05$ for 3D air sparging problem. Solution is shown for $t = 640[s]$ (Visualization with GRAPE).

The fluid properties are $\varrho_w = 1000$ [kg/m³], $\varrho_g = p_g/84149.6$ [kg/m³], $\mu_w = 10^{-3}$ [Pas], and $\mu_n = 1.65 \cdot 10^{-5}$ [Pas]. The initial values are $p_w(x, y) = 10^5 + (5 - y) \cdot 9810.0$ and $S_n = 0$.

The Brooks-Corey constitutive relations are used with the parameters:

Sand	Φ	k [m^2]	S_{wr}	S_{nr}	λ	p_d [Pa]	S_n^*
0	0.40	$5.04 \cdot 10^{-10}$	0.10	0.0	2.0	1600.0	–
1	0.39	$2.05 \cdot 10^{-10}$	0.10	0.0	2.0	1959.6	0.30
2	0.39	$5.62 \cdot 10^{-11}$	0.10	0.0	2.0	2565.7	0.55
3	0.41	$8.19 \cdot 10^{-12}$	0.10	0.0	2.0	4800.0	0.80

Figure 3 shows the coarsest mesh, consisting of 1492 tetrahedral elements. All internal boundaries are resolved by faces of the initial mesh. The mesh is generated with NETGEN [15]. By uniform refinement, a grid hierarchy of five levels is produced, with 6,111,232 elements and 1,040,129 nodes on the finest grid.

Figure 3 shows on the right an isosurface of non-wetting phase saturation $S_g = 0.05$ at final time $T = 640$ [s]. The initial time step size is $\Delta t = 8$ [s], so 80 time steps are needed (unless a time step reduction is enforced by the nonlinear solver). Visualization has been done with the graphics program GRAPE [14].

The results show that the finite volume method with interface conditions works with 3D unstructured meshes and captures correctly the discontinuities in the saturation that can especially be seen for the uppermost sand lens with a relative permeability that is two orders of magnitude smaller than the surrounding sand.

Thanks to parallel capabilities, the computations for this problem can be carried out on a Cray T3E. Performance results can be found in the table below for up to a million nodes (i.e., 2 million unknowns) mapped to 128 processors. The problem size is scaled with the number of processors, yielding constant workload for each processor for all configurations. Starting with two processors and scaling by 64 results in an almost fourfold increase in total computation time. This can be explained by the increased number of Newton steps on the finest mesh. We believe that this is due to the very thin layers of air under the sand lenses.

P	S	# elements	T	it_{nl}	it_{lin}	it_{avg}	t_{mg}
2	80	95488	10771	247	1355	5.5	3.44
16	81	763904	15201	320	1909	6.0	3.76
128	83	6111232	37297	693	4684	6.8	3.99

The multigrid method uses a point-block Gauß-Seidel smoother and performs very well as seen from the table above where we use the following abbreviations: P the number of processors, S the number of time steps, T the total execution time, it_{nl} the total number of newton iterations, it_{lin} the total number of multigrid cycles, it_{avg} the average number of multigrid cycles in each Newton step, and t_{mg} the time for one multigrid cycle in seconds. The multigrid method scales very well with respect to the average number of iterations and the time per iteration (parallel efficiency).

7 Discretization with Fractures

In the discretization of two-phase flow in fractured media we employ two fundamental properties of fractures. First, due to their small width, fractures can be discretized by means of finite elements or finite volumes of smaller dimension. Second, if we do not consider open fractures, the fractures can be treated as a porous medium.

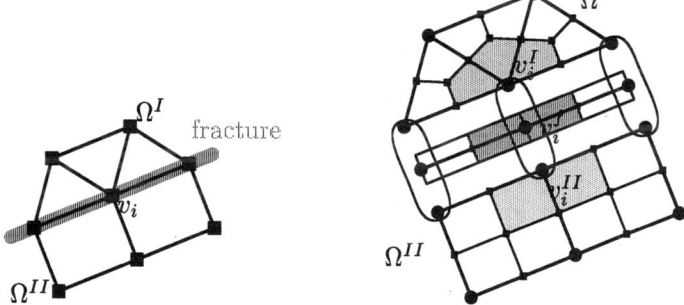

Figure 4: Finite volumes for fractures.

For the discretization we assume that interfaces between subregions and fractures are resolved by the grid. This situation is depicted in Fig. 4, where part of a fracture and the surrounding grid is shown. Since the interface conditions can be applied for vertices that lie on several subregions, we can apply the method to fractures as well. Vertex v_i in Fig. 4 is part of three subregions: two matrix regions Ω^I, Ω^{II} and the fracture.

To illustrate the discretization process, the grid detail is shown twice in Fig. 4, on the left with just the grid as it is stored and on the right with duplicated vertices at the fracture. Note that the vertex v_i still exists only once, but in the discretization the quantities associated with it may differ (dependent on the element we are viewing the vertex from). Also, shown is the secondary grid for the finite volume method. While the construction of the secondary grid for matrix elements remains the same, additional 1D finite volumes along the fracture in $\mathcal{R}^2$ and 2D finite volumes in $\mathcal{R}^3$ have to be considered. In Fig. 4, two 1D elements with v_i as one of their nodes are depicted.

The weak formulation of the problem is still as in (4.3). It is only in the evaluation of the terms $M_{\alpha h}$, $A_{\alpha h}$, and $Q_{\alpha h}$ that we have to take into account the increased number of finite volumes.

8 Gas Infiltration into Fractured Media

The effect of gas infiltration into a fractured medium is simulated on an example with five fractures. The domain size is $80m \times 100m$; the position of the fractures can be seen from Fig. 5.

The fluid properties are $\varrho_w = 1000$, $\mu_w = 10^{-3}$, $\varrho_g = p_g/84149.6$, and $\mu_g = 1.65 \cdot 10^{-5}$. The Brooks-Corey constitutive relations are used with the parameters $\Phi = 0.4$, $k = 10^{-8}$, and $p_d = 1000$ in the matrix and $\Phi = 0.39$, $k = 10^{-6}$, and $p_d = 2000$ in the fracture. $\lambda = 2$ and $S_{gr} = S_{wr} = 0$ is used for both matrix and fracture. The fracture width is $0.04m$.

The initial values are $S_g = 0$ and $p_w = (100-y) \cdot 9810.0$, and the boundary conditions are $S_g = 0$, $\phi_w = 0$ on the north boundary, $\phi_g = 0$, and hydrostatic pressure on the sides, and $\phi_g = 0$, $\phi_w = 0$ on the south boundary except for the intervals $[24m, 32m]$ to $[48m, 56m]$, where $S_g = -0.06 \ [kg/(sm^2)]$.

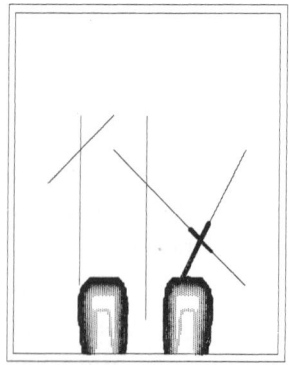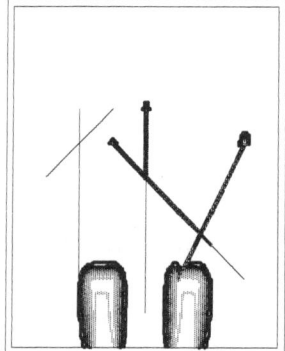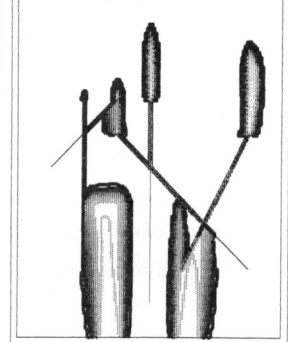

Figure 5: Gas infiltration into fractured domain at $t = 44s$, $55s$ and $85s$.

The problem is solved with a Newton method for the nonlinear equations and a multigrid-preconditioned Bi-CGstab on a grid hierarchy of five levels. The coarse grid has 91 elements and the finest grid has 27776 elements, which is created by three uniform refinement steps and one adaptive refinement step that only refines elements neighboring a fracture. ILU is used as a smoother for the multigrid method. The time step size is $\Delta t = 1s$.

Fig. 5 shows how gas infiltrates the water saturated medium and that gas moves very fast inside the fractures once the gas phase has reached them. At the top end of the fractures the gas phase is trapped until enough gas is there in order for the entry pressure for the matrix to be reached; then the matrix is infiltrated. The parameters for matrix and fractures have been chosen such that these effects can be observed. In natural rocks, material parameter differences between matrix and fractures are often so large that no infiltration from the fractures into the matrix takes place.

9   Comparison of Different Approaches

The strong influence of the topography of fractures on the permeability and on the multiphase flow behavior has been realized only in the recent years. Recent studies of naturally fractured media [11, 12] have shown that the description of the relative permeability must account for the roughness of the fracture walls, the fracture aperture, and the contact areas. The geostatistical model in [13] assumes that both phases can only flow simultaneously if the fracture apertures are correlated anisotropically. A survey of other approaches for the description of relative permeability-saturation relations in fractures can be found in [6].

In [13] the rough fractures are discretized as a field of *parallel plates* with different averaged apertures $\bar{a}_{ij}$ (see Fig. 6). The permeability of each parallel fracture is given by $k = b^2/12$. Normalizing the fracture domain to unit thickness, the permeability of each averaged aperture $\bar{a}_{ij}$ gets $\bar{k} = a^3/12$. The *cut-off-aperture* is defined by the capillary pressure $p_c = (2\sigma \cos\alpha)/a_c$, with the surface tension σ. The contact angle α is assumed to be zero.

In [12] the water phase is assumed to fill the void space between all plates with apertures lower than the cut-off-aperture a_c and the gas phase is assumed to fill the spaces between all plates with apertures bigger than a_c. In [9] this approach is referred to as the *separation assumption*.

In [9] a new aperture distribution based model is formulated. It assumes that a fraction of the plates with $\bar{a}_{ij} > a_c$ is still filled with the water phase.

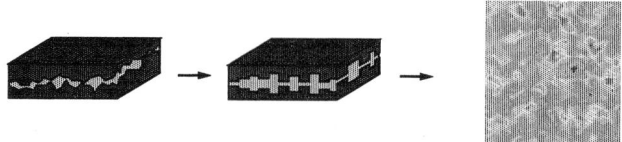

Figure 6: Approximation of rough fractures by parallel plates.

This fraction is given by the factor α. This assumption is called the *mix assumption*. Using the mix assumption, [9] reports better results for the relative transmissibilities and the gas saturation when the results of laboratory observations of a degassing experiment [10] are compared with the predictions based on the different assumptions.

Jarsjö developed two sets of constitutive relationships based on the separation assumption and on the mix assumption with $\alpha = 0.2$ (see Fig. 7). Both given relative permeability saturation functions have a very high gradient at $S_w \approx 1$. At this point little changes of the saturation have a big influence on hydraulic conditions within the fracture. The factor $\alpha = 0.2$ for the mix assumption results in a residual saturation for the water phase of $S_{wr} = 0.2$.

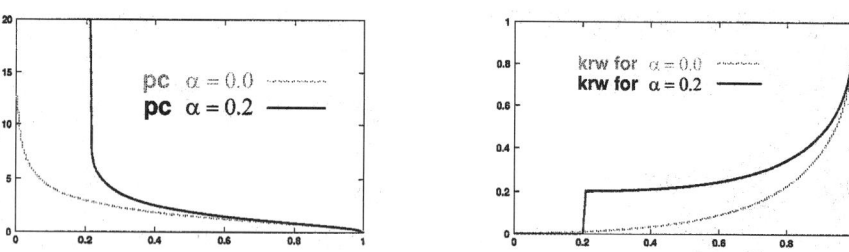

Figure 7: Comparison of capillary pressures (left) and relative permeabilities (right) based on separation assumption (dotted) and mix assumption (line).

Based on the data in [10], different permeability fields have been generated with the geostatistical tool SIMSET (Prof. Bardòssy, Institute for Hydraulic Engineering, Stuttgart University; best fit shown at Fig. 8). SIMSET uses the turning band method as the basic approach to generate geostatistical data.

The entering of gas from the bottom of the domain is simulated based on the generated permeability fields. The length of the inlet is set to be 0.02 m. The capillary pressure is computed for each plate via the Leverett

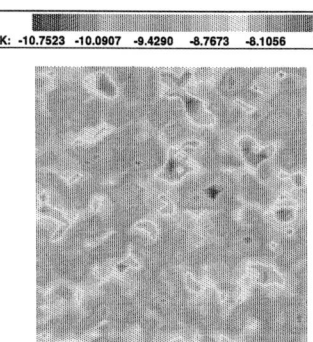

Figure 8: Permeability field based on stochastically generated aperture distribution for an area of $1\ m \times 1\ m$.

condition $p_c^{elem} = p_c^{avg}\sqrt{K^{avg}/K^{elem}}$; for the relative permeability function no upscaling concept is established.

For both assumptions the main flow paths are given by the areas of high permeability which are slightly connected. Neither the gas velocities shown differ much, nor the results for the effective permeability or the pressure fields. The most relevant differences occur in the saturations. The simulation using the constitutive relationships based on the mix assumption results in a wider spreading of the gas phase and in higher gradients for the saturation (see Fig. 9).

10   Conclusions

Parallel multigrid on unstructured grids can be efficiently used for multiphase flow problems in porous media. The MUFTE/UG program has proven to be successful for the simulation of realistic problems. While the current implementation focuses on a correct representation of the capillary pressure effect and fluid behavior at media discontinuities, a wide range of transport problems has been little explored so far. In a cooperation with the Institute for Scientific Computation of Texas A&M University, Department of Mathematics of Southern Methodist University, and the Exxon-Mobil Technology Company, we hope to expand the MUFTE/UG program to be able to handle transport problems. In particular, our future work will focus on the application of MUFTE/UG to multiphase, multicomponent, multi-dimensional flows with mass interchange between phases. Toward that end, a successful combination of MUFTE/UG and a phase behavior package will be critical.

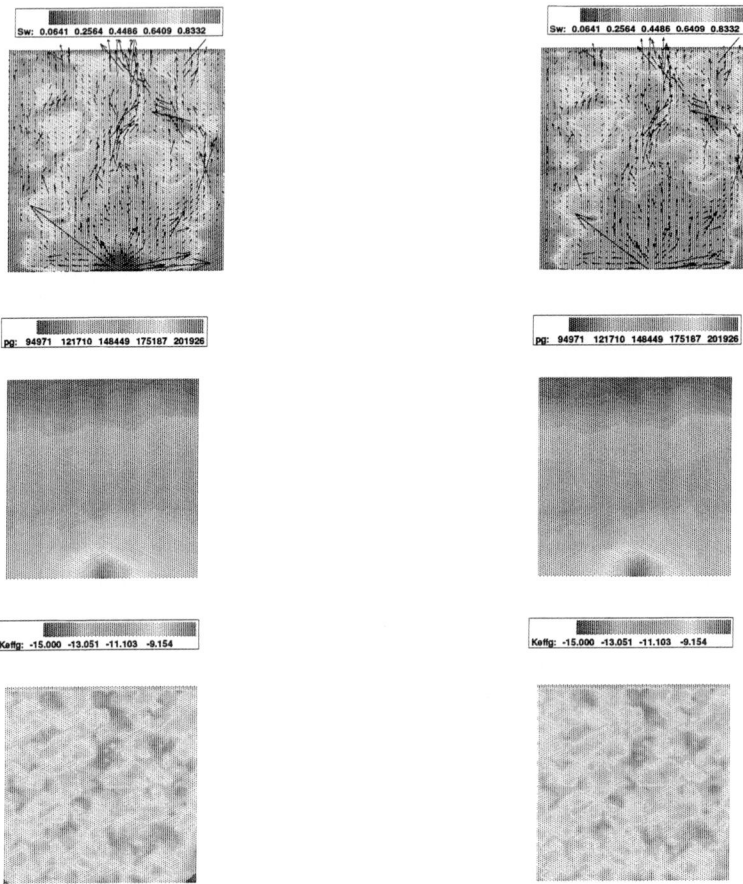

Figure 9: Water saturation (top), gas pressure (middle) and effective permeability (bottom) for relationships based on separation assumption (left) and mix assumption (right) at $t = 3.5s$.

Acknowledgments. We gratefully acknowledge that the constitutive relationships for the simulations shown in §9 were provided by Jerker Jersjö, Division of Water Resources Engineering, KTH Stockholm. Z. Chen and R. E. Ewing's research is supported in part by National Science Foundation grants DMS-9626179, DMS-9972147, and INT-9901498, by EAP grant R825207-01, and by a gift grant from the Mobil Technology Company.

References

[1] Bastian, P., Numerical computation of multiphase flows in porous media, Habilitation Thesis, Christian-Albrechts-Universität Kiel, submitted, 1999.

[2] Bastian, P., Birken, K., Lang, S., Johannsen, K., Neuß, N., Rentz-Reichert, H., and Wieners, C., UG: A flexible software toolbox for solving partial differential equations, *Computing and Visualization in Science* 1 (1997), 27–40. http://www.iwr.uni-heidelberg.de/~techsim.

[3] Bastian, P. and Helmig, R., Efficient fully coupled solution techniques for two-phase flow in porous media, *Adv. Water Res.* 23 (1999), 199–216.

[4] Birken, K., Ein modell zur effizienten parallelisierung von algorithmen auf komplexen, dynamischen datenstrukturen, Ph.D. Thesis, Universität Stuttgart, 1998.

[5] de Neef, M. J. and Molenaar, J., Analysis of DNAPL infiltration in a medium with a low permeable lense, *Comput. Geos.* 1 (1997), 191–214.

[6] Helmig, R., Theorie und numerik der mehrphasenströmungen in geklüftet-porösen Medien, Technical Report, Institut für Strömungsmechanik und Elektronisches Rechnen im Bauwesen, Universität Hannover, 1993.

[7] Helmig R., *Multiphase Flow and Transport Processes in the Subsurface- A Contribution to the Modeling of Hydrosystems*, Springer-Verlag, 1997.

[8] Hendrickson, B. and Leland, R., The CHACO user's guide 1.0, Technical Report SAND93-2339, Sandia National Laboratories, 1993.

[9] Jarsjö, J. and Destouni, G., Hydraulic conductivity relations in soil and fractured rock: Fluid component and phase interaction effects, SKB Report, Swedish Nuclear Fuel and Waste Management Company, 1998.

[10] Jarsjö, J. and Geller, J. T., Groundwater degassing: Laboratory experiments in rock fracture replicas with radial flow, SKB äspö hrl Progress Report, Swedish Nuclear Fuel and Waste Management Company, 1996.

[11] Persoff, P., Pruess, K., and Myer, L., Two-phase flow visualization and relative permeability measurement in transparent replicas of rough-walled rock fractures, Technical Report, University of California, 1991.

[12] Pruess, K., Cox, B. L., and Persoff, P., *A Casting and Imaging Technique for Determining Void Geometry and Relative Permeability Behavior of a Single Fracture Specimen*, Lawrence Berkeley Laboratoy, 1990.

[13] Pruess, K. and Tsang, Y. W., On two-phase relative permeability capillary pressure of rough-walled Rock Fractures, *Water Resour. Res.* **10** (1990).

[14] Rumpf, M., Neubauer, R., Ohlberger, M., and Schwörer, R., Efficient visualization of large-scale data on hierarchical meshes, in *Visual. Scient. Computing '97*, W. Lefer and M. Grave, eds., Springer-Verlag, 1997.

[15] Schöberl, J., A rule-based tetrahedral mesh generator, *Computing and Visualization in Science* **1** (1997), 1–26.

[16] van Duijn, C. J., Molenaar, J., and de Neef, M. J., Effects of capillary forces on immiscible two-phase flow in heterogeneous porous media, *Transport in Porous Media*, **21** (1995), 71–93.

The Modified Method of Characteristics for Compressible Flow in Porous Media

AIJIE CHENG   GAOHONG WANG

Abstract

Error estimates are derived for a finite element modified method of characteristics for a coupled system of partial differential equations modelling compressible flow in porous media. Some new techniques are introduced to conduct a convergence analysis. Optimal convergence rate is derived in the case of molecular diffusion and dispersion. One contribution of this paper is the demonstration of how molecular dispersion can be treated.

KEYWORDS: compressible displacement, MMOC, FEM, convergence.

1   Introduction

The modified method of characteristics (MMOC) was first formulated for a scalar parabolic equation by Douglas and Russell [6] and then extended by Russell [7] to nonlinear coupled systems modelling incompressible flow in porous media. Some improved error estimates were derived in [4]. Convergence of finite element methods (FEM) with MMOC for compressible flow was given by Yuan [9], but only molecular diffusion was considered.

In this paper we consider the model for compressible miscible displacement in porous media, which includes both of molecular diffusion and dispersion. A scheme of FEM with MMOC is established and analyzed.

Consider the single-phase miscible displacement of one compressible fluid by another in a porous medium. A set of equations modelling the pressure $p(x,t)$ and the concentration $c(x,t)$ are given by Douglas and Roberts [5]

$$
\begin{aligned}
d(c)\frac{\partial p}{\partial t} + \nabla \cdot u &= d(c)\frac{\partial p}{\partial t} - \nabla \cdot (a(c)\nabla p) = q, \\
\phi\frac{\partial c}{\partial t} + b(c)\frac{\partial p}{\partial t} + u \cdot \nabla c - \nabla \cdot (D(c,p_1,p_2)\nabla c) &= (\bar{c} - c)q,
\end{aligned}
\tag{1.1}
$$

for $x \in \Omega$ and $t \in J$, where Ω is a bounded domain in R^2 and $J = (0, T]$. We assume that no flow occurs across the boundary

$$u \cdot \nu = D(c, p_1, p_2)\nabla c \cdot \nu = 0, \quad (x, t) \in \partial\Omega \times J,$$
$$p(x, 0) = p_0(x), \quad c(x, 0) = c_0(x), \quad x \in \Omega, \tag{1.2}$$

where ν is the outer normal to $\partial\Omega$, $c_i = c_i(x, t)$, $i = 1, 2$, the concentration of ith component, $c = c_1 = 1 - c_2$, u Darcy's velocity, $k = k(x)$ the permeability of the rock, μ the local viscosity of the mixed fluid, $\phi = \phi(x)$ the porosity of the rock, q the source-sink term, and $\bar{c}$ the specified concentration at injection wells and $\bar{c} = c$ at production wells. D is a 2×2 diffusion matrix, which reflects the molecular diffusion and dispersion:

$$D = (D_{ij})_{2 \times 2}, \quad D_{ij} = D_{ij}(c, p_1, p_2) \equiv D_{ij}(x, c, p_1, p_2),$$

where $p_1 = \partial p/\partial x_1$ and $p_2 = \partial p/\partial x_2$. If only molecular diffusion is considered, then $D = \phi(x)d_m I$, where d_m is the molecular diffusivity and I is the 2×2 unit matrix.

Denote by z_1 and z_2 the compressibilities of the two components and set

$$a(c) = k(x)/\mu(c), \quad d(c) = \phi(x)(z_1 c_1 + z_2 c_2), \quad b(c) = \phi(x)c_1 c_2(z_1 - z_2).$$

For convenience, we assume the system above is Ω-periodic. Thus the no-flow boundary conditions above can be dropped. Additionally, the following assumptions are needed in the error analysis:

(1) There exist positive constants ϕ_*, ϕ^*, a_*, a^*, d_*, and d^* such that

$$\phi_* \le \phi(x) \le \phi^*, \quad a_* \le a(x, c) \le a^*, \quad d_* \le d_2(x, c) \le d^*,$$

for arbitrary $x \in \Omega$, $c \in [-\varepsilon', 1 + \varepsilon']$, and ε' is a small positive constant.

(2) D is positive-definite; i.e., there exists a constant D_* such that

$$e^T D(x, c, p_1, p_2)e \ge D_*|e|^2,$$

for arbitrary $x \in \Omega$, $c \in [-\varepsilon', 1 + \varepsilon']$, $(p_1, p_2) \in R^2$, and $e \in R^2$.

(3) There exist derivatives up to second order for $a(c)$ and $D_{ij}(c, p_1, p_2)$, which are bounded uniformly for $x \in \Omega$, $c \in [-\varepsilon', 1 + \varepsilon']$, $(p_1, p_2) \in R^2$.

For a finite element analysis for compressible two-phase flow, see [2].

2 The Procedure of FEM with MMOC

Let $M_{h_c} = M_h$ be a family of finite-dimensional subspaces of $H^1(\Omega)$ with the properties

$$\inf_{\psi \in M_h} \{\|v - \psi\| + h_c\|v - \psi\|_1 + h_c(\|v - \psi\|_{L^\infty(\Omega)} + \|v - \psi\|_{W^1_\infty(\Omega)})\}$$
$$\leq M\|v\|_s h_c^s, \quad v \in W^s_2(\Omega), \ 1 \leq s \leq l + 1,$$
$$\|\psi\|_{j,\infty} \leq M h_c^{-1}\|\psi\|_j, \quad \psi \in M_h, \ j = 0, 1,$$
$$\|\psi\|_1 \leq M h_c^{-1}\|\psi\|, \quad \psi \in M_h,$$
$$\|\nabla\psi\|_{0,\infty} \leq M h_c^{-1}\|\nabla\psi\|, \quad \psi \in M_h,$$

where M is independent of h_c. Similarly, we define another family of finite-dimensional subspaces of $H^1(\Omega)$, $N_{h_p} = N_h$, which satisfies the same properties as M_{h_c} with l and h_c replaced by k and h_p, respectively.

Let $s(x, t)$ be the unit vector in the characteristic direction and note that

$$\frac{\partial c}{\partial s(x,t)} = \frac{1}{\sqrt{|u|^2 + \phi^2}}(\phi\frac{\partial c}{\partial t} + u \cdot \nabla c). \tag{2.1}$$

Multiplying (1.1) by test functions $v, z \in H^1(\Omega)$ and integrating by parts, we obtain the variational form

$$(d(c)\frac{\partial p}{\partial t}, v) + (a(c)\nabla p, \nabla v) = (q, v), \quad v \in H^1(\Omega),$$
$$(\sqrt{|u|^2 + \phi^2}\frac{\partial c}{\partial s}, z) + (D(c, p_1, p_2)\nabla c, \nabla z) + (b(c)\frac{\partial p}{\partial t}, z) \tag{2.2}$$
$$= ((\bar{c} - c)q, z), \quad z \in H^1(\Omega).$$

Denote by Δt the time increment and let $t^n = n\Delta t$, $n = 0, 1, \ldots, N$, and $N = T/\Delta t$. If the approximation of $p(x, t)$ and $c(x, t)$ have been known at t^{n-1}, $(P_h^{n-1}, C_h^{n-1}) \in N_h \times M_h$, we define

$$u_h^{n-1} = -a(C_h^{n-1})\nabla P_h^{n-1}, \quad d_t f^n = \frac{f^n - f^{n-1}}{\Delta t},$$
$$\hat{x} = x - \frac{u_h^{n-1}}{\phi(x)}\Delta t, \quad \hat{f}^{n-1}(x) = f^{n-1}(\hat{x}).$$

Denote the approximate characteristic direction from (x, t^n) to $(\hat{x}, t^{n-1})$ as $\tau(x, t^n)$. The numerical scheme of FEM with MMOC can be defined by

$$(d(C_h^{n-1})\frac{P_h^n - P_h^{n-1}}{\Delta t}, v) + (a(C_h^{n-1})\nabla P_h^n, \nabla v) = (q^n, v), \quad v \in N_h,$$
$$(\phi\frac{C_h^n - \hat{C}_h^{n-1}}{\Delta t}, z) + (D(C_h^{n-1}, P_{h1}^{n-1}, P_{h2}^{n-1})\nabla C_h^n, \nabla z) \tag{2.3}$$
$$+(b(C_h^{n-1})\frac{P_h^n - P_h^{n-1}}{\Delta t}, z) = ((\bar{C}_h^n - C_h^n)q^n, z), \quad z \in M_h,$$

where $P_{h1} = \partial P_h/\partial x_1$, $P_{h2} = \partial P_h/\partial x_2$, and $\bar{C}_h^n = \begin{cases} \bar{c}^n, & q^n > 0, \\ C_h^n, & q^n < 0. \end{cases}$ First, $P_h^n \in N_h$ can be solved from (2.3); then $C_h^n \in M_h$ can be obtained by (2.3).

Before the error analysis, we define the elliptic projections $\tilde{p} : [0,T] \to N_h$ and $\tilde{c} : [0,T] \to M_h$ by

$$(a(c)\nabla(p - \tilde{p}), \nabla v) + \sigma(p - \tilde{p}, v) = 0, \quad v \in N_h,$$
$$(D(c, p_1, p_2)\nabla(c - \tilde{c}), \nabla z) + \lambda(c - \tilde{c}, z) = 0, \quad z \in M_h, \tag{2.4}$$

where λ and σ are positive constants large enough to ensure the coercivity of the bilinear form over $H^1(\Omega)$. Let

$$\eta \equiv p - \tilde{p}, \quad \pi \equiv \tilde{p} - P_h, \quad \zeta \equiv c - \tilde{c}, \quad \xi \equiv \tilde{c} - C_h.$$

Then,

$$p - P_h \equiv \pi + \eta, \quad c - C_h \equiv \xi + \zeta.$$

Lemma 2.1 [1, 3, 8] *If k, $l \geq 1$, there exists a constant M independent of h such that*

$$\|\zeta\| + h_c\|\zeta\|_1 + \|\frac{\partial\zeta}{\partial t}\| + h_c\|\frac{\partial\zeta}{\partial t}\|_1 \leq Mh_c^{l+1},$$

$$\|\eta\| + h_p\|\eta\|_1 + \|\frac{\partial\eta}{\partial t}\| + h_p\|\frac{\partial\eta}{\partial t}\|_1 \leq Mh_p^{k+1},$$

$$\|\tilde{p}\|_{W_\infty^1(J;W_\infty^1(\Omega))} + \|\tilde{c}\|_{W_\infty^1(J;W_\infty^1(\Omega))} \leq M,$$

$$\|\frac{\partial^2\eta}{\partial t^2}\|_1 \leq Mh_p^k, \quad \|\frac{\partial^3\eta}{\partial t^3}\|_{L^\infty(\Omega)} \leq M.$$

Let us select the initial approximation $P_h^0 = \tilde{p}^0$ and $C_h^0 = \tilde{c}^0$. Then,

$$\pi^0 = 0, \quad \xi^0 = 0. \tag{2.5}$$

Throughout the analysis, the symbol M denotes a generic constant, not the same at different places, which is independent of all mesh parameters. Similarly, ε denotes a generic small positive constant.

3 Convergence Analysis

Firstly, we establish error equations. Subtracting (2.3) from (2.2) and using the definition of projection, we obtain

$$(d(C_h^{n-1})d_t\pi^n, v) + (a(C_h^{n-1})\nabla\pi^n, \nabla v)$$
$$= -([d(c^n)\frac{\partial p^n}{\partial t} - d(C_h^{n-1})d_t p^n], v) - (d(C_h^{n-1})d_t\eta^n, v) \tag{3.1}$$
$$-([a(c^n) - a(C_h^{n-1})]\nabla\tilde{p}^n, \nabla v) + \sigma(\eta^n, v).$$

Then, let $n = 1$ and $v = d_t\pi^1$. It is shown that

$$(d(C_h^0)d_t\pi^1, d_t\pi^1) + (a(C_h^0)\nabla\pi^1, \nabla d_t\pi^1)$$
$$= -([d(c^1)\frac{\partial p^1}{\partial t} - d(C_h^0)d_t p^1], d_t\pi^1) - (d(C_h^0)d_t\eta^1, d_t\pi^1) \qquad (3.2)$$
$$- ([a(c^1) - a(C_h^0)]\nabla(\tilde{p}^1 - p^1), \nabla d_t\pi^1)$$
$$- ([a(c^1) - a(C_h^0)]\nabla p^1, \nabla d_t\pi^1) + \sigma(\eta^1, d_t\pi^1).$$

By $\xi^0 = \pi^0 = 0$ and inverse properties, we have

$$\|d_t\eta^1\|^2 \le M h_p^{2k+2},$$
$$([a(c^1) - a(C_h^0)]\nabla\eta^1, \nabla d_t\pi^1) \le M\|c^1 - \tilde{c}^0\|\,\|\nabla\eta^1\|\,\|\nabla d_t\pi^1\|_{L^\infty}$$
$$\le M\|c^1 - \tilde{c}^0\|h_p^k h_p^{-2}\|d_t\pi^1\| \le M(\Delta t^2 + \|\zeta^0\|^2) + \varepsilon\|d_t\pi^1\|^2, \quad k \ge 2.$$

Using integration by parts, we obtain

$$|([a(c^1) - a(C_h^0)]\nabla p^1, \nabla d_t\pi^1)|$$
$$\le |([a(c^1) - a(C_h^0)]\Delta p^1, d_t\pi^1)| + |\nabla([a(c^1) - a(C_h^0)] \cdot \nabla p^1, d_t\pi^1)|$$
$$\le M(\Delta t^2 + \|\zeta^0\|^2 + \|\nabla\zeta^0\|^2) + \varepsilon\|d_t\pi^1\|^2.$$

By the boundedness of d and a, we have

$$\|d_t\pi^1\|^2 + \frac{1}{\Delta t}\|\nabla\pi^1\|^2 + \Delta t\|\nabla d_t\pi^1\|^2 \le M(\Delta t^2 + h_c^{2l} + h_p^{2k+2}). \qquad (3.3)$$

If we try to derive a reasonable evolutionary error inequality for $p(x,t)$ from (3.1) which can be coupled suitably with that for $c(x,t)$ given later, some essential difficulties would occur unavoidably. Differently from the case of molecular diffusion, we now establish another error equation from a new viewpoint. Operating (3.1) by d_t and taking $v = d_t\pi^n$, we obtain the error equation

$$(d_t[d(C_h^{n-1})d_t\pi^n], d_t\pi^n) + (d_t[a(C_h^{n-1})\nabla\pi^n], \nabla d_t\pi^n)$$
$$= -(d_t[d(c^n)\frac{\partial p^n}{\partial t} - d(c^{n-1})d_t p^n], d_t\pi^n)$$
$$- (d_t[(d(c^{n-1}) - d(C_h^{n-1}))d_t p^n], d_t\pi^n) \qquad (3.4)$$
$$- (d_t[d(C_h^{n-1})d_t\eta^n], d_t\pi^n) + \sigma(d_t\eta^n, d_t\pi^n)$$
$$- (d_t[(a(c^n) - a(C_h^{n-1}))\nabla\tilde{p}^n], \nabla d_t\pi^n),$$

where $n \ge 2$. Denote the terms on the left side as A_1 and A_2, and the terms on the right side as $A_3, \ldots, A_7$ in order.

The induction hypotheses are useful

$$\|d_t\pi^r\|_{L^\infty(\Omega)} \le 1, \quad \|\nabla\pi^r\|_{L^\infty(\Omega)} \le 1, \quad \|\xi^r\|_{W_\infty^1(\Omega)} \le 1. \qquad (3.5)$$

Let $\Delta t = o(h_p)$, $h_c^l = o(h_p)$, and $k \geq 1$. From (3.3) and the inverse properties

$$\|d_t\pi^1\|_{L^\infty(\Omega)} \leq M h_p^{-1}\|d_t\pi^1\| \leq M h_p^{-1}(\Delta t + h_c^l + h_p^{k+1}) \leq 1,$$
$$\|\nabla\pi^1\|_{L^\infty(\Omega)} \leq M h_p^{-1}\|\nabla\pi^1\| \leq M h_p^{-1}(\Delta t + h_c^l + h_p^{k+1}) \leq 1.$$

Then from (2.5) and the above inequalities, the first equation in (3.5) is true for $r = 1$, the second is true for $r = 0, 1$, and the third is true for $r = 0$. Now, we suppose that the first and second equations in (3.5) are true for $r = 1, 2, \ldots, m-1$ and the third is true for $r = 0, 1, \ldots, m-1$. We demonstrate that (3.5) is true for $r = m$.

By expansion, we know

$$A_1 \geq d_t(d(C_h^{n-1})d_t\pi^n, d_t\pi^n) + (d_t d(C_h^{n-1})d_t\pi^{n-1}, d_t\pi^n)/2$$
$$-(d_t d(C_h^{n-1})d_t\pi^{n-1}, d_t\pi^{n-1})/2 \equiv A_1^1 + A_1^2 + A_1^3.$$

By the induction hypotheses and the linearity of $d(c)$ on c, we have

$$|A_1^2 + A_1^3| \leq M(h_c^{2l+2} + \|d_t\xi^{n-1}\|^2 + \|d_t\pi^{n-1}\|^2 + \|d_t\pi^n\|^2),$$

where $J^{n-1} = [t^{n-2}, t^{n-1}]$. For A_2, A_3, A_4, and A_6, we have

$$A_2 = (a(C_h^{n-1})d_t\nabla\pi^n, \nabla d_t\pi^n) + (d_t a(C_h^{n-1})\nabla\pi^{n-1}, \nabla d_t\pi^n) \equiv A_2^1 + A_2^2,$$
$$A_2^1 \geq a_*\|\nabla d_t\pi^n\|^2,$$
$$|A_2^2| \leq M(h_c^{2l+2} + \|d_t\xi^{n-1}\|^2 + \|\nabla\pi^{n-1}\|^2) + \varepsilon\|\nabla d_t\pi^n\|^2,$$
$$|A_3 + A_4 + A_6| \leq M(\Delta t^2 + h_c^{2l+2} + h_p^{2k+2} + \|\xi^{n-1}\|^2 + \|d_t\xi^{n-1}\|^2 + \|d_t\pi^n\|^2).$$

Now, only A_5 and A_7 are left. Divide A_5 into two parts

$$A_5 = -(d_t d(C_h^{n-1})d_t\eta^n, d_t\pi^n) - (d(C_h^{n-2})d_t^2\eta^n, d_t\pi^n) \equiv A_5^1 + A_5^2.$$

By inverse properties,

$$|A_5^1| \leq M(\|d_t\xi^{n-1}\| \, \|d_t\eta^n\| \, \|d_t\pi^n\|_{L^\infty(\Omega)}$$
$$+ \|d_t\zeta^{n-1}\| \, \|d_t\eta^n\| \, \|d_t\pi^n\|_{L^\infty(\Omega)} + \|d_t\eta^n\| \, \|d_t\pi^n\|).$$

By the mean value theorem and the boundedness of $\partial^3\eta/\partial t^3$ (Lemma 2.1), we have

$$A_5^2 = -(d(C_h^{n-2})\Delta t^{-2}\int_0^{\Delta t}[\tfrac{\partial\eta}{\partial t}(t^{n-1} + t) - \tfrac{\partial\eta}{\partial t}(t^{n-2} + t)]dt, d_t\pi^n)$$
$$= -(d(C_h^{n-2})\Delta t^{-1}\int_0^{\Delta t}[\tfrac{\partial^2\eta}{\partial t^2}(t^{n-2} + t) + \Delta t\tfrac{\partial^3\eta}{\partial t^3}(t^{n-2} + \theta\Delta t + t)]dt, d_t\pi^n),$$
$$|A_5^2| \leq M(\Delta t^{-1}\|\tfrac{\partial^2\eta}{\partial t^2}\|_{L^2(J^{n-1};L^2(\Omega))}^2 + \Delta t^2 + \|d_t\pi^n\|^2).$$

We can divide A_7 into three parts

$$
\begin{aligned}
A_7 \quad &= -(d_t[a(c^n) - a(c^{n-1})]\nabla\tilde{p}^n, \nabla d_t\pi^n) \\
&\quad -(d_t[a(c^{n-1}) - a(C_h^{n-1})]\nabla\tilde{p}^n, \nabla d_t\pi^n) \\
&\quad -([a(c^n) - a(C_h^{n-1})]\nabla d_t\tilde{p}^n, \nabla d_t\pi^n) \equiv A_7^1 + A_7^2 + A_7^3, \\
|A_7^1 + A_7^3| &\leq M(\Delta t^2 + h_c^{2l+2} + \|\xi^{n-1}\|^2) + \varepsilon\|\nabla d_t\pi^n\|^2.
\end{aligned}
$$

There exist $c^* \in [c^{n-2}, c^{n-1}]$ and $C_h^* \in [C_h^{n-2}, C_h^{n-1}]$ such that

$$
d_t[a(c^{n-1}) - a(C_h^{n-1})] = \frac{\partial a}{\partial c}(c^*)d_t c^{n-1} - \frac{\partial a}{\partial c}(C_h^*)d_t C_h^{n-1},
$$
$$
|c^* - C_h^*| \leq |c^{n-1} - c^{n-2}| + |c^{n-1} - C_h^{n-1}| + |c^{n-2} - C_h^{n-2}|.
$$

Then,

$$
|A_7^2| \leq M(\Delta t^2 + h_c^{2l+2} + \|\xi^{n-1}\|^2 + \|\xi^{n-2}\|^2 + \|d_t\xi^{n-1}\|^2) + \varepsilon\|\nabla d_t\pi^n\|^2.
$$

Multiplying (3.4) by Δt and summing that from $n = 2$ to $n = m$, if ε is taken to be small enough, by the above estimates and (3.3), we obtain

$$
\begin{aligned}
\|d_t\pi^m\|^2 + \sum_{n=1}^{m} \|\nabla d_t\pi^n\|^2\Delta t &\leq M\big(\Delta t^2 + h_c^{2l} + h_p^{2k} \\
\sum_{n=1}^{m}(\|\xi^n\|^2 + \|\nabla\pi^n\|^2 + \|d_t\xi^n\|^2 + \|d_t\pi^n\|^2)\Delta t\big),
\end{aligned}
\tag{3.6}
$$

for $m \geq 1$. This is the evolutionary error inequality for $p(x, t)$.

Now, we turn to the concentration equation. Subtracting (2.3) from (2.2) and taking $z = d_t\xi^n$, we see that

$$
\begin{aligned}
&(\phi d_t\xi^n, d_t\xi^n) + (D(C_h^{n-1}, P_{h1}^{n-1}, P_{h2}^{n-1})\nabla\xi^n, \nabla d_t\xi^n) \\
&= -([\phi\frac{\partial c^n}{\partial t} + u_h^{n-1} \cdot \nabla c^n - \phi\frac{c^n - \hat{c}^{n-1}}{\Delta t}], d_t\xi^n) \\
&\quad -([a(C_h^{n-1})\nabla P_h^{n-1} - a(c^n)\nabla p^n] \cdot \nabla c^n, d_t\xi^n) \\
&\quad -(\phi\frac{\zeta^n - \zeta^{n-1}}{\Delta t}, d_t\xi^n) - (\phi\frac{\xi^{n-1} - \hat{\xi}^{n-1}}{\Delta t}, d_t\xi^n) \\
&\quad -(\phi\frac{\zeta^{n-1} - \hat{\zeta}^{n-1}}{\Delta t}, d_t\xi^n) + \lambda(\xi^n + \zeta^n, d_t\xi^n) - ((\xi^n + \zeta^n)q_+, d_t\xi^n) \\
&\quad -([D(c^n, p_1^n, p_2^n) - D(C_h^{n-1}, P_{h1}^{n-1}, P_{h2}^{n-1})]\nabla\tilde{c}^n, d_t\xi^n) \\
&\quad -(b(c^n)\frac{\partial p^n}{\partial t} - b(C_h^{n-1})d_t P_h^n, d_t\xi^n),
\end{aligned}
$$

where $q_+ = \max(0, q)$. Multiplying this equation by Δt and summing over $1 \leq n \leq m$, we denote the resulting terms as $B_1, B_2, \ldots, B_{11}$ from the left to the right side. Obviously,

$$
B_1 \geq \phi_* \sum_{n=1}^{m} \|d_t\xi^n\|^2.
$$

B_2 can be decomposed as follows:

$$B_2 \geq \tfrac{1}{2} \sum_{n=1}^{m} d_t (D(C_h^{n-1}, P_{h1}^{n-1}, P_{h2}^{n-1}) \nabla \xi^n, \nabla \xi^n)$$
$$- \tfrac{1}{2} \sum_{n=1}^{m} (d_t D(C_h^{n-1}, P_{h1}^{n-1}, P_{h2}^{n-1}) \nabla \xi^{n-1}, \nabla \xi^{n-1}) \equiv B_2^1 + B_2^2.$$

By $\xi^0 = 0$ and

$$B_2^1 = \frac{1}{2} (D(C_h^{m-1}, P_{h1}^{m-1}, P_{h2}^{m-1}) \nabla \xi^m, \nabla \xi^m) \geq \frac{1}{2} D_* \|\nabla \xi^m\|^2,$$

we see that

$$\phi_* \sum_{n=1}^{m} \|d_t \xi^n\|^2 \Delta t + \frac{1}{2} D_* \|\nabla \xi^m\|^2 \leq B_3 + \ldots + B_{11} - B_2^2.$$

By estimating negative norms and integral transformations [2], we have

$$|B_3 + B_6 + B_7| \leq M \Delta t^2 \left\| \frac{\partial^2 \tau}{\partial \tau^2} \right\|_{L^2(J; L^2(\Omega))}^2$$
$$+ M \sum_{n=1}^{m} (\|\nabla \xi^n\|^2 + \|\nabla \zeta^n\|^2) \Delta t + \varepsilon \sum_{n=1}^{m} \|d_t \xi^n\|^2 \Delta t,$$

where τ is the approximate characteristic direction from (t^n, x) to $(t^{n-1}, \hat{x})$. A direct computation leads to

$$|B_4 + B_5 + B_8 + B_9 + B_{11}| \leq M\{\Delta t^2 + h_c^{2l+2} + h_p^{2k}$$
$$+ \sum_{n=1}^{m} (\|\xi^n\|^2 + \|d_t \pi^n\|^2 + \|\nabla \pi^n\|^2) \Delta t\} + \varepsilon \sum_{n=1}^{m} \|d_t \xi^n\|^2 \Delta t.$$

The crucial is to analyze B_2^2 and B_{10}. We find that

$$|d_t D_{ij}(C_h^{n-1}, P_{h1}^{n-1}, P_{h2}^{n-1})|$$
$$\leq M(|d_t \xi^{n-1}| + |d_t \zeta^{n-1}| + |d_t c^{n-1}| + |\nabla d_t \eta^{n-1}| + |\nabla d_t \pi^{n-1}| + |\nabla d_t p^{n-1}|).$$

By the induction hypotheses, we have

$$|B_2^2| \leq M(h_c^{2l+2} + h_p^{2k} + \sum_{n=1}^{m-1} \|\nabla \xi^n\|^2 \Delta t) + \varepsilon \sum_{n=1}^{m} (\|d_t \xi^n\|^2 + \|\nabla d_t \pi^n\|^2) \Delta t.$$

To handle B_{10}, by summation by parts and $\xi^0 = 0$, we obtain

$$B_{10} = \sum_{n=1}^{m} (d_t [D(c^n, p_1^n, p_2^n) - D(c^{n-1}, p_1^{n-1}, p_2^{n-1})] \nabla \tilde{c}^n, \nabla \xi^{n-1}) \Delta t$$
$$+ \sum_{n=1}^{m} (d_t [D(c^{n-1}, p_1^{n-1}, p_2^{n-1}) - D(C_h^{n-1}, P_{h1}^{n-1}, P_{h2}^{n-1})] \nabla \tilde{c}^n, \nabla \xi^{n-1}) \Delta t$$
$$+ \sum_{n=1}^{m} ([D(c^n, p_1^n, p_2^n) - D(C_h^{n-1}, P_{h1}^{n-1}, P_{h2}^{n-1})] d_t \nabla \tilde{c}^n, \nabla \xi^{n-1}) \Delta t$$
$$- ([D(c^m, p_1^m, p_2^m) - D(C_h^{m-1}, P_{h1}^{m-1}, P_{h2}^{m-1})] \nabla \tilde{c}^m, \nabla \xi^m)$$
$$\equiv B_{10}^1 + B_{10}^2 + \ldots + B_{10}^4.$$

It is easy to show that

$$|B_{10}^1 + B_{10}^3| \le M\{\Delta t^2 + h_c^{2l+2} + h_p^{2k} + \sum_{n=1}^{m}(\|\xi^n\|^2 + \|\nabla\pi^n\|^2 + \|\nabla\xi^n\|^2)\Delta t\},$$
$$|B_{10}^4| \le M(\Delta t^2 + h_c^{2l+2} + h_p^{2k} + \|\xi^{m-1}\|^2 + \|\nabla\pi^{m-1}\|^2) + \varepsilon\|\nabla\xi^{m-1}\|^2.$$

For B_{10}^2, using the mean value theorem and smoothness of D_{ij}, we have

$$d_t[D_{ij}(c^{n-1}, p_1^{n-1}, p_2^{n-1}) - D_{ij}(C_h^{n-1}, P_{h1}^{n-1}, P_{h2}^{n-1})]$$
$$\le M(\Delta t + |d_t\xi^{n-1}| + |d_t\zeta^{n-1}| + |\xi^{n-1}| + |\zeta^{n-1}| + |\xi^{n-2}| + |\zeta^{n-2}|$$
$$+ |\nabla d_t\pi^{n-1}| + |\nabla d_t\eta^{n-1}| + |\nabla\pi^{n-1}| + |\nabla\pi^{n-2}| + |\nabla\eta^{n-1}| + |\nabla\eta^{n-2}|).$$

Then,

$$|B_{10}^2| \le M\{\Delta t^2 + h_p^{2k} + h_c^{2l+2} + \sum_{n=1}^{m}(\|\xi^n\|_1^2 + \|\nabla\pi^n\|^2)\Delta t\}$$
$$+ \varepsilon\sum_{n=1}^{m}(\|d_t\xi^n\|^2 + \|\nabla d_t\pi^n\|^2)\Delta t.$$

Combining the above estimates and taking ε small enough, we have

$$\sum_{n=1}^{m}\|d_t\xi^n\|^2\Delta t + \|\nabla\xi^m\|^2 \le M\{\Delta t^2 + h_p^{2k} + h_c^{2l} + \|\xi^{m-1}\|^2 + \|\nabla\pi^{m-1}\|^2$$
$$+ \sum_{n=1}^{m}(\|\xi^n\|_1^2 + \|\nabla\pi^n\|^2)\Delta t\} + \varepsilon\sum_{n=1}^{m}\|\nabla d_t\pi^n\|^2\Delta t.$$

Multiplying (3.6) by $(M+1)$, adding the result to the above inequality, and taking ε small enough, we obtain

$$\sum_{n=1}^{m}(\|d_t\xi^n\|^2 + \|\nabla d_t\pi^n\|^2)\Delta t + \|\nabla\xi^m\|^2 + \|d_t\pi^m\|^2$$
$$\le M\{\Delta t^2 + h_p^{2k} + h_c^{2l} + \|\xi^{m-1}\|^2 + \|\nabla\pi^{m-1}\|^2$$
$$+ \sum_{n=1}^{m}(\|\xi^n\|_1^2 + \|\nabla\pi^n\|^2 + \|d_t\pi^n\|^2)\Delta t\}.$$

Observing that

$$\|f^m\|^2 - \|f^0\|^2 \le M\sum_{n=1}^{m}\|f^n\|^2\Delta t + \varepsilon\sum_{n=1}^{m}\|d_t f^n\|^2\Delta t,$$

we now have

$$\sum_{n=1}^{m}(\|d_t\xi^n\|^2 + \|d_t\pi^n\|_1^2)\Delta t + \|\xi^m\|_1^2 + \|d_t\pi^m\|^2 + \|\pi^m\|_1^2$$
$$\le M\{\Delta t^2 + h_p^{2k} + h_c^{2l} + \sum_{n=1}^{m}(\|\xi^n\|_1^2 + \|\pi^n\|_1^2 + \|d_t\pi^n\|^2)\Delta t\}.$$

Using Bellman's inequality, it is shown that

$$
\max_{1\leq n\leq m} \|d_t\pi^m\|^2 + \max_{1\leq n\leq m} \|\pi^m\|_1^2 + \max_{1\leq n\leq m} \|\xi^m\|_1^2
$$
$$
+ \sum_{n=1}^{m} (\|d_t\pi^n\|_1^2 + \|d_t\xi^n\|^2)\Delta t \leq M(\Delta t^2 + h_p^{2k} + h_c^{2l}). \tag{3.7}
$$

Let the parameters satisfy that

$$
\Delta t = o(h_p), \ \Delta t = o(h_c), \ k \geq 2, \ l \geq 2, \ h_c^l = o(h_p), \ h_p^k = o(h_c), \tag{3.8}
$$

and let Δt and h be small enough that

$$
Mh_p^{-1}(\Delta t + h_p^k + h_c^l) \leq 1, \ \ Mh_c^{-1}(\Delta t + h_p^k + h_c^l) \leq 1.
$$

Now, we demonstrate the induction hypotheses (3.5) is true for $r = m$. By (3.7), (3.8), and inverse properties, we find that

$$
\|d_t\pi^m\|_{L^\infty(\Omega)} \leq Mh_p^{-1}\|d_t\pi^m\| \leq Mh_p^{-1}(\Delta t + h_p^k + h_c^l) \leq 1,
$$
$$
\|\nabla\pi^m\|_{L^\infty(\Omega)} \leq Mh_p^{-1}\|\nabla\pi^m\| \leq Mh_p^{-1}(\Delta t + h_p^k + h_c^l) \leq 1,
$$
$$
\|\xi^m\|_{W_\infty^1(\Omega)} \leq h_c^{-1}\|\xi^m\|_1 \leq Mh_c^{-1}(\Delta t + h_p^k + h_c^l) \leq 1,
$$

so (3.5) is true for $r = m$.

Theorem 3.1 *Suppose that the assumptions given in §1 hold and the space and time discretizations satisfy (3.8). Then there exists a constant M such that, for (h_c, h_p) sufficiently small,*

$$
\max_{0\leq n\leq T/\Delta t} \|c^n - C_h^n\|_1^2 + \max_{0\leq n\leq T/\Delta t} \|p^n - P_h^n\|_1^2 + \max_{1\leq n\leq T/\Delta t} \|d_t(p^n - P_h^n)\|^2
$$
$$
+ \sum_{n=1}^{T/\Delta t} \|d_t(p^n - P_h^n)\|_1^2\Delta t + \sum_{n=1}^{T/\Delta t} \|d_t(c^n - C_h^n)\|^2\Delta t \leq M(\Delta t^2 + h_c^{2l} + h_p^{2k}),
$$

where M is independent of Δt, h_c, and h_p.

Remark. In addition to the optimal estimates in the $L^\infty(J; H^1(\Omega))$ norm, we also obtain an estimate of $d_t(p^n - P_h^n)$ in the $L^\infty(J; L^2(\Omega))$ norm and in the $L^2(J; H^1(\Omega))$ norm, which cannot be found previously. This should attribute to the special technique for establishing error equation (3.4).

References

[1] Cheng, A. J., Optimal error estimate of finite element method in $L^\infty(J; H^1(\Omega))$ norm for a model for miscible compressible displacement in porous media, *Numerical Mathematics, A Journal of Chinese Universities (in Chinese)* 16 (1994), 134–144.

[2] Chen, Z. and Ewing, R. E., Fully-discrete finite element analysis of multi-phase flow in groundwater hydrology, *SIAM J. Numer. Anal.* **34** (1997), 2228–2253.

[3] Ciarlet, P. G., *The Finite Element Method for Elliptic Problems*, North-Holland, Amsterdam, New York, 1978.

[4] Dawson, C. N., Russell, T. F., and Wheeler, M. F., Some improved error estimates for the modified method of characteristics, *SIAM J. Numer. Anal.* **26** (1989), 1487–1512.

[5] Douglas, J., Jr. and Roberts, J. E., Numerical methods for a model for compressible miscible displacement in porous media, *Math. Comp.* **41** (1983), 441–459.

[6] Douglas, J., Jr. and Russell, T. F., Numerical methods for convection-dominated diffusionproblem based on combining the method of characteristics with finite element or difference procedures, *SIAM J, Numer. Anal.* **19** (1982), 871–885.

[7] Russell, T. F., Time stepping along characteristics with incomplete iteration for a Galerkin approximation of miscible displacement in porous media, *SIAM J. Numer. Anal.* **22** (1985), 970–1013.

[8] Wheeler, M. F., A priori L^2 error estimates for Galerkin approximations to parabolic partial differential equations, *SIAM J. Numer. Anal.* **10** (1973), 723–759.

[9] Yuan, Y. R., Time stepping along characteristics for the finite element approximation of compressible miscible displacement in porous media, *Mathematica Numerica Sinica (in Chinese)* **14** (1992), 385–400.

A Numerical Algorithm for Single Phase Fluid Flow in Elastic Porous Media

HONGSEN CHEN   RICHARD E. EWING
STEPHEN L. LYONS   GUAN QIN   TONG SUN
DAVID P. YALE

Abstract

In this paper we consider an integrated model for single-phase fluid flow in elastic porous media. The model and mathematical formulation consist of mass and momentum balance equations for both fluid and porous media. We propose a mixed finite element scheme to solve simultaneously for the porous media displacement, fluid mass flux, and pore pressure. A prototype simulator for solving the integrated problem has been built based on a finite element object library that we have developed. We will present numerical and sensitivity results for the solution algorithm.

KEYWORDS: geomechanics, fluid flow, elastic deformation, porous media

1   Introduction

Mechanical processes in porous media involve two basic elements: fluid flow and rock deformation. It is important to address the coupling of fluid flow and rock deformation for stress-sensitive reservoirs in reservoir simulation. In stress-sensitive reservoirs, the change in fluid pressure, which is caused by fluid injection or production, perturbs the stress state of the rock. The change in stress state triggers the rock deformation which, in turn, affects the fluid flow processes by changing the volumetric behavior of the reservoir fluid and rock properties. In most conventional reservoir simulations, one only considers the impact of rock deformation on fluid flow through the pore compressibility which is expressed as a function of the fluid pressure. This simplistic approach is not adequate for simulating the coupling effects in stress-sensitive reservoirs. Moreover, reservoir simulations which are based on this simplified model can lead to wrong predictions for reservoir performance. Consequently, it is necessary to develop a coupled model where the dependence of flow and deformation on each other can be modeled simultaneously.

The key issue in a coupled model is how to describe the interaction between the flow and deformation. The concept of "effective stress", which is defined as the total stress of rock minus the fluid pressure, was first proposed by Terzaghi to take into account the effect of fluid pressure on the rock deformation [18]. Then this concept was generalized for the coupled model of fluid flow and rock deformation by Biot in a number of papers [1, 2, 3, 4], into what is called "poro-elasticity theory". The poro-elasticity theory has been widely used in civil, mining, petroleum, and environmental engineering for several decades. While considering the impact of the rock deformation on the processes of fluid flow, the rock properties, such as porosity and permeability, may change significantly with the variation of both fluid pressure and the stress state [7, 16]. The constitutive relations which describe the dependency of the rock properties on the deformation are derived based on various experimental observations and physical assumptions [5, 7, 8, 9, 13, 16]. Primarily, the effect on the volumetric behavior of fluid flow is widely used in developing coupled models, in which the pore compressibility depends on not only fluid pressure but also the mean stress of the rock [5, 20].

Generally, the mathematical models for solving the coupled problems are derived from physical principles based on certain physical assumptions [4, 5, 10, 11, 12, 19]. The mathematical formulations are composed of a system of mixed-type nonlinear partial differential equations because of the coupling terms in the equations which govern flow and deformation. Numerical solution algorithms for solving such a system are difficult because the algorithms need to address the mathematical characteristics of both flow and deformation problems and, at the same time, to handle the coupling terms properly. Standard finite element methods have been used to solve the coupled system [10, 11, 12, 19]. However, the application of standard Galerkin methods in solving fluid flow problems raises concerns because of the loss of local mass conservation which is an important issue in reservoir simulation. Osorio used a finite difference method to solve the same problem in a sequential fashion, which preserved local mass conservation [14]. However, the simultaneous solution procedure is more favorable compared with the sequential solution procedure due to computational efficiency.

In this paper we employ the mixed finite element method to discretize the governing equations in which the Galerkin method is used for the deformation equation and the lowest order Raviart-Thomas element is used for the fluid flow equations to conserve mass locally [6]. The numerical problems are solved in a fully coupled and fully implicit fashion, which requires an efficient iterative solver. The research simulator has been built up based on the finite element objects which have been developed in an object-oriented fashion [17]. The simulator has solved several test problems and field-scale simulation is ongoing. The primary numerical results show the strength of the solution procedure. The rest of the paper is organized as follows: In §2, we derive a mathematical formulation for fluid flow in elastic porous media from basic physical principles. In §3, we propose numerical schemes to solve mathematical problem. In §4, we briefly discuss our simulator development strategy. In §5, we show numerical results for solving an axis-symmetric problem by using the simulator. In §6, we draw some conclusions and identify the directions for future work.

2   Mathematical Model

In this paper, we make the following basic assumptions: (1) temperature of the reservoir is constant, (2) there is no mass exchange between the rock phase and fluid phase, and (3) the rock is elastic material. The governing equations of coupled rock deformation and fluid flow are derived from the following physical principles: mass conservation and momentum balance for both fluid and rock phases. Since there is no mass exchange between the rock and fluid, the mass is conserved in terms of phases:

$$\frac{\partial (\phi \rho_f)}{\partial t} + \nabla \cdot (\rho_f \phi \underline{v}_f) = q_f, \tag{2.1}$$

and

$$\frac{\partial ((1 - \phi)\rho_r)}{\partial t} + \nabla \cdot (\rho_r (1 - \phi)\underline{v}_r) = 0. \tag{2.2}$$

Equations (2.1) and (2.2) are the mass conservation for fluid and solid phase, respectively. The subscripts f stands for fluid phase and r for rock phase. In equations (2.1) and (2.2), ϕ stands for the porosity, ρ stands for the mass density and q_f stands for the flow rate of a source/sink term. The momentum balance for fluid is interpreted as Darcy's law in terms of fluid velocity relative to the rock velocity:

$$\underline{v} = \phi \left(\underline{v}_f - \underline{v}_r \right) = -\frac{\underline{\underline{k}}}{\mu_f} \left(\nabla p + \rho_f g \nabla D \right), \tag{2.3}$$

where μ stands for fluid viscosity, $\underline{k}$ for the rock permeability which is a tensor and g denotes the gravitational acceleration. The momentum balance for the solid phase is expressed by the steady state force balance:

$$\nabla \cdot \underline{\sigma} = \underline{q}_u, \qquad (2.4)$$

where $\underline{\sigma}$ stands for the stress state of the solid phase and $\underline{q}_u$ is a body force.

According to the poroelastic theory [4, 5], the effective stress is used in describing the stress-strain relationship:

$$\underline{\sigma}' = 2G\underline{\epsilon} + \lambda\nabla \cdot \underline{u}, \qquad (2.5)$$

where $\underline{\sigma}'$ stands for the effective stress, $\underline{u}$ for the displacement, $\underline{\epsilon}$ for the strain, G for shear moduli and λ for the Lamé constant. The elastic strain-displacement relationship is defined as

$$\underline{\epsilon} = \left(\nabla\underline{u} + \nabla\underline{u}^T\right)/2. \qquad (2.6)$$

The effective stress $\underline{\sigma}'$, is defined as follows according to [3, 5]:

$$\underline{\sigma}' = \underline{\sigma} - \alpha p\underline{\delta}, \qquad (2.7)$$

where p is fluid pressure, α is a coefficient which is positive but less than 1 and $\underline{\delta}$ is Kronecker δ. Generally, α could be a function of the fluid pressure and rock deformation. However, it is reasonable to assume that α is a constant for single phase fluid flow in elastic porous media. Combining equations (2.4)–(2.7), we obtain the governing equation for the rock deformation in terms of rock displacement $\underline{u}$ and fluid pressure p:

$$\nabla \cdot [G(\nabla\underline{u} + \nabla\underline{u}^T)] + \nabla(\lambda\nabla \cdot \underline{u}) + \nabla(\alpha p) = \underline{q}_u. \qquad (2.8)$$

In this paper, the effect of the rock deformation on the flow processes are treated through: the dependency of the porosity on both fluid pressure and mean stress, and Darcy's law is valid for the fluid velocity relative to the solid velocity. Substituting equation (2.3) into (2.1), we obtain the following form for the fluid mass balance equation:

$$\frac{d(\rho_f\phi)}{dt} + \rho_f\phi\nabla \cdot \underline{v}_r + \nabla \cdot (\rho_f\underline{v}) = q_f, \qquad (2.9)$$

where the total derivative is defined as $d/dt = \partial\partial t + \underline{v}_r \cdot \nabla$. From equation (2.2), we obtain the following relationship:

$$\nabla \cdot \underline{v}_r = \frac{1}{V_b}\frac{dV_b}{dt}, \qquad (2.10)$$

where V_b stands for the bulk volume of the rock phase. Since the porosity is defined as the ratio of pore volume V_p to the bulk volume V_b, we obtain the following relationship:

$$\frac{d\phi}{dt} = \frac{d}{dt}\left(\frac{V_p}{V_b}\right) = \phi\left(\frac{1}{V_p}\frac{dV_p}{dt} - \frac{1}{V_b}\frac{dV_b}{dt}\right). \tag{2.11}$$

By substituting equations (2.10) and (2.11) into equation (2.9), we obtain

$$\rho_f\phi\left(\frac{1}{\rho_f}\frac{d\rho_f}{dt} + \frac{1}{V_p}\frac{dV_p}{dt}\right) + \nabla \cdot (\rho_f\underline{v}) = q_f. \tag{2.12}$$

Generally the fluid density is a function of fluid pressure and the compressibility of the fluid is defined as $c_f = \frac{1}{\rho_f}\frac{d\rho_f}{dp}$. The change of the pore volume V_p is caused by both fluid pressure and mean stress of the rock phase when considering the coupling effects. Therefore, we obtain the following relationship:

$$\frac{1}{V_p}\frac{dV_p}{dt} = \frac{1}{V_p}\left(\frac{\partial V_p}{\partial \sigma}\right)_p\frac{d\sigma}{dt} + \frac{1}{V_p}\left(\frac{\partial V_p}{\partial p}\right)_\sigma\frac{dp}{dt}, \tag{2.13}$$

where the mean stress σ is defined as

$$\sigma = tr(\underline{\sigma})/3. \tag{2.14}$$

Then we define the compressibilities as follows:

$$c_{p\sigma} = -\frac{1}{V_p}\left(\frac{\partial V_p}{\partial \sigma}\right)_p \quad \text{and} \quad c_{pp} = \frac{1}{V_p}\left(\frac{\partial V_p}{\partial p}\right)_\sigma. \tag{2.15}$$

In equation (2.15), $c_{p\sigma}$ and c_{pp} are compressibilities of the pore volume. The first letter in subscripts denotes the type of volume (e.g. "p" for pore volume or "b" for bulk volume) and the second letter refers to the independent variable (e.g. "p" for fluid pressure and "σ" for the mean stress of the rock phase). We can relate these compressibilities to the experimentally measurable compressibilities: the compressibility of the solid grains c_r and the compressibility of the bulk volume with respect to mean stress $c_{b\sigma}$ [20]:

$$c_{p\sigma} = \frac{c_{b\sigma} - c_r}{\phi} \quad \text{and} \quad c_{pp} = \frac{c_{b\sigma} - (1 + \phi)c_r}{\phi}. \tag{2.16}$$

Combining equations (2.13) and (2.16) with equation (2.12), we obtain

$$c_t\frac{\partial p}{\partial t} - c_1\frac{\partial \sigma}{\partial t} + \nabla \cdot (\rho_f\underline{v}) = q_f, \tag{2.17}$$

where c_t and c_1 are given as follows:

$$c_t = \rho_f \left(c_{b\sigma} - (1 + \phi)c_r + \phi c_f \right), \tag{2.18}$$

and

$$c_1 = \rho_f \left(c_{b\sigma} - c_r \right). \tag{2.19}$$

We can obtain a relationship between volumetric strain and mean stress from equations (2.7) and (2.5):

$$\sigma = \frac{\nabla \cdot \underline{u}}{c_{b\sigma}} + \alpha p. \tag{2.20}$$

Because we are only considering elastic (small) deformation, it is reasonable to neglect the higher order term in the total derivative $\frac{d}{dt}$, then $\frac{d}{dt} = \frac{\partial}{\partial t}$. Combining equations (2.9), (2.18), (2.19), and (2.20), we obtain the governing equation for fluid flow as follows:

$$\underline{f} + \underline{K} \left(\nabla p + \rho_f g \nabla D \right) = 0,$$
$$c_1 \frac{\partial}{\partial t} \left(\frac{\nabla \cdot \underline{u}}{c_{bc}} \right) + c_1 \frac{\partial (\alpha p)}{\partial t} - \nabla \cdot \underline{f} - c_t \frac{\partial p}{\partial t} = -q_f, \tag{2.21}$$

where $\underline{K} = \rho_f \underline{k} / \mu_f$.

Equations (2.8) and (2.21) form the system of governing equations for single phase fluid flow through elastic rock. The primary variables are the displacement $\underline{u}$, mass flux of the fluid $\underline{f}$ and fluid pressure p.

To determine the solution of equations (2.8) and (2.21), we define the following boundary and initial conditions:

$$
\begin{aligned}
\underline{u} &= \underline{u}_d & &\text{on } \Gamma_{rd}, \\
G(\nabla \underline{u} + \nabla \underline{u}^T)\underline{n} + \lambda \nabla \cdot \underline{u}\underline{n} &= \underline{u}_n & &\text{on } \Gamma_{rn}, \\
p &= p_d & &\text{on } \Gamma_{fd}, \\
\underline{K}(\nabla p + \rho_f g \nabla D)\,\underline{n} &= \underline{f}_n & &\text{on } \Gamma_{fn}, \\
p &= p_0 & &\text{in } \Omega_0 \text{ at } t = 0, \\
\underline{u} &= \underline{u}_0 & &\text{in } \Omega_0 \text{ at } t = 0,
\end{aligned}
$$

where Ω stands for the solution domain, $\partial\Omega$ stands for the boundary of the solution domain. $\Gamma_{1d} \cup \Gamma_{1n} = \Gamma_{2d} \cup \Gamma_{2n} = \partial\Omega$. The subscripts r and f stand for the boundary condition in terms of the rock deformation and fluid flow, respectively. The subscripts d and n denote the Dirichlet and Neumann boundary conditions, respectively.

3   Numerical Solution Algorithm

To solve the system (2.8) and (2.21) numerically by using the mixed finite element method, we need to write the system in a weak formulation. To this end, we shall use the following standard notation for Sobolev spaces and their norms:

$$\tilde{\mathcal{U}} = \{\underline{u} \in H^1(\Omega)^3 \mid \underline{u} = \underline{u}_d \text{ on } \Gamma_{1d}\}, \quad \mathcal{U} = \{\underline{u} \in H^1(\Omega)^3 \mid \underline{u} = 0 \text{ on } \Gamma_{1d}\},$$
$$\tilde{\mathcal{V}} = \{\underline{f} \in L^2(\Omega)^3 \mid \nabla \cdot \underline{f} \in L^2(\Omega), \ \underline{f} \cdot \underline{n} = \underline{f}_n \text{ on } \Gamma_{2n}\},$$
$$\mathcal{V} = \{\underline{f} \in L^2(\Omega)^3 \mid \nabla \cdot \underline{f} \in L^2(\Omega), \ \underline{f} \cdot \underline{n} = 0 \text{ on } \Gamma_{2n}\},$$
$$\mathcal{W} = L^2(\Omega).$$

Now, the weak formulation for the coupled system (2.8) and (2.21) seeks $(\underline{u}, \underline{f}, p) \in \mathcal{U} \times \mathcal{V} \times \mathcal{W}$ satisfying

$$a(\underline{u}, \underline{v}) + (\alpha p, \nabla \cdot \underline{v}) - \langle \alpha p, \underline{v} \cdot \underline{n} \rangle_{\Gamma_{1n} \cap \Gamma_{2n}} = (\underline{q}_u, \underline{v}) + \langle \underline{u}_n, \underline{v} \rangle_{\Gamma_{1n}},$$
$$(\underline{\underline{K}}^{-1}\underline{f}, \underline{\psi}) - (p, \nabla \cdot \underline{\psi}) = -(\rho_f g \nabla D, \underline{\psi}) - \langle p_d, \underline{\psi} \cdot \underline{n} \rangle_{\Gamma_{1d}},$$
$$\left(c_1 \frac{\partial}{\partial t}\left(\frac{\nabla \cdot \underline{u}}{c_{bc}}\right), \kappa\right) + \left(c_1 \frac{\partial(\alpha p)}{\partial t}, \kappa\right) - (\nabla \cdot \underline{f}, \kappa) - \left(c_2 \frac{\partial p}{\partial t}, \kappa\right) = -(q_f, \kappa),$$

$$(3.1)$$

for all $(\underline{v}, \underline{\psi}, \kappa) \in \mathcal{U} \times \mathcal{V} \times \mathcal{W}$.

Let $h \in (0, 1)$ and $\mathcal{T}_h$ be a family of regular triangulations of Ω. Let $\mathcal{U}_h \in \mathcal{U}$ be the piecewise continuous linear finite element spaces and $(\mathcal{V}_h, \mathcal{W}_h) \in (\mathcal{V}, \mathcal{W})$ the lowest order Raviart-Thomas mixed finite element spaces [15]. The finite element approximation $(\underline{u}_h, \underline{f}_h, p_h) \in \mathcal{U}_h \times \mathcal{V}_h \times \mathcal{W}_h$ is the solution of the discretized version of (3.1):   Find $(\underline{u}_h, \underline{f}_h, p_h) \in \mathcal{U}_h \times \mathcal{V}_h \times \mathcal{W}_h$ such that for all $(\underline{v}_h, \underline{\psi}_h, \kappa_h) \in \mathcal{U}_h \times \mathcal{V}_h \times \mathcal{W}_h$

$$a(\underline{u}_h, \underline{v}_h) + (\alpha p_h, \nabla \cdot \underline{v}_h) - \langle \alpha p_h, \underline{v}_h \cdot \underline{n} \rangle_{\Gamma_{1n} \cap \Gamma_{2n}} = (\underline{q}_u, \underline{v}_h) + \langle \underline{u}_n, \underline{v}_h \rangle_{\Gamma_{1n}},$$
$$(\underline{\underline{K}}^{-1}\underline{f}_h, \underline{\psi}_h) - (p_h, \nabla \cdot \underline{\psi}_h) = -(\rho_f g \nabla D, \underline{\psi}_h) - \langle p_d, \underline{\psi}_h \cdot \underline{n} \rangle_{\Gamma_{1d}},$$
$$\left(c_1 \frac{\partial}{\partial t}\left(\frac{\nabla \cdot \underline{u}_h}{c_{bc}}\right), \kappa_h\right) + \left(c_1 \frac{\partial(\alpha p_h)}{\partial t}, \kappa_h\right) - (\nabla \cdot \underline{f}_h, \kappa_h) - \left(c_2 \frac{\partial p_h}{\partial t}, \kappa_h\right)$$
$$= -(q_f, \kappa_h).$$

$$(3.2)$$

A backward Euler scheme is employed for the time derivatives in (3.2) and a Newton iteration is used at each time step due to the nonlinearities in the governing equations. Here we present the simultaneous solution algorithm for solving the system as follows:

1. At time $t = 0$, the increment in displacement $\Delta \underline{u}$ is given as 0 and the

fluid pressure is known from the initial conditions for the model. The compressibilities are given as constants in this paper.

2. Update time to $t = t + \Delta t$.

3. Compute the porosity, fluid density and viscosity based on pressure and temperature.

4. Apply Newton's method for system (3.2) to generate Jacobian matrix.

5. Solve the linear system iteratively and then update the primary variables.

6. Repeat steps 3–5 until solution converges.

7. Go back to step 2 to update the porosity and the fluid properties and repeat the above process until time $t = T$.

4 Simulator Development

In the development of numerical simulation technologies, a flexible numerical tool is necessary to quickly test proposed numerical methods and algorithms for solving mathematical problems describing different physical problems. Therefore, the tool should be flexible in the choice of problems and finite element methods and its software components should be reusable. Moreover, the tool should also be simple to interface with other software packages. Based on the above considerations, we divided the simulator development into two steps: (1) develop object-oriented finite element objects (FEO) which can be utilized to solve various model problems by using different finite element schemes, and (2) build a simulator for solving specific physical problem, such as the coupled problem, by using these finite element objects.

In developing FEO, we focused on the implementation of the finite element formulation and interfaces to the mesh generator and linear solver, in addition to the geometry and partial differential equations [17]. There are two basic components in implementing the finite element method: (1) variational formulations which consist of a set of linear and bilinear forms; and (2) shape functions which generate the approximate solution. Consequently, we identified the objects as the bilinear forms, linear forms, piecewise polynomial spaces, boundary integral forms and essential boundary conditions. Object-oriented programming techniques, especially polymorphism, are used to uniformly process these different objects. A user interface specifies these objects and combines them to form a finite element problem.

Based on FEO, we need three steps to construct the simulator for solving the coupled problem. First, we need to develop a class which implements the constitutive relationships for updating the fluid and rock properties. Then we form a finite element problem for the coupled problem by using FEO. Finally, we need to develop a driver which implements the Newton iteration.

5   Computational Examples

A simulator for solving the coupled fluid flow and rock deformation problems has been developed. The simulator has been tested with various problems in order to debug and to improve the overall efficiency. In this paper, we show the numerical results for solving an axis-symmetric problem with known analytical solution. It is of practical importance to solve the axis-symmetric problem because the axis-symmetric problem is of great interest to study the single well injection or production case where the coupling effect can have a large impact on reservoir performance. Moreover, it is numerically efficient to use a 2-dimensional simulator to simulate 3-dimensional reservoir behavior. In the following test problem, we check the transformation which transforms the 3-dimensional problem into an axis-symmetric problem and to check the convergence rate of the proposed numerical algorithm.

By assuming that all coefficients and the solution are axis-symmetric, i.e., $\frac{\partial}{\partial \theta} = 0$ and $u_\theta = 0$, we can reformulate the system (2.8) and (2.21) in cylindrical coordinates. First, we have the following transformation formulas:

$$\nabla \cdot \underline{u} = \frac{1}{r}\left(\frac{\partial(ru_r)}{\partial r} + \frac{\partial(ru_z)}{\partial z}\right) = \frac{1}{r}\nabla_{rz} \cdot (r\underline{u}) \quad \text{if } u_\theta = 0,$$

$$\nabla \psi = \frac{\partial \psi}{\partial r}\underline{e}_r + \frac{\partial \psi}{\partial z}\underline{e}_z = \nabla_{rz}\psi \quad \text{if } \frac{\partial \psi}{\partial \theta} = 0,$$

where $\underline{e}_r = (\cos\theta, \sin\theta, 0)$, $\underline{e}_z = (0, 0, 1)$. Thus we obtain

$$\nabla \cdot [G(\nabla\underline{u} + \nabla\underline{u}^T)] = \frac{1}{r}\left[\frac{\partial}{\partial r}(2rG\frac{\partial u_r}{\partial r}) + \frac{\partial}{\partial z}(rG\frac{\partial u_r}{\partial z} + rG\frac{\partial u_z}{\partial r}) - \frac{2}{r}Gu_r\right]\underline{e}_r$$

$$+\frac{1}{r}\left[\frac{\partial}{\partial r}(rG\frac{\partial u_r}{\partial z} + rG\frac{\partial u_z}{\partial r}) + \frac{\partial}{\partial z}(2rG\frac{\partial u_z}{\partial z})\right]\underline{e}_z$$

$$= \frac{1}{r}\nabla_{rz} \cdot (rG(\nabla_{rz}\underline{u} + \nabla_{rz}\underline{u}^T)) - \frac{2}{r^2}Gu_r\underline{e}_r,$$

$$\nabla(\lambda\nabla \cdot \underline{u}) = \frac{\partial}{\partial r}\left[\frac{\lambda}{r}\left(\frac{\partial(ru_r)}{\partial r} + \frac{\partial(ru_z)}{\partial z}\right)\right]\underline{e}_r$$

$$+\frac{\partial}{\partial z}\left[\frac{\lambda}{r}\left(\frac{\partial(ru_r)}{\partial r} + \frac{\partial(ru_z)}{\partial z}\right)\right]\underline{e}_z = \nabla_{rz}\left(\frac{\lambda}{r}\nabla_{rz} \cdot (ru)\right).$$

Multiplying system (2.8) and (2.21) by r, we have

$$-\nabla_{rz} \cdot \left(rG(\nabla_{rz}\underline{u} + \nabla_{rz}\underline{u}^T)\right) + \tfrac{2}{r}G\underline{e}_1 u,$$

$$-r\nabla_{rz}\left(\tfrac{\lambda}{r}\nabla_{rz} \cdot (r\underline{u})\right) - r\nabla_{rz}(\alpha p) = r\underline{q}_u,$$

$$r\underline{f} + r\underline{K}\left(\nabla_{rz}p + \rho_f g \nabla_{rz}D\right) = 0,$$

$$c_1 r \frac{\partial}{\partial t}\left(\frac{\nabla_{rz} \cdot (r\underline{u})}{c_{bc}}\right) + c_1 r \frac{\partial(\alpha p)}{\partial t} - \nabla_{rz} \cdot (r\underline{f}) - c_2 r \frac{\partial p}{\partial t} = -rq_f. \tag{5.1}$$

Since the axis-symmetric problem defined in 2-dimensional rz coordinates is the transformation of the 3-dimensional problem in xyz Cartesian coordinates, there is no boundary conditions specified on $r = 0$.

We specify the solutions for $\underline{u}$, $\underline{f}$, and p as follows:

$$\underline{u} = te^{-t}\left(x(x^2 + y^2)e^z, y(x^2 + y^2)e^z, (x^2 + y^2)^2 z^2\right)^T,$$

$$p = (x^2 + y^2)^2 e^z \sin t, \tag{5.2}$$

$$\underline{f} = -\left(4x(x^2 + y^2), 4y(x^2 + y^2), (x^2 + y^2)^2\right)^T e^z \sin t.$$

Assuming all coefficients of system (2.8) and (2.21) are constants, $(\underline{u}, \underline{f}, p)$ is the solution of the system with the following right-hand sides:

$$\underline{q}_u = -G \begin{pmatrix} 16xe^z + x(x^2 + y^2)(e^z + 8z) \\ 16ye^z + y(x^2 + y^2)(e^z + 8z) \\ 4(x^2 + y^2)e^z + 16(x^2 + y^2)z^2 + 4(x^2 + y^2)^2 \end{pmatrix} te^{-t}$$

$$-\lambda \begin{pmatrix} 8xe^z + 8x(x^2 + y^2)z \\ 8ye^z + 8y(x^2 + y^2)z \\ 4(x^2 + y^2)e^z + 2(x^2 + y^2)^2 \end{pmatrix} te^{-t} - \alpha \begin{pmatrix} 4x(x^2 + y^2) \\ 4y(x^2 + y^2) \\ (x^2 + y^2)^2 \end{pmatrix} e^z \sin t,$$

$$-q_f = \tfrac{c_1}{c_{bc}}\left[4(x^2 + y^2)e^z + 2(x^2 + y^2)^2 z\right](1 - t)e^{-t}$$

$$+(c_1\alpha - c_2)(x^2 + y^2)^2 e^z \cos t + \left[16 + (x^2 + y^2)\right](x^2 + y^2)e^z \sin t.$$

The initial conditions for this analytical solution are:

$$\underline{u}_0 = 0 \quad \text{and} \quad p_0 = 0.$$

The analytical solution (5.2) is transformed to the axis-symmetric case:

$$\underline{u} = \begin{pmatrix} u_r \\ u_z \end{pmatrix} = \begin{pmatrix} r^3 e^z te^{-t} \\ r^4 z^2 te^t \end{pmatrix}, \tag{5.3}$$

and

$$p = r^4 e^z \sin t \quad \text{and} \quad \underline{f} = \begin{pmatrix} f_r \\ f_z \end{pmatrix} = \begin{pmatrix} -4r^3 e^z \sin t \\ -r^4 e^z \sin t \end{pmatrix}. \tag{5.4}$$

Substituting (5.3) and (5.4) into (5.1), we obtain the following right-hand sides $\underline{q}_u$ and q_f:

$$\underline{q}_u = -G \begin{pmatrix} 16re^z + r^3(e^z + 8z) \\ 4r^2e^z + 16r^2z^2 + 4r^4 \end{pmatrix} te^{-t}$$

$$-\lambda \begin{pmatrix} 8re^z + 8r^3z \\ 4r^2e^z + 2r^4 \end{pmatrix} te^t - \alpha \begin{pmatrix} 4r^3 \\ r^4 \end{pmatrix} e^z \sin t,$$

$$-q_f = \frac{c_1}{c_{bc}} \left[4r^2e^z + 2r^4z\right] (1-t)e^{-t} + (c_1\alpha - c_2)r^4e^z \cos t$$

$$+ \left[16 + r^2\right] r^2 e^z \sin t.$$

The resulting stress boundary condition is of Neumann type:

$$\left[rG(\nabla_{rz}\underline{u} + \nabla_{rz}\underline{u}^T) + \lambda\nabla_{rz} \cdot (r\underline{u})\right] \cdot \underline{n} =$$

$$Gte^{-t} \begin{pmatrix} 6r^3e^z & r^4e^z + 4r^4z^2 \\ r^4e^z + 4r^4z^2 & 4r^5z \end{pmatrix} \cdot \underline{n} + \lambda(4r^3e^z + 2r^5z)te^{-t}\underline{n}.$$

We have solved problem (5.1) by using the simultaneous solution algorithm with both triangular and quadrilateral elements. The L_2 average truncation error is calculated with respect to the analytical solution (5.3) and (5.4). We carried out four runs by refining the mesh size by a factor of 2 and refining the time step size by a factor of 4. Table 1 lists the L_2 average truncation errors for each primary variable in four runs using quadrilateral elements. It is clear that the proposed numerical algorithm is 2nd-order accurate.

6   Summary and Future Work

A mixed finite element method has been employed to solve single phase fluid flow through elastic porous media. A simultaneous solution algorithm has been implemented using FEO to solve this coupled problem in a fully coupled and fully implicit fashion. The simulator has been tested with various test problems and the numerical results presented in this paper show the overall numerical accuracy by the proposed algorithm is second order. The iterative solution procedure which can handle the resulting linear system from the coupled problem needs to be further studied and developed in order to carry out field-scale reservoir simulations.

	u_r	u_z	p	$f \cdot \underline{n}$
2×2	0.098	0.155	2.556	0.218
4×4	0.023	0.036	0.667	0.059
16×16	0.005	0.008	0.126	0.015
32×32	0.001	0.002	0.038	0.004

Table 1. Average truncation error.

References

[1] Biot, M. A., Nonlinear and semilinear rheology of porous solids', *J. Geophy. Res.* **73** (1973), 4924–4937.

[2] Biot, M. A. and Willis, D. G., The elastic coefficients of the theory of consolidation, *J. Appl. Mech.* **24** (1957), 594–601.

[3] Biot, M. A., General theory of three-dimensional consolidation, *J. Appl. Mech.* **25** (1956), 91–96.

[4] Biot, M. A., General theory of three-dimensional consolidation, *J. Appl. Phys.* **12** (1941), 155–164.

[5] Chen, H. Y., Teufel, L. W., and Lee, R. L., Coupled fluid flow and geomechanics in Reservoir Study - I. Theory and governing equations, presented in *The Proceding of SPE Annual Technical Conference & Exhibition*, Dallas, Oct. 1995, 22–25.

[6] Ewing, R. E., Problems Arising in the Modeling of Processes for Hydrocarbon Recovery, (Ewing, ed.) *The Mathematics of Reservoir Simulation*, 1983, SIAM, Philadelphia, PA, pp. 3-34.

[7] Fatt, I. and Davis, D. H., Reduction in permeability with overburden pressure, *Trans AIME* **195** (1952), 329-341.

[8] Holt, R. M., Permeability reduction induced by a nonhydrostatic stress field, *SPEFE* **Dec.** (1990), 444–448.

[9] Jones, F. O. and Owens, W. W., A laboratory study of low-permeability gas sands, *JPT* **Sept.** (1980), 1631–1640.

[10] Koutsabeloulis, N. C., Numerical modeling of soft reservoir behavior during fluid production, *Geotechnical Engineering in Hard Soil-Soft Rocks*, 1993.

[11] Koutsabeloulis, N. C., Heffer, K. J., and Wong, S., Numerical geomechanics in reservoir engineering, *Computer Methods and Advances in Geomechanics*, 1994.

[12] Lewis, R. W. and Sukirman, Y., Finite element modeling of three-phase flow in deforming saturated oil reservoirs, *Int. J. Num. & Analy. Methods Gemech.* **17** (1993), 577–598.

[13] Morita, N., *et al.*, Rock-property changes during reservoir compaction, *SPEFE* **Sept.** (1992), 197–205.

[14] Osorio, J. G., Numerical modeling of coupled fluid-flow/geomechanical behavior of reservoirs with stress-sensitive permeability, *Ph.D dissertation*, New Mexico Institute of Mining and Technology, Socorro, NM, 1998.

[15] Raviart, P. A. and Thomas, J. M., A Mixed Finite Element Method for 2nd Order Elliptic Problems, *Lecture Notes in Math.* **606**, Springer-Verlag, Berlin, 1977.

[16] Rhett, D. W. and Teufel, L. W., Effect of reservoir stress path on compressibility and permeability of sandstones, paper SPE 24756 presented at the *SPE Annual Technical Conference and Exhibition* Washington, DC, Oct. 4-7, 1990.

[17] Sun, T., Ewing, R. E., Chen, H., Lyons, S. L. and Qin, G., Object-oriented programming for general mixed finite element methods, *Object Oriented Methods for Interoperable Scientific And Engineering Computing; The Proceeding of the 1998 SIAM Workshop*, 1998.

[18] Terzaghi, K., Die berechnung der durchlassigkeitsziffer des tones aus dem verlauf der hydrodynamischen spannungsercheinungen, *Akademi der Wissenschaften in Wien, Sitzungsherichte, Mathematisch-naturwissenschaftliche Klasse Part IIa*, 1923.

[19] Zienkiewicz, O. C., Basic formulation of static and dynamic behavior of soil and other porous media, *Numerical Methods in Geomechanics*, 1982.

[20] Zimmerman, R. W., Somerton, W.H. and King, M.S., Compresibility of porous rocks, *Journal of Geophysical Research* **91** (1986), 12765–12777.

On the Discretization of Interface Problems with Perfect and Imperfect Contact

TATIANA CHERNOGOROVA   RICHARD E. EWING
OLEG ILIEV   RAYTCHO LAZAROV

Abstract

A second-order difference scheme for a first-order elliptic system with discontinuous coefficients is derived and studied. This approximation can be viewed as an improvement of the well-known scheme with harmonic averaging of the coefficients for a second order elliptic equation, which is first-order accurate for the gradient of the solution. The numerical experiments confirm the second order convergence for the scaled gradient, and demonstrate the advantages of the new discretization, compared with the older ones.

KEYWORDS: interface problems, second order discretization

1   Introduction

The single-phase fluid flow in a fully-saturated inhomogeneous porous media that occupies a bounded domain $\Omega \subset R^n$, $n = 1, 2, 3$ is often modeled by first-order system

$$\nabla \cdot \mathbf{u} = f(x), \quad \mathbf{u} = -K\nabla p \text{ for } x \in \Omega, \tag{1.1}$$

subject to various boundary conditions. Here $\mathbf{u}$ is the Darcy velocity, p is the pressure, and $K(x)$ is the permeability tensor. The first equation is the continuity equation, while the second one expresses the relation between the pressure gradient and the velocity by the linear Darcy law. We assume that $K(x)$ is a diagonal matrix with positive elements which may have jump discontinuities across given surfaces Γ called interfaces. In this paper we consider two types of conditions on the interface Γ: (a) a perfect contact:

$$[p] = 0, \quad [\mathbf{u} \cdot \mathbf{n}] = 0 \quad \text{for } \xi \in \Gamma; \tag{1.2}$$

and (b) imperfect contact:

$$(\mathbf{u} \cdot \mathbf{n})_+ = (\mathbf{u} \cdot \mathbf{n})_- = \alpha_\xi(p_+ - p_-) \quad \text{for } \xi \in \Gamma. \tag{1.3}$$

Here **n** is the normal to the interface Γ unit vector (with fixed direction), $g_\pm(\xi)$ denotes the right and left limits of the function g at point ξ, and $[g] = g_+ - g_-$.

Often the velocity is eliminated so that the system (1.1) is reduced to a second-order elliptic equation for the pressure. A modified finite volume discretization for such a class of elliptic problems has been introduced and studied in [2, 3] and numerical experiments, confirming second-order convergence for the pressure, have been presented. In the present paper we continue the research in [2, 3] by introducing a new approximation of the velocity and studying its properties. We also study the relative accuracy for the approximate pressure near the interface, and discuss its superconvergent behavior in a particular case. We restrict our consideration to a class of interface problems for which: (1) each interface is parallel to a co-ordinate axes, and (2) the velocity is continuously differentiable in the normal direction to the interfaces. Under these two assumptions, we construct a second-order approximation of the system (1.1) on a staggered grid so that the pressure values are computed at the centers of the control volumes, while the values of the normal component of the velocity are computed at faces of the cells. The numerical experiments show that the derived scheme has second order convergence for both, the pressure and the velocity, and is superior to the schemes with harmonic averaging, which exhibit second-order convergence for the pressure and first order convergence for the velocity. The computations also show that for a smooth velocity the proposed discretization gives a considerably more accurate pressure, compared to the results of the scheme with harmonic averaging of the coefficients.

2   Discretization of the Interface Problem

Here we present in detail the discretization of the one-dimensional (1-D) problems with perfect and imperfect contacts. The discretization of the multidimensional problem on a tensor-product grid is just a tensor product of 1-D discretizations.

2.1   Discretization of the Perfect Contact Problem

In order to illustrate our approach, we first consider the perfect contact problem (1.1) - (1.2) in the one-dimensional case. We introduce a standard uniform cell–centered grid, $x_0 = 0, x_1 = h/2, x_i = x_{i-1} + h$, $i =$

$2, \ldots, N, x_{N+1} = 1$, where $h = 1/N$. The internal grid points can be considered as centered around the control volumes $V_i = (x_{i-\frac{1}{2}}, x_{i+\frac{1}{2}})$ where $x_{i+\frac{1}{2}} = x_i + \frac{1}{2}h$, $x_{i-\frac{1}{2}} = x_i - \frac{1}{2}h$. The values of the pressure and of the right hand side are defined at the grid points x_i and are denoted by p_i, f_i. The values of the velocity are defined at the points $x_{i+\frac{1}{2}}$ and are denoted by $u_{i+\frac{1}{2}}$. Non-uniform grids can be treated in a similar way. Note, our approach is defined locally, at a particular control volume level, and it can work with standard vertex-based grids as well.

The second-order discretization of the continuity equation in (1.1) is straightforward:

$$u_{i+\frac{1}{2}} - u_{i-\frac{1}{2}} = h\, \varphi_i, \quad \varphi_i = \frac{1}{h} \int_{x_{i-\frac{1}{2}}}^{x_{i+\frac{1}{2}}} f(x)dx, \quad i = 1, 2, \ldots, N. \qquad (2.1)$$

Next, consider the Darcy law, rewritten in the form

$$-\frac{\partial p}{\partial x} = \frac{u(x)}{k(x)},$$

and integrate this expression over the interval (x_i, x_{i+1}):

$$-(p_{i+1} - p_i) = -\int_{x_i}^{x_{i+1}} \frac{\partial p}{\partial x} dx = \int_{x_i}^{x_{i+1}} \frac{u(x)}{k(x)} dx. \qquad (2.2)$$

We assume that the velocity $u(x)$ is two-times continuously differentiable on the interface, so it can be expanded around the point $x_{i+\frac{1}{2}}$ in Taylor series:

$$u(x) = u_{i+\frac{1}{2}} + (x - x_{i+\frac{1}{2}})\frac{\partial u_{i+\frac{1}{2}}}{\partial x} + \frac{(x - x_{i+\frac{1}{2}})^2}{2}\frac{\partial^2 u(\eta)}{\partial x^2}, \quad \eta \in (x_i, x_{i+1}). \quad (2.3)$$

After replacing the first derivative of the velocity at $x_{i+\frac{1}{2}}$ by a two-point backward difference, we get the following approximation of (2.2):

$$-(p_{i+1}-p_i) = u_{i+\frac{1}{2}} \int_{x_i}^{x_{i+1}} \frac{dx}{k(x)} + \frac{u_{i+\frac{1}{2}} - u_{i-\frac{1}{2}}}{h} \int_{x_i}^{x_{i+1}} \frac{(x - x_{i+\frac{1}{2}})}{k(x)} dx + O(h^3). \qquad (2.4)$$

Finally, we rewrite this equation in the following basic form:

$$-k_{i+\frac{1}{2}}^H \frac{p_{i+1} - p_i}{h} = u_{i+\frac{1}{2}} + a_{i+\frac{1}{2}}(u_{i+\frac{1}{2}} - u_{i-\frac{1}{2}}) + \psi_i, \qquad (2.5)$$

where $\psi_i = O(h^2)$ and

$$k_{i+\frac{1}{2}}^H = \left(\frac{1}{h}\int_{x_i}^{x_{i+1}} \frac{dx}{k(x)}\right)^{-1}, \quad a_{i+\frac{1}{2}} = k_{i+\frac{1}{2}}^H \frac{1}{h^2}\int_{x_i}^{x_{i+1}} \frac{x - x_{i+\frac{1}{2}}}{k(x)} dx. \qquad (2.6)$$

Here $k_{i+\frac{1}{2}}^H$ is the well-known harmonic averaging of the coefficient $k(x)$ over the cell (x_i, x_{i+1}), which has played a fundamental role in deriving accurate schemes for discontinuous coefficients (see, e.g. [4, 5, 6, 7]). This presentation of the velocity $u(x)$ is a starting point for our discretization. Since we have assumed that the velocity is smooth, then the consecutive terms in the right-hand side in (2.5) are $O(1)$, $O(h)$ and $O(h^2)$, respectively. Truncation of this sum after the first term produces the well-known scheme of Samarskii [5] with harmonic averaging of the coefficient. This scheme is $O(h)$-consistent for the velocity at the interface points and second-order accurate for the pressure in the discrete H^1-norm. Further in the text, we call this scheme *harmonic averaging* or HA scheme.

Now we derive an $O(h^2)$ consistent scheme for the velocity by disregarding only the ψ_i-term in (2.5). We denote by P_i the approximate pressure at the grid points in order to distinguish it from the exact values p_i. Similarly, the approximate velocity is denoted by $U_{i+\frac{1}{2}}^-$, $U_{i-\frac{1}{2}}^+$ etc., where the sign $\pm$ indicates the right and left values of the flux at the point. Note, that the exact fluxes are continuous, while the approximate fluxes may have different $U_{i+\frac{1}{2}}^\pm$ values. Thus, we get

$$-k_{i+\frac{1}{2}}^H \frac{P_{i+1} - P_i}{h} = U_{i+\frac{1}{2}}^- + a_{i+\frac{1}{2}}(U_{i+\frac{1}{2}}^- - U_{i-\frac{1}{2}}^+), \qquad (2.7)$$

$$-k_{i-\frac{1}{2}}^H \frac{P_i - P_{i-1}}{h} = U_{i-\frac{1}{2}}^+ + a_{i-\frac{1}{2}}(U_{i+\frac{1}{2}}^- - U_{i-\frac{1}{2}}^+). \qquad (2.8)$$

In summary, *the equations (2.1),(2.7), and (2.8) approximates the first-order system (1.1) with local truncation error $O(h^2)$.* Further, we refer to this approximation as a scheme with *improved harmonic averaging* or IHA scheme.

Now we transform the discretization to a more suitable form. Subtracting (2.8) from (2.7) we get

$$(1 + a_{i+\frac{1}{2}} - a_{i-\frac{1}{2}})(U_{i+\frac{1}{2}}^- - U_{i-\frac{1}{2}}^+) = -k_{i+\frac{1}{2}}^H \frac{P_{i+1} - P_i}{h} + k_{i-\frac{1}{2}}^H \frac{P_i - P_{i-1}}{h}. \qquad (2.9)$$

Combining this with the discretization of the continuity equation (2.1), we obtain

$$-\left(1 + a_{i+\frac{1}{2}} - a_{i-\frac{1}{2}}\right)^{-1} \frac{1}{h}\left(k_{i+\frac{1}{2}}^H \frac{P_{i+1} - P_i}{h} - k_{i-\frac{1}{2}}^H \frac{P_i - P_{i-1}}{h}\right) = \varphi_i. \qquad (2.10)$$

On the other hand, solving the system (2.7), (2.8) for $U_{i+\frac{1}{2}}^{-}$ and $U_{i-\frac{1}{2}}^{+}$, we get

$$
\begin{aligned}
U_{i+\frac{1}{2}}^{-} &= \frac{-k_{i+\frac{1}{2}}^{H} \dfrac{P_{i+1}-P_i}{h}\left(1-a_{i-\frac{1}{2}}\right) - k_{i-\frac{1}{2}}^{H} \dfrac{P_i - P_{i-1}}{h} a_{i+\frac{1}{2}}}{1 + a_{i+\frac{1}{2}} - a_{i-\frac{1}{2}}}, \\[2mm]
U_{i-\frac{1}{2}}^{+} &= \frac{k_{i+\frac{1}{2}}^{H} \dfrac{P_{i+1}-P_i}{h} a_{i-\frac{1}{2}} - k_{i-\frac{1}{2}}^{H} \dfrac{P_i - P_{i-1}}{h}\left(1+a_{i+\frac{1}{2}}\right)}{1 + a_{i+\frac{1}{2}} - a_{i-\frac{1}{2}}}.
\end{aligned}
\tag{2.11}
$$

The new scheme approximates the velocity with second-order accuracy, *independently* of the positions of the discontinuity of the coefficient $k(x)$. The price we paid is the necessity to evaluate the expressions $k_{i+\frac{1}{2}}^{H}, k_{i-\frac{1}{2}}^{H}$ and $a_{i+\frac{1}{2}}, a_{i-\frac{1}{2}}$ with an error no larger than $O(h^2)$. Let a point ξ where the coefficient $k(x)$ is discontinuous be in the form $\xi = x_i + \theta h$ for some i and $0 \le \theta \le 1$. Now we consider particular realizations of this scheme. The approximation of the integral in $k_{i+\frac{1}{2}}^{H}$ is done by splitting it into integrals over (x_i, ξ) and (ξ, x_{i+1}) and then applying the trapezoidal rule for each integral. This approach will produce an accurate enough evaluation of $k_{i+\frac{1}{2}}^{H}$:

$$
k_{i+\frac{1}{2}}^{H} \approx \left[\frac{\theta}{2}\left(\frac{1}{k_i} + \frac{1}{k_{\xi-0}}\right) + \frac{1-\theta}{2}\left(\frac{1}{k_{i+1}} + \frac{1}{k_{\xi+0}}\right) \right]^{-1}.
\tag{2.12}
$$

Note, that $k_{\xi-0}$ and $k_{\xi+0}$ are known from the interface condition.

Further, we continue with the second integral in (2.6). We again split the integral into two integrals and apply the trapezoidal rule for each of the two integrals:

$$
\begin{aligned}
\frac{1}{h^2}\int_{x_i}^{x_{i+1}} \frac{(x - x_{i+1})}{k(x)}\,dx &= \frac{1}{h^2}\int_{x_i}^{\xi} \frac{(x - x_{i+\frac{1}{2}})}{k(x)}\,dx + \frac{1}{h^2}\int_{\xi}^{x_{i+1}} \frac{(x - x_{i+\frac{1}{2}})}{k(x)}\,dx \\[2mm]
&= \frac{\theta}{2}\left(\frac{\theta - 0.5}{k_{\xi-0}} - \frac{0.5}{k_i}\right) + \frac{1-\theta}{2}\left(\frac{0.5}{k_{i+1}} + \frac{\theta - 0.5}{k_{\xi+0}}\right) + O(h^2).
\end{aligned}
\tag{2.13}
$$

In the case of a piece-wise constant coefficient $k(x)$ these formulas are exact and reduce to

$$
k_{i+\frac{1}{2}}^{H} = \left(\frac{\theta}{k_i} + \frac{1-\theta}{k_{i+1}}\right)^{-1} \quad \text{and} \quad a_{i+\frac{1}{2}} = \frac{1}{2}\frac{\theta(1-\theta)(k_i - k_{i+1})}{(1-\theta)k_i + \theta k_{i+1}}.
\tag{2.14}
$$

Obviously, if the point of discontinuity ξ is a midpoint of the grid, i.e. $\xi = x_{i+\frac{1}{2}}$, then $\theta = 1/2$ so that (2.14) reduces to

$$
k_{i+\frac{1}{2}}^{H} = 2\left(\frac{1}{k_i} + \frac{1}{k_{i+1}}\right)^{-1} \quad \text{and} \quad a_{i+\frac{1}{2}} = \frac{1}{4}\left(\frac{k_i - k_{i+1}}{k_i + k_{i+1}}\right).
$$

Remark 2.1 *Note that if $f(x) \equiv 1$ then $u''(x) = 0$ and the local truncation error is zero. Thus, HA scheme reproduces exactly piecewise linear solutions, while the IHA scheme reproduces exactly piecewise quadratic solutions.*

2.2 Discretization of the Imperfect Contact Problem

Discretization of imperfect contact problem in the case when interfaces are aligned with grid nodes, is studied in [6]. A harmonic averaging type of discretization for interfaces aligned with control volume faces has been discussed in [1]. Below we derive improved discretization for the case when interfaces are orthogonal to a co-ordinate axis. Consider the 1-D imperfect contact problem (1.1), (1.3). The discretization in this case is derived in a similar way as in the case of perfect contact, so we shall only list the final results. The second-order discretization of the continuity equation in (1.1) is almost the same as in the perfect contact case:

$$U_{i+\frac{1}{2}}^- - U_{i-\frac{1}{2}}^+ = h\,\varphi_i, \quad \varphi_i = \frac{1}{h}\int_{x_{i-\frac{1}{2}}}^{x_{i+\frac{1}{2}}} f(x)dx, \; i = 1, 2, ..., N. \qquad (2.15)$$

A second order consistent discretization to Darcy law is given by:

$$-\beta_{i+\frac{1}{2}}\frac{P_{i+1} - P_i}{h} = U_{i+\frac{1}{2}}^- + b_{i+\frac{1}{2}}(U_{i+\frac{1}{2}}^- - U_{i-\frac{1}{2}}^+). \qquad (2.16)$$

For piecewise constant coefficients we have

$$b_{i+\frac{1}{2}} = (\theta - 0.5) + \frac{h\alpha_\xi}{2A_\xi}\left((1 - \theta)^2 k_i - \theta^2 k_{i+1}\right),$$

where

$$A_\xi = k_i(1 - \theta)h\alpha_\xi + k_{i+1}(k_i + \theta h\alpha_\xi), \quad k_{i+\frac{1}{2}}^H = \left[\frac{1-\theta}{k_{i+1}} + \frac{\theta}{k_i}\right]^{-1},$$

$$\beta_{i+\frac{1}{2}} = k_{i+\frac{1}{2}}^H\left(\frac{k_{i+\frac{1}{2}}^H}{h\alpha_{i+\frac{1}{2}}} + 1\right)^{-1}.$$

Note, in this particular case, we have second order of approximation for the velocity near the interface when $\theta \neq 0.5$, and third order of approximation for the velocity on the interface when $\theta = 0.5$.

3 Numerical Experiments

Numerical experiments were performed in order to study the behavior of the proposed discretization (IHA) and to compare it with the widely used discretization with harmonic averaging of the coefficients (HA). We approx-imately solve model problems with known analytical solution in order to assess the accuracy of the numerical solution. Three test problems have been considered. First, we solve a 2-D perfect contact problem with different per-meabilities in 4 subregions, and investigate the cases when the interfaces are aligned with the control volume faces (i.e. $\theta = 0.5$), and when the grid is not aligned with the interfaces (i.e. $\theta \neq 0.5$). We show that IHA ensures second order convergence for the velocity in both cases, while HA is first-order accu-rate for the velocity in both cases. Both schemes are second-order accurate for the pressure, however the constant of convergence in IHA is much smaller. In the second test we solve a 2-D perfect contact problem with different per-meabilities in 16 subregions. Superconvergence for the pressure (namely, accuracy of $O(h^3)$) is observed for this particular problem in the case of a constant permeability in the whole domain Ω. This is due to some symmetry and some cancellation of reminder terms. Numerical experiments show that for smooth velocity the IHA scheme preserves this superconvergence for the pressure in the case of discontinuous piecewise constant coefficients, while HA does not preserve it. Again, IHA ensures second-order accuracy for the velocity, while HA is only first-order accurate for the velocity. In the third example we solve a imperfect contact problem.

We should note that these computations do not guarantee high order of convergence for solutions with low regularity, for example, those produced by discontinuous coefficients and $f(x) \equiv 1$. Such solutions belong to the Sobolev space $H^{1+\gamma}(\Omega)$, where $\gamma > 0$ depends on the ratios of the jumps at the intersection points of the interfaces and can be very small.

Example 1. 2-D perfect contact problem in a unit square with interfaces at $x = \xi$, $y = \eta$ (4 subregions). The permeability is a piecewise constant in each sub-domain and takes values $k = \{10^{-2}, 1, 10^{-4}, 10^{+6}\}$ in the 4 subre-gions, counting from left to right and from bottom to top. The exact solution in this case is: $p^{ex} = \frac{1}{k} sin\left(\frac{\pi x}{2}\right)(x - \xi)(y - \eta)\left(1 + x^2 + y^2\right)$. Results from computations are presented in Table 1 and Table 2. The following notations are used: HA and IHA denote harmonic averaging based discretization and improved HA discretization for the case of aligned interfaces (i.e. $\theta = 0.5$), respectively. The notation "$\theta-$" in front of the scheme's notation is used when $\theta \neq 0.5$. Maximum errors for the pressure and their ratios are pre-sented in Table 1. It is seen that both schemes converge with second order in both cases: aligned and non-aligned interfaces. The non-monotonic behavior of the convergence in the non-aligned case can be explained by the fact that θ varies from grid to grid, taking values larger or smaller than 0.5. In this way, the maximum value of the solution in different subregions is involved in the error estimate. Table 2 summarizes maximum errors for the velocity at the vertical interface (or at the nearest vertical line $x = x_{i+\frac{1}{2}}$ in the non-aligned case). It is clearly seen that IHA is second-order accurate for the velocity, while HA is only first-order accurate for it. Again non-monotone behavior of the convergence, related to alternating values for θ on the consecutive grids, is observed. A monotone convergence will be observed for fixed θ, for example, on grids contains $12 \times 12, 42 \times 42, 162 \times 162$ nodes.

EXAMPLE 1.2-D perfect contact problem with 4 subregions.

| Grid | $\xi = \frac{1}{2}, \eta = \frac{1}{2}$, (aligned) | | | | $\xi = \frac{1}{3}, \eta = \frac{1}{3}$, (non-aligned) | | | |
	HA scheme		IHA scheme		θ-HA scheme		θ-IHA scheme	
12x12	1.75d-2	-	3.34d-4	-	1.91d-2	-	4.48d-4	-
22x22	5.97d-3	2.9	7.64d-5	4.4	9.56d-3	2.0	1.45d-4	3.1
42x42	1.80d-3	3.3	1.94d-5	3.9	2.10d-3	4.6	4.02d-5	3.6
82x82	5.03d-4	3.6	4.97d-6	3.9	7.03d-4	3.0	1.10d-5	3.7
162x162	1.36d-4	3.7	1.26d-6	3.9	1.64d-4	4.3	2.80d-6	3.9
322x322	3.56d-5	3.8	3.17d-7	4.0	4.69d-5	3.5	7.19d-7	3.9

Table 1. Maximum errors for pressure and their ratios.

EXAMPLE 1.2-D perfect contact problem with 4 subregions.

| Grid | $\xi = \frac{1}{2}, \eta = \frac{1}{2}$, (aligned) | | | | $\xi = \frac{1}{3}, \eta = \frac{1}{3}$, (non-aligned) | | | |
	HA scheme		IHA scheme		θ-HA scheme		θ-IHA scheme	
12x12	4.20d-2	-	1.18d-3	-	2.19d-2	-	5.26d-4	-
22x22	2.14d-2	1.96	2.73d-4	4.3	2.83d-2	0.77	1.27d-4	4.1
42x42	1.08d-2	1.98	8.07d-5	3.4	5.62d-3	5.03	5.30d-5	2.4
82x82	5.40d-3	2.00	2.20d-5	3.7	7.11d-3	0.79	5.44d-6	9.7
162x162	2.70d-3	2.00	5.77d-6	3.8	1.41d-3	5.04	3.93d-6	1.4
322x322	1.35d-3	2.00	1.50d-6	3.8	1.78d-3	0.79	3.77d-7	10.

Table 2. Maximum errors for velocity and their ratios.

Example 2. 2-D perfect contact problem in a unit square with interfaces at $x = \xi_1 = 0.2$, $x = \xi_2 = 0.5$, $x = \xi_3 = 0.7$ and $y = \eta_1 = 0.3$, $y = \eta_2 = 0.6$, $y = \eta_3 = 0.8$ (16 subregions). The permeability is a piecewise constant function and in each sub-domain and takes values $k = 10 \times \{10, 10^{-3}, 1, 10^{-4}, 10^{-2}, 10, 10^{-3}, 1, 10, 10^{-3}, 1, 10^{-3}, 10^{-3}, 10, 10^{-2}, 1\}$ in the 16 subregions, counting from left to right and from bottom to top with

exact solution $p(x, y) = \frac{1}{k}(x - \xi_1)(x - \xi_2)(x - \xi_3)(y - \eta_1)(y - \eta_2)(y - \eta_3)$.
The computational results are presented in Tables 3 and 4. For comparison
we have included also the results for the constant coefficient $k(x, y) = 1$ in
the whole domain Ω. Notation "case B" is used to denote columns with these
results. One can observe superconvergence for the pressure in the continuous
case, which is also exhibited by IHA in the case of discontinuous coefficients.
Table 4 shows that IHA ensures second-order convergence for the velocity,
while HA is only first-order accurate. In our numerical experiments the exact
solution is zero at the interfaces, so the absolute values for the error there
might be small, compared to absolute errors far from interfaces, while the
relative errors can be very large. We can plot the relative error $\frac{p(x_i, y_j) - P_{i,j}}{p(x_i, y_j)}$
computed by HA and IHA, respectively. Here (x_i, y_j) are the centers of
the cells in the rectangular grid. Qualitatively, the results look similar, but
quantitatively they differ by orders of magnitude. The results show that IHA
produces a considerably more accurate solution near the interface.

Example 3. 1-D imperfect contact problem with interfaces at $\xi_1 = 0.2$
and $\xi_2 = 0.7$ with an exact solution:

$$p(x) = -\frac{d}{k(s+1)(s+2)}x^{s+2} - \frac{c}{2k}x^2 + ax + b.$$

The parameters k, c, d vary for the different subregions. The constants in the
imperfect contact interface conditions are $\alpha_1 = 10^2$, $\alpha_2 = 10$. Also, $u(0) = 1$,
$u(1) = 0$, and $s = \{1, 1, 1\}$, $k = \{1, 10, 10^2\}$, $c = \{10, 1, 10^2\}$, in the three
subregions, counting from the left.

EXAMPLE 2. 2-D perfect contact problem with 16 subregions.

Grid	$k(x, y) \equiv 1$ case B		aligned			
			HA scheme		IHA scheme	
12x12	1.75d-3	-	3.91d-3	-	1.65d-3	-
22x22	2.89d-4	6.06	1.49d-3	2.62	2.87d-4	5.75
42x42	4.18d-5	6.91	5.07d-4	2.93	4.17d-5	6.88
82x82	5.64d-6	7.41	1.55d-4	3.27	5.63d-6	7.41
162x162	7.33d-7	7.69	4.39d-5	3.53	7.32d-7	7.69
322x322	9.34d-8	7.85	1.19d-5	3.69	9.35d-8	7.83

Table 3. Maximum errors for pressure and their ratios.

EXAMPLE 2. 2-D perfect contact problem with 16 subregions.

Grid	$k(x,y) \equiv 1$ case B		aligned			
			HA scheme		IHA scheme	
12x12	1.74d-4	-	2.02d-3	-	9.37d-4	-
22x22	5.85d-5	2.97	1.17d-3	1.73	2.91d-4	3.21
42x42	1.67d-5	3.50	6.20d-4	1.89	8.22d-5	3.54
82x82	4.42d-6	3.77	3.18d-4	1.95	2.39d-5	3.44
162x162	1.14d-6	3.88	1.60d-4	1.99	6.56d-6	3.62
322x322	2.89d-7	3.94	8.06d-5	1.99	1.75d-6	3.75

Table 4. Maximum errors for velocity and their ratios.

EXAMPLE 3. 1-D imperfect contact problem with 3 subregions.

Grid	PRESSURE				VELOCITY			
	HA scheme		IHA scheme		HA scheme		IHA scheme	
12x12	4.56d-3	-	2.47d-3	-	3.87d-2	-	2.49d-2	-
22x22	1.33d-3	3.4	5.12d-4	4.8	9.70d-3	4.0	5.46d-3	4.6
42x42	3.58d-4	3.7	9.77d-5	5.2	2.43d-3	4.0	1.11d-3	4.9
82x82	9.24d-5	3.9	1.83d-5	5.3	6.07d-4	4.0	2.50d-4	5.4
162x162	2.35d-5	3.9	3.27d-6	5.6	1.52d-4	4.0	3.55d-5	5.8
322x322	5.92d-6	4.0	5.84d-7	5.6	3.80d-5	4.0	6.04d-6	5.9

Table 5. Maximum errors and their ratios.

The results from computations are presented in Table 5. As it was mentioned in Section 2.2 the IHA scheme approximates velocity at the interface with third-order accuracy in the case when the imperfect contact interface is aligned with cell faces. In this case, the HA scheme approximates the velocity at the interface to second order. This may explain the second-order convergence for the HA scheme, and the superconvergence for the IHA scheme.

Acknowledgments. The work of the first and the third authors has been supported partially by Bulgarian FSI under grant MM-811. Part of the work of the second and the fourth author has been supported by the USA EPA Grant R 825207-01-1.

References

[1] Angot, Ph., Modeling and visualization of thermal fields inside electronic systems under operating conditions, *IBM Technical Report*, **TR-47095**, 70 (1989).

[2] Chernogorova, T. and Iliev, O., A 2nd order discretization of imperfect contact problems with piecewise constant coefficients on cell-centered

grids, to appear in: *Notes in Numerical Fluid Mechanics*, M. Griebel et al (Eds), Proc. of Workshop on Large Scale Scientific Computations, June 1999, Sozopol, Bulgaria.

[3] Ewing, R., Iliev, O., and Lazarov, R., A modified finite volume approximation of second order elliptic equations with discontinuous coefficients, TR ISC-99-01-MATH, Texas A&M University, 1999 (submitted).

[4] Marchuk, G. I., *Methods of computational mathematics*, Moscow, Nauka, 1980.

[5] Samarskii, A. A., *Theory of difference schemes*, Moscow, Nauka, 1977.

[6] Samarskii, A. A. and Andréev, V. B., *Finite difference methods for elliptic problems*, Nauka Moscow, 1976 (in Russian), *Méthodes aux Différences pour Équations Elliptiques*, and MIR Moscou, 1978 (in French).

[7] Wesseling, P., *An Introduction to Multigrid Methods*, Wiley, N.Y., 1991.

Finite Element Analysis for Pseudo Hyperbolic Integral-Differential Equations

XIA CUI

Abstract

The finite element method and its analysis for pseudo-hyperbolic integral-differential equations with nonlinear boundary conditions is considered. A new projection is introduced to obtain optimal L^2 convergence estimates. The present techniques can be applied to treat elastic wave problems with absorbing boundary conditions in porous media.

KEYWORDS: pseudo-hyperbolic integral-differential equation, finite element, Sobolev-Volterra projection, convergence analysis

1 Introduction

We consider the finite element method for the pseudo-hyperbolic integral-differential equation subject to the initial and nonlinear boundary conditions

$$
\begin{aligned}
q(u)u_{tt} = \nabla \cdot (a(u)\nabla u_t + b(u)\nabla u) & \\
+ \int_0^t c(u(\tau))\nabla u(\tau)d\tau + f(u), & \quad x \in \Omega,\ t \in J, \\
a(u)\tfrac{\partial u_t}{\partial \gamma} + b(u)\tfrac{\partial u}{\partial \gamma} + \int_0^t c(u(\tau))\tfrac{\partial u(\tau)}{\partial \gamma}d\tau = g(u), & \quad x \in \partial\Omega,\ t \in J, \\
u(x,0) = \Phi(x), u_t = \Psi(x), & \quad x \in \Omega,
\end{aligned}
\tag{1.1}
$$

where $J = [0, T]$, $\Omega \subset R^d$ ($d \geq 1$ is the dimension of the space) is an open bounded domain with smooth boundary $\partial\Omega$, γ denotes the outer-normal direction to $\partial\Omega$, $\phi(u) = \phi(x, t, u)$, $c(u(\tau)) = c(t, \tau, u(\tau)) = c(t, \tau, x, u(x, \tau))$, and q, a, b, c, f, g, Φ, and Ψ are known functions. We also assume that the functions q, a, b, c, f, and g are smooth with bounded derivatives and there exist constants $q_* > 0$ and $a_* > 0$ such that $q(x, t, \psi) \geq q_*$ and $a(x, t, \psi) \geq a_*$, $\forall x \in \Omega$, $t \in J$, $\psi \in R$.

Pseudo-hyperbolic integral-differential equations are often used in the fields of visco-elasticity, nuclear physics, and biological mechanics. For existence, uniqueness, and continuous dependence of solutions, there are some

investigations carried out in [1, 5]; however, for their numerical approximation, there is few works available [4]. Since the antihunt, diffusion, and memory (Volterra) term emerge at the same time in their formulation, together
with nonlinear boundary conditions, the traditional Ritz projection [6], Ritz-
Volterra projection [2], and Sobolev projection [3] can no longer reflect this
global property, so it is difficult to do an error analysis. In this paper we
introduce a new projection to treat these difficulties, use integration by parts
and an induction hypothesis reasoning to deal with the nonlinearity of these
problems, and do a thorough and successful finite element (FE) numerical
analysis for (1.1).

An outline of the paper is as follows. In §2, a new projection is introduced and its approximation properties are studied. With the help of this
projection, the FE method and its numerical analysis is investigated and the
optimal L^2 convergence is established in §3.

In this paper K is a generic positive constant and may be different each
time it is used and ϵ is an arbitrarily small constant. Let $(\phi, \psi) = \int_\Omega \phi\psi dx$
and $\langle \phi, \psi \rangle = \int_{\partial\Omega} \phi\psi dx$; the norms in Banach spaces follow those in [4, 5].
We use the inequalities

$$ab \le \epsilon a^2 + \tfrac{1}{4\epsilon}b^2, \quad |\phi|^2_{L^2(\partial\Omega)} \le \epsilon\|\nabla\phi\|^2 + K(\epsilon)\|\phi\|^2,$$
$$|\phi|_r \le K(T,r)\|\phi\|_{r+\frac{1}{2}}, \quad 0 < r \le 3/2, \, r \ne 1.$$

2   A New Projection

Let $\mu \subset H^1(\Omega)$ be a finite element space associated with a quasi-regular
partition of Ω such that the elements have diameters bounded by h. Let
the index of μ be the integer k. It is frequently valuable to decompose the
convergence analysis of the FE method by passing through a FE projection of
the solution of the differential problem. To treat our problem, we introduce
a new projection: $\tilde{u}(t) : [0, T] \to \mu$ such that

$$(a(u)\nabla(\tilde{u}_t - u_t) + b(u)\nabla(\tilde{u} - u) + \int_0^t c(u(\tau))\nabla(\tilde{u} - u)(\tau)d\tau, \nabla v)$$
$$+\lambda(\tilde{u}_t - u_t, v) - \langle g(\tilde{u}) - g(u), v \rangle = 0 \quad \forall v \in \mu, \, t \in J, \tag{2.1}$$
$$\tilde{u}(0) = \Phi_h,$$

where λ is a positive constant assuring coercivity of the form and Φ_h is an
appropriate approximation of Φ in μ.

Lemma 2.1 *There exists a unique $\tilde{u}(t) \in \mu$ such that (2.1) being satisfied.*

Proof: We use an iterative procedure. Let $V^0 \in \mu$ and define $\{V^p\}$ by

$$(a(u)\nabla(V_t^p - u_t) + b(u)\nabla(V^p - u) + \int_0^t c(u(\tau))\nabla(V^{p-1} - u)(\tau)d\tau, \nabla v)$$
$$+\lambda(V_t^p - u_t, v) - \langle g(V^{p-1}) - g(u), v \rangle = 0 \quad \forall v \in \mu, t \in J,$$
$$V^p(0) = \Phi_h, \qquad \forall p \geq 1.$$

(2.2)

For $p \geq 1$, the unique existence of V^p comes from the general theory of initial value problems of ordinary differential equations. Setting $Z^p = V^{p+1} - V^p$ and $v = Z_t^p$, it is natural to see from the above relation that

$$a_* \|\nabla Z_t^p\|^2 + \lambda \|Z_t^p\|^2 \leq K[\|\nabla Z^p\|^2 + \|Z^{p-1}\|_1^2 + \int_0^t \|\nabla Z^{p-1}(\tau)\|^2 d\tau]$$
$$+ a_* \|\nabla Z_t^p\|^2 + \tfrac{1}{2}\|Z_t^p\|^2/2.$$

Using $\|Z^p\|_1^2 \leq K \int_0^t \|Z_t^p(\tau)\|_1^2 d\tau$ and Gronwall's lemma, there exists a constant $K_* > 0$ such that

$$\|Z_t^p\|_1^2 \leq K_* \int_0^t \|Z_t^{p-1}\|_1^2(\tau)d\tau \leq \frac{(K_*T)^p}{p!}, \qquad \forall p \geq 1.$$

Similarly, we have a upper bound for $\|V_t^p\|_1$. Hence, $\{V^p\}$ is a Cauchy sequence in μ and so there exists a unique $\tilde{u} \in \mu$ such that $V^p \to \tilde{u}$ and $V_t^p \to \tilde{u}_t$, as $p \to \infty$. Lemma 2.1 is completed by letting $p \to \infty$ in (2.2). $\square$

Let condition (P_1) denote this: $\|\Phi - \Phi_h\| + h\|\Phi - \Phi_h\|_1 \leq Kh^{k+1}\|\Phi\|_{k+1}$ and condition (P_2): $|\Phi - \Phi_h|_{-\frac{1}{2}} \leq Kh^{k+1}\|\Phi\|_{k+1}$. Let $\eta = u - \tilde{u}$ and $G = \int_0^1 g_u(su + (1-s)\tilde{u})ds$. (2.1) is represented as

$$(a(u)\nabla\eta_t + b(u)\nabla\eta + \int_0^t c(u(\tau))\nabla\eta(\tau)d\tau, \nabla v)$$
$$+\lambda(\eta_t, v) - <G\eta, v> = 0 \qquad \forall v \in \mu.$$

(2.3)

Lemma 2.2 *Under condition* (P_1), *there exists a* $K = K(u)$ *such that:* (1) *if* $\frac{\partial^j u}{\partial t^j} \in H^{k+1}(\Omega)$, $j = 0, 1$, $\sum_{j=0}^{1} \|\frac{\partial^j \nabla \eta}{\partial t^j}\| \leq K(\|u\|_{1,k+1} + \|\Phi\|_{k+1})h^k$; (2) *if*
$\frac{\partial^j u}{\partial t^j} \in H^{k+1}(\Omega)$, $j = 0, 1, 2$, $\sum_{j=0}^{2} \|\frac{\partial^j \nabla \eta}{\partial t^j}\| \leq K(\|u\|_{2,k+1} + \|\Phi\|_{k+1})\, h^k$; (3) *if*
$\frac{\partial^j u}{\partial t^j} \in H^{k+1}(\Omega)$, $j = 0, \ldots, 3$, $\sum_{j=0}^{3} \|\frac{\partial^j \nabla \eta}{\partial t^j}\| \leq K(\|u\|_{3,k+1} + \|\Phi\|_{k+1})h^k$; (4) *if*
$\frac{\partial^j u}{\partial t^j} \in H^{k+1}(\Omega)$, $j = 0, \ldots, 4$, $\sum_{j=0}^{4} \|\frac{\partial^j \nabla \eta}{\partial t^j}\| \leq K(\|u\|_{4,k+1} + \|\Phi\|_{k+1})h^k$, *where*
$\|\phi(t)\|_{r,s}^2 = \sum_{j=0}^{r} (\|\frac{\partial^j \phi(t)}{\partial t^j}\|_s^2 + \int_0^t \|\frac{\partial^j \phi(\tau)}{\partial t^j}\|_s^2 d\tau), r \geq 0.$

Proof: Let $Ru(t) : [0, T] \to \mu$ be the Ritz projection of u. From (2.3) and the inequalities in §1, we see

$$a_*\|\nabla\eta_t\|^2 + \lambda\|\eta_t\|^2 \leq (a(u)\nabla\eta_t + b(u)\nabla\eta + \int_0^t c(u(\tau))\nabla\eta(\tau)d\tau, \nabla(u_t - Ru_t))$$
$$+\lambda(\eta_t, u_t - Ru_t) - \langle G\eta, u_t - Ru_t\rangle$$
$$-(b(u)\nabla\eta + \int_0^t c(u(\tau))\nabla\eta(\tau)d\tau, \nabla\eta_t) + < G\eta, \eta_t >$$
$$\leq \tfrac{a_*}{2}\|\nabla\eta_t\|^2 + \tfrac{1}{2}\|\eta_t\|^2 + \epsilon\|\eta_t\|_1^2$$
$$+K[\|\eta\|_1^2 + \int_0^t \|\nabla\eta(\tau)\|^2 d\tau + \|u_t - Ru_t\|_1^2].$$

Noticing that $\eta(t) = \eta(0) + \int_0^t \eta_t(\tau)d\tau$ and $\eta(0) = \Phi - \Phi_h$ and using Gronwall's lemma, we get the first conclusion of Lemma 2.2.

Differentiate in time (2.3) and set $A = a_t(t, u(t)) + b(u)$, $B = b_t(t, u(t)) + c(t, t, u(t))$, $C = c_t(t, \tau, u(\tau))$, $D = g_u(t, \tilde{u})$, and $E = \int_0^1 g_u(t, su + (1 - s)\tilde{u})ds + u_t \int_0^1 g_{uu}(t, su + (1 - s)\tilde{u})ds$. Then we derive at

$$(a(u)\nabla\eta_{tt} + A\nabla\eta_t + B\nabla\eta + \int_0^t C\nabla\eta(\tau)d\tau, \nabla v)$$
$$+\lambda(\eta_{tt}, v) - < D\eta_t + E\eta, v >= 0 \quad \forall v \in \mu. \tag{2.4}$$

Similarly,

$$a_*\|\nabla\eta_{tt}\|^2 + \lambda\|\eta_{tt}\|^2 \leq \tfrac{a_*}{2}\|\nabla\eta_{tt}\|^2 + \tfrac{\lambda}{2}\|\eta_{tt}\|^2 + \epsilon\|\eta_{tt}\|_1^2$$
$$+K[\|\eta_t\|_1^2 + \|\eta\|_1^2 + \int_0^t \|\nabla\eta(\tau)\|^2 d\tau + \|u_{tt} - Ru_{tt}\|_1^2].$$

Hence the second conclusion holds by using the first one. $\quad$ [] Differentiating in t (2.4) similarly, we can obtain the third conclusion. The forth conclusion can be gained by a further differentiation sequence. $\qquad$ []

Lemma 2.3 *Under conditions* (P_1) *and* (P_2)*, there exists a* $K = K(u)$ *such that* (1) *if* $\frac{\partial^j u}{\partial t^j} \in H^{k+1}(\Omega)$, $j = 0, 1$, $\sum_{j=0}^{1} \|\frac{\partial^j \eta}{\partial t^j}\| \leq K(\|u\|_{1,k+1} + \|\Phi\|_{k+1}) h^{k+1}$;

(2) *if* $\frac{\partial^j u}{\partial t^j} \in H^{k+1}(\Omega)$, $j = 0, 1, 2$, $\sum_{j=0}^{2} \|\frac{\partial^j \eta}{\partial t^j}\| \leq K(\|u\|_{2,k+1} + \|\Phi\|_{k+1}) h^{k+1}$;

(3) *if* $\frac{\partial^j u}{\partial t^j} \in H^{k+1}(\Omega)$, $\sum_{j=0}^{3} \|\frac{\partial^j \eta}{\partial t^j}\| \leq K(\|u\|_{3,k+1} + \|\Phi\|_{k+1}) h^{k+1}$, $j = 0, \ldots, 3$;

(4) *if* $\frac{\partial^j u}{\partial t^j} \in H^{k+1}(\Omega)$, $\sum_{j=0}^{4} \|\frac{\partial^j \eta}{\partial t^j}\| \leq K(\|u\|_{4,k+1} + \|\Phi\|_{k+1}) h^{k+1}$, $j = 0, \ldots, 4$.

Proof: Define $\alpha \in H^1(\Omega)$ in the way

$$(a(u)\nabla\alpha + b(u)\nabla\eta + \int_0^t c(u(\tau))\nabla\eta(\tau)d\tau, \nabla v)$$
$$+\lambda(\alpha, v) - \langle G\eta, v\rangle = 0 \quad \forall v \in H^1(\Omega). \tag{2.5}$$

By subtracting (2.5) from (2.3), we see that $\tilde{u}_t$ is the standard Ritz projection of $u_t - \alpha$ onto μ. Hence, by the elliptic regularity of (2.5) and Lemma 2.2, we see that

$$\|\eta_t\| \le \|\alpha\| + Kh(\|\eta_t\|_1 + \|\alpha\|_1) \le \|\alpha\| + Kh^{k+1}(\|u\|_{1,k+1} + \|\Phi\|_{k+1}). \quad (2.6)$$

It remains to estimate $\|\alpha\|$. To do this, let $\beta \in H^1(\Omega)$ be defined by

$$\begin{aligned}
&(a(u)\nabla\beta + b(u)\nabla\eta + \textstyle\int_0^t c(u(\tau))\nabla\eta(\tau)d\tau, \nabla v) \\
&+\lambda(\beta, v) - <G\eta, v> = (\alpha, v) \quad \forall v \in H^1(\Omega).
\end{aligned} \quad (2.7)$$

Setting $v = \beta$ in (2.5) and $v = \alpha$ in (2.7) and denoting $z = \alpha - \beta$, we find that

$$\begin{aligned}
\|\alpha\|^2 &= \left\langle \eta, b(u)\frac{\partial z}{\partial\gamma} \right\rangle - (\eta, \nabla \cdot [b(u)\nabla z]) + \int_0^t \{< \eta(\tau), c(u(\tau))\frac{\partial z(t)}{\partial\gamma} > \\
&\quad -(\eta(\tau), \nabla \cdot [c(u(\tau))\nabla z(t)])\}d\tau - <G\eta, z> \\
&\le K\{|\eta|_{-\frac{1}{2}}^2 + \|\eta\|^2 + \int_0^t [|\eta(\tau)|_{-\frac{1}{2}}^2 + \|\eta(\tau)\|^2]d\tau\} + \epsilon(|z|_{\frac{3}{2}}^2 + \|z\|_2^2 + |z|_{\frac{1}{2}}^2),
\end{aligned}$$

while subtracting (2.7) from (2.5) leads to $\|z\|_2 \le K\|\alpha\|$, so we obtain

$$\|\alpha\| \le K\{|\eta(0)|_{-\frac{1}{2}} + \|\eta(0)\| + \int_0^t [|\eta_t(\tau)|_{-\frac{1}{2}} + \|\eta_t(\tau)\|]d\tau\}. \quad (2.8)$$

Now, we turn to estimate $|\eta_t|_{-\frac{1}{2}}$. For this purpose, we define $\vartheta \in H^1(\Omega)$ in such a way

$$\begin{aligned}
&(a(u)\nabla\vartheta + b(u)\nabla\eta + \textstyle\int_0^t c(u(\tau))\nabla\eta(\tau)d\tau, \nabla v) \\
&+\lambda(\vartheta, v) - <G\eta, v> = <\delta, v> \quad \forall v \in H^1(\Omega),
\end{aligned} \quad (2.9)$$

where $\delta \in H^{\frac{1}{2}}(\partial\Omega)$, $|\delta|_{\frac{1}{2}} = |\eta_t|_{-\frac{1}{2}}$, and $<\delta, \eta_t> = |\eta_t|_{-\frac{1}{2}}^2$. Let $\varphi = \vartheta - \alpha$. It follows from (2.5) and (2.9) that

$$(a(u)\nabla\varphi, \nabla v) + \lambda(\varphi, v) = <\delta, v>, \quad \forall v \in H^1(\Omega). \quad (2.10)$$

Hence, $\|\varphi\|_2 \le K|\eta_t|_{-\frac{1}{2}}$. Setting $v = \eta_t$ in (2.10) and using integration by parts, we obtain

$$\begin{aligned}
|\eta_t|_{-\frac{1}{2}}^2 &= (a(u)\nabla\eta_t + b(u)\nabla\eta + \textstyle\int_0^t c(u(\tau))\nabla\eta(\tau)d\tau, \nabla(\varphi - R\varphi)) \\
&\quad +\lambda(\eta_t, \varphi - R\varphi) - <G\eta, \varphi - R\varphi> - <\eta, b(u)\frac{\partial\varphi}{\partial\gamma}> +(\eta, \nabla \cdot [b(u)\nabla\varphi]) \\
&\quad - \int_0^t \{< \eta(\tau), c(u(\tau))\frac{\partial\varphi(t)}{\partial\gamma} > -(\eta(\tau), \nabla \cdot [c(u(\tau))\nabla\varphi(t)])\}d\tau + <G\eta, \varphi> \\
&\le Kh^2[\|\eta_t\|_1^2 + \|\eta\|_1^2 + \int_0^t \|\nabla\eta(\tau)\|^2 d\tau] + K[|\eta(0)|_{-\frac{1}{2}}^2 + \|\eta(0)\|^2] \\
&\quad +K\int_0^t [|\eta_t(\tau)|_{-\frac{1}{2}}^2 + \|\eta_t(\tau)\|^2 d\tau + \epsilon\|\varphi\|_2^2.
\end{aligned}$$

Applying Lemma 2.2 and Gronwall's lemma to above inequality, we see that

$$|\eta_t|_{-\frac{1}{2}} \le Kh^{k+1}(\|u\|_{1,k+1} + \|\Phi\|_{k+1}) + K \int_0^t \|\eta_t(\tau)\|^2 d\tau. \qquad (2.11)$$

Thus (2.8) yields a upper bound for $\|\alpha\|$. Substituting this bound into (2.6), we see again from Gronwall's lemma that

$$\|\eta_t\| \le Kh^{k+1}(\|u\|_{1,k+1} + \|\Phi\|_{k+1}).$$

Thus the first conclusion of Lemma 2.3 is valid as a consequence of $\|\eta\| \le K[\|\eta(0)\| + \int_0^t \|\eta_t(\tau)\| d\tau]$.

To demonstrate the second conclusion, we introduce the auxiliary equation

$$\begin{aligned}
&(a(u)\nabla\sigma + A\nabla\eta_t + B\nabla\eta + \int_0^t C\nabla\eta(\tau)d\tau, \nabla v) \\
&+\lambda(\sigma, v) - < D\eta_t + E\eta, v \ge 0 \quad \forall v \in H^1(\Omega).
\end{aligned} \qquad (2.12)$$

Subtracting (2.12) from (2.4) implies that $\tilde{u}_{tt}$ is the Ritz projection of $u_{tt} - \sigma$ in μ. Hence, by the elliptic regularity of (2.12) and Lemma 2.2, we have

$$\|\eta_{tt}\| \le \|\sigma\| + Kh(\|\eta_{tt}\|_1 + \|\sigma\|_1) \le \|\sigma\| + Kh^{k+1}(\|u\|_{2,k+1} + \|\Phi\|_{k+1}). \quad (2.13)$$

To estimate $\|\sigma\|$, we define $\zeta \in H^1(\Omega)$ such that

$$\begin{aligned}
&(a(u)\nabla\zeta + A\nabla\eta_t + B\nabla\eta + \int_0^t C\nabla\eta(\tau)d\tau, \nabla v) \\
&+\lambda(\zeta, v) - < D\eta_t + E\eta, v \ge (\sigma, v) \quad \forall v \in H^1(\Omega).
\end{aligned} \qquad (2.14)$$

Setting $v = \zeta$ in (2.12) and $v = \sigma$ in (2.14), denoting $y = \sigma - \zeta$, and using integration by parts, we derive at

$$\begin{aligned}
\|\sigma\|^2 \le K\{&|\eta_t|^2_{-\frac{1}{2}} + |\eta|^2_{-\frac{1}{2}} + \|\eta_t\|^2 + \|\eta\|^2 \\
&+ \int_0^t [|\eta(\tau)|^2_{-\frac{1}{2}} + \|\eta(\tau)\|^2] d\tau\} + \epsilon\|y\|_2^2.
\end{aligned} \qquad (2.15)$$

Note that subtracting (2.14) from (2.12) results in $\|y\|_2 \le K\|\sigma\|$, so from (2.15), (2.11), and the first conclusion, we can show that

$$\|\sigma\| \le K(\|u\|_{1,k+1} + \|\Phi\|_{k+1}).$$

Now, by (2.13), we obtain the second conclusion of Lemma 2.3. Differentiating in time (2.4) for one and two times, we can derive the third and forth conclusions. □

It is easy to show that

Lemma 2.4 *Under conditions* (P_1), (P_2), *and* $k \geq d/2$, *there exists a* $K = K(u)$ *such that:* (1) *(1) if* $\frac{\partial^j u}{\partial t^j} \in H^{k+1}(\Omega)$, $j = 0,1$, *then* $\|\frac{\partial^j u}{\partial t^j}\|_{L^\infty(L^\infty)} + \|\frac{\partial^j \nabla \tilde{u}}{\partial t^j}\|_{L^\infty(L^\infty)} \leq K$; (2) *if* $\frac{\partial^j u}{\partial t^j} \in H^{k+1}(\Omega)$, $j = 0,1,2$, *then* $\|\frac{\partial^j u}{\partial t^j}\|_{L^\infty(L^\infty)} + \|\frac{\partial^j \nabla \tilde{u}}{\partial t^j}\|_{L^\infty(L^\infty)} \leq K$; (3) *if* $\frac{\partial^j u}{\partial t^j} \in H^{k+1}(\Omega)$, $j = 0,\dots,3$, *then* $\|\frac{\partial^j u}{\partial t^j}\|_{L^\infty(L^\infty)} + \|\frac{\partial^j \nabla \tilde{u}}{\partial t^j}\|_{L^\infty(L^\infty)} \leq K$; (4) *if* $\frac{\partial^j u}{\partial t^j} \in H^{k+1}(\Omega)$, $j = 0,\dots,4$, *then* $\|\frac{\partial^j u}{\partial t^j}\|_{L^\infty(L^\infty)} + \|\frac{\partial^j \nabla \tilde{u}}{\partial t^j}\|_{L^\infty(L^\infty)} \leq K$.

Remark. If we set $c \equiv 0$ in (2.1), then the projection becomes a Sobolev projection. Since formulation (2.1) also includes the Volterra term, we call the new projection Sobolev-Volterra (SV) projection. We see, on one hand, from the above lemmas, that the SV projection has the same priori approximation properties as the Ritz, Ritz-Volterra, and Sobolev projections. On the other hand, it reflects the global property of the combination of antihunt term, diffusion term, and memory term, so it may be very convenient to the FE numerical analysis for the pseudo-hyperbolic integral-differential equation and pseudo-parabolic integral-differential equation. In the next section, we apply the SV projection to simplify the FE analysis for problem (1.1).

3   The FE Analysis

Define the continuous-time finite element scheme: finding $U(t) \in \mu$, such that

$$(q(U)U_{tt}, v) + (a(U)\nabla U_t + b(U)\nabla U + \int_0^t c(U(\tau))\nabla U(\tau)d\tau, \nabla v)$$
$$= (f(U), v) + <g(U), v> \quad \forall v \in \mu, t \in J, \qquad (3.1)$$
$$U = \Phi_h, \quad U_t = \Psi_h, \qquad t = 0,$$

where Ψ_h is an approximation of Ψ in μ, $c(U(\tau)) = c(t, \tau, x, U(x, \tau))$, and $q(U) = q(x, t, U)$.

Theorem 3.1 *Under conditions* (P_1), (P_2), $k \geq d/2$, *and* $\frac{\partial^j u}{\partial t^j} \in H^{k+1}(\Omega)$, $j = 0,1,2$, *for scheme* (3.1), *then*

$$\|U_t - u_t\|_{L^\infty(L^2)} + \|U - u\|_{L^\infty(L^2)}$$
$$+ h[\|U_t - u_t\|_{L^2(H^1)} + \|U - u\|_{L^\infty(H^1)}] = O(h^{k+1}).$$

Proof: Let $\xi = U - \tilde{u}$. Then $U - u = \xi - \eta$. Manipulation of (1.1), (3.1), and (2.1) leads to the error equation

$$
\begin{aligned}
&(q(U)\xi_{tt}, v) + (a(U)\nabla\xi_t + b(U)\nabla\xi \\
&+ \int_0^t c(U(\tau))\nabla\xi(\tau)d\tau, \nabla v) = ([q(u) - q(U)]u_{tt} \\
&+ q(U)\eta_{tt} + [f(U) - f(u)] - \lambda\eta_t, v) \\
&+ ([a(u) - a(U)]\nabla\tilde{u}_t + [b(u) - b(U)]\nabla\tilde{u} + \int_0^t [c(u(\tau)) \\
&- c(U(\tau))]\nabla\tilde{u}(\tau)d\tau, \nabla v) + \ <g(U) - g(\tilde{u}), v> \quad \forall v \in \mu, \ t \in J, \\
&\xi(0) = 0, \xi_t(0) = \Psi_h - \tilde{u}_t(0).
\end{aligned}
\tag{3.2}
$$

Let $v = \xi_t$ and estimate (3.2) term by term; we see that

$$
\begin{aligned}
&\tfrac{1}{2}\tfrac{d}{dt}(q(U)\xi_t, \xi_t) + (a_* - \epsilon)\|\nabla\xi_t\|^2 \leq K[\|\eta\|^2 + \|\eta_t\|^2 + \|\eta_{tt}\|^2 \\
&+ \int_0^t \|\eta(\tau)\|^2 d\tau] + K[\|\xi\|_1^2 + \|\xi_t\|^2 + \int_0^t \|\xi(\tau)\|_1^2 d\tau] + \epsilon\|\xi_t\|_1^2.
\end{aligned}
$$

Integrating this inequality for t from 0 to s, $0 \leq s \leq T$, and using Gronwall's lemma, we obtain Theorem 3.1. □

Divide $[0, T]$ into M equal intervals, and let $\Delta t = T/M$, $t_l = l\Delta t$, $t_{l+\frac{1}{2}} = (1 + \frac{1}{2})\Delta t$, $\phi^l = \phi(t_l)$, $d_t\phi^l = \frac{1}{\Delta t}(\phi^{l+1} - \phi^l)$, $\partial_t\phi^l = \frac{1}{2\Delta t}(\phi^{l+1} - \phi^{l-1})$, $\partial_{tt}\phi^l = \frac{1}{(\Delta t)^2}(\phi^{l+1} - 2\phi^l + \phi^{l-1})$, $\phi^{l+\frac{1}{2}} = \frac{1}{2}(\phi^{l+1} + \phi^l)$, and $\overline{\phi}^l = \frac{1}{2}(\phi^{l+1} + \phi^{l-1})$. We have

$$
\int_{t_l}^{t_{l+1}} \phi(\tau)\psi(\tau)d\tau = \Delta t\phi(t_{l+\frac{1}{2}})\psi^{l+\frac{1}{2}} + \varepsilon_l(\phi, \psi).
$$

A discrete-time finite element scheme can be defined: Find $U^{n+1} \in \mu$ such that

$$
\begin{aligned}
&(q^n(U)\partial_{tt}U^n, v) + (a^n(U)\nabla\partial_t U^n + b^n(U)\nabla U^n + \Delta t \sum_{l=0}^{n-1} c_{nl}(U)\nabla U^{l+\frac{1}{2}}, \nabla v) \\
&= (f^n(U), v) + \langle g^n(U), v\rangle \quad \forall v \in \mu, \ n = 1, 2, \ldots,
\end{aligned}
\tag{3.3}
$$

where $c_{nl}(U) = c(t_n, t_{l+\frac{1}{2}}, x, U^{l+\frac{1}{2}})$ and $q^n(U) = q(x, t_n, U^n)$. Let $\xi^n = U^n - \tilde{u}^n$. Then $U^n - u^n = \xi^n - \eta^n$; we have such an approximation result.

Theorem 3.2 *Under conditions* (P_1), (P_2), $k \geq d/2$, $\frac{\partial^j u}{\partial t^j} \in H^{k+1}(\Omega)$, $j = 0, \ldots, 4$, *and*

$$
\|d_t\xi^0\|_1 + \|\xi^0\|_1 = O(h^{k+1} + (\Delta t)^2),
\tag{3.4}
$$

for scheme (3.3), *then,*

$$
\begin{aligned}
&\|\partial_{tt}(U - u)\|_{L^2(L^2)} + \|d_t(U - u)\|_{L^\infty(L^2)} + \|U - u\|_{L^\infty(L^2)} \\
&+ h[\|d_t(U - u)\|_{L^\infty(H^1)} + \|U - u\|_{L^\infty(H^1)}] = O(h^{k+1} + (\Delta t)^2).
\end{aligned}
$$

Proof: We see from (1.1), (3.3), and (2.1) that ξ satisfies that, for $n \geq 1$ and $\forall v \in \mu$,

$$
\begin{aligned}
\sum_{i=1}^{2} L_i^n &= (q^n(U)\partial_{tt}\xi^n, v) + (a^n(U)\nabla\partial_t\xi^n, \nabla v) \\
&= ([q^n(u) - q^n(U)]u_{tt} + q^n(U)[(u_{tt}^n - \partial_{tt}u^n) + \partial_{tt}\eta^n] \\
&\quad + [f^n(u) - f^n(U)] - \lambda\eta_t^n, v) \\
&\quad + ([a^n(u) - a^n(U)]\nabla\tilde{u}_t^n + a^n(U)(\nabla\tilde{u}_t^n - \nabla\partial_t\tilde{u}^n), \nabla v) \\
&\quad - (b^n(U)\nabla\xi^n - [b^n(u) - b^n(U)]\nabla\tilde{u}^n, \nabla v) \\
&\quad - (\Delta t \sum_{l=0}^{n-1} \hat{c}_{nl}(U)\nabla\xi^{l+\frac{1}{2}} + \Delta t \sum_{l=0}^{n-1} [\hat{c}_{nl}(u) - \hat{c}_{nl}(U)]\nabla\tilde{u}^{l+\frac{1}{2}} \\
&\quad - \sum_{l=0}^{n-1} \varepsilon_l(c, \nabla\tilde{u})|_{t=t_n}, \nabla v) + <g^n(U) - g^n(\tilde{u}), v> = \sum_{i=1}^{5} R_i^n,
\end{aligned}
\tag{3.5}
$$

where $\hat{c}_{nl}(u) = c(t_n, t_{l+\frac{1}{2}}, x, u(x, t_{l+\frac{1}{2}}))$. Setting $v = \partial_{tt}\xi^n$ in (3.5), we see that

$$
\begin{aligned}
2\Delta t \sum_{n=1}^{N-1} L_2^n &\geq a_* \|\nabla d_t\xi^{N-1}\|^2 - K\|\nabla d_t\xi^0\|^2 \\
&\quad - K\Delta t \sum_{n=0}^{N-2} (1 + h^{-\frac{d}{2}}\|d_t\xi^n\|)\|\nabla d_t\xi^n\|^2.
\end{aligned}
\tag{3.6}
$$

and

$$
\begin{aligned}
2\Delta t \sum_{n=1}^{N-1} \sum_{i=2}^{5} R_i^n &\leq K[\|\xi^{N-1}\|_1^2 + \|\xi^1\|_1^2 + \|d_t\xi^0\|_1^2 \\
&\quad + \|\eta\|_{L^\infty(L^2)}^2 + \|\eta_t\|_{L^2(L^2)}^2 + (\Delta t)^4] + \epsilon\|d_t\xi^{N-1}\|_1^2 \\
&\quad + K\Delta t \sum_{n=1}^{N-2} [\|d_t\xi^n\|_1^2 + (1 + h^{-\frac{d}{2}}\|d_t\xi^n\|)\|\nabla d_t\xi^n\|^2] + \sum_{n=1}^{N-1} \|\xi^n\|_1^2,
\end{aligned}
\tag{3.7}
$$

where $\partial_{tt}\xi^n = \frac{1}{\Delta t}(d_t\xi^n - d_t\xi^{n-1})$ and $\sum_{n=1}^{N-1} (\phi^n, \psi^n - \psi^{n-1}) = (\phi^{N-1}, \psi^{N-1}) - (\phi^1, \psi^0) + \sum_{n=1}^{N-2} (\phi^n - \phi^{n+1}, \psi^n)$ have been used to accomplish the steps of (3.7). It is easy to estimate other terms. Applying these relations to (3.5) and using $\|\xi^n\|_1^2 \leq K\|\xi^0\|_1^2 + K\Delta t \sum_{l=0}^{n-1} \|d_t\xi^l\|_1^2$ and $\|d_t\xi^{N-1}\|^2 \leq K\|d_t\xi^0\|^2 + K\Delta t \sum_{n=0}^{N-1} \|d_t\xi^n\|^2 + \epsilon\Delta t \sum_{n=1}^{N-1} \|\partial_{tt}\xi^n\|^2$, we obtain, for $N \geq 3$,

$$
\begin{aligned}
\Delta t \sum_{n=1}^{N-1} \|\partial_{tt}\xi^n\|^2 + \|d_t\xi^{N-1}\|_1^2 &\leq K[\|d_t\xi^0\|_1^2 + \|\xi^0\|_1^2 \\
&\quad + (\Delta t)^4 + \|\eta\|_{L^\infty(L^2)}^2 + \|\eta_t\|_{L^2(L^2)}^2 + \|\eta_{tt}\|_{L^2(L^2)}^2] \\
&\quad + K_2\Delta t \sum_{n=0}^{N-2} \|d_t\xi^n\|_1^2 + K_3\Delta t \sum_{n=0}^{N-2} (1 + h^{-\frac{d}{2}}\|d_t\xi^n\|)\|\nabla d_t\xi^n\|^2.
\end{aligned}
\tag{3.8}
$$

For $N = 2$, we obtain (3.8) by estimating $R_i^n (i = 2, \ldots, 5)$ directly. Under condition (3.4), the first term on the right side of (3.8) is less than $K_1[h^{2k+2} + (\Delta t)^4]$. We obtain Theorem 3.2 using an induction hypothesis procedure, together with Lemmas 2.2 and 2.3.   ☐

Letting $v = \partial_{tt} U^n$ in (3.3), we have the stability result.

Theorem 3.3 *Under the conditions of Theorem 3.2 for scheme (3.3), $N \geq 2$, then*

$$\Delta t \sum_{n=1}^{N-1} \|\partial_{tt} U^n\|^2 + \|d_t U^{N-1}\|_1^2 + \|U^N\|_1^2$$

$$\leq K[\|d_t U^0\|_1^2 + \|U^0\|_1^2 + \Delta t \sum_{n=1}^{N-1} \|f^n(U)\|^2 \tag{3.9}$$

$$+\Delta t \sum_{n=1}^{N-1} |d_t g^{n-1}(U)|_{L^2(\partial\Omega)}^2 + |g^{N-1}(U)|_{L^2(\partial\Omega)}^2 + |g^0(U)|_{L^2(\partial\Omega)}^2].$$

Another time-discrete finite element scheme is defined: Find $U^{n+1}, W^{n+1} \in \mu$ such that

$$(q^n(U)\partial_t W^n, v) + (a^n(U)\nabla \overline{W}^n + b^n(U)\nabla U^n + \Delta t \sum_{l=0}^{n-1} c_{nl}(U)\nabla U^{l+\frac{1}{2}}, \nabla v)$$
$$= (f^n(U), v) + <g^n(U), v> \quad \forall v \in \mu, \, n = 1, 2, \ldots,$$
$$d_t U^n = W^{n+\frac{1}{2}}, \quad n = 0, 1, 2, \ldots.$$
$$\tag{3.10}$$

Denote $w = u_t$, $\tilde{w} = \tilde{u}_t$, $\rho = w - \tilde{w}$, and $\theta^n = W^n - \tilde{w}^n$. Then we have

Theorem 3.4 *Under conditions (P_1), (P_2), $k \geq d/2$, $\frac{\partial^j u}{\partial t^j} \in H^{k+1}(\Omega)$, $j = 0, \ldots, 4$, and*

$$\|\theta^0\|_1 + \|\theta^1\|_1 + \|\xi^0\|_1 = O(h^{k+1} + (\Delta t)^2), \tag{3.11}$$

for scheme (3.10), then

$$\|\partial_t(W - w)\|_{L^2(L^2)} + \|W - w\|_{L^\infty(L^2)} + \|U - u\|_{L^\infty(L^2)}$$
$$+h[\|W - w\|_{L^\infty(H^1)} + \|U - u\|_{L^\infty(H^1)}] = O(h^{k+1} + (\Delta t)^2).$$

Proof: By (1.1), (3.10), and (2.1), we have the error equation

$$(q^n(U)\partial_t \theta^n, v) + (a^n(U)\nabla \overline{\theta}^n, \nabla v) = ([q^n(u) - q^n(U)]w_t$$
$$+q^n(U)[(w_t^n - \partial_t w^n) + \partial_t \rho^n] + [f^n(u) - f^n(U)] - \lambda \eta_t^n, v)$$
$$+([a^n(u) - a^n(U)]\nabla \tilde{w}^n + a^n(U)(\nabla \tilde{w}^n - \nabla \overline{\tilde{w}}^n), \nabla v) + \sum_{i=3}^{5} R_i^n, \tag{3.12}$$

where $R_i^n (i = 3, 4, 5)$ follows the form in (3.5). Setting $v = \partial_t \theta^n = (\theta^{n+\frac{1}{2}} - \theta^{n-\frac{1}{2}})/\Delta t$ in (3.12) and estimating the terms in a similar way as that for Theorem 3.2, we obtain

$$\Delta t \|\partial_t \theta^1\|^2 + \|\theta^2\|_1^2 \leq K[\|\theta^0\|_1^2 + \Delta t \|\theta^1\|_1^2 + \|\xi^0\|_1^2 + \Delta t h^{2k+2} + (\Delta t)^4],$$

$$\Delta t \sum_{n=1}^{N-1} \|\partial_t \theta^n\|^2 + \|\theta^N\|_1^2 + \|\theta^{N-1}\|_1^2 \leq K[\|\theta^0\|_1^2 + \|\theta^1\|_1^2 + \|\xi^0\|_1^2$$

$$+ (\Delta t)^4 + \|\eta\|_{L^\infty(L^2)}^2 + \|\eta_t\|_{L^2(L^2)}^2 + \|\rho_t\|_{L^2(L^2)}^2] + K_2 \Delta t \sum_{n=1}^{N-1} \|\theta^n\|_1^2$$

$$+ K_3 \Delta t \sum_{n=2}^{N-1} [1 + h^{-\frac{d}{2}} (\|\theta^{n-\frac{1}{2}}\| + \|\theta^{n-\frac{3}{2}}\|)](\|\theta^n\|_1^2 + \|\theta^{n-1}\|_1^2), \ N \geq 3.$$

With an analogous inductive hypothesis reasoning procedure to Theorem 3.2, Theorem 3.4 can be proved. ☐

Theorem 3.5 *Under the conditions of Theorem 3.4 for scheme* (3.10), $N \geq 2$, *then*

$$\Delta t \sum_{n=1}^{N-1} \|\partial_t W^n\|^2 + \|W^N\|_1^2 + \|W^{N-1}\|_1^2 + \|U^N\|_1^2 + \|U^{N-1}\|_1^2$$

$$\leq K[\|W^0\|_1^2 + \|W^1\|_1^2 + \|U^0\|_1^2 + \Delta t \sum_{n=1}^{N-1} \|f^n(U)\|^2$$

$$+ \Delta t \sum_{n=1}^{N-1} |d_t g^{n-1}(U)|_{L^2(\partial\Omega)}^2 + |g^{N-1}(U)|_{L^2(\partial\Omega)}^2 + |g^0(U)|_{L^2(\partial\Omega)}^2].$$

From all above results, we see that the FE schemes proposed in this paper are all uniquely resolvable and have optimal H^1 and L^2 convergent properties to the original problem.

Remark 1. In the theorems in §3, we assumed $k \geq d/2$, which is needed in Lemma 2.4. However, in the case of semilinear equations, e.g., $a = a(x, t)$, $b = b(x, t)$, and $c = c(t, \tau, x)$, since Lemma 2.4 is no longer necessary in the numerical analysis, this restriction on k can be removed.

Remark 2. In almost all of the lemmas and theorems in this paper, we used conditions (P_1) and (P_2). Practically, these assumptions are not difficult to satisfy, for example, provided that $\Phi \in H^{k+1}(\Omega)$ and Φ_h is given by

$$(\nabla(\Phi - \Phi_h), \nabla v) + (\Phi - \Phi_h, v) = 0, \ \forall v \in \mu.$$

It is trivial that (P_1) and (P_2) are available [3].

Remark 3. The idea in this paper can be used to a more generalized case, e.g., when the $\nabla \cdot (a\nabla u_t + b\nabla u + \int_0^t c\nabla u(\tau)d\tau)$ term in equation (1.1) takes a more generic form $\sum_{i,j=1}^{d} \frac{\partial}{\partial x_i}(a_{ij}\frac{\partial u_t}{\partial x_j} + b_{ij}\frac{\partial u}{\partial x_j} + \int_0^t c_{ij}\frac{\partial u(\tau)}{\partial x_j}d\tau)$, where $a_{ij} = a_{ij}(x,t,u)$, $b_{ij} = b_{ij}(x,t,u)$, $c_{ij} = c_{ij}(t,\tau,x,u(x,\tau))$, and $A = (a_{ij})$ is a $d \times d$ symmetric and positive-defined matrix. We can present similar schemes and obtain similar conclusions as we have done here.

Acknowledgments. The author thanks Professor Yirang Yuan for his helpful suggestions, and Professors Longjun Shen and Xijun Yu for their warmly encouragement.

References

[1] Cui, S., Global solutions for a class of nonlinear integral-differential equations, *Acta Math. Appl. Sinica* **2** (1993), 191–200.

[2] Lin Y., Thomèe, V., and Wahlbin, L., Ritz-Volterra projections to finite-element spaces and applications to integral-differential and related equations, *SIAM. J. Numer. Anal.* **4** (1991), 1047–1070.

[3] Lin, Y. and Zhang, T., Finite element methods for nonlinear Sobolev equations with nonlinear boundary conditions, *J. Math. Anal. Appl.* **165** (1992), 180–191.

[4] Wang H., Finite element method and error estimates for semilinear pseudo-hyperbolic integral-differential equation's initial boundary value problem, Master Degree's Thesis, Shandong University, Jinan, China, 1998.

[5] Wang, S., On the initial boundary value problem and initial value problem for the semilinear pseudo-hyperbolic integral-differential equation, *Acta Math. Appl. Sinica* **4** (1995), 567–578.

[6] Wheeler, M. F., A priori L^2 error estimates for Galerkin approximations to parabolic partial differential equations, *SIAM. J. Numer. Anal.* **4** (1973), 723–759.

A CFL-Free Explicit Scheme with Compression for Linear Hyperbolic Equations

RONALD A. DEVORE   HONG WANG   JIANG-GUO LIU
HONG XU

Abstract

We develop an unconditionally stable, explicit numerical scheme for linear hyperbolic equations, which arises as an advection-reaction equation in porous medium flows. The derived scheme generates accurate numerical solutions even if large time steps are used, and conserves mass. Furthermore, this scheme has the capability of performing adaptive compression while maintaining the accuracy of the compressed solution and mass conservation. Numerical results show the strong potential of the method.

KEYWORDS: characteristic methods, hyperbolic equations, multiresolution analysis, wavelet decomposition

1 Introduction

Hyperbolic partial differential equations describe the displacement of oil by injected fluid in petroleum recovery, the subsurface contaminant transport and remediation, aerodynamics, and many other applications. Because of the moving steep fronts present in their solutions, the numerical treatment of these equations often presents severe difficulties. In late 1920's, Courant, Friedrichs, and Lewy proved the famous result that there are no explicit, unconditionally stable, consistent finite difference schemes for hyperbolic partial differential equations [2]. Consequently, most numerical methods developed for hyperbolic equations are explicit but subject to the well-known Courant-Friedrichs-Lewy (CFL) condition. Very small time steps often have to be used in numerical simulations to meet the stability requirement of the numerical methods. Implicit methods could be unconditionally stable, but an algebraic system must be solved at each time step in order to obtain numerical solutions.

In this paper we develop an unconditionally stable, explicit numerical scheme for linear hyperbolic equations. The derived scheme generates accurate numerical solutions even if large time steps are used, and conserves mass. Furthermore, this scheme has the capability of carrying out adaptive compression while maintaining the accuracy of the numerical solution and mass conservation. Computational results are presented to show the strong potential of the method developed.

2   A Reference Equation

We consider the initial-value problem for the linear hyperbolic equation

$$
\begin{aligned}
u_t + \nabla \cdot (\mathbf{v}u) + Ku &= f(\mathbf{x},t), \quad (\mathbf{x},t) \in \mathbb{R}^d \times (0,T], \\
u(\mathbf{x},0) &= u_0(\mathbf{x}), \qquad \mathbf{x} \in \mathbb{R}^d,
\end{aligned}
\tag{2.1}
$$

where $\mathbf{v}(\mathbf{x},t) := (V_1(\mathbf{x},t), V_2(\mathbf{x},t), \ldots, V_d(\mathbf{x},t))$ is a fluid velocity field, $K(\mathbf{x},t)$ is a first-order reaction coefficient, $f(\mathbf{x},t)$ is a given source function, $u(\mathbf{x},t)$ is the unknown function, and $u_0(\mathbf{x})$ is a given initial condition. $\mathbf{x} := (x_1, \ldots, x_d)$, $u_t := \frac{\partial u}{\partial t}$, $\nabla := (\frac{\partial}{\partial x_1}, \frac{\partial}{\partial x_2}, \ldots, \frac{\partial}{\partial x_d})$. We assume that $u_0(\mathbf{x})$ and $f(\mathbf{x},t)$ have compact support, so the exact solution $u(\mathbf{x},t)$ also has compact support for any finite time $t > 0$.

We define a uniform partition on the temporal interval $[0,T]$ by $t_n := n\Delta t$ with $\Delta t = T/N$. If we choose the test functions $w(\mathbf{x},t)$ to be of compact support in space, to vanish outside the interval $(t_{n-1}, t_n]$, and to be discontinuous in time at time t_{n-1}, the weak formulation for Eq. (2.1) is

$$
\begin{aligned}
\int_{\mathbb{R}^d} u(\mathbf{x},t_n)w(\mathbf{x},t_n)d\mathbf{x} &- \int_{t_{n-1}}^{t_n} \int_{\mathbb{R}^d} u(w_t + \mathbf{v}\cdot\nabla w - Kw)(\mathbf{x},t)d\mathbf{x}dt \\
&= \int_{\mathbb{R}^d} u(\mathbf{x},t_{n-1})w(\mathbf{x},t_{n-1}^+)d\mathbf{x} + \int_{t_{n-1}}^{t_n} \int_{\mathbb{R}^d} f(\mathbf{x},t)w(\mathbf{x},t)d\mathbf{x}dt,
\end{aligned}
\tag{2.2}
$$

where $w(\mathbf{x},t_{n-1}^+) := \lim_{t \to t_{n-1}^+} w(\mathbf{x},t)$ takes into account the fact that $w(\mathbf{x},t)$ is discontinuous in time at time t_{n-1}.

Motivated by the ELLAM framework of Celia *et al* [1], the test functions w in Eq. (2.2) are chosen from the solution space of the adjoint equation of Eq. (2.1). Let $\mathbf{y} = \mathbf{r}(\theta; \bar{\mathbf{x}}, \bar{t})$, with $\bar{t} \in [t_{n-1}, t_n]$, be the characteristic determined by

$$
\frac{d\mathbf{y}}{d\theta} = \mathbf{v}(\mathbf{y},\theta), \quad \text{with} \quad \mathbf{y}|_{\theta=\bar{t}} = \bar{\mathbf{x}}.
\tag{2.3}
$$

Then the adjoint equation of Eq. (2.1) is rewritten as

$$-\frac{d}{d\theta}w(\mathbf{r}(\theta;\bar{\mathbf{x}},\bar{t}),\theta) + K(\mathbf{r}(\theta;\bar{\mathbf{x}},\bar{t}),\theta)w(\mathbf{r}(\theta;\bar{\mathbf{x}},\bar{t}),\theta) = 0,$$
$$w(\mathbf{r}(\theta;\bar{\mathbf{x}},\bar{t}),\theta)|_{\theta=\bar{t}} = w(\bar{\mathbf{x}},\bar{t}), \tag{2.4}$$

leading to the following expression for w

$$w(\mathbf{r}(\theta;\bar{\mathbf{x}},\bar{t}),\theta) = w(\bar{\mathbf{x}},\bar{t})e^{-\int_\theta^{\bar{t}} K(\mathbf{r}(\gamma;\bar{\mathbf{x}},\bar{t}),\gamma)d\gamma}. \tag{2.5}$$

Therefore, once the test functions $w(\mathbf{x}, t_n)$ are specified at time t_n, they are determined completely and in fact vary exponentially along the characteristic $\mathbf{r}(\theta;\mathbf{x}, t_n)$ for $\theta \in [t_{n-1}, t_n]$.

To avoid confusion in the derivation, we replace the dummy variables $\mathbf{x}$ and t in the second term on the right-hand side of Eq. (2.2) by $\mathbf{y}$ and θ. For any $\mathbf{y} \in \mathbb{R}^d$, there exists an $\mathbf{x} \in \mathbb{R}^d$ such that $\mathbf{y} = \mathbf{r}(\theta;\mathbf{x}, t_n)$. We obtain

$$\begin{aligned}
&\int_{t_{n-1}}^{t_n} \int_{\mathbb{R}^d} f(\mathbf{y}, \theta)w(\mathbf{y}, \theta)d\mathbf{y}d\theta \\
&= \int_{\mathbb{R}^d} \int_{t_{n-1}}^{t_n} f(\mathbf{r}(\theta;\mathbf{x}, t_n),\theta)w(\mathbf{r}(\theta;\mathbf{x}, t_n),\theta)\left|\frac{\partial \mathbf{r}(\theta;\mathbf{x}, t_n)}{\partial \mathbf{x}}\right|d\theta d\mathbf{x} \\
&= \int_{\mathbb{R}^d} F(\mathbf{x}, t_n)w(\mathbf{x}, t_n)\left[\int_{t_{n-1}}^{t_n} e^{-K(\mathbf{x},t_n)(t_n-\theta)}d\theta\right]d\mathbf{x} + E_1(f, w) \\
&= \int_{\mathbb{R}^d} \Lambda(\mathbf{x}, t_n)f(\mathbf{x}, t_n)w(\mathbf{x}, t_n)d\mathbf{x} + E_1(f, w).
\end{aligned} \tag{2.6}$$

Here $\Lambda(x, t_n) = (1 - e^{-K(\mathbf{x},t_n)\Delta t})/K(\mathbf{x}, t_n)$ if $K(\mathbf{x}, t_n) \neq 0$, or Δt otherwise. $E_1(f, w)$ is the local truncation error.

Incorporating Eq. (2.6) into Eq. (2.2), we obtain a reference equation

$$\begin{aligned}
&\int_{\mathbb{R}^d} u(\mathbf{x}, t_n)w(\mathbf{x}, t_n)d\mathbf{x} \\
&= \int_{\mathbb{R}^d} u(\mathbf{x}, t_{n-1})w(\mathbf{x}, t_{n-1}^+)d\mathbf{x} + \int_{\mathbb{R}^d} \Lambda(\mathbf{x}, t_n)(fw)(\mathbf{x}, t_n)d\mathbf{x} + E(w),
\end{aligned} \tag{2.7}$$

where $E(w) = \int_{t_{n-1}}^{t_n} \int_{\mathbb{R}^d} u[w_t + \mathbf{v} \cdot \nabla w - Kw](\mathbf{x}, t)d\mathbf{x}dt + E_1(f, w)$.

3   Multiresolution and Wavelet Decomposition

To develop a numerical scheme based on the reference equation (2.7), we need to define the trial and test functions at time t_n. To do so, we recall the notions of multiresolution analysis and wavelet decompositions.

In the standard Fourier analysis, L^2-functions are represented by linear combinations of sines and cosines. In 1910, Haar studied the representation of L^2-functions by step functions taking values ± 1 [5]. In the 1980's, these ideas were explored and developed further into the theory of wavelets. The first wavelets were introduced in early 1980's by Stromberg [10] and Morlet *et al* [8]. One of the best ways of constructing wavelets is multiresolution analysis. This approach began in image-processing [9, 11] and was introduced into mathematics by Mallat [6]. Multiresolution analysis was used by Daubechies [3] to construct compactly supported orthogonal wavelets with arbitrary smoothness, which include the Haar wavelets as the simplest case.

Definition A sequence of closed subspaces $\{\mathcal{V}_j\}_{j \in \mathbf{Z}}$ ($\mathbf{Z}$-the set of all integers) of $L^2(\mathbb{R})$ is a *Multiresolution Analysis* if

(a) these spaces are nested: $\mathcal{V}_j \subset \mathcal{V}_{j+1}$ for $\forall j \in \mathbf{Z}$;

(b) these spaces are dense in $L^2(\mathbb{R})$: $\overline{\cup}_{j \in \mathbf{Z}} \mathcal{V}_j = L^2(\mathbb{R})$ and $\cap_{j \in \mathbf{Z}} \mathcal{V}_j = \emptyset$;

(c) $\mathcal{V}_0$ is invariant under integer shifts: $f(\cdot) \in \mathcal{V}_0 \Longrightarrow f(\cdot - k) \in \mathcal{V}_0$, $\forall k \in \mathbf{Z}$;

(d) $\mathcal{V}_j$ is obtained from $\mathcal{V}_0$ by dilation: $f(\cdot) \in \mathcal{V}_j \Longleftrightarrow f(2^{-j} \cdot) \in \mathcal{V}_0$, $\forall j \in \mathbf{Z}$;

(e) $\mathcal{V}_0$ is generated by a single (the so-called scaling) ϕ and its translates: $\{\phi_{0,k} : k \in \mathbf{Z}\}$ is an orthonormal basis in $\mathcal{V}_0$, where

$$\phi_{j,k}(x) := 2^{j/2} \phi(2^j x - k), \quad j, k \in \mathbf{Z}. \tag{3.1}$$

The conditions (d) and (e) in the definition implies that the family $\{\phi_{j,k} : k \in \mathbf{Z}\}$ forms an orthonormal basis for $\mathcal{V}_j$. Let $\mathcal{P}_j : L^2(\mathbb{R}) \longrightarrow \mathcal{V}_j$ be the orthogonal projection operator, then the conditions (a) and (b) in the definition concludes that for any $f \in L^2(\mathbb{R})$

$$\lim_{j \to +\infty} \|\mathcal{P}_j f - f\|_{L^2(\mathbb{R})} = 0, \tag{3.2}$$

i.e.,

$$\mathcal{P}_j f = \sum_{k \in \mathbf{Z}} (f, \phi_{j,k}) \phi_{j,k} \longrightarrow f, \quad \text{as } j \to +\infty. \tag{3.3}$$

Let $\mathcal{W}_{j-1}$ be the orthogonal complement of $\mathcal{V}_{j-1}$ in $\mathcal{V}_j$

$$\begin{aligned} \mathcal{V}_j &= \mathcal{V}_{j-1} \oplus \mathcal{W}_{j-1} = \dots \\ &= \mathcal{V}_J \oplus \mathcal{W}_J \oplus \mathcal{W}_{J+1} \oplus \dots \oplus \mathcal{W}_{j-1}, \quad \text{for } j > J. \end{aligned} \tag{3.4}$$

Then, if $f \in L^2(\mathbb{R})$, the telescoping sum

$$P_j f = P_J f + \sum_{i=J}^{j-1} (P_{i+1} f - P_i f) \tag{3.5}$$

represents $P_j f \in V_j$ as an element of $V_J \oplus W_J \oplus W_{J+1} \oplus \ldots \oplus W_{j-1}$. By (3.2), Eq. (3.5) provides an approach of approximating arbitrary L^2-functions by sequences of functions from V_J and W_i for $J \leq j < +\infty$. More importantly, the spaces W_j can be generated by a single (the so-called wavelet) function ψ. In other words, ψ and its integer translates $\psi_{0,k}$, with $\psi_{j,k}$ being defined by

$$\psi_{j,k}(x) := 2^{j/2} \psi(2^j x - k), \quad j, k \in \mathbf{Z}, \tag{3.6}$$

constitute an orthonormal basis for W_0. For each fixed j, the $\psi_{j,k}$ ($k \in \mathbf{Z}$) form an orthogonal basis for W_j.

4 A CFL-Free Explicit Scheme with Conservative Compression

In this section we develop a CFL-free, explicit scheme for the initial-value problem (2.1). We define the finite-dimensional space $S_j(\mathbb{R}^d)$ to be the tensor product of the one-dimensional space V_j

$$S_j(\mathbb{R}^d) := (V_j(\mathbb{R}))^d = \operatorname{span} \{\Phi_{j,\mathbf{k}}(\mathbf{x})\}_{\mathbf{k}=(k_1,k_2,\ldots,k_d)\in\mathbf{Z}^d}, \tag{4.1}$$

where $\Phi_{j,\mathbf{k}}(\mathbf{x}) := \phi_{j,k_1}(x_1)\phi_{j,k_2}(x_2)\ldots\phi_{j,k_d}(x_d)$ and $\phi_{j,k}(x) \in V_j(\mathbb{R})$ is defined by Eq. (3.1).

With these preparations, we replace the exact solution u in Eq. (2.7) by the trial functions $U(\mathbf{x}, t_n) \in S_j(\mathbb{R}^d)$ and drop the error term $E(w)$ in Eq. (2.7), leading to the following numerical scheme: find $U(\mathbf{x}, t_n) \in S(\mathbb{R}^d)$ such that

$$\int_{\mathbb{R}^d} U(\mathbf{x}, t_n) w(\mathbf{x}, t_n) dx$$
$$= \int_{\mathbb{R}^d} U(\mathbf{x}, t_{n-1}) w(\mathbf{x}, t_{n-1}^+) dx + \int_{\mathbb{R}^d} \Lambda(\mathbf{x}, t_n) f(\mathbf{x}, t_n) w(\mathbf{x}, t_n) dx, \tag{4.2}$$
$$\forall w(\mathbf{x}, t_n) = \Phi_{j,\mathbf{k}}(\mathbf{x}) \in S_j(\mathbb{R}^d).$$

Although the integrals in Eq. (4.2) are formally defined on the space $\mathbb{R}^d$, they are in fact supported locally since the test functions $w(\mathbf{x}, t_n) = \Phi_{j,\mathbf{k}}(\mathbf{x})$ have compact support. Moreover, because the solution $U(\mathbf{x}, t_{n-1})$ and $f(\mathbf{x}, t_n)$

have compact support, Eq. (4.2) only needs a finite number of operations at each time level. Thirdly, because the scaling functions $\Phi_{j,\mathbf{k}}(\mathbf{x})$, $\mathbf{k} \in \mathbf{Z}^d$, form an orthonormal system for $\mathcal{S}_j(\mathbb{R}^d)$, the scheme (4.2) is explicit. It can be proven that the scheme (4.2) is unconditionally stable.

It is well known that in practice the solutions of linear transport equations are fairly smooth outside some very small (dynamic) regions, where the solutions can develop steep fronts or even shock discontinuities. Therefore, some kind of local refinement or adaptive techniques should be used. Extensive research has been conducted in this regard in the finite difference, finite element, and finite volume methods. In the current context, this can be realized in terms of compression in a very natural way. For example, we can apply a thresholding to the scheme (4.2) to minimize the number of equations that need to be formulated and solved, and so significantly improve the computational efficiency. However, a direct application to the scaling functions $\Phi_{j,\mathbf{k}}$ will introduce mass balance error. In contrast, because of their vanishing moments, applying such a technique to the wavelets does not affect mass balance. Hence, we use the wavelet decomposition (3.4) to choose another basis for the space $\mathcal{V}_j(\mathbb{R})$

$$V_j(\mathbb{R}) = \mathrm{span} \left\{ \phi_{l,k}^{(\alpha)}(x) \right\}_{k \in \mathbf{Z}, \; J \le l \le j, \; \alpha = 0,1}, \tag{4.3}$$

where $\phi_{l,k}^{(0)}(x) := \phi_{l,k}(x)$ and $\phi_{l,k}^{(1)}(x) := \psi_{l,k}(x)$. This leads to another expression for the space $\mathcal{S}_j(\mathbb{R}^d)$

$$S_j(\mathbb{R}^d) = \mathrm{span} \left\{ \Phi_{l,\mathbf{k}}^{(\alpha)}(\mathbf{x}) \right\}_{\mathbf{k} \in \mathbf{Z}^d, \; J \le l \le j, \; \boldsymbol{\alpha} = (\alpha_1, \alpha_2, \ldots, \alpha_d) \in \{0,1\}^d}. \tag{4.4}$$

In this case, the numerical scheme is still defined by Eq. (4.2), but using a different basis for $\mathcal{S}_j(\mathbb{R}^d)$. We begin by projecting the initial condition $u_0(\mathbf{x})$ onto the space $\mathcal{S}_j(\mathbb{R}^d)$ to obtain an approximant $\tilde{U}_0(\mathbf{x}, t_0)$. For $n = 1, 2, \ldots, N$, we apply a thresholding technique to compress $\tilde{U}_0(\mathbf{x}, t_{n-1})$ to obtain $U(\mathbf{x}, t_{n-1})$. Then, using its wavelet expansion as an error indicator, we solve the reduced scheme (4.2) at time step t_n with the minimal number of basis functions used to obtain $\tilde{U}(\mathbf{x}, t_n)$. Next, we compress $\tilde{U}(\mathbf{x}, t_n)$ to obtain $U(\mathbf{x}, t_n)$. We continue this process until we reach the final time step $t_N = T$.

5   Numerical Experiments

In this section we present numerical results to show the potential of the numerical scheme developed in this paper. The spatial domain is $\Omega = (0, 5)$, the time interval is $[0, T] = [0, 2]$, a velocity field $V = 1 + 0.1x$ is chosen, $K = f = 0$. The initial condition $u_0(x)$ is

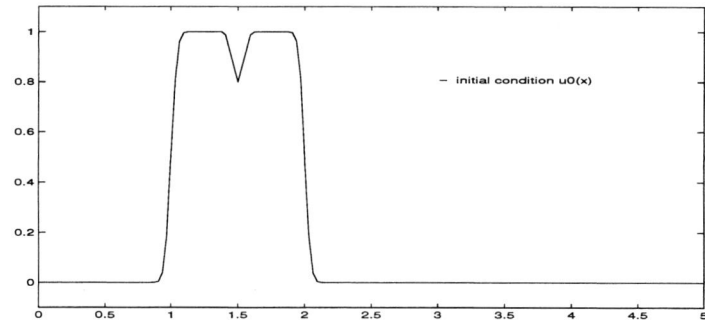

Figure 1:The initial condition $u_0(x)$.

$$u_0(x) := \text{Imp}(x; 1, 2, 0.05) - 0.2\, \text{Tri}(x; 1.5, 0.1), \tag{5.1}$$

with the hat function $\text{Tri}(x; a, b)$ and the impulse function $\text{Imp}(x; a, b, c)$ being defined by

$$\text{Tri}(x; a, b) := \begin{cases} 1 + \dfrac{x - a}{b} & \text{if } a - b \leq x \leq a, \\ 1 - \dfrac{x - a}{b} & \text{if } \phantom{a - b \leq{}} a \leq x \leq a + b, \\ 0 & \text{otherwise,} \end{cases} \tag{5.2}$$

and

$$\text{Imp}(x; a, b, c) := \frac{1}{2}\left[\text{erf}\left(\frac{x - 1}{0.05}\right) - \text{erf}\left(\frac{x - 2}{0.05}\right)\right], \tag{5.3}$$

where the error function $\text{erf}(x) := (2/\sqrt{\pi}) \int_0^x e^{-z^2} dz$. The initial condition $u_0(x)$ is plotted in Figure 1.

In the numerical experiments, we use a time step of $\Delta t = 0.4$. The coarse and fine spatial levels are $J = 3$ and $j = 7$, respectively. This leads to a maximal Courant number of 192. At the final time $T = 2$, the L^1 error of the uncompressed solution is 0.00477 while the L^1 error of the compressed solution is 0.00515. The compression ratio is 30.19. We plot the uncompressed and compressed solutions against the analytical solution in Figure 2.

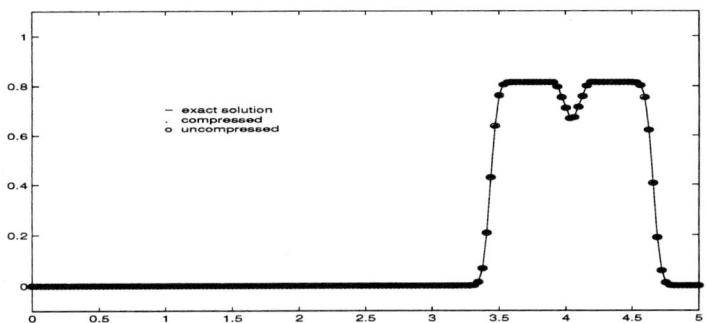

FIGURE 2:The analytical, uncompressed, and compressed numerical solutions.

References

[1] Celia, M. A., Russell, T. F., Herrera, I., Ewing R. E., An Eulerian-Lagrangian localized adjoint method for the advection-diffusion equation, *Advances in Water Resources* **13** (1990), 187–206.

[2] Courant, R., Friedrichs, K. O., and Lewy, H., Uber die partiellen differenzen-gleichungen der mathematisches physik, *Math. Annalen* **100** (1928), 32–74.

[3] Daubechies, I., Orthogonal bases of compactly supported wavelets, *Commun. Pure and Appl. Math.* **41** (1988), 909–996.

[4] I. Daubechies: *Ten Lectures on Wavelets*, **61** of CBMS-NSF Regional Conference Series in Applied Mathematics, SIAM, Philadelphia, 1992.

[5] Haar, A., Zur Theorie der Orthogonalen Funktionen-Systeme,*Math. Ann.* **69** (1910), 331–371.

[6] Mallet, S. G., Multiresolution and wavelet orthonormal bases in $L^2(\mathbb{R})$, *Trans. Amer. Math. Soc.* **315** (1989), 69–87.

[7] Meyer, Y., *Ondelettes et Opérateurs*, **1** and **2**, Hermann, Paris, 1992.

[8] Morlet, J., Arens, G., Fourgeau, I., and Giard, D., Wave propagation and sampling theory, *Geophysics* **47** (1982), 203–236.

[9] Smith, M. J. and Barnwell, D. P., Exact reconstruction for tree-structured subband coders, *IEEE Trans. ASSP* **34** (1986), 434–441.

[10] Stromberg, J. O., A modified Franklin system and higher order spline on $\mathbb{R}^n$ as unconditional bases for Hardy spaces, Becker W. *et al* (eds.), Wadsworth Math. Series, Belmont CA, 1982, 475–493.

[11] Vetterli, M., Filter banks allowing perfect reconstruction,*Signal Processing* **10** (1986), 219–244.

Maximizing Cache Memory Usage for Multigrid Algorithms for Applications of Fluid Flow in Porous Media

CRAIG C. DOUGLAS   JONATHAN HU

MOHAMED ISKANDARANI   MARKUS KOWARSCHIK

ULRICH RÜDE   CHRISTIAN WEISS

Abstract

Computers today rely heavily on good utilization of their cache memory subsystems. Compilers are optimized for business applications, not scientific computing ones, however. Automatic tiling of complex numerical algorithms for solving partial differential equations is simply not provided by compilers. Thus, absolutely terrible cache performance is a common result.

Multigrid algorithms combine several numerical algorithms into a more complicated algorithm. In this paper, an algorithm is derived that allows for data to pass through cache exactly once per multigrid level during a V cycle before the level changes. This is optimal cache usage for large problems that do not fit entirely in cache. The numerical techniques and algorithms discussed in this paper can be easily applied to numerical simulation of fluid flows in porous media.

KEYWORDS: multigrid, cache, threads, sparse matrix, iterative methods, domain decomposition, compiler optimization

1   Introduction

Multigrid methods are widely known as the fastest methods for solving elliptic partial differential equations. This belief was derived when computers were designed very differently than today. Accessing one word of data took a set amount of time due to computers having one level of memory.

Since the early 1980's, processors have sped up 5 times faster per year than memory. Multilevel memories, using *memory caches* were developed to compensate for the uneven speed ups in the hardware. Essentially all computers today, from laptops to distributed memory supercomputers use cache memories to keep the processors busy.

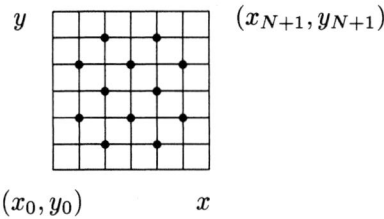

Figure 1: Simple grid with red points marked

By the term cache, we mean a fast memory unit closely coupled to the processor. In the interesting cases, the cache is further divided into a great many *cache lines*, which hold copies of contiguous locations of main memory. The cache lines may hold data from quite separate parts of main memory.

Tiling is the process of decomposing a computation into smaller blocks and doing all of the computing in each block one at a time. Tiling is an attractive method for improving data locality. In some cases, compilers can do this automatically. However, this is rarely the case for realistic scientific codes. In fact, even for simple examples, manual help from the programmers is, unfortunately, necessary.

Language standards interfere with compiler optimizations. Due to the requirements about loop variable values at any given moment in the computation, compilers are not allowed to fuse nested loops into a single loop. In part, it is due to coding styles that make very high level code optimization (nearly) impossible [11].

Before transforming the standard Gauss-Seidel algorithms into cache aware versions, let us define two operations for updating the approximate solution on either one or two rows of a grid:

$$Update(\text{ row, color } [, \text{ direction}])$$

and

$$UpdateRedBlack(\text{ row, } [, \text{ direction}])$$

We implicitly assume that both *Update* and *UpdateRedBlack* only compute on rows that actually exist.

The operation *Update* does a Gauss-Seidel update in each of the columns of a single row in the grid. The color is one of red, black, or all. The direction

Standard		Cache aware	
1.	Do $j = 1, N$	1.	$Update(1, \text{red})$.
1a.	$Update(j, \text{red})$.	2.	Do $j = 2, N$
2.	Do $j = 1, N$	2a.	$UpdateRedBlack(j)$.
2a.	$Update(j, \text{black})$.	3.	$Update(N, \text{black})$.

Figure 2: Standard and cache aware Gauss-Seidel

is optional and can be natural or reverse. The natural order is assumed unless a direction is given. Hence, symmetric Gauss-Seidel is easily implemented.

The operation *UpdateRedBlack* operates on all of the red points (x_i, y_j) in row j of the grid and on all of the black points $(x_i, y_{j\pm1})$ in the preceding row $j\pm1$ (the updates are paired). The $\pm$ depends on the choice of direction. This fuses the red-black calculation so that we do a red-black ordered Gauss-Seidel with only one sweep across the grid instead of the standard two passes.

The update operations can be modified for SOR, SSOR, or ADI relaxation methods. Within the update operations we can further optimize the process by including Linpack style optimizations like loop unrolling to get $2 - 7$ updates per iteration through the loop.

Some of the motivation behind this paper can be summarized in a simple example. Consider the grid in Figure 1, where the boundary points are included in the grid. Both standard and cache aware algorithms for the red-black ordered Gauss-Seidel iteration are given in Figure 2.

Each iteration of the standard algorithm, all of the data passes through the cache twice. Hence, with a small change in the algorithm's implementation, the data only passes through cache once. Unfortunately, no compiler for commonly used languages (e.g., Fortran, C, or C++) seems to exist that can optimize the first form of the red-black ordered Gauss-Seidel algorithm automatically into the second form.

Multigrid algorithms combine a number of operations in order to work. These include iterative methods (typically relaxation methods), residual computation, projection of residuals onto a coarser grid, and interpolation of corrections onto a finer grid. These are typically programmed as separate routines, which makes the components easy to replace and modify.

However, a number of components re-use data in a manner that is suitable for algorithms that are *cache aware*. Algorithms will be developed such that data passes through the memory cache once while computing on a given level before a level change.

This paper is concerned with algorithmic changes that are highly portable. Such techniques as loop unrolling, though mentioned, are not really stressed. The intention is that codes written using the algorithms in this paper will work well on anything from a PC to a high end RISC processor based supercomputer with only trivial tuning (one parameter). This means that we are not trying to get every last floating and fixed point operation out of a computer, just an integer factor speed up for a modest amount of work.

2   Relaxation Methodology and Notation

Consider solving the following set of problems:

$$A_i u_i = f_i, \qquad 1 \le i \le k,$$

where $u_i \in \mathbb{R}^{n_i}$ and $n_i > n_{i+1}$. Level 1 is the real problem that the solution is wanted on. All other levels are smaller, or coarser, approximations to level 1. The linear systems result from discretizing a partial differential equation over a given grid Ω_i. The discretization can be any standard finite element, difference, volume, or wavelet approach (see [1] and [2] for examples of this approach). Further, the discrete grids Ω_i will be assumed to be structured and regular.

2.1   Once through cache naturally ordered Gauss-Seidel

Consider the naturally ordered Gauss-Seidel first. Let us restrict our attention to matrices A_j which are based on discretization methods which are local to only 3 neighboring rows of the grid (e.g., a 5 or 9 point discretization). We have to assume that $\ell + m - 1$ rows of a $N \times N$ grid G fit entirely into cache simultaneously and that $m < \ell$.

Figure 3 contains the complete algorithm for passing data through cache only once for the naturally ordered Gauss-Seidel. There are two special cases to the cache aware algorithm: the first block of rows and the rest of the blocks.

The first case is for the first ℓ rows of the grid. At the end of step 1 in Figure 3, the data associated with rows 1 to ℓ has been brought into cache only once, not m times. The data in rows 1 to $\ell - m + 1$ has been updated m times. The data in rows j, $\ell - m + 2 \le j \le \ell$ has been updated $\ell - j + 1$ times.

Once the first block of grid rows is partially updated, we have a new block to update and must also finish updating the previous block of grid rows. This

Algorithm Cache-GSNat
1.   Do $it = 0, m - 1$
    1a.   Do $i = 1, \ell - it$
        1a1. $Update(i, \text{all})$.
2.   Do $j = \ell + 1, N, \ell$.
    2a.   Do $it = 0, m - 1$
        2a1.   Set $j_1 = \min(j + \ell, N)$.
        2a2.   If $j + \ell < N$, then
            2a2a. $j_2 = \max(j_1 - it - 1, j)$.
        2a3.   Else
            2a3a. $j_2 = N$
        2a4.   Do $i = j, j_2$
            2a4a. $Update(i, \text{all})$.
        2a5.   Do $i = j - 1, j + it - m + 1, -1$
            2a5a. $Update(i, \text{all})$.

Figure 3: Once through cache naturally ordered Gauss-Seidel algorithm

corresponds to step 2 in Figure 3. Note that the second inner loop (step 2a3) runs in the opposite order as the first inner loop (step 2a2), which ensures that the updates are bitwise identical to the standard algorithm. In effect, we have applied a domain decomposition methodology to the standard iteration. Each iteration requires a new, slightly smaller domain.

The cache aware algorithm can be generalized to grids which are too large to fit entire rows into cache. In this case, a multidimensional approach is necessary when dropping rows and columns from the subdomains. In each of the three pictures below, computation occurs in the smallest subdomain only (the outer parts represent the buffers that have been added). The buffers grow by one line in each of x and y each iteration.

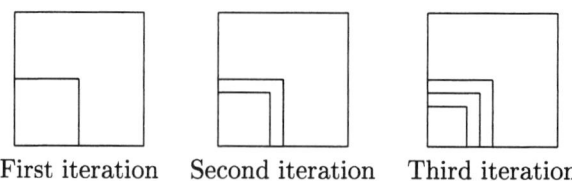

First iteration   Second iteration   Third iteration

While computing in neighboring subdomains, the rest of the updates are done carefully in order to maintain bitwise identical updates. In the current subdomain, the calculation is done using the natural ordering. In the trailing subdomain, the calculation is done in the opposite order. In the three pictures below, the shrinking subdomain on the left is the trailing subdomain.

Algorithm Cache-GSRedBlack
1.   Do $j = 1, N, \ell$.
   1a.   Set $j_1 = \min(j + \ell - 1, N)$.
   1b.   Do $it = 0, m - 1$
      1b1.   Do $i = j, j_1 - 2 * it$
         1b1a.   *Update*(i, red).
         1b1b.   *Update*$(i - 1, \text{black})$.
      1b2.   Do $i = j - 1, j - m + 2 * it, -2$
         1b2a.   If $\mod(j, 2) = 1$ then
                  1b2a1.   *UpdateRedBlack*(i)
                  1b2a2.   *UpdateRedBlack*$(i - 1)$
         1b2b.   If $\mod(j, 2) = 0$ then
                  1b2b1.   *UpdateRedBlack*$(i - 1)$
                  1b2b2.   *UpdateRedBlack*(i)

Figure 4: Once through cache red-black ordered Gauss-Seidel algorithm

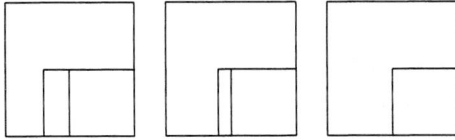

2.2   Once through cache red-black ordered Gauss-Seidel

Red-black ordered Gauss-Seidel is significantly more complex to code in the once through cache style rather than in the normal style. Let us restrict our attention to matrices A_j which are based on discretization methods which are local to only 3 neighboring rows of the grid (e.g., a 5 or a 9 point discretization). For a discussion of a 9 point discretization on this topic, see [9] and [12].

Figure 4 contains the complete algorithm for passing data through cache only once for the red-black ordered Gauss-Seidel. In §2.1, the naturally ordered Gauss-Seidel was shown geometrically to be similar to a domain decomposition method where the size of the subdomains shrank each iteration (forming an increasingly larger buffer each iterations). For the red-black ordered Gauss-Seidel, the buffer is saw tooth shaped on the top and grows each iteration by points from two grid rows instead of one. On the right side, the buffer grows each iteration by only one grid column (similar to the naturally ordered Gauss-Seidel case).

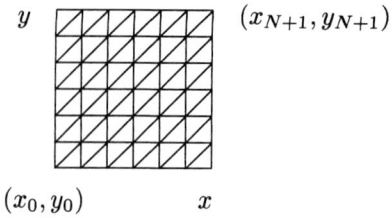

Figure 5: Simple triangular grid

2.3 Once through cache ADI and line relaxation methods

ADI and line relaxation algorithms are easily made cache aware using similar techniques to the ones given in §§2.1-2.2. One difference is that entire lines of unknowns must fit in cache at once in order to guarantee bitwise compatibility with the standard algorithm implementations.

2.4 Once through cache relaxation methods on triangular grids

Triangular grids are also easily made cache aware. The grids must be quasi-uniform and highly structured and the ordering of A_i must lead to blocks of diagonal submatrices. A simple example of a suitable grid is given in Figure 5. A combination of the methods given in §§2.1-2.2 with the nonzero graph of the A_i's is used.

Grids that are unstructured, including ones that have been chosen through an adaptive gridding procedure, require a different approach than is covered in this paper. §4 briefly discusses an unstructured grid approach.

2.5 Unstructured and quasi-structured grids

Unstructured grids presents a challenge that is not addressed in this paper. Quasi-unstructured grids (see Figure 6) can frequently be accommodated using techniques similar to structured grids. In particular, the number of graph connections in the matrices A_i are usually predictable, just like in the structured grid case. Both of these cases are considered in [5], [6], and [7].

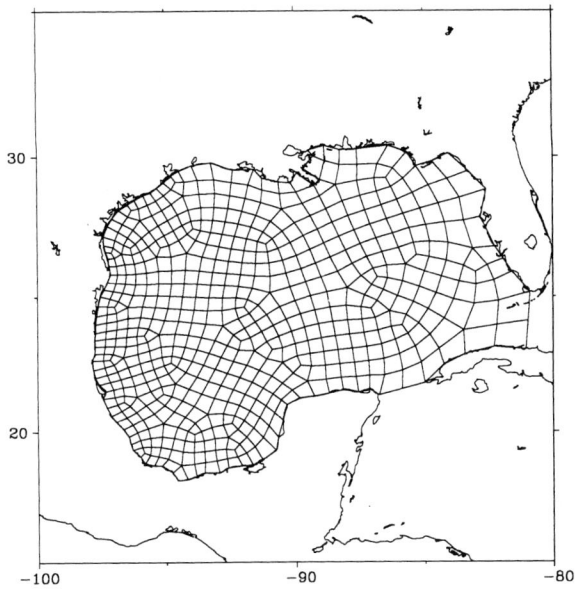

Figure 6: Quasi-structured grid.

Algorithm Vcycle(k, $\{A_i, u_i, f_i, r_i\}_{i=1}^k$)
1. Do $i = 1, 2, \cdots, k-1$
    1a.   Approximately solve $A_i u_i = f_i$.
    1b.   Compute a residual $r_i \leftarrow f_i - A_i u_i$.
    1c.   Set $f_{i+1} \leftarrow R_i r_i$ and $u_{i+1} \leftarrow 0$.
2. Do $i = k, k-1, \cdots, 1$
    2a.   Approximately solve $A_i u_i = f_i$.
    2b.   If $i > 1$, then set $u_{i-1} \leftarrow u_{i-1} + P_i u_i$.

Figure 7: V cycle definition

3 Combining Multigrid Components

A typical multigrid method is based on a V cycle multigrid method (see Figure 7). Implementing a W or F Cycle (or any other correction cycle) is a trivial extension.

All multigrid correction algorithms are a simple combination of two distinct parts: the *pre-correction step* and the *post-correction step*. These are referred to by McCormick [10] as slash cycles. While both steps may have an approximate solve step included, the change of level step at the end of each has quite different cache effects.

There are 3 major operations in the V cycle:

1. Approximate solves: steps 1a and 2a. This is typically a relaxation method for simple problems, but can be any iterative method.

2. Restriction of residuals: steps 1b and 1c. This is typically a weighted average of nearby residuals. It is used to compute a residual correction problem's right hand side (represented by the operator R_i) on the next coarser grid.

3. Prolongation of corrections: step 2b. This is typically a second or fourth order interpolation method (represented by the operator P_i).

In a typical multigrid code, a V cycle is implemented very similarly to the description given here. By using a structured language methodology, different algorithms can be substituted for the default ones trivially.

As was shown in [3], when a grid Ω_i gets too large, a change in the algorithm is required which changes the global ordering. In essence, we use a domain decomposition approach to find disjoint two dimensional subdomains Ω_{ij} whose union is Ω_i. Further, the data associated with the Gauss-Seidel operation on each subdomain must fit entirely in cache. The size of the subdomains depends heavily on the number of nonzeroes per row in the matrix A_i, the sparse matrix storage method, and what iteration of the Gauss-Seidel iteration we are on. Hence, we really have a set of disjoint subdomains $\Omega_{ij}^{(\ell)}$, where $\ell = 1, \cdots, m$.

Data passes through cache once almost everywhere each time a level is reached with this transformation. Due to connections between subdomains, sometimes a very few points (rather than whole subdomains) have data pass through cache twice. However, we can do better, as will be described in this and the next section.

Algorithm Cache-Vcycle(k, $\{A_i, u_i, f_i, r_i\}_{i=1}^{k}$)
1. Do $i = 1, 2, \cdots, k-1$
    1a.   Do $\ell = 1, 2, \cdots, m_i$
            1a1.   Determine $\Omega_{ij}^{(\ell)}$.
            1a2.   For each $\Omega_{ij}^{(\ell)}$,
                    1a2a.   If $\ell = 1$, then $u_i \big|_{\Omega_{ij}^{(\ell)}} \leftarrow 0$.
                    1a2b.   Do 1 iteration of approximate solve of $A_i u_i = f_i$.
                    1a2c.   If $\ell = m_i$, then compute as much of r_i as is possible.
            1a3.   If $\ell = m_i$, then complete the calculation of r_i and calculate $f_{i+1} = R_i r_i$.
2. Do $i = k, k-1, \cdots, 1$
    2a.   Do $\ell = 1, 2, \cdots, m_i$
            2a1.   Determine $\Omega_{ij}^{(\ell)}$.
            2a2.   For each $\Omega_{ij}^{(\ell)}$,
                    2a2a.   If $\ell = 1$ and $k = 1$, then $u_i \big|_{\Omega_{ij}^{(\ell)}} \leftarrow 0$.
                    2a2a.   If $\ell = 1$ and $k > 1$, then
$$u_i \big|_{\Omega_{ij}^{(\ell)}} \leftarrow u_i \big|_{\Omega_{ij}^{(\ell)}} + P_{i+1} u_{i+1} \big|_{\Omega_{ij}^{(\ell)}}.$$
                    2a2c.   Do 1 iteration of approximate solve of $A_i u_i = f_i$.

Figure 8: Cache aware V cycle definition

The computational subdomains $\Omega_{ij}^{(\ell)}$ must be further refined in order to use as large of subdomains as possible each iteration of the relaxation algorithm. The last iteration of the relaxation algorithm must be treated differently due to the projection or interpolation steps that must be done. As is noted in [11], only 50-60% of the cache is actually available for use by a given program. This is a side effect of multitasking operating systems.

Determining the sizes of the $\Omega_{ij}^{(\ell)}$'s per iteration ℓ can be done as a preprocessing step and is inexpensive. In order to efficiently do loop unrolling and/or tiling, we need a small amount of information about the computer that we are going to use. First, we need to know how big the usable part of the cache actually is. Knowing the size of usable cache and the number of points in a line we calculate

- $\Omega_{ij}^{(1)}$ for either a precorrection or a postcorrection step.

- $\Omega_{ij}^{(\ell)}$ for $\ell = 2, \cdots, m_i - 1$.

- $\Omega_{ij}^{(m_i)}$ for either a precorrection or a postcorrection step.

The second item we need to know is how many points to unroll in the loops

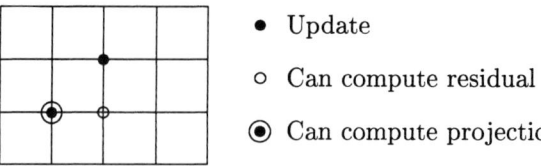

Figure 9: Five point discretization, point relaxation

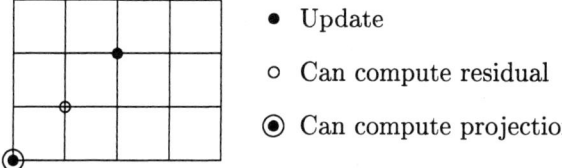

Figure 10: Nine point discretization, point relaxation

in the *Update* and *UpdateRedBlack* code.

Three steps occur during the pre-correction step: smoothing, residual calculation, and projection. Two sets of computational subdomains $\Omega_{ij}^{(\ell)}$ are necessary. One is for when just the relaxation method is used as a smoother and the other is when all three components are used at once.

There is a scheduling issue when implementing cache aware multigrid algorithms. Figures 9-11 graphically show when the residual can be computed based on the last update in the relaxation method. Where a projection can be centered is also shown based on which residuals have been computed. Figure 12 shows when interpolation to the next finer level can be scheduled. What is expressed is simply a "compute as soon as you can" principal.

For simplicity, point relaxation methods should probably be written assuming a nine point operator (see Figure 11). In order to write highly efficient code with no special cases, two extra "ghost" rows and columns are needed

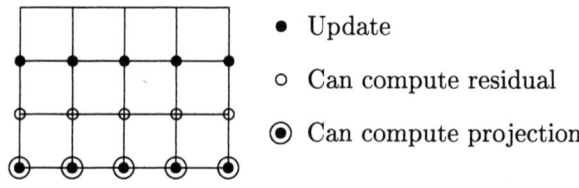

Figure 11: Five or nine point discretization, line relaxation

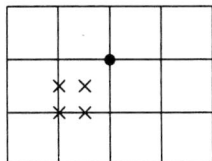

- Update

× Can compute interpolants on finer grid

Figure 12: Five or nine point discretization, interpolation

in the computational grids. Values there are set to zero and no relaxation updates occur (requiring a minor post processing). While for tiny grids this adds a significant amount of storage, we are only interested in problems that are much larger than would fit in the cache. Hence, the padding on the coarser grids only amounts to a trivial increase in the overall storage requirements (and is useful for parallel computing versions of the algorithms).

Two steps occur in the post-correction step: smoothing and interpolation. In order for this half of a correction scheme to be optimally cache aware, the last step must use a somewhat smaller $\Omega_{ij}^{(m_i)}$ than the rest of the iterations. This is because the interpolant on level $i + 1$ is added to much larger vector, which is typically four times the size of the vector on level i.

4   Numerical Results and Conclusions

In [5], [6], and [7] a collection of problems are solved on structured grids (2D and 3D) and unstructured grids (in 2D). Speedups range from 100% to 300% over using standard, well coded implementations.

Reducing the number of times data passes through cache to the absolute minimum eliminates one of the major areas where multigrid does not take full advantage of the hardware that it is implemented on. The algorithms in this paper show how to reach this minimum.

When using cache aware implementations, there are two issues that must be faced: is true portability wanted and how much performance is demanded? By true portability, all that must be determined for a given machine are two numbers: the number of elements in a cache line and the number of cache lines that can be guaranteed to be usable.

If ultra high performance is demanded, true portability becomes much harder. Loop unrolling, relaxation updates along diagonals, specialty BLAS for short vectors (in machine language), and other machine specific optimizations are necessary. One interesting aspect is that most RISC based processors are very similar. Hence, the non-machine language optimizations will work well on similar processors.

Using cache aware multigrid algorithms requires a different programming style than is traditional (see [4] and [8]). In order to make the codes viable, a rigid programming style must be maintained for all components of the code. This includes a uniform subscripting method for variables, an isolation of the minimal number of lines of code that is necessary for a give discretized problem to do major components (e.g., updating a point by the relaxation method), and a decision on how portable the code will be.

If only true portability is demanded, a general code can be constructed quite easily for 5-9 point operators A_i (see [4] for example codes). For the examples in §3, only four quite small pieces of code have to be written as separate "include" files or macro definitions. Only the interpolation file will typically have more than one target point to modify.

The ideas in this paper can be applied not just to natural and red-black ordered Gauss-Seidel, but to SOR variants, line relaxation methods, and ADI. Applying the ideas to typical parallel smoothers is also a straight forward extension.

Acknowledgments. This research was supported in part by the Deutsche Forschungsgemeinschaft (project Ru 422/7-1), the National Science Foundation (grants DMS-9707040 and ACR-9721388), NATO (grant CRG 971574), and the National Computational Science Alliance (grant OCE980001N and utilized the NCSA SGI/Cray Origin2000).

References

[1] Bank, R. E. and Douglas, C. C., Sharp estimates for multigrid rates of convergence with general smoothing and acceleration, *SIAM J. Numer. Anal.* **22** (1985), 617–633.

[2] Douglas, C. C., Multi–grid algorithms with applications to elliptic boundary–value problems, *SIAM J. Numer. Anal.* **21** (1984), 236–254.

[3] Douglas, C. C., Caching in with multigrid algorithms: problems in two dimensions. *Paral. Alg. Appl.* **9** (1996), 195–204.

[4] Douglas, C. C., Reusable cache memory object oriented multigrid algorithms. See http://www.ccs.uky.edu/~douglas under *preprints*, 1999.

[5] Douglas, C. C., Hu, J., and Iskandarani, M., Preprocessing costs of cache based multigrid, in *Proceeding of ENUMATH99: Third European Conference on Numerical Methods for Advanced Applications*, 8 pages, Singapore, World Scientific, 2000.

[6] Douglas, C. C., Hu, J., Karl, W., Kowarschik, M., Rüde, U., and Weiss, C., Fixed and adaptive cache aware algorithms for multigrid methods, in *European Multigrid VI*, Lecture Notes in Computational Science and Engineering, 7 pages, Springer, Berlin, 2000.

[7] Douglas, C. C., Hu, J., Kowarschik, M., Rüde, U., and Weiss, C., Cache optimization for structured and unstructured grid multigrid, *Electron. Trans. Numer. Anal.*, 9, 2000.

[8] Douglas, C. C., Hu, J., Rüde, U., and Bittencourt, M, Cache based multigrid on unstructured two dimensional grids. Notes on Numerical Fluid Mechanics, 11 pages, Vieweg, Braunschweig, 1999, Proceeding of the 14th GAMM-Seminar Kiel on 'Concepts of Numerical Software', January, 1998.

[9] Hellwagner, H., Weiß, C., Stals, L., and Rüde, U., Efficient implementation of multigrid on cache based architectures. Notes on Numerical Fluid Mechanics. Vieweg, Braunschweig, 1999, Proceeding of the 14th GAMM-Seminar Kiel on 'Concepts of Numerical Software, January, 1998.

[10] McCormick, S. F., Multigrid methods for variational problems: general theory for the V-cycle, *SIAM J. Numer. Anal.* **22** (1985), 634–643.

[11] Philbin, J., Edler, J., Anshus, O. J., Douglas, C. C., and Li, K., Thread scheduling for cache locality, in *Proceedings of the Seventh ACM Conference on Architectural Support for Programming Languages and Operating Systems*, pages 60–73, Cambridge, MA, 1996, ACM.

[12] Stals, L. and Rüde, U., Techniques for improving the data locality of iterative methods. Technical Report MRR 038–97, Australian National University, 1997.

A Locally Conservative Eulerian-Lagrangian Method for Flow in a Porous Medium of a Mixture of Two Components Having Different Densities

Jim Douglas, Jr.   Felipe Pereira   Li-Ming Yeh

Abstract

The object of this paper is to develop an efficient, conservative, Eulerian-Lagrangian numerical method for the differential system describing miscible displacement of one incompressible fluid by another of different density in a porous medium. The method will be a variant of the "Locally Conservative Eulerian-Lagrangian Method" that has been studied for immiscible displacement.

KEYWORDS: LCELM, MMOC, MMOCAA

1   Introduction

There have been many discussions of numerical methods for the simulation of miscible displacement in porous media over the past two decades; see, *e.g.*, [2, 10, 13, 14, 15, 17, 18, 27, 28, 29, 30, 34, 36, 37]. However, almost all of this work has been related to a model for miscible flow that is consistent with the conservation of momentum only if the two components in a binary mixture have equal densities [20]. The object here is to treat the more realistic and complex system [20] that properly describes the flow of a binary mixture when the components are allowed to have significantly different densities; in particular, the numerical approximation of this system will be addressed by an efficient and *locally conservative* Eulerian-Lagrangian method ($LCELM$) that takes into account the dominance of transport over diffusion in the physics. Our method is based on an $LCELM$ treatment of immiscible displacement [24]; a somewhat similar locally conservative method was analyzed by Arbogast and Wheeler [3] for linear problems; they and Chilikapati [2] had applied their method to the older miscible displacement model, though not without some difficulties. It is clear from the title of [3] that their derivation of their method was based on different concepts from those underlying our derivation.

The authors consider the $LCELM$ in [24] and in this paper to be a natural development in the family of modified method of characteristics ($MMOC$) procedures originally introduced by Douglas and Russell [25] and applied first by Russell [36, 37] and then by many others to (equal density) miscible displacement and to immiscible displacement [13, 14, 26, 27, 28, 29]. The $MMOC$ has several, by now well-known, advantages (including, in particular, computational efficiency) and one fundamental flaw, which in the miscible displacement problem is the failure to preserve as an algebraic identity the total masses of the two components, an important physical requirement. Recently, for the immiscible displacement problem, Douglas, Furtado, and Pereira [19] formulated a variant of the $MMOC$ called the *modified method of characteristics with adjusted advection* ($MMOCAA$) which does conserve the component masses globally and also preserves the conceptual and computational advantages of the $MMOC$; however, it does not necessarily conserve the component masses locally. There is a collection of finite element methods, called generically $ELLAM$, that can address conservation problems locally, though not all implementations of these methods do exhibit local conservation; see [2, 4, 8, 9, 31]. We believe that $ELLAM$ methods have not been applied to the system for miscible flow with differing densities. The cost of $ELLAM$ procedures appears to be significantly greater than for the $MMOCAA$ procedure or the $LCELM$ technique to be described in this paper.

As mentioned above, an $LCELM$ procedure has been described in detail by the authors in [24] for approximating the solution of immiscible displacement in porous media; the experimental evidence reported in [24] clearly indicates the superiority of the $LCELM$ over either the $MMOC$ or the $MMOCAA$. A convergence proof for a simpler application of the $LCELM$ concept has been obtained by Douglas and Huang [21]. There is a fundamental distinction between the various $MMOC$ procedures and $LCELM$ techniques. All $MMOC$ schemes relate to approximating transport-dominated parabolic problems beginning from nondivergence forms of the equations; $LCELM$ schemes are derived from the divergence forms of the equations. It is the use of the divergence form of a parabolic equation that allows relatively easy localization of desired conservation principles in a form amenable to the application of finite element or finite difference methods. Both $MMOC$ and $LCELM$ procedures are properly viewed as operator-splitting techniques, with the transport being separated from the diffusion in the parabolic equation. The two families of methods treat the diffusive part of a time step in like manner; the difference lies in the treatment of the transport. In $MMOC$ procedures, certain characteristics are traced back from the advanced time level to the previous level; normally, these characteristics are associated with node points for finite differences or quadrature points for finite elements. The trace-backs arising in $LCELM$ procedures lead to predecessor *sets*, not predecessor *points*. In the immiscible displacement problem, the integral curves that generate the predecessor sets are not the characteristics of the first-order transport operator coming from the nondivergence form of the saturation equation; for the miscible displacement problem the relevant integral curves do coincide with characteristics of the first-order transport operator, though they are usually associated with points that would not be employed in $MMOC$ schemes.

2   Unequal Density Miscible Displacement

A model [20] for unequal-density miscible displacement can be described as follows. The model assumes that the components are incompressible and that there is no change in volume resulting from the mixing of the components; i.e., the volume of the mixture is the sum of the partial volumes. Let ρ_α and ρ_β denote the densities of the αth and β^{th} pure components in the binary mixture and assume that $\rho_\alpha > \rho_\beta$. Set

$$\sigma = (\rho_\alpha - \rho_\beta)/\rho_\beta.$$

Let $c = c_\alpha = 1 - c_\beta$; then the density of the mixture is given by

$$\rho = \rho(c) = \rho_\beta(1 + \sigma c).$$

Denote the pressure by p and the Darcy velocity by u. The *pressure equation* is given by the equations

$$\nabla \cdot \left(u + \frac{\sigma}{1 + \sigma c} D(u) \nabla c \right) = q, \qquad x \in \Omega,$$

$$u = -\frac{1}{\mu(c)} k(x) \left(\nabla p - \rho(c) g \nabla d \right), \qquad x \in \Omega,$$

$$(2.1)$$

where $D(u)$ is the dispersion tensor given by

$$D(u) = \phi(x)\big(d_{mol}I + |u|(d_{long}E(u) + d_{trans}E^{\perp}(u))\big), \qquad (2.2)$$

$\phi(x)$ is the porosity of the medium, d_{mol} is the (small) molecular diffusion coefficient, d_{long} and d_{trans} are the longitudinal and transverse dispersion coefficients, $E(u)$ is the projection along the Darcy velocity, and $E^{\perp}(u) = I - E(u)$ is its orthogonal compliment. Also, q is the total external flow rate, with $q > 0$ denoting injection and $q < 0$ production; $k(x)$ is the permeability tensor; $\mu(c)$ is the viscosity, frequently quite strongly dependent on the concentration; g is the gravitational constant and, finally, $d(x)$ is the depth function in the reservoir Ω. For simplicity, a "no physical transport" boundary condition

$$u \cdot \nu = 0, \qquad x \in \partial\Omega, \qquad (2.3)$$

will be assumed; then, compatibility to incompressibility requires that

$$\int_{\Omega} q(x,t)\, dx = 0, \qquad \forall t.$$

The *concentration equation*, in divergence form, is given by

$$\phi\frac{\partial c}{\partial t} + \nabla \cdot \left(cu - \frac{1}{1+\sigma c}D(u)\nabla c\right) = \tilde{c}q, \qquad x \in \Omega, \qquad (2.4)$$

where $\tilde{c}$ is the concentration in the fluid in the external source;

$$\tilde{c}(x,t) = \begin{cases} \tilde{c}(x,t), & \text{a specified function, if } q > 0, \\ c(x,t), & \text{the resident concentration, if } q < 0. \end{cases} \qquad (2.5)$$

A "no diffusive transport" boundary condition will be assumed for the concentration, and an initial concentration must be specified:

$$(D\nabla c)\cdot \nu = 0, \qquad x \in \partial\Omega, \qquad (2.6)$$

and

$$c(x,0) = c_{init}(x), \qquad x \in \Omega. \qquad (2.7)$$

The model above reduces to the older model if $\sigma = 0$; *i.e.*, when the two components have equal densities.

Recall [20] that the Darcy flux u is a mass-averaged flux, not the volumetric flow rate. It is very convenient to introduce the volumetric flow rate; let

$$w = u + \frac{\sigma}{1+\sigma c}D(u)\nabla c. \qquad (2.8)$$

Note that it follows from the two boundary conditions (2.3) and (2.6) that

$$w \cdot \nu = 0, \qquad x \in \partial\Omega. \tag{2.9}$$

A small amount of manipulation shows that the concentration equation can be rewritten in the form

$$\nabla_{t,x} \cdot \begin{pmatrix} \phi c \\ cw \end{pmatrix} - \nabla_x \cdot (D\nabla_x c) = \tilde{c}q = \tilde{c}q^+ - cq^-. \tag{2.10}$$

It is also convenient to write (2.10) in mixed form by introducing a concentration flux; let

$$z = -D(u)\nabla_x c. \tag{2.11}$$

Then, the concentration equation takes the mixed form

$$\begin{aligned} \nabla_{t,x} \cdot \begin{pmatrix} \phi c \\ cw \end{pmatrix} + \nabla_x \cdot z &= \tilde{c}q^+ - cq^-, \qquad x \in \Omega, \\ z + D(u)\nabla_x c &= 0, \qquad\qquad\qquad x \in \Omega. \end{aligned} \tag{2.12}$$

If the pressure equation is written in terms of w and p and the concentration flux z is used to replace $D\nabla c$ in the equation, then the pressure equation can be seen to be given, also in mixed form, by

$$\begin{aligned} \mu(c)k(x)^{-1}w + \nabla_x p &= \rho(c)g\nabla_x d - \frac{\sigma\mu(c)}{1+\sigma c}k(x)^{-1}z, \quad x \in \Omega, \\ \nabla_x \cdot w &= q, \qquad\qquad\qquad\qquad\qquad\qquad x \in \Omega, \\ u &= w + \frac{\sigma}{1+\sigma c}z, \qquad\qquad\qquad\qquad x \in \Omega. \end{aligned} \tag{2.13}$$

The boundary conditions are expressed by the relations

$$z \cdot \nu = w \cdot \nu = 0, \qquad x \in \partial\Omega. \tag{2.14}$$

The introduction of the volumetric flux w will allow the system to be treated in a manner that closely resembles the way the older model has been approximated; in particular, we shall be able to use essentially the same operator splitting for the new system as for the old. If w had not been introduced, it is not clear that the pressure and concentration equations could be effectively separated as they are in the procedure described below.

3 The Basic Operator Splitting Algorithm

Assume that we shall approximate both $\{w, p\}$ and $\{z, c\}$ in the lowest index Raviart-Thomas mixed finite element space $\mathcal{RT} = \mathcal{V} \times \mathcal{W}$ [35] over the same partition $\{\Omega_j\}$. (There is no logical constraint that the partitions for the pressure and concentration be the same; in fact, we shall introduce a local refinement in the pressure partition in order to obtain a better approximation of a point source or sink. There can be other advantages to the use of different partitions for the two equations; see [16, 17, 18].) Let

$$\Delta t_p = i_1 \Delta t_c = i_1 i_2 \Delta t_{ct}, \tag{3.1}$$

where i_1 and i_2 are positive integers; in their numerical experiments the authors have frequently chosen i_1 to be one or two and i_2 to be about ten. Set

$$t^m = m\Delta t_p, \quad t_n = n\Delta t_c, \quad t_{n,\kappa} = t_n + \kappa\Delta t_{ct},$$

where normally $0 \leq \kappa \leq i_2$. The pressure equation will be approximated at time levels t^m; the concentration will be approximated through transport microsteps corresponding to times $t_{n,\kappa}$ and diffusive steps corresponding to times t_n. We shall indicate a function f evaluated at these times by $f^m = f(t^m)$, $f_n = f(t_n)$, and $f_{n,\kappa} = f(t_{n,\kappa})$; all functions will be assumed single-valued (*i.e.*, if $t^m = t_n$, then $f^m = f_n$).

Define an extrapolation operator as follows:

$$Ef(t) = \begin{cases} f^0, & 0 < t \leq t^1, \\ \dfrac{t - t^{m-1}}{t^m - t^{m-1}} f^m - \dfrac{t - t^m}{t^m - t^{m-1}} f^{m-1}, & t^m < t \leq t^{m+1}. \end{cases}$$

Then, the algorithm can be described as follows.

1° Given $C^0 = c_{init}(x)$, find $\{W^0, P^0, U^0\}$, where $\{W^0, P^0\} \in \mathcal{RT}$ and

$$U^0 = W^0 + \frac{\sigma}{1 + \sigma C^0} Z^0.$$

If c_{init} vanishes, so that the simulation begins when the injection of the solvent commences, then $Z^0 = 0$. If c_{init} is not identically zero, then the initial pressure problem is nonlinear, since $Z^0 = -D(U^0)\nabla_x C^0$ is unknown. A suggestion for working the nonlinear elliptic problem is to apply the method of continuity to it in the following form. First, set $\sigma = 0$, so that the elliptic problem is linear (and corresponds to the old miscible model). Solve for a first approximation $U^{0,0}$ to U^0:

$$\left(\mu(C^0)k^{-1}W^{0,0}, \gamma\right) - (\nabla_x \cdot \gamma, P^{0,0}) = \left(\rho(C^0)g\nabla_x d, \gamma\right), \quad \gamma \in \mathcal{V},$$
$$(\nabla_x \cdot W^{0,0}, \eta) = (q, \eta), \qquad\qquad\qquad\qquad\qquad\qquad \eta \in \mathcal{W},$$
$$U^{0,0} = W^{0,0}.$$

$$(3.2)$$

Now, let $\sigma^m = m\delta\sigma$, where $\sigma^M = \sigma$. Then, for $m = 1, \ldots, M$, set $Z^{0,m-1} = -D(U^{0,m-1})\nabla_x C^0$ and solve the linear problem

$$\left(\mu(C^0)k^{-1}W^{0,m}, \gamma\right) - (\nabla_x \cdot \gamma, P^{0,m})$$

$$= \left(\rho(C^0)g\nabla_x d, \gamma\right) - \left(\frac{\sigma\mu(C^0)}{1+\sigma C^0}k^{-1}Z^{0,m-1}, \gamma\right), \quad \gamma \in \mathcal{V}, \quad (3.3)$$
$$(\nabla_x \cdot W^{0,m}, \eta) = (q, \eta), \qquad\qquad\qquad\qquad\qquad\qquad \eta \in \mathcal{W},$$

for $\{W^{0,m}, P^{0,m}\}$ and set

$$U^{0,m} = W^{0,m} + \frac{\sigma}{1+\sigma C^0}Z^{0,m-1};$$

$U^{0,M}$ should give a very good approximation of U^0, which should then be obtainable in a small number of Newton iterations starting from $Z^{0,M} = -D(U^{0,M})\nabla_x C^0$.

If C^0 is constant, then $\{W^0, P^0\} \in \mathcal{RT}$ satisfies (3.2).

2° Assume $\{U^m, W^m, P^m, Z^m, C^m\}$ known for some $m \geq 0$. Then, carry out a full pressure time step employing the operator-splitting procedure outlined in 3° − 5° below.

3° Let $t^m \leq t_n < t^{m+1}$. Split the concentration time step (t_n, t_{n+1}) into calculations for transport and diffusion. Associate transport with the system

$$\nabla_{t,x} \cdot \begin{pmatrix} \phi c \\ cw \end{pmatrix} = \tilde{c}q^+ - cq^-, \qquad x \in \Omega, \qquad (3.4)$$
$$w \cdot \nu = 0, \qquad\qquad\qquad\qquad\qquad x \in \partial\Omega,$$

and apply the *LCELM* transport microstepping procedure (to be detailed below in §5) over the microsteps $(t_{n,\kappa-1}, t_{n,\kappa})$, $\kappa = 1, \ldots, i_2$, so that $t_{n,i_2} = t_{n+1}$, to obtain the transport-approximation $\overline{C}_{n+1} \in \mathcal{W}$.

4° Given $\overline{C}_{n+1}$, complete the concentration time step by approximating the solution of the diffusive system

$$\phi\frac{C_{n+1} - \overline{C}_{n+1}}{\Delta t_c} - \nabla_x \cdot (D((EU)_{n+1})\nabla_x C_{n+1}) = 0, \quad x \in \Omega, \qquad (3.5)$$
$$(D((EU)_{n+1})\nabla_x C_{n+1}) \cdot \nu = 0, \qquad\qquad\qquad\qquad x \in \partial\Omega;$$

i.e., find $\{Z_{n+1}, C_{n+1}\} \in \mathcal{RT}$ such that

$$
\left(D((EU)_{n+1})^{-1} Z_{n+1}, \gamma\right) - (\nabla_x \cdot \gamma, C_{n+1}) = 0, \quad \gamma \in \mathcal{V},
$$

$$
\left(\phi \frac{C_{n+1} - \overline{C}_{n+1}}{\Delta t_c}, \eta\right) + (\nabla_x \cdot Z_{n+1}, \eta) = 0, \qquad \eta \in \mathcal{W}. \tag{3.6}
$$

If $t_{n+1} = t^{m+1}$, set $\{Z^{m+1}, C^{m+1}\} = \{Z_{n+1}, C_{n+1}\}$.

5° Given $\{Z^{m+1}, C^{m+1}\}$, find $\{W^{m+1}, P^{m+1}\} \in \mathcal{RT}$ such that

$$
\left(\mu(C^{m+1}) k^{-1} W^{m+1}, \gamma\right) - (\nabla_x \cdot \gamma, P^{m+1})
$$

$$
= \left(\rho(C^{m+1}) g \nabla_x d, \gamma\right) - \left(\frac{\sigma \mu(C^{m+1})}{1 + \sigma C^{m+1}} k^{-1} Z^{m+1}, \gamma\right), \quad \gamma \in \mathcal{V},
$$

$$
(\nabla_x \cdot W^{m+1}, \eta) = (q, \eta), \qquad\qquad\qquad \eta \in \mathcal{W}, \tag{3.7}
$$

and set

$$
U^{m+1} = W^{m+1} + \frac{\sigma}{1 + \sigma C^{m+1}} Z^{m+1}. \tag{3.8}
$$

6° Repeat pressure time steps to the desired final time.

The partial differential system is nonlinear, since the concentration enters the pressure equation nonlinearly through the viscosity and the term including the concentration flux, which in turn involves the dispersion tensor $D(U)$. The dispersion tensor's appearance in the concentration equation is another nonlinearity in the system. However, the operator-splitting procedure outlined above produces a numerical method that is linear in every part, with the exception that a nontrivial initial concentration can induce a single nonlinear elliptic problem for the initial pressure distribution.

4   Discretization of the Pressure Equation

The $LCELM$ procedure to be employed herein is based on that introduced by the authors [24] to treat immiscible displacement in porous media. It was assumed above that the solutions of the pressure and concentration equations are being approximated in Raviart-Thomas spaces $\mathcal{RT}$ of lowest index over the same rectangular partition (in two or three spatial dimensions). Consequently, the pressure and the concentration variables are approximated by piecewise constant functions, while each component of the fluxes W and Z is approximated on each element by a function that is linear in the corresponding component of the space variable and is constant in the other components of the space variable. In the model for miscible displacement being considered in this paper, the Darcy velocity u plays a secondary rôle to the flux w, in that w is the natural choice for the flux in the pressure equation and is the velocity variable appearing in the transport part of the concentration equation. If the external flow is restricted to a few elements (as is usual in practice), then in almost all of the domain w will be approximated in a discrete divergence-free manner by the mixed method flux variable W. In the $LCELM$ procedure that will be used to approximate transport in the concentration equation, it will be necessary to evaluate W (or, more precisely, EW) at the vertices of the elements; however, W, as produced by (3.7), is discontinuous at vertices, and some averaging or interpolation procedure must be applied at such points. It was observed in [24] that it is strongly advisable to devise an evaluation at vertices that preserves the divergence-free nature of the approximation in the neighborhood of these points. The method used in [24] and that will be repeated here is to define a bilinear or trilinear interpolation $\widetilde{W}$ of element-center values of W. It can be seen by a short calculation that, if $q = 0$ on all of the elements having a point $\mathbf{x}$ as vertex, $\widetilde{W}$ is divergence free on boxes centered at $\mathbf{x}$ and contained in these elements. The extrapolation of $\widetilde{W}$ remains divergence free in this sense and will be employed in the transport calculation.

An alternative to the interpolation would be to offset the pressure partition from that for the concentration. If the element centers for the concentration partition are taken as the vertices of the (rectangular) elements in the pressure partition, then the flux W would be continuous at the vertices of the concentration partition and W would be discrete divergence free about vertices in pressure elements where $q = 0$. On the other hand, $D(U)$ would be discontinuous in concentration elements, which would complicate the diffusive part of a concentration time step.

5   Transport Microsteps by an $LCELM$ Technique

Let us derive the local conservation law that will provide the basis for our $LCELM$ approach to the miscible problem. Recall the transport equation

$$\nabla_{t,x} \cdot \begin{pmatrix} \phi c \\ cw \end{pmatrix} = \tilde{c}q^+ - cq^-, \quad x \in \Omega, \tag{5.1}$$

$$w \cdot \nu = 0, \qquad\qquad\qquad x \in \partial\Omega,$$

Following [24], we consider the space-time slice $Q = \Omega \times [t_{n,\kappa}, t_{n,\kappa+1}]$. Let $\mathcal{K}$ be a reasonably shaped, simply-connected subset of Ω, and define a subset $\mathcal{D} = \mathcal{D}_{n,\kappa}(\mathcal{K})$ of Q as follows. For each $x \in \partial\mathcal{K}$, construct the solution $y(x; t)$ of the final value problem

$$\frac{dy}{dt} = \frac{cw}{\phi c} = \frac{w}{\phi}, \quad t_{n,\kappa+1} > t \geq t_{n,\kappa},$$
$$y(x; t_{n,\kappa+1}) = x, \tag{5.2}$$

and set

$$\overline{x}_{n,\kappa}(x) = y(x; t_{n,\kappa}). \tag{5.3}$$

Then, let $\overline{\mathcal{K}} = \overline{\mathcal{K}}_{n,\kappa}$ be the interior of the set $\{\overline{x}_{n,\kappa}(x) : x \in \partial\mathcal{K}\}$, and let $\mathcal{D}$ be the tube determined by $\mathcal{K}$, $\overline{\mathcal{K}}$, and the integral curves (5.2). (For Δt_{ct} sufficiently small, the map $x \to \overline{x}_{n,\kappa}$ is one-to-one, so that this construction can be carried out.) Now, denote the outward normal to $\partial\mathcal{D}$ by $\vartheta(x, t)$ and note that it is orthogonal to the vector $(\phi c, cw)^t$ on the lateral surface of $\mathcal{D}$. Then, integrate (5.1) over $\mathcal{D}$:

$$\int_{\mathcal{D}} \nabla_{t,x} \cdot \begin{pmatrix} \phi c \\ cw \end{pmatrix} dx\, dt = \int_{\partial\mathcal{D}} \begin{pmatrix} \phi c \\ cw \end{pmatrix} \cdot \vartheta\, dA$$
$$= \int_{\mathcal{K}} \phi c_{n,\kappa+1}\, dx - \int_{\overline{\mathcal{K}}} \phi c_{n,\kappa}\, dx \tag{5.4}$$
$$= \int_{\mathcal{D}} (\tilde{c}q^+ - cq^-)\, dx\, dt.$$

Thus, mass is conserved locally in the transport step if

$$\int_{\mathcal{K}} \phi c_{n,\kappa+1}\, dx = \int_{\overline{\mathcal{K}}} \phi c_{n,\kappa}\, dx + \int_{\mathcal{D}} (\tilde{c}q^+ - cq^-)\, dx\, dt. \tag{5.5}$$

The no-flow boundary condition is handled in a natural way in (5.4), since the integral curves (5.2) do not exit Ω in this case. In fact, if $x \in \partial\Omega$, then the integral curve remains in $\partial\Omega$ and $\mathcal{D}$ has a portion of its lateral surface contained in $\partial\Omega \times [t_{n,\kappa}, t_{n,\kappa+1}]$. Hence, no special cases arise for subsets $\mathcal{K}$ close to the boundary for these boundary conditions.

Before proceeding to the discretization of (5.5), let us discuss the relations between (5.5) and the corresponding conservation law for immiscible displacement and that employed in [2] for the older miscible displacement model. It is easy to see that the nondivergence form of the concentration equation is given by

$$\phi\frac{\partial c}{\partial t} + w \cdot \nabla_x c - \nabla_x \cdot (D(u)\nabla_x c) = (\tilde{c} - c)q^+. \tag{5.6}$$

Thus, the integral curves determined by (5.2) coincide with the characteristics of the first order part of (5.6), whereas in the immiscible problem treated in [24] they do not. In the miscible displacement model treated by Arbogast, Chilipataki, and Wheeler [2] and in the paper by Arbogast and Wheeler [3] on linear convection-diffusion equations, our volumetric flow rate w is replaced by the Darcy velocity u, which is the volumetric flow rate of the incompressible mixture when $\sigma = 0$. Thus, the tubes $\mathcal{D}$ represent, both for them and for us, actual flow tubes, with the consequence that the volume of $\overline{\mathcal{K}}$ must equal that of $\mathcal{K}$ in the absence of external flow. This seems to have introduced some difficulties in their calculations, which we believe (on the basis of our experience with the immiscible displacement problem [24] and that reported in [1] for the older miscible model) to be related to the way they evaluated u at the vertices of their elements. The constraint itself does not arise in the immiscible problem.

Now, let us turn to the discretization of the conservation law (5.5). It was seen in [24] that an acceptable choice for choosing the sets $\{\mathbf{K}\}$ for conservation is to use exactly the concentration partition, and we shall make this choice here. So, let $\mathbf{K} = \Omega_j$ for some j. We shall define $\overline{\mathbf{K}}$ as the box obtained as follows. First, approximate the integral curves (5.2) at the vertices of $\mathbf{K}$. Since the microstep Δt_{ct} is usually small (*i.e.*, $i_2 = 10$ or so) with respect to Δt_c, we shall assume that it suffices to follow the tangent to the integral curve through each vertex of $\mathbf{K}$ back to the time level $t_{n,\kappa}$, using $(E\widetilde{W})_{n,\kappa+1}$ as defined above. Thus, if $\mathbf{x}$ is a vertex of $\mathbf{K}$, then its predecessor point $\overline{\mathbf{x}}_{n,\kappa}(\mathbf{x})$ is given by

$$\overline{\mathbf{x}}_{n,\kappa}(\mathbf{x}) = \mathbf{x} - (E\widetilde{W})_{n,\kappa+1}(\mathbf{x})\Delta t_{ct}/\phi(x). \tag{5.7}$$

If Ω is two-dimensional, then for sufficiently small Δt_{ct}, the four points $\overline{\mathbf{x}}_{n,\kappa}(\mathbf{K})$ form the vertices of a quadrilateral, $\overline{\mathbf{K}} = \overline{\mathbf{K}}_{n,\kappa}$, which intersects at most nine elements. (In theory, we could take steps long enough that the predecessor set would intersect elements other than the immediate neighbors of Ω_j, but it is our intention to treat very inhomogeneous media, for which the longer microsteps would lead to rather inaccurate simulations, at least if $\overline{\mathbf{K}}$ is defined using only the vertices of $\mathbf{K}$.) If Ω lies in three-space, the eight points $\overline{\mathbf{x}}_{n,\kappa}$ determine a box with ruled, almost quadrilateral surfaces for faces. The common face of two adjacent elements Ω_i and Ω_j is mapped to a common face between the corresponding predecessor sets. Denote the tube generated by $\mathbf{K}$ and $\overline{\mathbf{K}}$ by $\mathbf{D} = \mathbf{D}_{n,\kappa}(\mathbf{K})$.

Next, we must discretize the conservation relation (5.5) as nearly exactly as we can. First, given the uncertainties of real reservoir data, it suffices to assume that the porosity ϕ is constant on each element of the partition $\{\Omega_j\}$. Then, the integral over $\mathbf{K}$ can be computed exactly. If Ω is two-dimensional, an algorithm for computing the integral over $\overline{\mathbf{K}}$ was given in §10 of [24]. The argument for approximating the integral over the tube $\mathbf{D}$ by a trapezoidal rule in time given in §7 of [24] is independent of dimension. The remaining question is how to approximate almost exactly the integral over $\overline{\mathbf{K}}$ in the three-dimensional case. Perhaps the simplest suggestion is to compute the coordinates of the image of the center of each face of $\mathbf{K}$ under the map specified above and to approximate the image of a face of $\mathbf{K}$ by the four triangles determined by the images of the vertices and center of the face. Then, an algorithm generalizing that of §10 in [24] in the two-dimensional case can be written down and applied. Now, assuming that each integral in (5.5) can be evaluated exactly or to high precision, let us replace (5.5) by the approximation

$$\int_{\mathbf{K}} \phi \overline{C}_{n,\kappa+1}\, dx = \int_{\overline{\mathbf{K}}} \phi \overline{C}_{n,\kappa}\, dx + \frac{\Delta t_{ct}}{2}\int_{\mathbf{K}} (\widetilde{C}_{n,\kappa+1}q^+ - \overline{C}_{n,\kappa+1}q^-)\, dx$$
$$+ \frac{\Delta t_{ct}}{2}\int_{\overline{\mathbf{K}}_{n,\kappa}} (\widetilde{C}_{n,\kappa}q^+ - \overline{C}_{n,\kappa}q^-)\, dx, \quad \kappa = 0, \ldots, i_2 - 1,$$

where

$$\overline{C}_{n,0} = C_n, \quad x \in \Omega.$$

Let $|\mathbf{K}|$ denote the volume of $\mathbf{K}$. Then, $\overline{C}_{n,\kappa+1} = \overline{C}_{j,n,\kappa+1} = \overline{C}_{n,\kappa+1}(\Omega_j)$ satisfies the local, linear equation

$$\left(\phi|\mathbf{K}| + \frac{\Delta t_{ct}}{2}\int_{\mathbf{K}} q^-\, dx\right)\overline{C}_{n,\kappa+1} = \int_{\overline{\mathbf{K}}_{n,\kappa}} \phi \overline{C}_{n,\kappa}\, dx + \frac{\Delta t_{ct}}{2}\int_{\mathbf{K}} \widetilde{C}_{n,\kappa+1}q^+\, dx$$
$$+ \frac{\Delta t_{ct}}{2}\int_{\overline{\mathbf{K}}_{n,\kappa}} \left(\widetilde{C}_{n,\kappa}q^+ - \overline{C}_{n,\kappa}q^-\right)\, dx, \quad \kappa = 0, \ldots, i_2 - 1. \tag{5.8}$$

The transport part of the concentration time step is completed by setting

$$\overline{C}_{n+1} = \overline{C}_{n,i_2}, \quad x \in \Omega. \tag{5.9}$$

The diffusive part of the concentration time step has been adequately defined by equations (3.6) above. Thus, the complete concentration time step has been specified. And, since the pressure time step has been determined by (3.7) and (3.8), the entire computational algorithm has been specified.

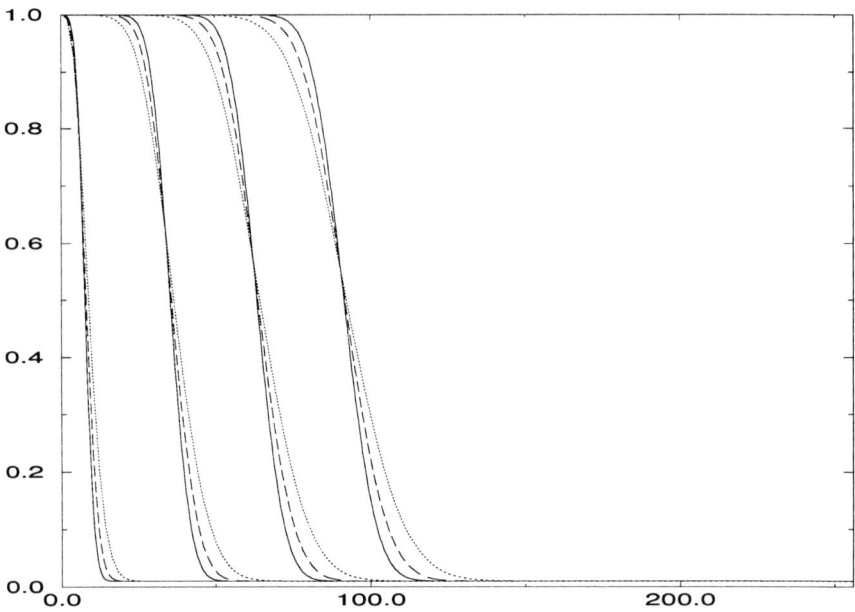

Figure 1: Comparison of $LCELM$ simulations of a homogeneous reservoir using 128, 256, and 512 elements at 50, 250, 450, 650 days. Refinement produces a sharper front.

6   Numerical Experiments

Viscosity	$\mu_\alpha = .5$ cP	$\mu_\beta = 10$ cP
Density	$\rho_\alpha = 1 \ g/\,cm^3$	$\rho_\beta = .7 \ g/\,cm^3$
Porosity	$\phi = .2$	
Absolute permeability	$k = 2$ mdarcy	

Viscosity function
$$\mu(c) = \mu_\beta \left(1 + c((\frac{\mu_\beta}{\mu_\alpha})^{0.25} - 1)\right)^{-4}$$

Here, we present numerical results to show convergence of the $LCELM$ method. Assume the reservoir to be $256\,m \times 256\,m$. Inject fluid (with $\tilde{c}^n = 1$) uniformly along the left edge of the reservoir and let the (total) production rate be uniform along the right edge; no flow is allowed along the top and bottom edges of the reservoir as they appear in the graphics. Gravity will be neglected. The injection rate is taken to be one pore–volume every five years, and data below are held fixed for the computational results discussed here.

Figure 1 shows a mesh refinement study for a homogeneous reservoir (corresponding to a one-dimensional problem). Three different discretizations, with $h = h_x = 2\,m$, $1\,m$, and $.5\,m$, were used for the $LCELM$ simulation. The concentration c obtained for the three grid sizes are almost the same, indicating convergence for the concentration. Figure 2 is a mesh refinement study for a heterogeneous reservoir at 200 and 300 days; the permeability for this test was defined on a 64×64 grid. In this case, three different discretizations (64×64, 128×128, 256×256 grids) were again used. It is observed that flow patterns from the three different grids are almost the same; if the mesh size is smaller, the concentration front is sharper.

References

[1] Almeida, C., Douglas, J., Jr., and Pereira, F., From nonconservative to locally conservative Eulerian-Lagrangian methods for miscible displacement in heterogeneous formations, to appear.

[2] Arbogast, T., Chilikapati, A., and Wheeler, M. F., A characteristics-mixed finite element for contaminant transport and miscible displacement, "Computational Methods in Water Resources IX, vol. 1: Numerical Methods in Water Resources," T. F. Russell *et al.*, eds., Elsevier Applied Science, London New York, 1992, 77–84.

[3] Arbogast, T. and Wheeler, M. F., A characteristics-mixed finite element for advection-dominated transport problems, *SIAM J. Numer. Anal.* **32** (1995), 404–424.

[4] Binning, P. and Celia, M. A., Two-dimensional Eulerian Lagrangian localised adjoint method for the solution of the contaminant transport equation in the saturated and unsaturated zones, "Proceedings, Tenth International Conference on Computational Methods in Water Resources," A. Peters *et al.*, eds., Kluwer, Dordrecht, 1994, vol. 1, 165–172.

[5] Brezzi, F., Douglas, J., Jr., and Marini, L. D., Two families of mixed finite elements for second order elliptic problems, *Numer. Math.* **47** (1985), 217–235.

[6] Brezzi, F., Douglas, J., Jr., Durán, R., and Fortin, M., Mixed finite elements for second order elliptic problems in three variables, *Numer. Math.* **51** (1987), 237–250.

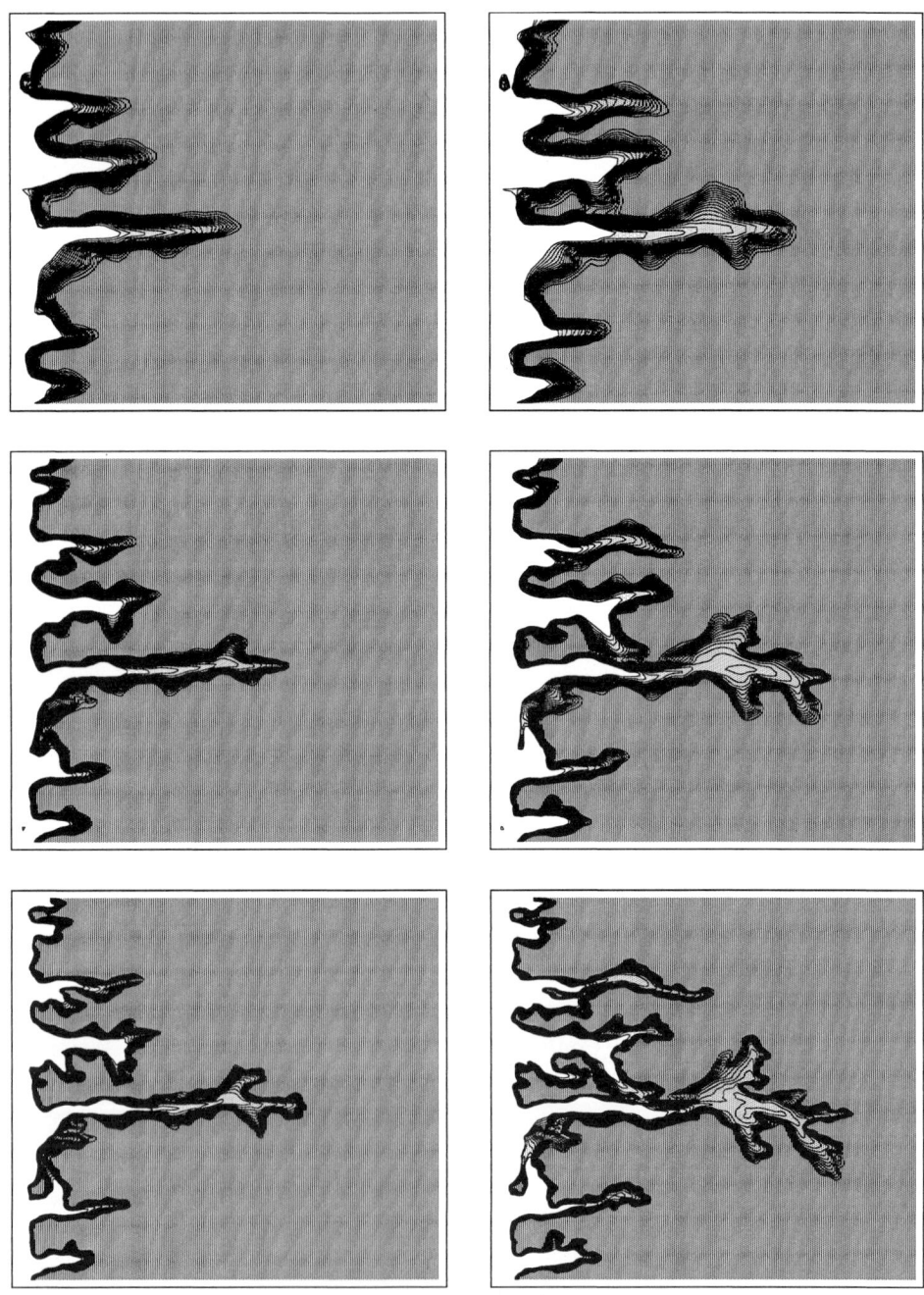

Figure 2: Comparison of the concentration for discretizations of 64×64 (top), 128×128 (middle), 256×256 (bottom) at 200 (left) and 300 (right) days.

[7] Brezzi, F., Douglas, J., Jr., Fortin, M., and Marini, L. D., Efficient rectangular mixed finite elements in two and three space variables, *R.A.I.R.O. Modélisation Mathématique et Analyse Numérique* **21** (1987), 581–604.

[8] Celia, M. A., Eulerian-Lagrangian localized adjoint methods for contaminant transport simulations, "Proceedings, Tenth International Conference on Computational Methods in Water Resources," A. Peters *et al.*, eds., Kluwer, Dordrecht, 1994, vol. 1, 207–216.

[9] Celia, M. A., Russell, T. F., Herrera, I., and Ewing, R. E., An Eulerian-Lagrangian local adjoint method for the advection-diffusion equation, *Adv. Water Resour.* **13** (1990), 187–206.

[10] Chavent, G. and Jaffré, J., *Mathematical Models and Finite Elements for Reservoir Simulation*, North-Holland, Amsterdam, 1986.

[11] Chen, Zhangxin and Douglas, J., Jr., Prismatic mixed finite elements for second order elliptic problems, *Calcolo* **26** (1989), 135–148.

[12] Ciarlet, P. G., *The Finite Element Method for Elliptic Equations*, North-Holland, Amsterdam, 1978.

[13] Douglas, J., Jr., Simulation of miscible displacement in porous media by a modified method of characteristics procedure, in "Numerical Analysis", Lecture Notes in Mathematics, volume 912, Springer-Verlag, Berlin, 1982.

[14] Douglas, J., Jr., Finite difference methods for two-phase incompressible flow in porous media, *SIAM J. Numer. Anal.* **20** (1983), 681–696.

[15] Douglas, J., Jr., Numerical methods for the flow of miscible fluids in porous media, in "Numerical Methods in Coupled Systems", pages 405–439, John Wiley and Sons, Ltd., London, R. W. Lewis, P. Bettess, and E. Hinton, eds., 1984.

[16] Douglas, J., Jr., Superconvergence in the pressure in the simulation of miscible displacement, *SIAM J. Numer. Anal.* **22** (1985), 962–969.

[17] Douglas, J., Jr., Ewing, R. E., and Wheeler, M. F., The approximation of the pressure by a mixed method in the simulation of miscible displacement, *R.A.I.R.O., Anal. Numér.* **17** (1983), 17–33.

[18] Douglas, J., Jr., Ewing, R. E., and Wheeler, M. F., A time-discretization procedure for a mixed finite element approximation of miscible displacement in porous media, *R.A.I.R.O., Anal. Numér.* **17** (1983), 249–265.

[19] Douglas, J., Jr., Furtado, F., and Pereira, F., On the numerical simulation of waterflooding of heterogeneous petroleum reservoirs, *Computational Geosciences* **1** (1997), 155–190.

[20] Douglas, J., Jr., Hensley, J. L., Wei, Y., Yeh, L., Jaffré, J., Paes Leme, P. J., Ramakrishnam, T. S., and Wilkinson, D. J., A derivation for Darcy's law for miscible fluids and a revised model for miscible displacement in porous media, in "Mathematical Modeling in Water Resources", T. F. Russell, R. E. Ewing, C. A. Brebbia, W. G. Gray, and G. F. Pinder, eds., vol. 2, 165–178, Computational Mechanics Publications, Elsevier Applied Science, Southhampton, Boston, 1992.

[21] Douglas, J., Jr. and Huang, C.-S., A convergence proof for an application of a locally conservative Eulerian-Lagrangian method, to appear.

[22] Douglas, J., Jr., Huang, C.-S., and Pereira, F., The modified method of characteristics with adjusted advection, to appear in Numerische Mathematik; currently available as Technical Report #298, Center for Applied Mathematics, Purdue University.

[23] Douglas, J., Jr., Huang, C.-S., and Pereira, F., The modified method of characteristics with adjusted advection for an immiscible displacement problem, in "Advances in Computational Mathematics", Lecture Notes in Pure and Applied Mathematics 202, Marcel Dekker, Inc., New York-Basel-Hong Kong, 1999, 53–73, Z. Chen, Y. Li, C. A. Micchelli, Y. Xu (eds.).

[24] Douglas, J., Jr., Pereira, F., and Yeh, L., A locally conservative Eulerian-Lagrangian numerical method and its application to nonlinear transport in porous media, to appear.

[25] Douglas, J., Jr. and Russell, T. F., Numerical methods for convection-dominated diffusion problems based on combining the method of characteristics with finite element or finite difference procedures, *SIAM J. Numer. Anal.* **19** (1982), 871–885.

[26] Douglas, J., Jr., and Yuan, Y., Numerical simulation of immiscible flow in porous media based on combining the method of characteristics with mixed finite element procedures, in "Numerical Simulation in Oil Recovery," The IMA Volumes in Mathematics and its Applications, vol. 11, 119–131, Springer-Verlag, Berlin and New York, 1988, M. F. Wheeler, ed.

[27] Ewing, R. E. and Russell, T. F., Efficient time-stepping methods for miscible displacement problems in porous media, *SIAM J. Numer. Anal.* **19** (1982), 1–67.

[28] Ewing, R. E., Russell, T. F., and Wheeler, M. F., Simulation of miscible displacement using mixed methods and a modified method of character-

istics, in "Proceedings, Seventh SPE Symposium on Reservoir Simulation," Paper SPE 12241, 71–81, Society of Petroleum Engineers, Dallas, Texas, 1983.

[29] Ewing, R. E., Russell, T. F., and Wheeler, M. F., Convergence analysis of an approximation of miscible displacement in porous media by mixed finite elements and a modified method of characteristics, *Comp. Meth. Appl. Mech. Eng.* **47** (1984), 73–92.

[30] Ewing, R. E. and Wheeler, M. F., Galerkin methods for miscible displacement problems in porous media, *SIAM J. Numer. Anal.* **17** (1980), 351–365.

[31] Healy, R. W. and Russell, T. F., A finite-volume Eulerian-Lagrangian localized adjoint method for solving the advection-diffusion equation, *Water Resour. Res.* **29** (1993), 2399–2413.

[32] Nedelec, J. C., Mixed finite elements in $\mathbf{R}^3$, *Numer. Math.* **35** (1980), 315–341.

[33] Peaceman, D. W., Improved treatment of dispersion in numerical calculation of multidimensional miscible displacement, *Soc. Petroleum Engr. J.* **6** (1966), 213–216.

[34] Peaceman, D. W., *Fundamentals of Numerical Reservoir Simulation*, Elsevier, New York, 1977.

[35] Raviart, P. A. and Thomas, J. M., A mixed finite element method for second order elliptic problems, in "Mathematical Aspects of the Finite Element Method", Lecture Notes in Mathematics, volume 606, 292–315, Springer-Verlag, Berlin, New York, 1977, I. Galligani and E. Magenes, eds.

[36] Russell, T. F., An incompletely iterated characteristic finite element method for a miscible displacement problem, Ph.D. thesis, University of Chicago, Chicago, 1980.

[37] Russell, T. F., Time stepping along characteristics with incomplete iteration for a Galerkin approximation of miscible displacement in porous media, *SIAM J. Numer. Anal.* **22** (1985), 970–1013.

[38] Wheeler, M. F., A priori L^2-error estimates for Galerkin approximations to parabolic partial differential equations, *SIAM J. Numer. Anal.* **10** (1973), 723–759.

Validation of Non-Darcy Well Models Using Direct Numerical Simulation

VLADIMIR A. GARANZHA   VLADIMIR N. KONSHIN
STEPHEN L. LYONS   DIMITRIOS V. PAPAVASSILIOU
GUAN QIN

Abstract

We describe discrete well models for 2-D non-Darcy fluid flow in anisotropic porous media. Attention is mostly paid to the well models and simplified calibration procedures for the control volume mixed finite element methods, including the case of highly distorted grids.

KEYWORDS: well models, non-Darcy flows, anisotropy, distorted grids

1   Introduction

Flow around high production rate gas wells deviates from Darcy's law. This phenomenon has been successfully modeled by the two-term Forchheimer law [7].

In reservoir simulation, the discrete well model is a relation between the production/injection rate of the well, the well-block pressure and the bottom-hole pressure. This relation is specific to the basic approximation scheme used for discretizing the governing equations. Such well models are well understood in the case of Darcy flow and are mostly based on various generalizations of the effective radius concept [1, 10]. In the case of Forchheimer flow in isotropic media, we find the effective radius as a function of the dimensionless Forchheimer number using the invariant behavior of discrete solutions near the well blocks. Such invariant properties are analyzed in [6] and are assessed numerically by solving a set of auxiliary problems which reproduce the known analytical solutions around a single isolated well in an infinite domain. Our numerical experiments show that this calibration procedure is very accurate and can also be applied on non-uniform and highly distorted grids.

In the general case of anisotropic media, there is no consensus on a specific formulation of Forchheimer's law that is backed by experiments or from

first principles. In this work, we derive well models only for the simplest formulation [9], which provides reasonably good fit to pore network simulation
results such as those obtained by Thauvin and Mohanty [14].

In this case direct numerical simulation becomes a critical tool for estimating the validity of a specific well model. Such verification requires
much higher accuracy compared to conventional reservoir simulation techniques and imposes strict requirements on the quality of the approximation
scheme. To this end we have developed a modification of the control volume mixed finite element (CVMFE) scheme [4] based on quadrilateral grids
in 2-D and on hexahedral grids in 3-D. This scheme is conservative, nearly
optimal among second order schemes, and naturally incorporates harmonic
averaging of reservoir properties. Moreover it allows us to obtain reliable
results on highly distorted grids and even on grids that are "degenerate" in
the conventional finite element sense (for example on grids with non-convex
cells in the plane.)

Numerical experiments in the 2-D case show that very fine grids are necessary in order to obtain grid independent results using direct simulation.
Such grids may be impractical in the 3-D case, especially in the case of devi-
ated wells. Hence there is a need for high order methods suitable for accurate
resolution of the flow in the near-well region using coarse grids.

2   Problem Formulation
   and Governing Equations

The governing equations that describe steady-state, single-component, single
phase, isothermal flow in porous media are

$$\nabla \cdot (\rho \mathbf{u}) = f, \quad \mu K^{-1} \mathbf{u} + \rho \beta |\mathbf{u}| \mathbf{u} + \nabla p = 0, \tag{2.1}$$

where ρ denotes the fluid density, $\mathbf{u}$ the velocity vector, μ the dynamic viscosity and p the pressure. The porous medium is characterized by the permeability tensor K, the porosity ϕ and the Forchheimer coefficient β, which
can be a tensor in some formulations. The right hand side f is associated
with the presence of wells-localized mass inflows or outflows in the reservoir.

Different Forchheimer law formulations are available for the anisotropic
case. Thauvin and Mohanty, [14], require that $\beta = \{b_{ij}\}$ in order to fit their
network simulation results. Knupp and Lage [9] have suggested an anisotropic
formulation using a variational approach. Their formulation can be written

as follows

$$\mu K^{-1}\mathbf{u} + \beta\rho\frac{(\mathbf{u}\cdot K^{-1}\mathbf{u})^{\frac{1}{2}}}{(\det K^{-1})^{\frac{3}{2n}}}K^{-1}\mathbf{u} + \nabla p = 0, \quad n = 2, 3, \ \beta \ \text{is scalar,} \quad (2.2)$$

where $n = 2, 3$ is the space dimension. This model has fewer degrees of freedom as compared to (2.1). Nevertheless, it provides a good fit to the data from [14]. However, these data are still not enough in order to choose a particular tensorial model.

3   CVMFE on Distorted Quadrilateral Grids

In order to describe the discrete approximation to system (2.1) on quadrilateral grids for each element we introduce the mapping $\mathbf{r} = \mathbf{r}(\xi_1, \xi_2)$ which maps the unit square on the quadrilateral in physical coordinates $\mathbf{r} = (x_1, x_2)^T$ and associated metric entities

$$\mathbf{g}_i = \frac{\partial\mathbf{r}}{\partial\xi_i}, \quad S = \{s_{ij}\}, \quad s_{ij} = \frac{\partial x_i}{\partial\xi_j}, \quad J = \det S, \quad \mathbf{g}_j^T\mathbf{g}^i = J\delta_{ij},$$

where $\mathbf{g}_i, \mathbf{g}^i$ are the covariant and scaled contravariant basis vectors and J is the Jacobian of the mapping. Using above notations the governing equations can be written as follows:

$$\sum_{i=1}^{n}\frac{\partial}{\partial\xi_i}\rho V^i = Jf, \tag{3.1}$$

$$\sum_{j=1}^{n}\frac{1}{J}\mathbf{g}_i^T\left(\mu I + \beta\rho\frac{(\mathbf{u}\cdot K^{-1}\mathbf{u})^{\frac{1}{2}}}{(\det K^{-1})^{\frac{3}{2n}}}\right)K^{-1}\mathbf{g}_j V^j + \frac{\partial}{\partial\xi_i}p = 0 \tag{3.2}$$

$$V^i = \mathbf{u}^T\mathbf{g}^i, \quad \mathbf{u} = \frac{1}{J}\sum_{i=1}^{n}\mathbf{g}_i V^i \ \text{(Piola mapping).}$$

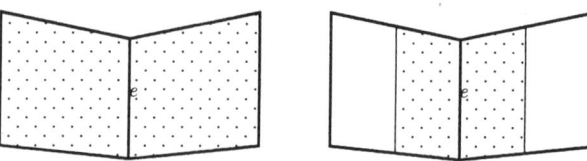

Figure 1: Support for the edge-centered flux base $\boldsymbol{\phi}_e$ (left) and for the flux test function $\boldsymbol{\psi}_e$ (right).

The Control Volume Mixed Finite Element methods (CVMFE) as introduced in [4] are based on the lowest order Raviart-Thomas (RT_0) flux basis functions and cell-based pressures. Equation (3.1) is integrated over the grid cell in the parametric space with mid-point quadrature rules for the contour integrals. The Forchheimer law (3.2) is integrated over the edge-centered control volumes in the parametric space and mid-point rules are used for pressure contour integrals. This derivation is equivalent to using the Raviart-Thomas flux-pressure bases, piecewise-constant pressure test functions and flux test functions $\boldsymbol{\psi}_e$ defined by (3.3), which is illustrated on Fig. 1 for the cases when the edge e locally coincides with the vector $\mathbf{g}_2$.

$$
\phi_e = \begin{cases} (1-\xi_1)\frac{1}{J}\mathbf{g}_1 & \text{in} \quad c_{\text{right}}(e) \\ \xi_1\frac{1}{J}\mathbf{g}_1 & \text{in} \quad c_{\text{left}}(e) \\ 0 & \text{elsewhere} \end{cases}
\qquad
\boldsymbol{\psi}_e = \begin{cases} \frac{1}{J}\mathbf{g}_1 & \text{in} \quad c_{\text{right}}(e) \\ \frac{1}{J}\mathbf{g}_1 & \text{in} \quad c_{\text{left}}(e) \\ 0 & \text{elsewhere} \end{cases}
\tag{3.3}
$$

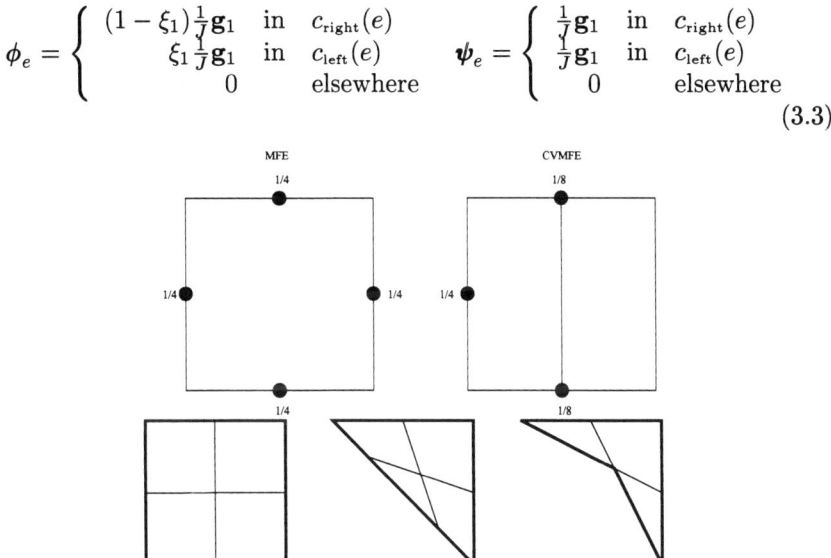

Figure 2: Equivalent first order quadrature rules and distorted grid cells.

In [4] the value $\frac{1}{J}$ is in fact approximated by a constant in each half-cell which allows the integrals over cells to be computed exactly. The basic advantage of CVMFE is the observed $O(h^2)$ convergence [4] in terms of pressure and fluxes on highly nonuniform grids and in the presence of strong coefficient jumps. Moreover it possesses optimal spectral resolution properties in a whole range of the discrete harmonics and provides accurate solutions on mildly distorted grids. The main drawbacks of CVMFE are the lack of the Linear Preservation (LP) property and nonsymmetric discrete metric tensor.

It is possible to derive CVMFE alternatively as a "low order" approximation to conventional mixed finite element (MFE) method [12]. Thorough analysis of relations between finite volume and finite element methods with different quadrature rules can be found in [2] where it was shown that most FE methods in primal and dual formulations can be written in the "factorized" form, or as a flux differences in terms of finite volume methods.

In [8] it was shown that there exist quadrature rules for the MFE integrals with first order algebraic accuracy which result in partial error cancellation property, namely they result in the discrete system which is identical to that resulting from RT_0 bases and flux test functions defined by (3.3) with certain first order quadrature rules, which are shown on Fig. 2.

The resulting discrete system coincides with CVMFE [4] on grids with affine cells, the discrete metric tensor is symmetric positive definite on admissible cells. Moreover the scheme is Linearity Preserving and is more accurate than the original CVMFE on distorted grids. The set of admissible grids is wider for this scheme as compared to conventional finite elements in the following sense: the necessary condition for the convergence of discrete solutions in the case $\beta = 0$ in fully discrete norms is that the Jacobian of the local mapping in the cell edge centers is bounded from below by a positive constant (see Fig. 2). The convergence proof is similar to that in [13] and the invertibility of local mapping for each element is not required, i.e., the cell shown on Fig. 2 (center), is admissible and the local discrete metric tensor is positive definite and has condition number of the order of unity in this case. An example of degenerate cell is shown on Fig. 2 (right).

Similar conclusions are valid in the primal formulation, i.e., for the bilinear finite element method and control volume finite element method.

4   Analytical Estimates for the Equivalent Well-Block Radius r_0 in Isotropic Darcy Flow

In the case of infinite uniform grid with square cells and isotropic Darcy flow the generic dimensionless discrete system which comprises several well known approximation schemes can be written as follows:

$$
\begin{aligned}
&-(2 + 2w)P_{ij} + w(P_{i-1\,j} + P_{i+1\,j} + P_{i\,j-1} + P_{i\,j+1}) \\
&+ \tfrac{1-w}{2}(P_{i-1\,j-1} + P_{i+1\,j-1} + P_{i-1\,j+1} + P_{i+1\,j+1}) = \\
&sQ_{ij} + (1-s)\left(\tfrac{36}{64}Q_{ij} + \tfrac{6}{64}(Q_{i-1\,j} + Q_{i+1\,j} + Q_{i\,j-1} + Q_{i\,j+1})\right. \\
&\left. + \tfrac{1}{64}(Q_{i-1\,j-1} + Q_{i+1\,j-1} + Q_{i-1\,j+1} + Q_{i+1\,j+1})\right),
\end{aligned}
\tag{4.1}
$$

where i, j are the grid node(cell) indices.

We seek the solution P_{ij} to (4.1) in the infinite computational domain $-\infty \leq i, j \leq +\infty$ with the following right hand side Q

$$Q_{00} = 1, \quad Q_{ij} = 0, \quad i^2 + j^2 > 0. \tag{4.2}$$

The problem (4.1),(4.2) is closed with the following condition

$$P_{00} = 0, \quad \lim_{r \to \infty} \frac{P_{ij}}{r} = 0, \quad r = \Delta x \sqrt{i^2 + j^2}, \tag{4.3}$$

where Δx is the side of the square grid cell.

The solution to problem (4.1)–(4.3) exists and is unique [1]. Similar to [1] our objective is to find the value r_0 such that the following equality is valid

$$\lim_{r \to \infty} \left(P_{ij} - \frac{1}{2\pi} \ln \frac{r_0}{r} \right) = 0, \quad r = \Delta x \sqrt{i^2 + j^2}. \tag{4.4}$$

Equation (4.1) leads to some popular schemes, in particular the values $s = 1, w = 1$ correspond to conventional finite difference scheme (FD), $s = 0, w = \frac{1}{3}$ is the bilinear finite element (BFE) scheme, while $s = 0, w = \frac{1}{2}$ correspond to the control volume finite element method (CVFE) and the control volume mixed finite element method (CVMFE). In the latter case equation (4.1) is deduced from the extended system by elimination of flux variables. In order to underline the difference between the single-cell production term and multiple-cell production term we include the schemes CVFE' and BFE' which are defined by the parameters $s = 1, w = \frac{1}{2}$ and $s = 1, w = \frac{1}{3}$, respectively.

Andreev [1] derived the asymptotic expansion for the case $s = 1$. It can be generalized for the general case $s \neq 1$ by adding a constant c,

$$P_{ij} = c + \frac{1}{2\pi} \left(\ln \frac{\Delta x}{r} - \frac{3}{2} \ln 2 - \gamma + \frac{1}{2} \ln w \right) + O\left(\frac{\Delta x^2}{r^2} \right), r = \Delta x \sqrt{i^2 + j^2},$$

where $\gamma = 0.57722156649\ldots$ is the Euler constant. The value of c for several approximation schemes was computed in [8] using the analytical solutions from [1] and superposition principle. Comparing the above equality with (4.4) we obtain that

$$\frac{r_0}{\Delta x} = e^{-\gamma - \frac{3}{2} \ln 2 + \frac{1}{2} \ln w + 2\pi c},$$

A simple approximate solution approach for finding r_0 was suggested in [10] using the observation that discrete solution in the near well cells is close to the analytical solution. Omitting the derivation details we obtain

$$\frac{r_0}{\Delta x} = e^{-\frac{\pi}{1+w}(s+(1-s)\frac{9}{16}) + \frac{1-w}{2(1+w)} \ln 2}.$$

All results are summarized in Table 1.

Scheme	$\frac{r_0}{\Delta x}$, exact value	exact value	$\frac{r_0}{\Delta x}$, Peaceman	error %
BFE	$\frac{(2+\sqrt{3})^{\frac{5\sqrt{3}}{16}} e^{-\gamma+\frac{3\pi}{32}}}{2\sqrt{6}}$	0.313833	$2^{1/4} e^{-27\pi/64}$	0.68
BFE'	$\frac{e^{-\gamma}}{2\sqrt{6}}$	0.114607	$2^{1/4} e^{-3\pi/4}$	1.65
FD	$\frac{e^{-\gamma}}{2\sqrt{2}}$	0.198506	$e^{-\pi/2}$	4.72
CVMFE	$\frac{1}{4} e^{\frac{1}{8}(4-8\gamma+\pi)}$	0.3427305	$2^{1/6} e^{-3\pi/8}$	0.83
CVFE'	$\frac{e^{-\gamma}}{4}$	0.140365	$2^{1/6} e^{-2\pi/3}$	1.52

Table 1. Equivalent radius for different numerical schemes and Darcy flow.

5 Calibration Procedure Based on the Solution of Auxiliary Problem

The equivalent radius does depend on the discrete representation of point sources/sinks. The most natural discrete approximation to the δ-function is by the piecewise-constant hat function, which is illustrated in Fig. 3. Rigorous analysis and convergence proofs for such approximations as applied to the Darcy law case can be found in [5]. We write the contribution to the right hand side f in (2.1) from a single well as follows

$$f = \frac{Q}{H}\phi(\mathbf{r} - \mathbf{r}_0),$$

$$\phi(\mathbf{r}) = \begin{cases} \frac{(\lambda_1+\lambda_2)^2}{4d^2\lambda_1\lambda_2}, & \mathbf{r} \in \Omega_h(K) \\ 0, & \mathbf{r} \notin \Omega_h(K) \end{cases}, \quad \Omega_h(K) = \{\mathbf{r} : |\mathbf{a}_i \cdot \mathbf{r}| < \frac{d\lambda_i}{\lambda_1 + \lambda_2}\}.$$

Here $\mathbf{r}_0$ is the well location, H is the height of the perforated zone (fully penetrating vertical wells are assumed), Q is the mass rate of the well, $\mathbf{a}_i, \lambda_i$ are the unit eigenvector and eigenvalue of K in the vicinity of the well, respectively, or

$$K = A\Lambda A^T, \quad A = (\mathbf{a}_1, \mathbf{a}_2), \quad \Lambda = \mathrm{diag}(\lambda_i), \quad A^T A = I,$$

while d is chosen such that the well block can be placed inside $\Omega_h(K)$, e.g., on square grid $d = \Delta x$.

The reason for this choice of hat function is that it is non-zero in the square in the transformed coordinates

$$\mathbf{r}' = \Lambda^{-\frac{1}{2}} A^T \mathbf{r}. \qquad (5.1)$$

This typically results in multiple-cell production terms for a single well and the estimates for r_0 differ from those in [1] and [11] since the superposition principle should be used for its computation in the linear case.

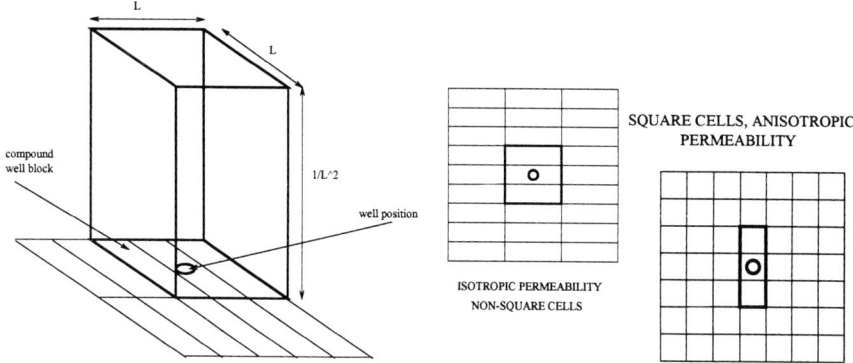

Figure 3: Discrete approximation to the production terms.

The basic motivation for this model of discrete sources and sinks is that numerical solutions in the vicinity of the well are more accurate compared to single-cell production term (typically by factor 4 to 7 on the grid with 1:3 aspect ratio shown on Fig. 3.

The well models in the case of isotropic Forchheimer flow are based on the analytical solution for radial flow around isolated well [3]

$$p(r) = p_R + \frac{\mu}{k\rho} \frac{Q}{2\pi H} \ln\left(\frac{r}{R}\right) + \frac{\beta Q|Q|}{\rho(2\pi H)^2} \left(\frac{1}{R} - \frac{1}{r}\right), \qquad (5.2)$$

Using the assumption that the finite difference solution in the cells near the well block is close to the analytical solution (5.2), it was shown in [6] that the local behavior of the discrete solution is described by the dimensionless Forchheimer number of the well block defined as $Fo = \frac{\beta k|Q|}{4\Delta x \mu H}$. The equivalent radius $\alpha = r_0/\delta x$ was found as a solution to the nonlinear equation

$$\frac{\pi}{2}(1 + Fo) = \ln(\frac{1}{\alpha}) + \frac{2}{\pi} Fo(\frac{1}{\alpha} - 1),$$

In [6] the well model was derived also for the bilinear finite element approximations, or BFE′ in our notations since single-cell production term was used. The result looks as follows

$$p_0 = p_w + \frac{\mu}{k\rho} \frac{Q}{2\pi H} \ln\left(\frac{r_w}{\alpha_1 \Delta x}\right) + \frac{\beta Q|Q|}{\rho(2\pi H)^2} \left(\frac{1}{\alpha_2 \Delta x} - \frac{1}{r_w}\right),$$

where p_0, p_w are the well block pressure and flowing well pressure, respectively, $\alpha_1 = 2^{\frac{1}{4}} e^{\frac{-3\pi}{4}}$ (see the same value in Table 1), $\alpha_2 = \frac{8(\Gamma + \sqrt{2})}{4(\Gamma + \sqrt{2})(1 + \frac{1}{\sqrt{2}}) + 9\pi^2}$, and $\Gamma \approx 1.35$ is the empirical calibration constant. The above results have provided insight into the problem, however they cannot be used in the case of irregular and distorted grids. To this end we suggest to find r_0 via solution of small auxiliary system using the following procedure:

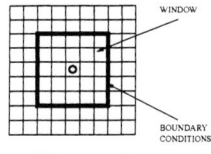

(a) choose a window around the well block; (b) write discrete approximation to governing equations in this window; (c) specify Dirichlet boundary conditions using the analytical solution (5.2); (d) solve discrete system, find the pressure in the well block p_0 and find r_0 via $\frac{r_0}{R} = e^{(p_0 - p_R)\frac{2\pi H k \rho}{\mu Q}}$

Figure 4.

Choosing the window size to be 3×3 cells results in accuracy which is comparable to that of the Peaceman method while using well block plus 2 cells in each direction typically allows to obtain 3 to 4 correct digits in r_0.

An attractive feature of this simple calibration procedure is that it can be used in the case of anisotropic permeability using the following analytical solution to (2.1) and (2.2)

$$p(\tilde{r}) = p_R + \frac{\mu}{(\det K)^{\frac{1}{2}}\rho}\frac{Q}{2\pi H}\ln\left(\frac{\tilde{r}}{R}\right) + \frac{\beta Q|Q|}{\rho(2\pi H)^2}\left(\frac{1}{R} - \frac{1}{\tilde{r}}\right), \quad \tilde{r} = \frac{(\mathbf{r}^T K^{-1}\mathbf{r})^{\frac{1}{2}}}{(\det K^{-1})^{\frac{1}{4}}}.$$

(5.3)

This analytical solution is derived from (5.2) using the transformation of space variables (5.1). In this case the well model can be written as follows

$$p_w = p_0 + \frac{\mu}{(\det K)^{\frac{1}{2}}\rho}\frac{Q}{2\pi H}\ln\left(\frac{\tilde{r}_w}{\tilde{r}_0}\right) + \frac{\beta Q|Q|}{\rho(2\pi H)^2}\left(\frac{1}{\tilde{r}_0} - \frac{1}{\tilde{r}_w}\right),$$

where $\tilde{r}_w$ is the mean well radius computed according to [11].

$$\tilde{r}_w = \frac{1}{2}\left(\mathrm{cond}(K)^{\frac{1}{4}} + \mathrm{cond}(K)^{-\frac{1}{4}}\right)r_w = \frac{r_w}{2\pi}\int_0^{2\pi}\left(\frac{(\mathbf{z}^T K^{-1}\mathbf{z})}{(\det K^{-1})^{\frac{1}{2}}}\right)^{\frac{1}{2}}d\phi,$$

where $\mathbf{z}^T = (\cos(\phi), \sin(\phi))$. Now the calibration procedure is modified as follows: (a) fix constants R and p_R; (b) solve discretized equations in a "window" around the well with Dirichlet BC specified by (5.3); (c) find pressure p_0 in the well block and find $\tilde{r}_0$ as a solution to nonlinear system (5.3) using the equality $p(\tilde{r}_0) = p_0$. It is important that the physical properties for the calibration procedure should be the same as for the reservoir simulation, i.e., the Forchheimer number or its generalizations should be the same in both cases.

6 Numerical Experiments

Typical well model validation scenario requires the following stages: a) derivation of the well model; b) numerical experiments with flow around isolated well; c) numerical simulation of 5-spot flow on Cartesian/distorted grids using well models; d) validation using direct simulation of the 5-spot flow on extremely refined radial grids near wells. Table 2 shows the comparison of computed data on 201×201 grid with analytical solutions which clearly shows that the multiple-cell production terms result in more accurate solutions.

Scheme	$\dfrac{\|p_h - p\|_{L_1}}{\|p\|_{L_1}}, \%$	$\dfrac{\|p_h - p\|_{L_2}}{\|p\|_{L_2}}, \%$	$\dfrac{\|p_h - p\|_C}{\|p\|_C}, \%$
BFE	0.0049	0.039	0.41
BFE'	0.0074	0.164	2.31
FD	0.0060	0.079	0.93
CVFE'	0.0032	0.054	0.74
CVMFE	0.0026	0.018	0.23

Table 2. Discrete norms of errors for different schemes. (201×201 cells)

Injection rate	0.05 mmscf/day
Number of injection wells	4
Injection well coordinates (ft)	(0, 0) (200, 0) (0, 200) (200, 200)
Number of production wells	1
Production rate	4×0.05 mmscf
Production well coordinates (ft)	(100, 100)
Well radius	0.35ft
Reservoir dimensions	200 ft× 200 ft× 1 ft
Initial pressure, p_I	5000 psia
Fluid density	1.783926×10^{-1} g/cm^3
Fluid density at 1atm, 60°F	$6.76361 \cdot 10^{-4}$ g/cm^3
Fluid viscosity	2.5574794×10^{-2} cp
$k_{11}, k_{22}(mD)$	10,10 or 10,100
$k_{12} = k_{21}$	0
Forchheimer coefficient β [ft^{-1}]	$0, 1.71 \cdot 10^{10}, 1.71 \cdot 10^{11}$
Boundary conditions	No-flow at all boundaries

Table 3. Simulation conditions.

Two different sample coarse grid configurations for the same 5-spot flow

are shown in Fig. 5

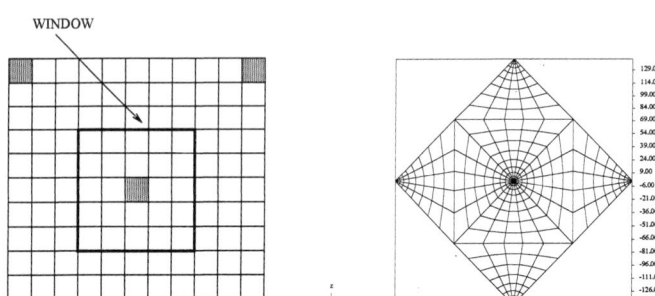

Figure 5: Different grids for 5-spot well configuration.

All data for this problem are presented in Table 3.

The well model validation results in the isotropic case for the CVMFE scheme are presented in Table 4. Production terms and well model are used on the 11×11 Cartesian grid while in direct simulation fluxes through well boundaries are specified. In this case the bottom-hole pressure p_w is the well model quality indicator. Very fine grids were used for direct simulation, typical cell size near well was about $r_w/50$.

β	$p_0 - p_I$	$r_0/\Delta x$	$p_w - p_I$	$p_w - p_I$, direct simulation	err %
0	-121.683	0.342926	-262.149	-262.602	0.17
$1.71 \cdot 10^{10}$	-132.067	0.35065	-490.176	-490.694	0.11
$1.71 \cdot 10^{11}$	-225.451	0.37137	-2542.4	-2543.5	0.04
$1.71 \cdot 10^{11}$			-2560 [6]		

Table 4. Comparison of simulation results.

b	β	$p_0 - p_I$	$r_0/\Delta x$	$p_w - p_I$	$p_w - p_I$, direct simulation	err %
0	$1.71 \cdot 10^{11}$	-387.46	0.376743	-2543.1	-2543.5	0.01
0.43	$1.71 \cdot 10^{11}$	-402.745	0.358036	-2543.1	-2543.5	0.01

Table 5. Results for the distorted grid simulations.

The numerical results illustrating the influence of the grid distortion on the accuracy of the calibration procedure in the isotropic case are presented in Table 5. Here $b\Delta x$ is the value of the quasi-random displacement for the grid nodes. The initial 23×23 Cartesian grid and the resulting distorted grid are shown on Fig. 5.

Both grids along with discrete pressure and pressure errors are presented on Fig. 6. The pressure errors for the 5-spot flow are computed via comparison with very fine grid results.

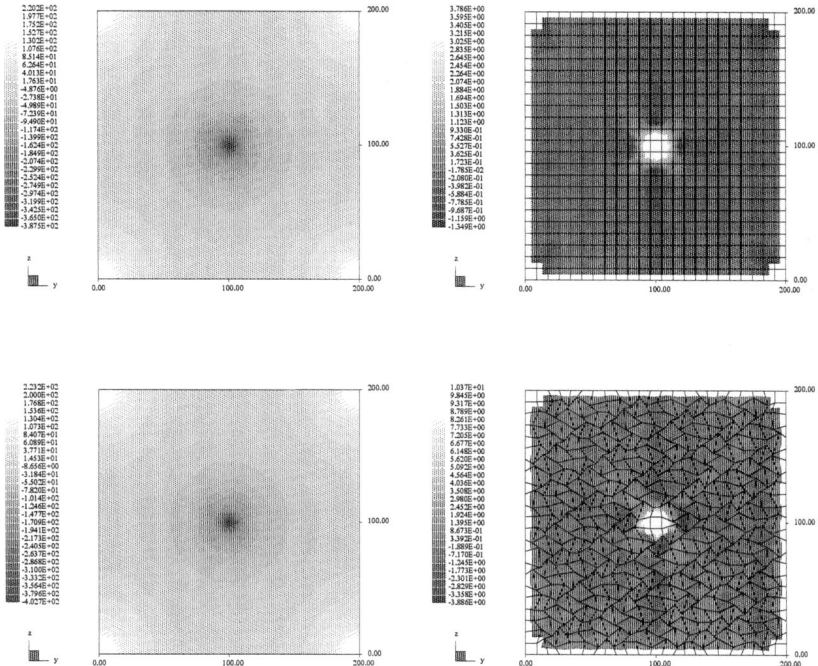

Figure 6: Pressure contour maps and error maps for regular and distorted grid simulations.

The pressure errors are only 3 times larger as compared to the results on square-cell grids. This is quite good, given that the distorted grid contains a lot of poorly shaped and non-convex cells.

The well model validation in the anisotropic case is more difficult since the normal flux distribution through the well boundary is not known. Hence in direct simulation p_w from the well model is specified while the predicted Q becomes the quality measure. Another observation is that as a rule of thumb the window around the well in the calibration procedure should be much larger as compared to the isotropic case for the same well model accuracy. The preliminary results of simulation on the 47×47 grid with square cells are presented in Table 6.

β	$r_0/\Delta x$	$p_w - p_I$	exact Q	predicted Q	err %
$1.71 \cdot 10^{11}$	0.46	-2072	0.2	0.20015	0.076

Table 6. Validation results for $k_{11} = 10, k_{22} = 100$.

7 Conclusions

Asymptotic methods and the superposition principle allow the derivation of exact expressions for the equivalent radius r_0 for various approximation schemes on Cartesian grids.

An inexpensive black-box calibration procedure allows r_0 to be computed in general grid configurations for non-Darcy flows, including anisotropic cases.

The CVMFE approximation scheme provides optimal resolution for near-well flow, including the case of distorted grids. The derivation of CVMFE via low-order "cancellation" quadrature rules for MFE integrals can improve accuracy on highly distorted grids and can make the admissible set of grids in 2-D and 3-D much wider.

Direct simulation of Forchheimer flows is very expensive with high order approximation schemes being desirable.

Acknowledgments: This work was supported by Mobil Technology Company. We are grateful to I.E. Kaporin for providing the linear solver used in this work and for drawing our attention to the papers by V.B. Andreev. Partial support of research of the first author by NWO grant 047003017 is gratefully acknowledged.

References

[1] Andreev, V. B. and Kryakvina, S. A., The fundamental solution of a one-parameter family of difference approximations of the Laplace operator in the plane, *Zh. Vychisl. Mat. Mat. Fiz* **13** (1973), 343–355.

[2] Axelsson, O. and Gustaffson, I., An efficient finite element method for nonlinear diffusion problems, *Bull. Greek Math. Soc.* **32** (1991), 45–61.

[3] Basniev, K. S., Kochina, I. N., and Maksimov, V. M., *Underground hydromechanics*, Moscow, Nedra. 1993.

[4] Cai, Z., Jones, J. E., McCormick, S. F., and Russel, T. F., Control-Volume Mixed Finite Element Methods, *Computational Geosciences* **1** (1997), 289–315.

[5] Chavent, G. and Jaffre, J., *Mathematical models and finite elements for reservoir simulation*, North-Holland, 1986.

[6] Ewing, R. E., Lazarov, R., Lyons, S. L., Papavassiliou, D. V., Pasciak, J., and Qin, G., Numerical Well Model for Non-Darcy Flow through Isotropic Media, *ISC Research Report ISC-98-11-MATH* (1998).

[7] Forchheimer, P., Wasserbewegung durch Boden *Zeit. Ver. Deut. Ing.* **45** (1901), 1781–1788.

[8] Garanzha, V. A. and Konshin, V. N., *Approximation schemes and discrete well models for the numerical simulation of the 2-D non-Darcy fluid flows in porous media* Comm. in Appl. Math., Computing Center RAS, 1999.

[9] Knupp, P. and Lage, J., Generalization of Forchheimer-extended Darcy flow model to tensor permeability case via a variational principle, *J. Fluid Mech.* **299** (1995), 97–104.

[10] Peaceman, D. W., Interpretation of well-block pressure in numerical reservoir simulation, *SPE Paper 6893, SPE Journal, Trans. AIME* **253** (1978), 183–194.

[11] Peaceman, D. W., Interpretation of well-block pressures in numerical reservoir simulation with non-square grid blocks and anisotropic permeability, *SPE Journal* (1983), 531–543.

[12] Raviart, P. A. and Thomas, J. M., *A mixed finite element method for 2nd order elliptic problems. Mathematical aspects of the finite element method. Lecture notes in Mathematics,* **606**, Springer, 1977.

[13] Samarskii, A. A., Koldoba, V. A., Poveshchenko, Yu. A., Tishkin V. F., and Favorskiy, A. P., *Difference Schemes on Irregular Grids*, Minsk, 1996, (in Russian).

[14] Thauvin, F. and Mohanty, K. K,, *Modeling of non-Darcy flow through porous media*, SPE paper 38017, presented at 1997 SPE Res. Simulation Symp., Dallas, Texas, USA, 8-11 June, 1997.

Mathematical Treatment of Diffusion Processes of Gases and Fluids in Porous Media

Norbert Herrmann

Abstract

The transport of fluids and gases in narrow pore systems is described by the transport equation and the material balance equation. In this paper we start with a typical example of such a process and develop the underlying parabolic partial differential equation and the corresponding initial and boundary conditions. Afterwards we describe how to reformulate the problem into a Fredholm integral equation of the first kind, which leads to the boundary element method. Whereas there exists an almost complete approach to the finite difference method and the finite element method, comparably little is known for the BEM. We use the collocation method to solve the Fredholm integral equation of the first kind and present a convergence theorem.

A computer program shows that the the predicted error is in good agreement with the calculated result.

KEYWORDS: parabolic equation, boundary element method, collocation method

1 The Physical Problem

For a given gas A the general gas law is well-known:

$$\text{gas law} \qquad p \cdot V = n \cdot k \cdot T, \qquad (1.1)$$

where p is the pressure, V the volume, n the number of molecules, k the gas constant, and T the temperature. The concentration c is defined as

$$c = \frac{n}{V} = \frac{p}{kT}. \qquad (1.2)$$

The transport of gases obeys the following equation:

$$\text{transport equation} \qquad \vec{j} = -D^{eff} \cdot \operatorname{grad} c, \qquad (1.3)$$

where we denote by $\vec{j}$ the flux of the gas and by D^{eff} the (effective) diffusion coefficient.

A complete description of the gas transport has also to bear in mind the effect of adsorption a, because for an isothermic process we have

$$a = f(c), \tag{1.4}$$

and this function f is normally a nonlinear one. If we would like to include f, we would get a nonlinear equation. Because we want to explain how the boundary element method could be applied, we omit the influence of adsorption ($f \equiv 0$) to have a linear equation. Later we will show where linearity is needed.

The change of the concentration c means that there are sinks or sources in the region. This is mathematically described by the divergence $\operatorname{div} \vec{j}$ of the flux. So we get the equation

$$\text{balance equation} \qquad \frac{\partial c}{\partial t} = -\operatorname{div} \vec{j}. \tag{1.5}$$

Combining (1.3) and (1.5) leads to the following partial differential equation:

$$\frac{\partial c}{\partial t} = \operatorname{div}(D^{eff} \operatorname{grad} c), \tag{1.6}$$

which is a parabolic partial differential equation.

At the start of the process we know the concentration from the experiment which gives us the initial condition

$$c = \begin{cases} c_i & \text{for } t = 0 \text{ in the interior,} \\ c_R & \text{for } t = 0 \text{ on the boundary.} \end{cases} \tag{1.7}$$

As the first boundary condition we demand that no change of concentration take place in the centre of symmetry, say the origin:

$$\operatorname{grad} c(0, t) = 0 \quad \text{in the centre of symmetry and for all } t. \tag{1.8}$$

The second boundary condition originates from a balance equation

$$D^{eff} \cdot \operatorname{grad} c = -C \cdot \frac{\partial c}{\partial t} \quad \text{on the boundary for all } t, \tag{1.9}$$

where C is a constant with respect to time and space.

The above consideration can be summarized in the following system of equations:

$$\frac{\partial c}{\partial t} = \operatorname{div}(D^{eff} \operatorname{grad} c),$$

$$c = \begin{cases} c_i, & \text{for } t = 0 \text{ in the interior} \\ c_R, & \text{for } t = 0 \text{ on the boundary.} \end{cases}$$

$$\operatorname{grad} c(0, t) = 0 \quad \text{in the centre of symmetry and for all } t.$$

$$D^{eff} \cdot \operatorname{grad} c = -C \cdot \frac{\partial c}{\partial t} \quad \text{on the boundary for all } t.$$

Two essential difficulties appear in this system:

1. The initial condition is not continuous.

2. The boundary conditions and are both of Neumann type.

2 The Boundary Integral Equation

Let us consider the following model problem:

$$
\begin{aligned}
\dot{u} - \Delta u &= f & \text{in } \Omega \times I, \\
u(x,t) &= g(x,t) & \text{on } \Gamma \times I, \\
u(x,0) &= 0 & \text{for } x \in \Omega.
\end{aligned}
$$

In the finite element method (FEM) we transform this model problem into a weak form by multiplying both sides with a test function v and integrating over the whole domain Ω. Here we do exactly the same. But since our problem is time dependent, we need a further integration over the time and so come to the equation

$$
\int_0^{t_o} \int_\Omega (\dot{u}(x,t) - \Delta u(x,t)) \cdot v(x,t_o - t) \, dx \, dt = \int_0^{t_o} \int_\Omega f(x,t) v(x,t_o - t) \, dx \, dt.
$$

In the next step we use the above Green's formula instead of Green's formula in the FEM:

$$
\int_0^{t_o} \int_\Omega [(\dot{u}(x,t) - \Delta u(x,t)) v(x,t_o - t) - u(x,t_o - t)(\dot{v}(x,t) - \Delta v(x,t))] \, dx \, dt
$$

$$
= \int_0^{t_o} \int_\Gamma \left[u(x,t) \cdot \frac{\partial v(x,t_o - t)}{\partial n} - \frac{\partial u(x,t)}{\partial n} \cdot v(x,t_o - t) \right] d\gamma \, dt.
$$

The next step consists in applying an appropriate test function. In the FEM we use a trial function from a certain Sobolev space, which fulfils the boundary conditions. Here we choose the Green's function for the (linear) differential operator $\dot{u} - \Delta u$, and this is the reason why we are so keen to have a linear problem. Otherwise, we have a trouble to find a Green function.

The Green function for the operator $\dot{u} - \Delta$ is well-known:

$$
G(x,t) = \begin{cases} \dfrac{1}{2\sqrt{\pi t}} e^{-\frac{x^2}{4t}} & t > 0, x \in \mathbb{R} \\ 0 & t \le 0, x \in \mathbb{R} \end{cases}
$$

This function is now used as the test function

$$
v(x_o, t_o - t) := G(x_o - x, t_o - t).
$$

The most important property of Green's function is the following identity:

$$u(x_o, t_o) = \int_0^{t_o} \int_\Omega u(x, t - t_o)(\dot{G} - \Delta G) \, dx \, dt.$$

With this formula, we get the following representation formula:

$$u(x_o, t_o) = \int_0^{t_o} \int_\Omega u(x, t - t_o)(\dot{G} - \Delta G) \, dx \, dt$$

$$= \int_0^{t_o} \int_\Gamma \left[\frac{\partial u(x, t)}{\partial n} \cdot G(x_o - x, t_o - t) \right.$$

$$\left. - u(x, t) \cdot \frac{\partial G(x_o - x, t_o - t)}{\partial n} \right] d\gamma \, dt$$

$$+ \int_0^{t_o} \int_\Omega f(x, t) \cdot G(x_o - x, t_o - t) \, dx \, dt.$$

In the theory of integral equations, parts of the right-hand side are well-known, so

$$K_o[u](x_o, t_o) := \int_0^{t_o} \int_\Gamma \frac{\partial u(x, t)}{\partial n} \cdot G(x_o - x, t_o - t) \, d\gamma \, dt \qquad (2.1)$$

is called the single layer potential and

$$K_1[u](x_o, t_o) := \int_0^{t_o} \int_\Gamma u(x, t) \cdot \frac{\partial G(x_o - x, t_o - t)}{\partial n} \, d\gamma \, dt \qquad (2.2)$$

is called the double layer potential. Indeed, if we consider the potential of a single plate the Gauß law gives exactly (2.1) and the same is true for the potential of a double layer (2.2).

We summarize a few theoretical results, which can be found in [1].

Theorem 2.1 *The single layer potential*

$$K_o : H^{-\frac{1}{2}, -\frac{1}{4}}(\Gamma) \to \tilde{H}_{loc}^{1, \frac{1}{2}}(\mathbb{R}^n \times I)$$

and the double layer potential

$$K_1 : H^{\frac{1}{2}, \frac{1}{4}}(\Gamma) \to \tilde{H}^{1, \frac{1}{2}}(\Omega, \partial_t - \Delta)$$

are both continuous operators.

Theorem 2.2 (Plemelj-Sochozky) *Let $\psi \in H^{-\frac{1}{2}, -\frac{1}{4}}(\Gamma), w \in H^{\frac{1}{2}, \frac{1}{4}}(\Gamma)$, and γ describe the restriction operator to the boundary Γ. We abbreviate by $[\gamma u]$ the difference of the boundary values on Γ, when coming from outside Ω^o and from inside Γ^i:*

$$[\gamma u] := \gamma(u|_{\Omega^o}) - \gamma(u|_{\Omega^i}).$$

Then we have the following jump relations:

$$[\gamma K_o \psi] = 0, \qquad [\gamma K_1 w] = w.$$

Note that the above equation means that the single layer potential does not jump, but in the double layer potential we have a jump when going from outside and from inside to the boundary.

3 Neumann Problem of a Diffusion Process

We now apply these theoretical considerations to our model problem. Let Ω be a bounded convex domain in $\mathbb{R}^2$ and let its boundary $\Gamma = \partial\Omega$ be sufficiently smooth. Then we consider the following Neumann boundary value problem:

$$\frac{\partial c(x,t)}{\partial t} = D^{eff} \cdot \Delta c(x,t), \quad (x,t) \in \Omega \times (0,T],$$

$$c(x,0) = c_o(x), \qquad\qquad x \in \overline{\Omega}, \tag{3.1}$$

$$\frac{\partial c(x,t)}{\partial n} = f(x,t), \qquad\qquad (x,t) \in \Gamma \times (0,T].$$

We know from [3] that if $\Gamma = \partial\Omega \in C^2$, f continuous, and $c_o \in C^1$ in an enviroment of Γ, then there exists a unique solution. The Green function is

$$G(x,t) := \frac{1}{4\pi D^{eff} t} \exp\left(-\frac{|x|^2}{4D^{eff} t}\right).$$

With this function, we get our representation formula

$$c(x_o,t_o) = -\int_0^t \int_\Gamma \frac{\partial}{\partial n} G(x-x_o,t-t_o) \cdot c(x,t)\,d\gamma_x\,dt$$

$$+ \int_0^t \int_\Gamma G(x-x_o,t-t_o)\frac{\partial}{\partial n} c(x,t)\,d\gamma_x\,dt$$

$$+ \int_\Omega G(x-x_o,t) \cdot c_o(x)\,dx, \qquad x_o \in \Omega,\, 0 < t_o < T.$$

For $x_o \to \Gamma$, we have to observe the jump relation and come to the following equation which is now valid for $x_o \in \Gamma$:

$$\tfrac{1}{2} c(x_o,t_o) = -p.v. \int_0^t \int_\Gamma \frac{\partial}{\partial n} G(x-x_o,t-t_o) \cdot c(x,t)\,d\gamma_x\,dt$$

$$+ \int_0^t \int_\Gamma G(x-x_o,t-t_o)\frac{\partial}{\partial n} c(x,t)\,d\gamma_x\,dt$$

$$+ \int_\Omega G(x-x_o,t) \cdot c_o(x)\,dx, \qquad x_o \in \Gamma,\, 0 < t_o < T.$$

Now, including the boundary conditions, we are led to an equation which could be called the mathematical model of our problem:

$$\tfrac{1}{2}c(x_o,t_o)+p.v.\int_0^t \int_\Gamma \tfrac{\partial}{\partial n}G(x-x_o,t-t_o)\cdot c(x,t)\,d\gamma_x\,dt$$

$$= \int_0^t \int_\Gamma G(x-x_o,t-t_o)\cdot f(x,t)\,d\gamma_x\,dt$$

$$+ \int_\Omega G(x-x_o,t)\cdot c_o(x)\,dx, \qquad x_o \in \Gamma,\, 0<t_o<T.$$

4 The Collocation Method

As a model problem, we consider a ball where our diffusion process should take place. Because of the symmetry we can restrict our attention to a two dimensional ball, i.e., a circle, so that the boundary Γ is the circle line shown below:

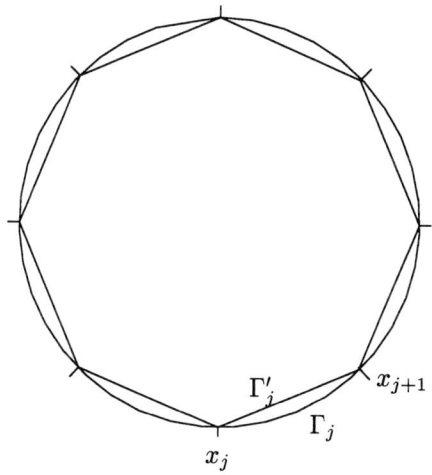

We divide this line into N parts by fixing the points $x_1, x_2, \ldots, x_{N+1} = x_1$. We call

$$\Gamma_j := \text{arc }(x_j, x_{j+1}), \quad \Gamma'_j := \overline{x_j, x_{j+1}}.$$

It is not necessary that all the parts are of equal size, but they should be quasiuniform, i.e., we demand:

$$\max_j |\Gamma_j|/\min_j |\Gamma_j| < const.$$

We introduce the abbreviations

$$h := \max_j |\Gamma'_j|, \quad k := T/M,\ t_p = p\cdot k,\ p=1,\ldots,M,$$

where M is the number of time steps. As the first approximation we consider the polygon Γ', formed by $\Gamma_1, \ldots, \Gamma_N$, instead of the circle line Γ, and solve the boundary integral on Γ'. In the same way we have to use the inner region Ω' of Γ' instead of Ω. Then we use the ansatz

$$\tilde{x}(x', t) := \sum_{p=1}^{M} \sum_{j=1}^{N} c_j^p \Phi_j^p(x', t),$$

where Φ consists of the following functions:

$$\Phi_j^p(x', t) = \varphi_j(x') \cdot \chi_p(t), \quad x' \in \Gamma', \ t \in (0, T]$$

with φ_j being the linear spline basis functions on Γ', $j = 1, \ldots, N$ χ_p the constant spline basis functions on $[0, T]$, $p = 1, \ldots, M$. We define the finite dimensional ansatz space

$$V^{N,M} := \text{span} \ \langle \Phi_j^p \rangle_{j,p}.$$

Now, our question consists in finding

$$(c_j^p)_{1 \leq j \leq N, 1 \leq p \leq M}.$$

such that

$$\tfrac{1}{2} c_j^p + p.v. \int_0^t \int_{\Gamma'} \tfrac{\partial}{\partial n'} G(x' - x_j, t - t_p) \cdot \tilde{c}(x', t) \, d\gamma_{x'} \, dt$$

$$= \int_0^t \int_{\Gamma'} G(x' - x_j, t - t_p) \cdot (P_h f)(x', t) \, d\gamma_x' \, dt$$

$$+ \int_{\Omega'} G(x' - x_j, t_p) \cdot c_o(x') \, dx'.$$

This equation looks complicated and indeed it is complicated. It is a simple system of $n \cdot M$ linear equations with the $N \cdot M$ unknowns c_j^p, but the entries of the system matrix are singular integrals where we have to use their principal values. On the other hand, this procedure is not as tedious as the application of the finite element method for solving the boundary integral equation. The FEM leads to double integrals over the boundary combined with a time integration.

5   Convergence

Important not only for the theoretical background but also for the numerical approximation is the following result, which states the stability of the collocation method:

Theorem 5.1 *There exists $\varepsilon_o > 0$ such that, for all grids on $\Gamma \times [0,T]$, with*

$$\varepsilon_o > \varepsilon := \sqrt{k}/h > 0, \tag{5.1}$$

the coefficient matrix K_N is diagonally dominant and we have

$$\|K_N^{-1}\|_\infty < C,$$

where the constant $C > 0$ is independent of h, k, and N.

The restriction (5.1) arises typically in the finite difference method. It means that we can not refine the space step h without refining the time step. In the finite difference method this is needed to ensure the stability, and here we also see that we get stability in that the inverse of the coefficient matrix is uniformly bounded independent of N.

From the above theorem we conclude the convergence result, which gives linear convergence of the collocation method, linear in space and time.

Theorem 5.2 *Let u be the exact solution of (3.1), $u_h := P_h u$ (the projection of u onto $V^{N,M}$), and $e(x',t) := u_h(x',t) - u(x,t)$, $x' \in \Gamma'$, $t \in I$. There exists a positive constant $C > 0$ such that*

$$\|e\|_\infty := \max_{x' \in \Gamma', t \in I} |e(x',t)| < C \cdot (h+k).$$

6 Numerical Results

With the help of a computer program we solve the following problem: Given a ball of radius one and an oval shown below, we choose $u \equiv 1$ to be the solution and adapt the boundary and initial conditions, respectively. Therefore, we are able to compare the exact solution with the approximate solution and it is possible to calculate the L^∞-norm for their difference. The result is shown in the figure below. It can be seen that we have indeed a good agreement with the theoretical result in Theorem 5.2.

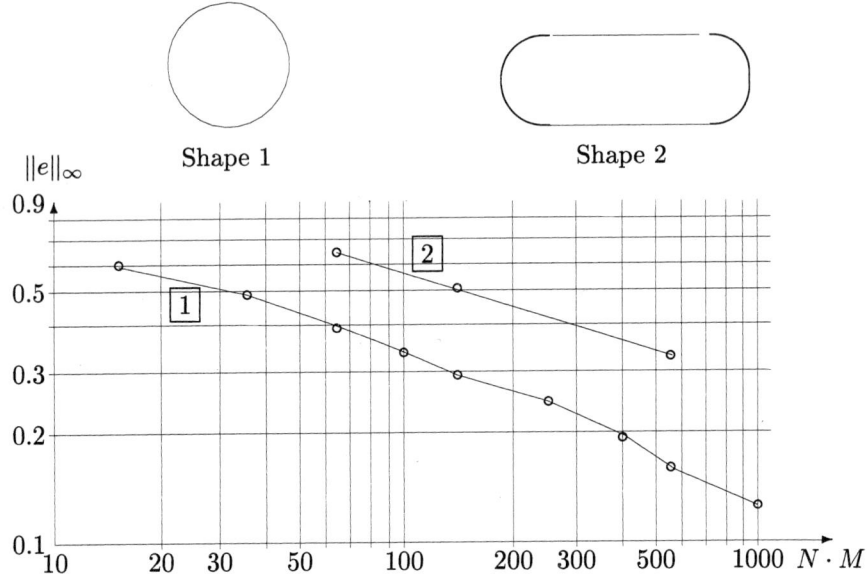

References

[1] Costabel, M., Boundary integral operators for the heat equation, *Integral Equations and Operator Theory* **13** (1990), 498–552.

[2] Costabel, M., Onoshi, K., and Wendland, W. L., *A Boundary Element Collocation Method for the Neumann Problem of the Heat Equation*, in Inverse and Ill–Posed Problems, Academic Press Inc., 1987.

[3] Friedman, A., *Partial Differential Equations of Parabolic Type*, Prentice–Hall. Inc., 1964.

[4] Herrmann, N., Time discretization of linear parabolic problems, *Hung. J. Ind. Chem.* **19** (1991), 275–281.

[5] Herrmann, N., Numerical problems in determining pore-size distribution in porous material, Workshop at the University of Budapest, Invited lecture, 1995.

[6] Herrmann, N., Siefer, J., Stephan, E. P., and Wagner, R., *Mathematik und Umwelt*, Edition Univ. Hannover, Theodor Oppermann Verlag, Hannover, 1994.

[7] Herrmann, N. and Stephan, E. P., *FEM und BEM, Einführung*, Eigendruck Inst. f. Angew. Math., Univ. Hannover, 1991.

[8] Iso, Y. Convergence of doundary element solutions for the heat equation, *Preprint*, 1991.

Implementation of a Locally Conservative Eulerian-Lagrangian Method Applied to Nuclear Contaminant Transport

CHIEH-SEN HUANG   ANNA M. SPAGNUOLO

Abstract

Recently, in the study of computational geosciences, a new Locally Conservative Euler-Lagrangian Method (LCELM) [3] was introduced by Douglas, Pereira, and Yeh. They have shown superior results to those using the Modified Method of Characteristics (MMOC) and the Modified Method of Characteristics with Adjust Advection (MMO-CAA) for the problem of two-phase, immiscible, incompressible flow in porous media. The object of this paper is to implement the LCELM applied to the transport of a high-level nuclear decay chain for the purpose of locally conserving the mass of each element in the chain. This method is coupled with mixed finite elements for the spatial discretization of each concentration equation and for the pressure equation. Computational results comparing the LCELM, the MMOC, and the MMOCAA are presented.

KEYWORDS: LCELM, MMOC, MMOCAA

1 Introduction

This work is an extension of the work done in [4], where the modified method of characteristics (MMOC) and a modified method of characteristics with adjusted advection (MMOCAA) were applied to approximate a finite number of concentrations corresponding to elements in a nuclear decay chain. See [8] for the development and proof of convergence of the MMOCAA in this miscible displacement problem.

High-level nuclear waste is contained in engineering barriers and then it is buried. For the purpose of assessing nuclear waste repository sites, we are interested in tracking the contaminants that might escape containment and enter saturated groundwater flows. In particular, an important chain, namely

$$^{234}U \rightarrow {}^{230}Th \rightarrow {}^{226}Ra,$$

is considered in [5, 6].

We let Ω denote our domain which represents the reservoir. Let $d(x)$, $x \in \Omega$ be the local elevation. Since we are considering contaminants that have half-lives on the orders of tens of thousands of years, a change in temperature will not affect any of the decay factors. Therefore, we assume that the phenomena occur at a constant temperature. Furthermore, we assume that we are dealing with very low concentrations of each contaminant. This implies that the viscosity μ and the density ρ of the fluid are independent of the concentrations. Let N_c be the number of contaminants in the decay chain, and let c_α, $\alpha = 1, \ldots, N_c$ be the αth element. Furthermore, denote by $q = q(x, t)$ a macroscopically distributed source term; q is positive at injection points and negative at production points. Let $J = (0, T]$ be the time interval of interest.

Recently, a fast, accurate, and stable numerical method [3], the locally conservative Eulerian-Lagrangian method (LCELM) for transport-dominated diffusive systems was derived. The LCELM retains the computational efficiency of the MMOC and its variant, the MMOCAA. Moreover, the LCELM preserves the desired conservation principles locally, while the MMOCAA conserves them globally, and the MMOC does not at all.

It should be noted that the restriction of the LCELM to miscible flow in porous media is essentially the same as the characteristics-mixed method in [1]. The fundamental difference between these new methods and the MMOC and the MMOCAA is that the new methods consider the partial differential system in divergence form and then split the transport from the diffusion, rather than relate to the nondivergence form and make use of the characteristics associated with the first-order transport part of the system in a fractional step procedure that splits the transport from the diffusive part of the system. It is the use of the divergence form that allows the localization of the transport so that the desired conservation property can also be localized.

This paper closely follows the computational ideas in [3]. There, immiscible displacement is considered; here, the flow is miscible and there is a finite number of contaminants to trace, giving a vector concentration. The LCELM applied to high-level nuclear contaminant transport is an alteration of the original MMOC and MMOCAA for miscible flow [4] that achieves local mass conservation of each component in the system. A set of comparisons showing the MMOC, the MMOCAA, and the LCELM are given for the first element in the decay chain.

The organization of the paper is as follows. The flow system is described in §2. The finite element spaces to be used for the spatial discretization are introduced in §3. In §4, the problem is discretized in time and the details of the LCELM procedure are given; in §4.1, the LCELM will be described in the differential setting, followed by the LCELM transport step in §4.2, the diffusive step in §4.3, and the pressure calculation in §4.4. Numerical results are given in §5; comparisons of the MMOCAA, MMOC, and LCELM will be presented.

2 The Flow System

For $\alpha = 1, \ldots, N_c$, the following equations govern the flow of a high-level nuclear decay chain through porous media:

$$u = -(\frac{k}{\mu}(\nabla p + \rho g \nabla d)), \quad x \in \Omega, \quad t \in J,$$
$$\nabla \cdot u = q, \quad\quad\quad\quad\quad x \in \Omega, \quad t \in J,$$
$$(2.1a)$$

$$\nabla_{t,x} \cdot \left(\frac{\phi r_\alpha c_\alpha}{c_\alpha u} \right) - \nabla \cdot (D_\alpha(u)\nabla c_\alpha)$$
$$= -\lambda_\alpha r_\alpha c_\alpha + \lambda_{\alpha-1} r_{\alpha-1} c_{\alpha-1} + \tilde{c}_\alpha q, \quad x \in \Omega, \quad t \in J,$$
$$c_\alpha u \cdot n_\Omega = (D_\alpha(u)\nabla c_\alpha) \cdot n_\Omega = 0, \quad\quad\quad x \in \partial\Omega, \quad t \in J,$$
$$(2.1b)$$

$$c_\alpha(x,0) = c_{\alpha,0}(x), \quad\quad\quad\quad\quad\quad\quad x \in \Omega,$$

where $\lambda_0 = r_0 = c_0 = 0$ and n_Ω is the outward unit normal to Ω. In (2.1a), $p = p(x,t)$ is the pressure, g is the gravitational constant, $k = k(x)$ is the permeability of the medium, and $\phi = \phi(x)$ is the porosity of the medium. Then, with $u = u(x,t)$ denoting the volumetric flow rate, the first equation represents Darcy's law, and the second reflects the incompressibility of the fluid. For $\alpha = 1, \ldots, N_c$, equations (2.1b) are the mass-balance equations for the concentration c_α of the αth contaminant in a chain that consists of N_c elements starting with c_1. Note that (2.1b) is in divergence form. Let λ_α, for $\alpha = 1, \ldots, N_c$ be the decay constant of the αth element. Also, since each species has the potential to adhere to the rock during the flow process, we denote by r_α the retardation factor of the αth contaminant for $\alpha = 1, \ldots, N_c$. For simplicity, the boundary conditions reflect the assumption that the periphery of the reservoir is impermeable. Compatibility to incompressibility requires that

$$\int_\Omega q\,dx = 0. \tag{2.2}$$

Following [7] for each contaminant, we assume $D_\alpha = D_\alpha(\phi, u)$, for $\alpha = 1, \ldots, N_c$, is the $\dim(\Omega) \times \dim(\Omega)$ matrix,

$$D_\alpha = \phi(x)[d_{\alpha,mol}I + |u|(d_{\alpha,long}E(u) + d_{\alpha,trans}E^\perp(u))], \tag{2.3}$$

representing diffusion. In (2.3), $d_{\alpha,mol}$ is the molecular diffusion coefficient, and $d_{\alpha,long}$ and $d_{\alpha,trans}$ are, respectively, the longitudinal and transverse dispersion coefficients for the αth contaminant. The matrix E is the projection along the direction of flow given by

$$E(u) = |u|^{-2}[u_i u_j], \tag{2.4}$$

and $E^\perp = I - E$. Usually, $d_{\alpha,long}$ is considerably larger than $d_{\alpha,trans}$. The term $\tilde{c}_\alpha$ is the specified concentration of the αth contaminant whenever q is positive, and it is the unknown concentration c_α when q is negative. Finally, the last equation in (2.1b) specifies the initial condition.

3 Spatial Discretization by Mixed Finite Elements

For simplicity, we do not take gravity into account from this point. Below, we rewrite the flow equations in terms of flux vectors, because we use mixed formulations for our space and time discretizations:

$$\nabla \cdot u = q, \quad u = -\frac{k}{\mu}\nabla p, \tag{3.1}$$

$$\nabla_{t,x} \cdot \begin{pmatrix} \phi r_\alpha c_\alpha \\ c_\alpha u \end{pmatrix} + \nabla \cdot v_\alpha = g_\alpha, \quad v_\alpha = -D_\alpha(u)\nabla c_\alpha, \tag{3.2}$$

where $g_\alpha = -\lambda_\alpha r_\alpha c_\alpha + \lambda_{\alpha-1}r_{\alpha-1}c_{\alpha-1} + (\tilde{c}_\alpha - c_\alpha)q^+$. Our numerical procedure is developed starting from equations (3.1) and (3.2).

Let

$$\Omega = [0, LX] \times [0, LY],$$

and set $H = \{HX, HY\}$, where $HX = LX/NX$ and $HY = LY/NY$. Then, let $X_i = iHX$ and $Y_j = jHY$, and define the elements of the partition $\mathcal{T} = \mathcal{T}(H) = \{M_{ij} : i = 1, \ldots, NX, j = 1, \ldots, NY\}$ by $M_{ij} = [X_{i-1}, X_i] \times$

$[Y_{j-1}, Y_j]$; $\mathcal{T}$ will serve for both the pressure and the concentration equations. Let

$$V = V(H) = \{\vec{v} \in H(\mathrm{div}, \Omega) : \vec{v}\,|_{M_{ij}} \in \mathcal{P}_{1,0} \times \mathcal{P}_{0,1} \text{ and } \vec{v} \cdot \vec{n} = 0 \text{ on } \partial\Omega\},$$
$$W = W(H) = \{w : w\,|_{M_{ij}} \in \mathcal{P}_0\} \subset L^2(\Omega),$$

where $\mathcal{P}_k$ denotes the set of polynomials of total degree k and $\mathcal{P}_{k,\ell}$ denotes the tensor product of polynomials of degree k in x by those of degree ℓ in y. Then, set

$$\mathcal{M} = \mathcal{M}(H) = V \times W;$$

i.e., the lowest index Raviart-Thomas mixed finite element space over the partition $\mathcal{T}$.

We shall seek an approximate solution to the system (3.1), (3.2).

4   Discretization in Time

Following [2, 4], we employ a time-discretization procedure based on operator splitting ideas. To do this, let

$$\Delta t_p = a_1 \Delta t_d, \qquad\qquad \Delta t_d = a_2 \Delta t_{tr}, \qquad\qquad (4.1)$$

where a_1 and a_2 are positive integers. Let $t^m = m\Delta t_p$ and denote by f^m a function f evaluated at time t^m. Similarly, let $t_n = n\Delta t_d$, $f_n = f(t_n)$, $f_{n,\kappa} = f(t_{n,\kappa})$, where $t_{n,\kappa} = t_n + \kappa\Delta t_{tr}$, and $f_{m,n} = f(t_{m,n})$, where $t_{m,n} = m\Delta t_p + n\Delta t_d$.

The pressure is approximated at times $t^m, m = 0, 1, \ldots,$. For each concentration equation, the LCELM procedure is applied. It splits into transport microsteps and diffusive fractional steps, in which the concentration is computed at times $t_{n,\kappa}$ and t_n, respectively, $n = 1, 2, \ldots, \kappa = 1, \ldots, a_2$. A detailed algorithm is given in the following subsections.

4.1   The Differential LCELM Procedure

Recall that the concentration equations can be written in the divergence form as

$$\nabla_{t,x} \cdot \left(\frac{\phi r_\alpha c_\alpha}{c_\alpha u}\right) - \nabla \cdot (D_\alpha(u)\nabla c_\alpha) = g_\alpha, \qquad\qquad (4.2)$$

where $g_\alpha = -\lambda_\alpha r_\alpha c_\alpha + \lambda_{\alpha-1} r_{\alpha-1} c_{\alpha-1} + \tilde{c}_\alpha q$ for $\alpha = 1, \ldots, N_c$.

Then, the fractional stepping procedure for it corresponds to the transport equation

$$\nabla_{t,x} \cdot \begin{pmatrix} \phi r_\alpha c_\alpha \\ c_\alpha u \end{pmatrix} = g_\alpha, \tag{4.3}$$

followed by the diffusive part given by

$$\phi r_\alpha \frac{\partial c_\alpha}{\partial t} + \operatorname{div} v_\alpha = 0. \tag{4.4}$$

Following the notation in [3], consider the space-time slice $Q = \Omega \times [t_{n,\kappa}, t_{n,\kappa+1}]$. Let K be a reasonably shaped, simply-connected subset of Ω, and define a subset $D = D_{n,\kappa}(K)$ of Q as follows. For each $x \in \partial K$, construct the solution $y(x; t)$ of the final value problem

$$\frac{dy}{dt} = \frac{u}{\phi r_\alpha}, \quad t_{n,\kappa+1} > t \geq t_{n,\kappa}, \quad y(x; t_{n,\kappa+1}) = x, \tag{4.5}$$

and set

$$\overline{x}_{n,\kappa}(x) = y(x; t_{n,\kappa}). \tag{4.6}$$

Then, let $\overline{K} = \overline{K}_{n,\kappa}$ be the interior of the set $\{\overline{x}_{n,\kappa}(x) \; : \; x \in \partial K\}$ and D the tube determined by K, $\overline{K}$, and the integral curves (4.5). (For Δt_{tr} sufficiently small, the map $x \to \overline{x}_{n,\kappa}$ is one-to-one, so that this construction can be carried out.) Now, denote the outward normal to ∂D by $\sigma(x, t)$, note that it is orthogonal to the vector $(\phi r_\alpha c_\alpha, c_\alpha u)^t$ on the lateral surface of D, and integrate (4.3) over D:

$$\int_D \nabla_{t,x} \cdot \begin{pmatrix} \phi r_\alpha c_\alpha \\ c_\alpha u \end{pmatrix} dx \, dt = \int_{\partial D} \begin{pmatrix} \phi r_\alpha c_\alpha \\ c_\alpha u \end{pmatrix} \cdot \sigma \, dA$$

$$= \int_K \phi r_\alpha c_\alpha(t_{n,\kappa+1}, x) \, dx - \int_{\overline{K}} \phi r_\alpha c_\alpha(t_{n,\kappa}, x) \, dx \tag{4.7}$$

$$= \int_D g_\alpha \, dx \, dt.$$

Thus, mass is conserved locally in the transport step, as defined in (4.3) above, if

$$\int_K \phi r_\alpha c_\alpha(x, t_{n,\kappa+1}) \, dx = \int_{\overline{K}} \phi r_\alpha c_\alpha(x, t_{n,\kappa}) \, dx + \int_D g_\alpha \, dx \, dt. \tag{4.8}$$

4.2 The LCELM Transport Microstep

Computation results show that letting $\mathbf{K}_{ij} = M_{ij}$ is a reasonable choice. Let's define $\overline{\mathbf{K}}_{\alpha, ij}$ as the quadrilateral obtained by approximating the integral curves (4.5) at the vertices $x_{i,j,k}$, $k = 1, \dots, 4$, of $\mathbf{K}_{ij}$. Let

$$\overline{\mathbf{x}}_{\alpha, i, j, k, n, \kappa} = x_{i,j,k} - \frac{I^{m,n,\kappa+1} u_i}{\phi r_\alpha} \Delta t_{tr} \tag{4.9}$$

where $I^{m,n,\kappa+1}u_i$ is the interpolation of the known velocity at times t_m and t_{m+1} at time $t_{m,n,\kappa+1}$ defined by

$$I^{m,n,\kappa+1}u_i = \left(1 - \frac{(\kappa+1)\Delta t_{tr}}{\Delta t_p}\right) u_i^m + \left(\frac{(\kappa+1)\Delta t_{tr}}{\Delta t_p}\right) u_i^{m+1}.$$

Note that there are N_c copies of the predecessor set of $\mathbf{K}_{ij}$, one for each α. Since the microstep Δt_{tr} is usually small with respect to Δt_d, in (4.9) we follow the tangent to the integral curve through $(x_{i,j,k}, t_{n,\kappa+1})$ back to the time level $t_{n,\kappa}$. Note that the trace back is in exactly the same direction as the characteristic; however, in the LCELM it associated with the vertices of K, rather than in the interior of K as in the MMOC and MMOCAA.

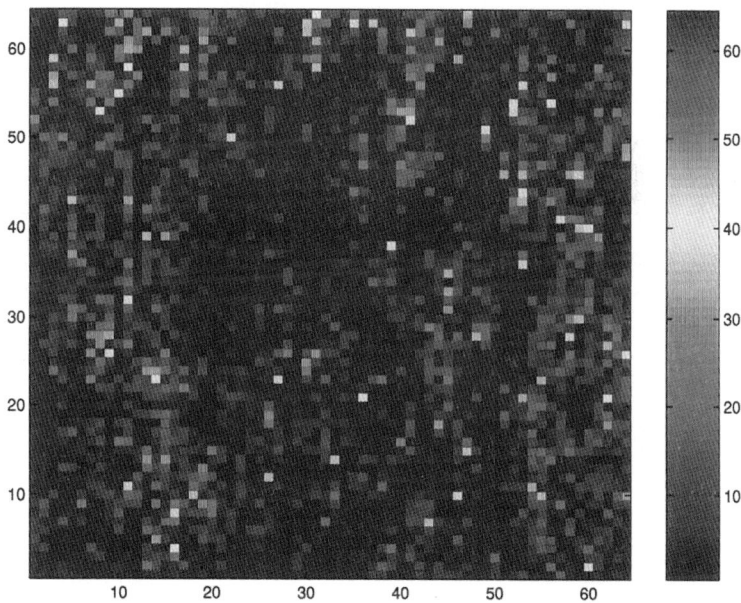

FIGURE 1: The heterogeneous permeability field (in millidarcies) with a coefficient of variation of 2.86 on a 64 × 64 grid .

Next, we must discretize the local conservation relation (4.8). Using the trapezoidal rule, we have

$$\int_{\mathbf{K}_{ij}} \phi r_\alpha z_{\alpha,n,\kappa+1} \, dx = \int_{\overline{\mathbf{K}}_{\alpha,ij}} \phi r_\alpha z_{\alpha,n,\kappa} \, dx + \frac{\Delta t_{tr}}{2} \left(\int_{\mathbf{K}_{ij}} g_\alpha \, dx + \int_{\overline{\mathbf{K}}_{\alpha,ij}} g_\alpha \, dx \right),$$

where $z_{\alpha,n,\kappa}$ is used to denote the solutions of concentration in the transport step. This is an explicit calculation and only needs to be performed when g_α does not vanish. Moreover, unlike the immiscible problem, it is a linear equation so the solutions can be solved directly.

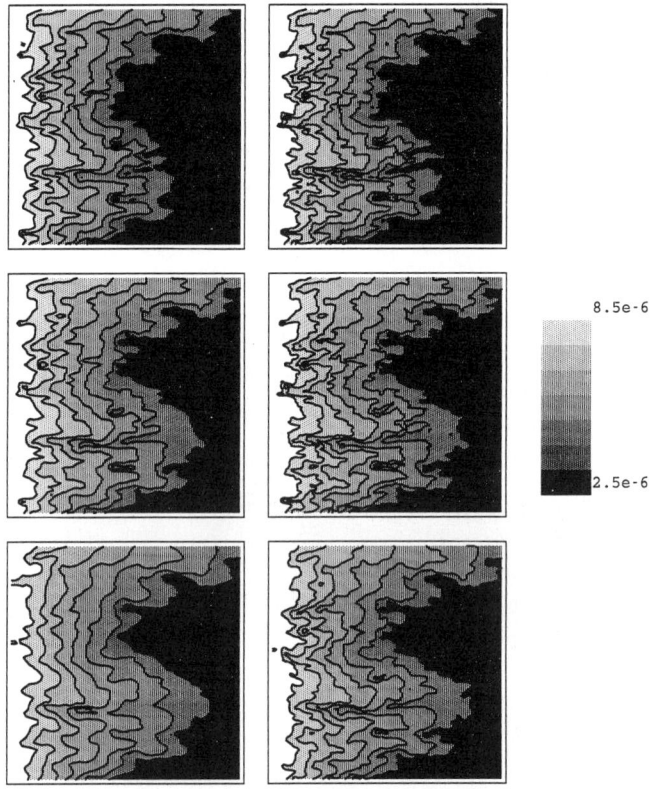

FIGURE 2: The pictures are results of a mesh refinement study.

4.3 The LCELM Diffusive Fractional Step for the Concentrations

After performing the transport microsteps, we have the functions

$$\bar{c}_{\alpha,n} = z_{\alpha,n,a_2-1}(t_{n+1}) \in W$$

which will be the initial conditions (one for each $\alpha = 1, \ldots, N_c$) at time t_n for the diffusive steps.

We shall apply the mixed finite element method to the equations

$$v_{\alpha,n+1} = -D_\alpha(u)(\nabla c_\alpha), \quad \phi r_\alpha \frac{c_{\alpha,n+1} - \bar{c}_{\alpha,n}}{\Delta t_d} + \operatorname{div} v_{\alpha,n+1} = 0, \quad (4.10)$$

subject to the boundary conditions

$$v_{\alpha,n+1} \cdot \vec{n} = 0. \quad (4.11)$$

Thus the mixed finite element equations take the form

$$\begin{aligned}
\left(\frac{1}{D_\alpha(u)} v_{\alpha,n+1}, \vec{v}\right) - (c_{\alpha,n+1}, \operatorname{div} \vec{v}) &= 0, & \vec{v} \in V, \\
\left(\phi r_\alpha \frac{c_{\alpha,n+1} - \bar{c}_{\alpha,n}}{\Delta t_d}, w\right) + (\operatorname{div} v_{\alpha,n+1}, w) &= 0, & w \in W.
\end{aligned} \quad (4.12)$$

4.4 The LCELM Pressure Calculation

The equations for $\{u^m, p^m\}$ are given by (see (3.1))

$$\nabla \cdot u^m = q, \quad u^m = -\frac{k}{\mu} \nabla p^m, \quad (4.13)$$

subject to the boundary condition $u^m \cdot \vec{n} = 0$ on $\partial\Omega$. The corresponding mixed finite element equations are

$$\begin{aligned}
\left(\frac{\mu}{k} u^m, \vec{v}\right) - (p^m, \operatorname{div} \vec{v}) &= 0, & \vec{v} \in V, \\
(\operatorname{div} u^m, w) &= (q, w), & w \in W.
\end{aligned} \quad (4.14)$$

5 Implementation

The algorithm described in this paper is applied to the decay chain $^{234}U \rightarrow^{230} Th \rightarrow^{226} Ra$. As a preliminary study, only limited numerical results are presented here; more intensive numerical simulations will be carried out elsewhere. A heterogeneous reservoir (see Figure 1) with a coefficient of variation of 2.86 is used. The reservoir is initially saturated with water. The leakage (or injection) of uranium from its container into the reservoir occurs at the left-hand side of the reservoir (slab geometry). The injection is at a constant rate of 500 years per pore volume with a concentration of uranium of 1e-5. Only the results of uranium are presented. The values in Table 1 are fixed in all of the computations.

The relative mass-balance errors using the MMOC in Fig. 2 are 11.9% and 10.6%, for the grids from left to right, respectively. Therefore, the MMOC is

Table 1: Parameter Values

Viscosity	$\mu(\text{poise}/((\text{g/cc})) = 0.01$
Porosity	$\phi = 0.01$
Decay Constant (^{234}U)	$\lambda_1 = 9 \times 10^{-14}(1/\text{sec})$
Retention Factor(^{234}U)	$r_1 = 12$
Dispersion(d_{long})	$d_{long} = 1 \times 10^{-3}(\text{cm}^2/\text{sec})$
Dispersion(d_{trans})	$d_{trans} = 1 \times 10^{-4}(\text{cm}^2/\text{sec})$
Dispersion(d_{mol})	$d_{mol} = 1 \times 10^{-7}(\text{cm}^2/\text{sec})$

more accurate when the grid is finer, which is what we expect. However, the MMOCAA and LCELM conserve mass exactly, regardless of grid size. Notice that in the MMOCAA and LCELM figures, the fingers are more defined and reach the well faster than those in the MMOC. However, the fingers in the LCELM results are not more defined than those in the MMOCAA results, but they do reach the well faster.

The pictures in Figure 2 are results of a mesh refinement study of uranium in a heterogeneous reservoir with a coefficient of variation of 2.86 after 5100 years of simulation using the MMOC (top row), the MMOCAA (middle row) and the LCELM (bottom row). The grids have elements of sizes 64 × 64 (left-hand side pictures) 128 × 128 (right-hand side pictures).

6   Acknowledgements

The authors express their sincere thanks to Jim Douglas, Jr. and Li-Ming Yeh for their advice throughout this work.

References

[1] Arbogast, T. and Wheeler, M. F., A characteristic-mixed finite element method for advection-dominated transport problems, *SIAM J. Numer. Anal.* **32** (1995), 404–424.

[2] Douglas, J., Jr., Furtado, F., and Pereira, F., On the numerical simulation of waterflooding of heterogeneous petroleum reservoirs, *Comput. Geosc.* **1** (1997), 155–190.

[3] Douglas, J., Jr., Pereira, F., and Yeh, L., A Locally Conservative Eulerian-Lagrangian Numerical Method and its Application to Nonlinear Transport in Porous Media, to appear.

[4] Huang, C.-S. and Spagnuolo, A. M., Implementation and Computation of the Modified Method of Characteristics with Adjusted Advection Applied to Nuclear Contaminant Transport: Part II, submitted.

[5] Kang, C. H., Chambré, P. L., Pigford, T. H., and Lee, W.-L., Near-field transport of radioactive chains, "Proceedings of the 2nd Annual International Conference on High Level Radioactive Waste Management," Las Vegas, 1991, 1054–1060.

[6] Kischinhevsky, M. and Paes-Leme, P. J., Modelling and numerical simulations of contaminant transport in naturally fractured porous media, *Transport in Porous Media* **26** (1997), 25–49.

[7] Peaceman, D. W., *Fundamentals of Numerical Reservoir Simulation*, Elsevier, New York, 1977.

[8] Spagnuolo, A. M., Convergence analysis of an approximation of contaminant transport in porous media by mixed finite elements and a modified method of characteristics with adjusted advection: Part I, submitted.

Application of a Class of Nonstationary Iterative Methods to Flow Problems

Xiuren Lei   Hong Peng

Abstract

Convergence of a certain class of nonstationary iterative methods applied to the numerical solution of algebraic linear systems arising in flow problems is studied. The iteration matrix of these methods can be expressed by a constant matrix plus a variable matrix tending to zero. The conclusions of convergence based on the matrix spectrum are given and applied to a class of semi-iterative methods.

KEYWORDS: algebraic linear system, iterative method, convergence, matrix spectrum

1   Introduction

Iterative methods for obtaining the numerical solution of algebraic linear systems can be classified into two broad groups: linear stationary methods and linear nonstationary methods. For the stationary case, there is a well-known sufficient and necessary condition for convergence; i.e., the spectral radius of the associated iteration matrix is less than one. On the other hand, the problem of convergence of the nonstationary methods is more complex. For this case, the result based on the spectrum radius analysis is not generally applicable.

However, as we prove in this paper, for a certain class of nonstationary iterative methods whose iteration matrix can be expressed by a constant matrix plus a variable matrix tending to zero, it is possible to establish a sufficient condition for convergence based on the spectral radius. Also, we apply the result to a class of semi-iterative methods [1], e.g., the Chebyshev semi-iterative methods, and obtain the relevant conditions for convergence of these methods. The result in [2] is a particular case of this paper.

2   Convergence of Nonstationary Methods

We consider the system of linear equations

$$Ax = b, \tag{2.1}$$

where A is an $N \times N$ nonsingular matrix and b is a vector of length N. Let a class of nonstationary iterative methods consistent with (2.1) be written as

$$x^{(n)} = G_n x^{(n-1)} + h_n, \qquad n \geq 1, \tag{2.2}$$

where the condition is satisfied

$$G_n = G + U_n, \tag{2.3}$$

G is a constant matrix independent of n, and

$$U_n \to 0, \qquad n \to \infty.$$

By (2.2), we can write

$$x^{(n)} = T_n x^{(0)} + K_n, \qquad n \geq 1,$$

where

$$T_n = G_n G_{n-1} \cdots G_1,$$
$$K_n = h_n + G_n h_{n-1} + G_n G_{n-1} h_{n-2} + \cdots + G_n G_{n-1} \cdots G_2 h_1.$$

Theorem 2.1 (Theorem 3.1 in [3]) *The method (2.2) converges if and only if*

$$\lim_{n \to \infty} T_n = 0.$$

Let $S(A)$ denote the spectral radius of a matrix A; i.e.,

$$S(A) = \max_{\lambda} |\lambda|,$$

where λ is an eigenvalue of A.

We need the following three lemmas from [4, pp. 141-146].

Lemma 2.1 *Given a matrix A of order N and $\epsilon > 0$, there exists a matrix M such that*

$$S(A) \leq \|MAM^{-1}\|_\infty \leq S(A) + \epsilon.$$

Lemma 2.2 *Let A be an $N \times N$ matrix and $\epsilon > 0$. There exist two positive constants $\mu_1 > 0$ and $\sigma > 0$, depending on A and ϵ, such that if for a sequence of $N \times N$ matrices U_k with*

$$\|U_k\|_\infty \leq \mu_1, \qquad k = 1, 2, \ldots,$$

where

$$\mu_1 = \epsilon/(2\sigma), \quad \sigma = \|M\|_\infty \|M^{-1}\|_\infty,$$

then we have

$$\left\| \prod_{k=1}^{m} (A + U_k) \right\|_\infty \leq \sigma(S(A) + \epsilon)^m, \qquad m = 1, 2, \ldots.$$

Lemma 2.3 *Let A be an $N \times N$ matrix such that $S(A) > 1$ and $\epsilon > 0$ such that $S(A) - \epsilon > 1$. If for a sequence of $N \times N$ matrices U_k there exists $\delta = \delta(A, \epsilon) > 0$ such that*

$$\|U_k\|_1 \leq \delta, \qquad k = 1, 2, \ldots,$$

then the sequence of matrices

$$\prod_m := \prod_{k=1}^{m} (A + U_k)$$

is not convergent when $m \to \infty$.

Theorem 2.2 *Let G be given by (2.3). Then, (a) the iterative method (2.2) is convergent if $S(G) < 1$; (b) it is not convergent if $S(G) > 1$.*

Proof: (a) The desired result can be obtained by Lemmas 2.1 and 2.2 and Theorem 2.1. (b) By Lemma 2.3 and Theorem 2.1, we obtain the desired result.   ☐

3   Convergence of Semi-Iterative Methods

Now, we apply Theorem 2.2 to a certain class of semi-iterative methods for obtaining the solution of (2.1). Let the iterative method for (2.1) be written as

$$x_{n+1} = \bar{G} x_n + k, \qquad n \geq 0. \tag{3.1}$$

By [5], we get the acceleration polynomial

$$Q_0(x) = 1, \quad Q_1(x) = \nu_1 x - \nu_1 + 1,$$
$$Q_{n+1}(x) = \rho_{n+1}(\nu_{n+1} x + 1 - \nu_{n+1}) Q_n(x) + (1 - \rho_{n+1}) Q_{n-1}(x), \quad n \geq 1,$$

where ρ_i and ν_i are real numbers. The corresponding semi-iterative methods can be obtained by the three-term recurrence relation

$$u^{(1)} = \nu_1(\bar{G}u^{(0)} + k) + (1 - \nu_1)u^{(0)},$$
$$u^{(n+1)} = \rho_{n+1}[\nu_{n+1}(\bar{G}u^{(n)} + k) + (1 - \nu_{n+1})u^{(n)}] + (1 - \rho_{n+1})u^{(n-1)},$$
$$(3.2)$$

for $n \geq 1$. We can rewrite (3.2) as

$$u^{(n)} = \rho_n\left\{[\nu_n\bar{G} + (1 - \nu_n)I]u^{(n-1)} + \nu_nk\right\} + (1 - \rho_n)u^{(n-2)}.$$

Then we have

$$\begin{pmatrix} u^{n-1} \\ u^n \end{pmatrix} = \begin{pmatrix} 0 & I \\ (1 - \rho_n)I & \rho_n[\nu_n\bar{G} + (1 - \nu_n)I] \end{pmatrix} \begin{pmatrix} u^{n-2} \\ u^{n-1} \end{pmatrix} + \rho_n\nu_n\begin{pmatrix} 0 \\ k \end{pmatrix};$$

i.e.,

$$V^{(n)} = G_n^{\star}V^{(n-1)} + h_n^{\star},$$

where

$$G_n^{\star} = \begin{pmatrix} 0 & I \\ (1 - \rho_n)I & \rho_n[\nu_n\bar{G} + (1 - \nu_n)I] \end{pmatrix}, \qquad (3.3)$$
$$h_n^{\star} = \rho_n\nu_n\begin{pmatrix} 0 \\ k \end{pmatrix}, \qquad V^{(n)} = \begin{pmatrix} u^{n-1} \\ u^n \end{pmatrix}.$$

Theorem 3.1 *Let* $\lim_{n\to\infty} \rho_n = \rho$, $\lim_{n\to\infty} \nu_n = \nu$, $\bar{G}$ *be given by* (3.1), *and*

$$\bar{G}_1 = \begin{pmatrix} 0 & I \\ (1 - \rho)I & \rho[\nu\bar{G} + (1 - \nu)I] \end{pmatrix}.$$

Then, (a) the method (3.2) *is convergent if* $S(\bar{G}_1) < 1$; *(b) it is not convergent if* $S(\bar{G}_1) > 1$.

Proof: Set

$$H_n = \rho_n[\nu_n\bar{G} + (1 - \nu_n)I], \qquad H_\infty = \rho[\nu\bar{G} + (1 - \nu)I].$$

Then (3.3) can be written as

$$G_n^{\star} = \bar{G}_1 + \bar{U}_n,$$

where

$$\bar{U}_n = \begin{pmatrix} 0 & 0 \\ (\rho - \rho_n)I & H_n - H_\infty \end{pmatrix}.$$

Since

$$\lim_{n\to\infty} \bar{U}_n = 0,$$

by Theorem 2.2, we complete the proof.   ▯

The Chebyshev semi-iterative methods, discussed in [2], is a particular case of the semi-iterative methods mentioned here. The results in [2] can be obtained from Theorem 3.1.

Acknowledgments. This work is supported in part by the Natural Science Fund of South China University of Technology (E5-121-025).

References

[1] Young, D. M., *Iterative Solution of Large Linear Systems*, Academic Press, New York, London, 1971.

[2] Lei, X., On convergence of Chebyshev semi-iterative methods, *J. Math. Research & Exposition* **9** (1989), 277–281 (in Chinese).

[3] Santos, N. R. and Linhares, O. L., Convergence of Chebyshev semi-iterative methods, *J. Comp. Appl. Math.* **16** (1986), 59–68.

[4] Ostrowsky, A. M., *Solution of Equation in Euclidean and Banach Spaces*, Academic Press, New York, London, 1973.

[5] Hagman, L. A. and Young, D. M., *Applied Iterative Methods*, Academic Press, New York, London, 1981.

Reservoir Thermal Recover Simulation on Parallel Computers

BAOYAN LI   YUANLE MA

Abstract

The rapid development of parallel computers has provided a hardware background for massive refine reservoir simulation. However, the lack of parallel reservoir simulation software has blocked the application of parallel computers on reservoir simulation. Although a variety of parallel methods have been studied and applied to black oil, compositional, and chemical model numerical simulations, there has been limited parallel software available for reservoir simulation. Especially, the parallelization study of reservoir thermal recovery simulation has not been fully carried out, because of the complexity of its models and algorithms. The authors make use of the message passing interface (MPI) standard communication library, the domain decomposition method, the block Jacobi iteration algorithm, and the dynamic memory allocation technique to parallelize their serial thermal recovery simulation software NUMSIP, which is being used in petroleum industry in China. The parallel software PNUMSIP was tested on both IBM SP2 and Dawn 1000A distributed-memory parallel computers. The experiment results show that the parallelization of I/O has great effects on the efficiency of parallel software PNUMSIP; the data communication bandwidth is also an important factor, which has an influence on software efficiency.

KEYWORDS: domain decomposition method, block Jacobi iteration algorithm, reservoir thermal recovery simulation, distributed-memory parallel computer

1 Introduction

The rapid development of parallel computers provides a solid hardware foundation for developing parallel software used in industries such as oil reservoir recovery and airplane manufacture. There are mainly three kinds of parallel computers, i.e., shared-memory, distributed-memory, and parallel computers of the mixed type. The typical vector computers are Cray series, such as

Cray90. The well-known IBM SP2 and Dawn 1000A are distributed-memory computers. As for shared-memory parallel computers, the scales of computation models are limited by the bandwidth of data bus and CPU number. But for a distributed-memory one, it has good extensibility and does not limit the computation scale. However, accessing to the memory of a neighbor CPU has to be realized by programmers. It increases the complexity of programs. It is noticeable that a mixed-type of computers are developed such as Oringin2000 of SGI Company. This kind of computers greatly decreases the complexity of programs.

In the past 20 years, the study of parallelization of serial software of oil reservoir recovery simulation has been carried out on all these kinds of parallel computers [1, 16, 15, 6, 9]. It focuses on parallelization strategy, linear equation solution and parallel software application [12, 9, 13, 10]. Wallis , *et al.* [17] developed a nested algorithm to solve linear equations and obtained high speedup in distribution calculation environment. Killough [7] applied the domain decomposition method (DDM) based on local grid refinement to the black oil and compositional models. Briens, *et al.* [1] studied the stability of parallel software of the compositional model. Mejerink [11] parallelized the black oil numerical simulation software Bosium by using the block Jacobi iteration method. Rame [16], *et al.* parallelized the chemical drive software UTCHEM by using the vector parallel technology. It can be seen that parallelization of software of the black oil, compositional, and chemical models has been deeply studied. In particular, the parallelization algorithms of linear equation solution have been applied to the calculation of the black oil model numerical simulation. However, there has been no literature about parallelization of the thermal recovery simulation.

Most of serial software are parallelized mainly in the part of linear equation setup and solution. For most of reservoir simulation software, the establishment of coefficient matrices and solution of linear equations, respectively, takes 30-40% and 50-60% calculation time; the PVT and rock physical property parameter calculation and the initialization of data fields take less than 10% calculation time. Therefore, parallelization of linear equation solution is the key to obtain high parallel efficiency. Under a shared-memory computation environment, a high parallel efficiency can be obtained in such mode. However, for a distributed-memory computer, the memory of a node is limited and it may be impossible to load data without parallelization of the data load. Besides, data output is also an important factor, which leads to communication between nodes. Especially, for large-scale reservoir thermal recovery simulation, such as history match and whole-scale reservoir simulation, the frequency of I/O operation is high and the scale of data stream is large, because of frequent operation of large number of injection and production wells. In general, data I/O is a bottom neck of reservoir simulation. In a distributed-memory computation environment, it is necessary to parallel the I/O part of simulation software.

To meet the increasing requirement for the model scale of refine reservoir simulation, the authors parallelize the thermal recovery simulation software NUMSIP by using DDM and test it on distributed-memory parallel computers. A parallelization strategy is designed to parallelize not only setup and solution of linear equations, but also I/O, PVT and rock property parameter calculation and data initialization to improve efficiency of programs. If the boundaries between sub-domains are along natural faults, there is no communication between such sub-domains. In such cases, the authors apply an irregular non-overlapping DDM to reduce communication between sub-domains. The linear equation solution algorithm is the block Jacobi iteration method. To guarantee the compatibility of software, the standard message interface MPI is used to realize the data communication. PNUMSIP is tested on both IBM SP2 and Dawn 1000A.

This paper consists of five parts. The model and algorithm of the serial software NUMSIP is introduced in §2. In §3, we give the parallelization strategy, data structure, and data communication mode of PNUMSIP. The test results are shown and analyzed in §4. We conclude with a few remarks in §5.

2   Software NUMSIP

NUMSIP is a simulation software of reservoir thermal recovery. It has been widely used in many oil fields in China, such as Liao He Oil Field, Sheng Li Oil Field, and Xing Jiang Oil Field. It can be used for designing reservoir recovery schemes of old and new oil fields, the history match, the production prediction, and the study of residual oil distribution.

The following physics factors and processes are presented in the model of NUMSIP:

a) viscosity, gravity and capillary force,

b) heat conduction and convection processes,

c) heat losses of over burden and under burden of reservoir,

d) effect of temperature on physical property parameters of oil, gas and water,

e) rock compression and expansion.

But kinetic, heat transfer, mass transfer, and the thermal cracking of hydro-carbons caused by molecule diffusion and dispersion are not considered.

The governing equations include the mass conservation equation, the energy conservation equation, and the mole fraction and saturation constraint equations.

1) The mass conservation equation

$$\sum_{j=1}^{N_p} \left(\int \int \int_V \frac{\partial}{\partial T} (\varphi S_j \rho_j \chi_{ij}) dv + \int \int_S (\rho_j \chi_{ij} v_j) ds \right)$$

$$+ q_i = 0, \qquad i = 1, \ldots, N_c, \tag{2.1}$$

where φ is the rock porosity, ρ_j the density of the material in phase j, S_j the saturation of the material in phase j, v_j the surface flow velocity of the material in phase j, χ_{ij} the mole fraction of composition i in phase j, q_i the source item of composition i, and N_p the number of phases.

2) The energy conservation equation

$$\int \int \int_V \frac{\partial}{\partial T} \left(\varphi \sum_{j=1}^{N_p} \rho_j S_j u_j + (1 - \varphi) \rho_{rock} C_p (T - T_i) \right) dv$$

$$+ \int \int_S (q_h + q_c) ds + Q_h + Q_c = 0, \tag{2.2}$$

where u_j is the inner energy of the material in phase j, ρ_{rock} the rock density, C_p the rock heat capacity, T_i the temperature of component i, q_h the enthalpy flow velocity, q_c the heat flow velocity, Q_c the heat source item, and Q_h the enthalpy source item.

3) The mole fraction constraint equation

$$\sum_{i=1}^{N_c} \chi_{ij} = 1. \tag{2.3}$$

4) The saturation constraint equation

$$\sum_{j=1}^{N_p} S_j = 1. \tag{2.4}$$

The finite difference method (FDM) is used to discrete the above partial differential equations. More advanced methods [4] will be used. The sorting methods, such as the natural, red-black, and D4 sorting can be selected by users to establish linear equations. Both the Gauss method and ORTHOMIN method can be used to solve linear equations.

The serial NUMSIP consists of 258 modules. Because it has a good program structure, it is convenient for organization of structure of its parallel version PNUMSIP. To spare space, NUMSIP stores the data fields in a compact array. It decreases readability of programs and data access efficiency and increases the difficulty to parallelize them. It is not suitable for parallelization under a shared-memory parallel calculation environment.

3 Parallelization of NUMSIP

The aims to parallelize NUMSIP are a) to increase the model scale, b) to obtain the speedup and efficiency of software, and c) to develop a transplantable and stable parallel software of thermal recovery simulation. Toward that end, various factors are synthetically taken into account.

3.1 Software parallelization

Because 60-80% percent of calculation time is used for establishment and solution of linear equations, the prevailing strategy is to parallelize only these parts. However, it results in a problem; e.g., the model scale is limited by the size of accessible memory of a CPU. This problem becomes prominent in a parallel computation environment consisting of a workstation cluster. To solve this problem, a global parallel scheme is designed to parallelize a whole calculation process in DDM. It will benefit for improving speedup and efficiency of a program. To decrease data communication and simplify data structure, a non-overlapping irregular DDM is adopted. It is particularly suitable for the case that boundaries between sub-domain are along a natural fault, which can decrease data communication. For every sub-domain, the matured serial algorithm is used.

3.2 Domain decomposition

To realize the domain decomposition, a 2D array $Id(i, j)$ is introduced to mark different sub-domains. Its definition is given as follows:

$$Id(i, j) = k \quad \text{for grid } (i, j) \text{ belonging to sub-domain } k. \quad (3.1)$$

The boundaries between sub-domains can be traced from $Id(i, j)$. Only the boundaries, which are not along natural faults, can be determined as valid boundaries. For a valid boundary, there will be communication between the two sub-domains divided by this boundary.

3.3 Data structure

On the basis of the original compact data array, two buffer arrays Bi and Be are introduced, respectively, to store the inner and outer boundary of a sub-domain. In the iteration calculation process, Bi is updated by the data field generated by the calculation at different iteration steps. It is transferred to the outer boundary buffer array Be of the neighboring sub-domain; see in Fig. 1.

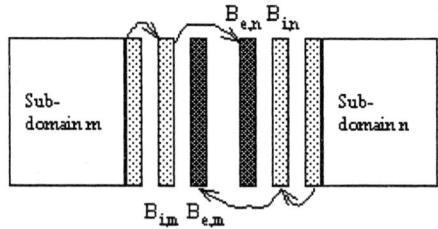

Fig.1 data structure and communication

1	2	3	4	5	6	7	8

Fig. 2. Domain decomposition of test problem T_1.

1	2	3	4	5	6	7	8
9	10	11	12	13	14	15	16

Fig. 3. Domain decomposition of test problem T_2.

3.4 Data communication

According to the time variant characteristic of data to be transferred, the communication data are divided into three types, i.e., static data and slow and fast transient data. The data describing the geometric model of reservoir and rock property parameters are static data. The base values of the data field at a time step in the iteration calculation process are slow transient data, and the increment values of the data field at an iteration step of the Newton-Raphson iteration are fast transient data. The block communication mode is used to transfer static data and slow transient data. But the non-block communication mode is adopted for transferring fast transient data to reduce communication overhead and improve communication efficiency.

3.5 Time step control

To ensure that the well data of all production periods can be safely loaded and satisfy the Newton-Raphson linearization condition, the minimum time step is selected. The domain is divided by using DDM and every sub-domain has a predicted next time step at every time step. To synchronize the calculation processes at different sub-domains, the time step control is determined as follows:

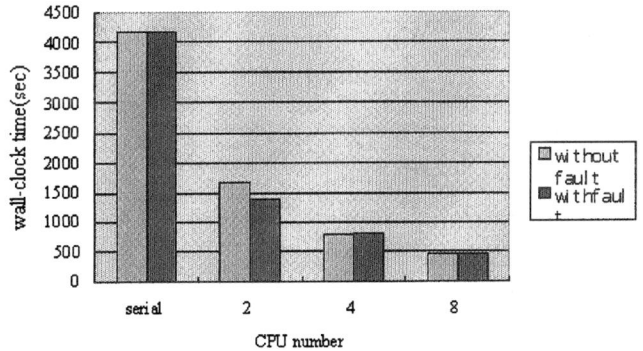

Fig.4   Calculation time of test problem T_1

a) predict time step $t_{i,k+1}$ at sub-domain i, $i = 1, 2, \ldots, N$,

b) determine the $(k + 1)$th synchronic time step t_{k+1} by

$$t_{k+1} = \min\{t_{1,k+1}, t_{2,k+1}, \ldots, t_{N,k+1}\},$$

c) determine the $(k + 1)$th time step $t_{i,k+1}$ of sub-domain i, $t_{i,k+1} = t_{k+1}$.

3.6 Linear equation solution algorithm

The block Jacobi iteration algorithm is adopted to solve linear equations. The linear equations are established in the FDM and Newton-Raphson method. They are presented as follows:

$$\begin{bmatrix} A_{11} & A_{12} & \cdots & A_{1N} \\ A_{21} & A_{22} & \cdots & A_{2N} \\ A_{31} & A_{32} & \cdots & A_{3N} \\ & & & \\ A_{N1} & A_{N2} & \cdots & A_{NN} \end{bmatrix} \begin{bmatrix} X_1 \\ X_2 \\ X_3 \\ \\ X_N \end{bmatrix} = \begin{bmatrix} B_1 \\ B_2 \\ B_3 \\ \\ B_N \end{bmatrix} \tag{3.2}$$

where $X_1, X_2, \ldots, X_N$, respectively, are solution vectors of data fields of sub-domains $G_1, G_2, \ldots, G_N$. To solve these linear equations, an initial solution vector is given as $X^0 = [X_1^0, \ldots, X_N^0]^t$. The kth iteration value of solution vector can be obtained from

$$A_{ii}X_i^k = B_i - \sum_{j=1}^{i-1} A_{ij}X_j^k - \sum_{j=i+1}^{N} A_{ij}X_j^{k-1}. \tag{3.3}$$

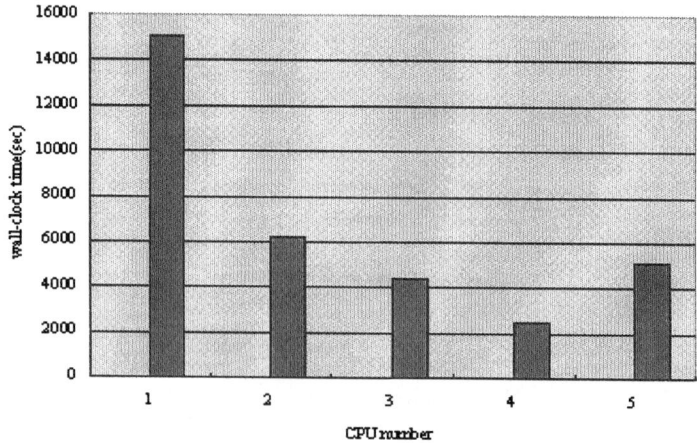

Fig.5 calculation time of test problem T2 without fault

The modification work of NUMSIP focuses on I/O. More than ten modules are modified and 24 new modules are added to realize parallelization of I/O, tracing of boundaries, establishment of linear equations, well distributions, and data communication.

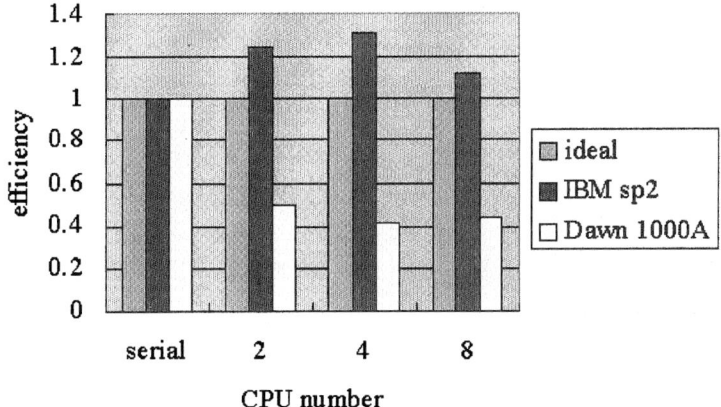

Fig.6 Comaration of efficiency of PNUMSIP tested
on IBM sp2 and Dawn 1000A

4 Test and Calculation Results

4.1 Hardware environment

The IBM SP2 used for testing PNUMZIP is a distributed-memory computer, which consists of IBM RISC system/6000 workstations, high performance switch board, control workstation, and SP ether network. Its CPU clock frequency is 66.7 MHz. The peak calculation speed is 226Mflops. The communication bandwidth of its high performance switch board is 40Mbps and the bandwidth of its SP ether network is 1.25Mbps. It has 28 calculation nodes. Four of them are wide nodes and the rests are narrow nodes. A wide node has 128M memory and a narrow one has 64M memory. Parallel calculation process can run on only narrow nodes.

Dawn 1000A used for testing PNUMSIP has a structure similar to that of IBM SP2. It has eight nodes used for calculation. The clock frequency of CPU chips is 200 MHz. A node has 256M memory and 4.3G hard disk. The communication bandwidth of its high performance switch board is 10Mbps.

4.2 Test Results and analysis

The test problems are cases used for study on well group schemes. One well group consists of 8 wells. Because of the limit of memory size of the used parallel computers, the data model size of test problems T_1 and T_2, respectively, consists of only 8 and 16 well groups. Their grid sizes, respectively, are $96 \times 12 \times 10$ and $96 \times 24 \times 10$ grids. The deposition of well groups of these two problems are given in Figs. 2 and 3. There is a fault between sub-domain 4 and 5.

Tables 1 and 2 give the experiment results of T_1 and T_2 tested on IBM SP2, where WOF and WF indicate without fault and with fault, respectively. Figs. 4 and 5 are the calculation time of these two test problems. It can be seen that a superlinear speedup can be obtained on IBM SP2, if the number of CPU is not more than ten. However, the efficiency of PNUMSIP decreases seriously for T_2, compared with T_1.

Case	CPU number	Serial	2	4	8
WOF	Wall-clock time (s)	4180.3	1674.9	797.6	464.4
WOF	Speed up	1.000	2.496	5.241	9.002
WOF	Efficiency	1.000	1.248	1.310	1.125
Case	CPU number	Serial	2	4	8
WF	Wall-clock time (s)	4180.3	1387.1	810.6	465.8
WF	Speed up	1.000	3.014	5.158	8.974
WF	Efficiency	1.000	1.507	1.289	1.122

Table 1. Experiment results of T_1 on IBM sp2.

CPU number	Serial	2	4	8	16
Wall-clock time (s)	15059.3	6223.9	4400.4	2492.4	5104.4
Speed up	1.000	2.420	3.422	6.0422	2.950
Efficiency	1.000	1.210	0.856	0.755	0.184

Table 2. Experiment results of T_2 on IBM sp2.

The reason for the superlinear speedup can be explained by the parallelization of I/O. For the serial software NUMSIP, all the well data are loaded by one CPU. But for the parallel software PNUMSIP, the well data are simultaneously loaded by a number of CPUs. The total cache buffer size used by PNUMSIP is much greater than that used by NUMZIP. Therefore, the input time for PNUMSIP will be much shorter than that for NUMSIP. Table 1 shows that the parallel efficiency is improved for a case that the boundaries between sub-domains are along natural faults, only if all the boundaries between sub-domains are not valid boundaries. The total length of boundaries between sub-domains of T_2 is generally greater than that of T_1 and it results in more data communication and more modifications of linear equation coefficient matrices, which leads to the increase of iteration times of the linear equation solution. Therefore, the parallel efficiency of T_2 is less than that of T_1. The rapid increase of data communication and iteration calculation times can also explain why the parallel efficiency drops seriously, as the number of CPUs becomes too large.

Case	CPU number	Serial	2	4	8
WOF	Wall-clock time (s)	1238.8	1244.9	741.6	351.3
WOF	Speed up	1.000	0.995	1.670	3.526
WOF	Efficiency	1.000	0.498	0.418	0.441

Table 3. Experiment results of T_1 on Dawn 1000A.

Table 3 is the results of T_1 without faults tested on Dawn 1000A. Fig. 4 is the comparison of the efficiency of PNUMSIP obtained on IBM SP2 and Dawn 1000A with case T_1. It shows that the speedup and efficiency of PNUM-SIP running on IBM SP2 is much higher than those of PNUMSIP running on Dawn 1000A with the same test problem. But the calculation time of PNUM-SIP running on Dawn 1000A is less than that of PNUMSIP running on IBM SP2. Because the clock frequency of CPUs on Dawn 1000A is faster than that of CPUs on IBM SP2, as seen in Section 4.1, the calculation speed of Dawn 1000A is faster than that of IBM sp2, so the calculation time of Dawn 1000A is less than that of IBM SP2. Compared with IBM SP2, the communication speed of Dawn 1000A is too slow, as seen in Section 4.1. It affects the iteration calculation process of PNUMSIP. Because the non-block communication is used to transfer fast transient data and the data communication delay is mainly caused by the low communication speed, the iteration solution values of neighboring sub-domains used for calculating new iteration solution values may not be updated in time. It slows down the iteration calculation speed and results in a low speedup and efficiency of parallel software.

5   Conclusions

The following conclusion can be drawn from the test results of PNUMSIP. The parallelization of benefits the efficiency improvement of parallel software of reservoir thermal recovery simulation. The data communication bandwidth of computers is an important factor which affects the speedup and efficiency of PNUMSIP. PNUMSIP is a stable and efficient parallel software of reservoir thermal recovery simulation.

References

[1] Briens, F. J. L., Wu, C. H., and Gazdag, J., Compositional reservoir simulation in parallel supercomputing environment, SPE Paper 21214, the 11th SPE Symp. on Reserv. Simul. in Anaheim, CA, February 17-20, 1997.

[2] Bhogeswara, R. and Killough, J. E., Parallel linear solvers for reservoir simulation: A generic approach for existing and emerging computer architectures, SPE Paper 25240, Presented at the 12th SPE Symposium on Reservoir Simulation in New Orleans, LA, Feb. 28-March 3, 1993.

[3] Chen, Zhangxin, Qin, G., and Ewing, R. E., Analysis of a compositional model for fluid flow in porous media, *SIAM J. Appl. Math.*, to appear.

[4] De Sturler, E., Incomplete block LU preconditioners on slightly overlapping subdomains for a massively parallel computer, *Applied Numerical Mathematics* **19** (1995) 129–146.

[5] Dfaz, J. C. and Shenei, K., Domain decomposition and Schur complement approaches to coupling the well equations in reservoir simulation, *SIAM J. SCI. Comput.* **16** (1995), 29–39.

[6] Ghori, S. G., Wang, C. H., and Lim, M. T., Compositional reservoir simulation on CM-5 and KSR-1 parallel machines, the SPE Paper 29110, Presented at the 13th SPE Symposium on Reservoir Simulation in San Antonio, TX, Feb. 12-15, 1995.

[7] Killough, J. E., Camilleri, D., Darlow, B. L., and Foster, J. A., A parallel reservoir simulator based on local grid refinement, SPE Paper 37978, Presented at the 1997 Reservoir Simulation in Dallas, TX, June 1997.

[8] Kokar, G. and Killaugh, J. E., An asynchronous parallel linear equation solution technique, SPE Paper 29142, Presented at the 13th SPE Symposium on Reserv. Simulation in San Antonio, TX, Feb. 12-15, 1995.

[9] Killough, J. E. and Bhogeswara, R., Simulation of compositional reservoir phenomena on a distributed memory parallel computer, SPE Paper 21208, Presented at the 11th SPE Symposium on Reservoir Simulation held in Anheim, California, February 17-20, 1997.

[10] Kortas, S. and Angot, P., A practical and portable model of programming for iterative solvers on distributed memory machine, *Parallel Computing* **22** (1996), 487-512.

[11] Meijerink, J. A., Van Daolen, D. T., and Hoogerbrugge, D. J., Towards a more effective parallel reservoir simulator, SPE Paper 21212, the 11th SPE Symposium on Reserv. Simul. in Anaheim, CA, Feb. 17-20, 1997.

[12] Naculand, U. E., and Lett, G. S., Under and over relaxation techniques for accelerating nonlinear domain decomposition method, SPE Paper 25244, Presented at the 12th SPE Symposium on Reservoir Simulation held in New Orleans, LA, U.S.A., February 28-March 3, 1993.

[13] Notay, Y., An efficient parallel discrete PDE solver, *Parallel Computing* **21** (1995), 1725–1748.

[14] Pommerell, C. and Fichter, W., Memory aspects and performance of iterative solvers, *SIAM J. Sci. Comput.* **15** (1994), 460–473.

[15] Quenes, A., Weiss, W., and Sulten, A. J., Parallel reservoir automatic history matching using a network of workstations and PVM, the SPE Paper 29107, Presented at the 13th SPE Symposium on Reservoir Simulation held in San Antonio, TX, U.S.A., February 12-15, 1995.

[16] Rame, M. and Delshad, M., A compositional reservoir simulation on distributed memory parallel computers, SPE Paper 29103, the 11th SPE Symp. on Reserv. Simul. in San Antonio, TX, Febr. 12-15, 1995.

[17] Wallis, J. R., Foster, J. A., and Kendall, R. P., A new parallel iterative linear solution method for large-scale reservoir simulation, SPE Paper 21209, Presented at the 11th SPE Symposium on Reservoir Simulation held in Anaheim, California, February 17-20, 1997.

A Class of Lattice Boltzmann Models with the Energy Equation

Yuanxiang Li   Shengwu Xiong   Xiufen Zou

Abstract

In this paper a class of lattice Boltzmann models with the energy equation for simulating fluid thermodynamics are studied. The features of this class of models are that the discrete velocity set consists of multi-speed velocities and the internal energy of fluid is introduced by a multi-speed. Therefore, the energy term appears in the local equilibrium distribution functions of these models. Two examples are given in this paper. One is a 1D model and the other is a 2D model, which are used to model a shock wave tube problem and the Benard convection problem, respectively.

KEYWORDS: lattice Boltzmann model, energy equation, shock wave tube, Benard convection

1 Introduction

The lattice gas automata (LGA) method has attracted considerable attention during the last several years in both modeling physical systems and solving partial differential equations. In more recent years there has been a trend of using the lattice Boltzmann (LB) scheme instead of the lattice gas automata method. Unlike the LGA method in which one keeps track of each individual particle, in the lattice Boltzmann approach we are only interested in the particle distribution function. While retaining the advantages existing in the LGA, such as parallel implementation, simple programming, and clear physical images, the LB method is more efficient and accurate in computation in an essentially noise-free manner [1] and has been applied in many areas such as flows in porous media and magneto-fluids. However, most of the models previously considered have not involved the energy equation. In this paper we discuss a class of lattice Boltzmann models with the energy equation by introducing multi-speed velocities to discrete velocity sets. Then particles can exchange internal energy during interactions. Therefore, the energy term appears in particle distribution functions (i.e., the local equilibrium distribution function). In §2, a general description of this class of models is given. In §3, two examples are given; one is a 1D model and the other is a 2D model. Numerical experiments are made in §4. The two models are used to model a shock wave tube problem and the Benard convection problem, respectively. The results show that the models can effectively simulate fluid thermodynamics.

2   A Class of Lattice Boltzmann Models

Divide the domain of a flowing field in a d-dimensional space into a regular lattice and the discrete velocity set $V = \{\mathbf{e}_i : i = 1, \ldots, b\}$. Let $f_i(x, t)$ be the particle distribution function with velocity $\mathbf{e}_i$ at site x and time t. The model evolves according to the lattice Boltzmann equation

$$f_i(x + \mathbf{e}_i, t + 1) = f_i(x, t) + \Omega_i, \tag{2.1}$$

and

$$\Omega_i = -\frac{1}{\tau}(f_i(x, t) - f_i^{eq}(x, t)), \tag{2.2}$$

where Ω_i is called the collision term or the collision function and τ is a relaxation factor ($\tau > 1/2$). According to the physical conservation law, Ω_i must satisfy the conservation law of mass, momentum and energy

$$\sum_{i=1}^{b} \Omega_i = 0, \quad \sum_{i=1}^{b} \Omega_i \mathbf{e}_i = 0, \quad \sum_{i=1}^{b} \Omega_i \frac{1}{2} \mathbf{e}_i^2 = 0. \tag{2.3}$$

In (2.2), f_i^{eq} is the local equilibrium distribution function depending on the dynamic quantities of density, momentum, and energy at site x and time t. In the classical gas dynamics, f_i^{eq} is the Maxwell-Boltzmann distribution. As an approximation of the Maxwell-Boltzmann distribution, let

$$f_i^{eq} = c_0 + c_1(\mathbf{e}_i \cdot u) + c_2(\mathbf{e}_i \cdot u)^2 + c_3 u^2 + c_4(\mathbf{e}_i \cdot u)^3 + \ldots, \tag{2.4}$$

where u is the velocity of fluid and $c_k (k = 0, 1, \ldots)$ are undetermined coefficients, which depend on ρ and ε that are the density and the specific internal energy of the fluid, respectively, defined as follows:

$$\rho = \sum_{i=1}^{b} f_i(x,t) = \sum_{i=1}^{b} f_i^{eq}(x,t),$$

$$\rho \mathbf{u} = \sum_{i=1}^{b} f_i(x,t)\mathbf{e}_i = \sum_{i=1}^{b} f_i^{eq}(x,t)\mathbf{e}_i, \tag{2.5}$$

$$\rho \varepsilon = \tfrac{1}{2} \sum_{i=1}^{b} f_i(x,t)(\mathbf{e}_i - \mathbf{u})^2 = \tfrac{1}{2} \sum_{i=1}^{b} f_i^{eq}(x,t)(\mathbf{e}_i - \mathbf{u})^2.$$

Taking the zeroth, first, and second order moments of $\mathbf{e}_i$ from (2.1) and using (2.3) and (2.5), we can obtain the mass, momentum, and energy equations of fluid

$$\frac{\partial \rho}{\partial t} + \nabla(\rho \mathbf{u}) = 0, \quad \frac{\partial(\rho \mathbf{u})}{\partial t} + \nabla \Pi = 0, \quad \frac{\partial(\rho E)}{\partial t} + \nabla Q = 0, \tag{2.6}$$

where Π is the momentum-flux tensor of fluid, $E = \varepsilon + \tfrac{1}{2}u^2$ (the energy), and Q is the energy-flux vector, which are defined as

$$\Pi_{\alpha\beta} = \sum_{i=1}^{b} e_{i\alpha} e_{i\beta} f_i(x,t), \quad Q_\alpha = \frac{1}{2} \sum_{i=1}^{b} (\mathbf{e}_i)^2 (\mathbf{e}_i)_\alpha f_i(x,t). \tag{2.7}$$

To derive the Navier-Stokes equation, Π must be isotropic. To obtain Π must be combined with some concrete models. In the next section we present two example models; Π is obtained approximately by using the Chapman-Enskog expansion [3] and properly selecting the local equilibrium distribution function like (2.4).

3   Thermodynamic Models

In the thermodynamic motion, fluid particles must exchange energy. Hence, to simulate the thermodynamic feature, modules of $\mathbf{e}_i(i = 1, 2, \ldots, b)$ in the set V of LB models should not be the same.

3.1   A 1D Model

For the problem of 1D flows, the straight line is divided into a uniform lattice and the step size is unit one. Then the set V is chosen as $V = \{-2, -1, 0, 1, 2\}$, which contains five velocities and three speeds, 0, 1, and 2. According to the speed $\sigma(\sigma = 0, 1, 2)$, $\mathbf{e}_i$ and f_i^{eq} are rewritten as $\mathbf{e}_{\sigma i}$ and $f_{\sigma i}^{eq}$, respectively.

Now, we choose $f_{\sigma i}^{eq}$ as follows:

$$f_0^{eq} = \rho(1 - \varepsilon) - \tfrac{1}{2}\rho u^2 + \tfrac{1}{4}\rho u^4,$$
$$f_{1i}^{eq} = \tfrac{1}{3}\rho\varepsilon + \tfrac{1}{3}\rho(1 - \varepsilon)(\mathbf{e}_{1i} \cdot \mathbf{u}) + \tfrac{1}{6}\rho u^2 - \tfrac{1}{6}\rho(\mathbf{e}_{1i} \cdot \mathbf{u})^3 - \tfrac{1}{6}\rho u^4, \qquad (3.1)$$
$$f_{2i}^{eq} = \tfrac{1}{6}\rho\varepsilon + \tfrac{1}{24}\rho(1 + 2\varepsilon)(\mathbf{e}_{2i} \cdot \mathbf{u}) + \tfrac{1}{12}\rho u^2 + \tfrac{1}{12}\rho(\mathbf{e}_{2i} \cdot \mathbf{u})^3 + \tfrac{1}{24}\rho u^4,$$

with $i = 1, 2$. Taking the Chapman-Enskog expansion of the particle distribution function $f_{\sigma i}$ by the local equilibrium distribution function $f_{\sigma i}^{eq}$, we have

$$f_{\sigma i} = f_{\sigma i}^{(0)} + f_{\sigma i}^{(1)} + f_{\sigma i}^{(2)} + \cdots, \qquad (3.2)$$

where $f_{\sigma i}^{(0)} = f_{\sigma i}^{eq}$. Likely, Π and Q possess the same expansion after using (2.7):

$$\Pi = \Pi^{(0)} + \Pi^{(1)} + \Pi^{(2)} + \cdots, \quad Q = Q^{(0)} + Q^{(1)} + Q^{(2)} + \cdots. \qquad (3.3)$$

Under the zeroth order approximation, we have

$$\Pi^{(0)} = 2\rho\varepsilon + \rho u^2 = p + \rho u^2, \quad Q^{(0)} = \rho E u + p u, \qquad (3.4)$$

with $p = 2\rho\varepsilon$ (the pressure). Consequently, the Euler equation is obtained:

$$\frac{\partial\rho}{\partial t} + \frac{\partial(\rho\mathbf{u})}{\partial x} = 0, \quad \frac{\partial(\rho\mathbf{u})}{\partial t} + \frac{\partial(p + \rho u^2)}{\partial x} = 0, \quad \frac{\partial(\rho E)}{\partial t} + \frac{\partial(\rho E u + p u)}{\partial x} = 0. \quad (3.5)$$

If taking the expansion to the first order term of Π and Q, we can have the Navier-Stokes equation and the heat conduction equation

$$\begin{aligned}
\frac{\partial(\rho u)}{\partial t} + \frac{\partial(\rho u^2)}{\partial x} &= -\frac{\partial p}{\partial x} + \frac{\partial}{\partial x}\left(\mu\frac{\partial u}{\partial x}\right), \\
\frac{\partial(\rho E)}{\partial t} + \frac{\partial(\rho E u)}{\partial x} &= -\frac{\partial(p u)}{\partial x} + \frac{\partial}{\partial x}\left(\kappa\frac{\partial\varepsilon}{\partial x}\right),
\end{aligned} \qquad (3.6)$$

with $\mu = 2\rho\varepsilon(2\tau - 1)$ and $\kappa = 7\rho\varepsilon(2\tau - 1)/3$, which are the viscosity and the heat conductivity respectively.

3.2 A 2D Model

As for a flow problem on the 2D plane, the field is divided into a uniform square mesh and the step size is unit one in both directions. Then the set V is chosen as $V = \{(0,0), (1,0), (\sqrt{2},\sqrt{2}), (0,1), (-1,\sqrt{2}), (-1,0), (-\sqrt{2},-\sqrt{2}),$ $(0,-1), (\sqrt{2},-\sqrt{2}), (2,0), (0,2), (-2,0), (0,-2)\}$, which contains thirteen velocities and four speeds, 0, 1, $\sqrt{2}$, and 2. Now, the $f_{\sigma i}^{eq}$'s are chosen as

$$f_0^{eq} = \rho(1 - \tfrac{5}{2}\varepsilon + \tfrac{9}{4}\varepsilon^2) - \tfrac{2}{3}\rho u^2,$$

$$f_{1i}^{eq} = \rho(\tfrac{2}{3}\varepsilon - \tfrac{5}{6}\varepsilon^2) + \rho(\tfrac{2}{3} - \varepsilon)(\mathbf{e}_{1i} \cdot \mathbf{u}) + \tfrac{1}{2}\rho(\mathbf{e}_{1i} \cdot \mathbf{u})^2 - \tfrac{1}{6}\rho u^2 + \tfrac{1}{3}\rho(\mathbf{e}_{1i} \cdot \mathbf{u})^3,$$

$$f_{\sqrt{2}i}^{eq} = \tfrac{1}{8}\rho\varepsilon^2 + \tfrac{1}{4}\rho\varepsilon(\mathbf{e}_{\sqrt{2}i} \cdot \mathbf{u}) + \tfrac{1}{8}\rho(\mathbf{e}_{\sqrt{2}i} \cdot \mathbf{u})^2 - \tfrac{1}{24}\rho u^2 + \tfrac{1}{8}\rho(\mathbf{e}_{\sqrt{2}i} \cdot \mathbf{u})^3,$$

$$f_{2i}^{eq} = \rho(\tfrac{7}{48}\varepsilon^2 - \tfrac{1}{24}\varepsilon) - \rho(\tfrac{1}{24} - \tfrac{1}{18}\varepsilon)(\mathbf{e}_{2i} \cdot \mathbf{u}) - \tfrac{1}{96}\rho u^2 + \tfrac{1}{96}\rho(\mathbf{e}_{2i} \cdot \mathbf{u})^3,$$

$$(3.7)$$

with $i = 1, 2, 3, 4$. By the same way as above, taking the Chapman-Enskog expansion to the zeroth order term and the first order term, we can obtain the Euler and Navier-Stokes equations, respectively,

$$\partial_t(\rho u_\alpha) + \partial_\beta(\rho u_\alpha u_\beta) = -\partial_\alpha p, \quad \partial_t(\rho\varepsilon) + \partial_\beta(\rho\varepsilon u_\beta) = -\partial_\beta(p u_\beta), \qquad (3.8)$$

and

$$\partial_t(\rho u_\alpha) + \partial_\beta(\rho u_\alpha u_\beta) = -\partial_\alpha p + \partial_\beta[\nu\partial_\beta(\rho u_\alpha) + \zeta\partial_\alpha(\rho u_\beta)],$$

$$\partial_t(\rho\varepsilon) + \partial_\beta(\rho\varepsilon u_\beta) = -\partial_\beta(p u_\beta) + \partial_\alpha(\kappa\rho\partial_\alpha\varepsilon) + \partial_\alpha[\nu u_\beta\partial_\beta(\rho u_\alpha) + \zeta\partial_\alpha(\rho u^2)],$$

$$(3.9)$$

where $\nu = \varepsilon(2\tau - 1)/2$, $\kappa = \varepsilon(2\tau - 1)$, and $\zeta = 2$, which are the shear viscosity, heat conductivity, and bulk viscosity, respectively.

3.3   Numerical Experiments

To examine the models presented above, we apply them to two problems, which are a shock wave tube problem and the Benard convection problem.

The 1D model in §3.1 is applied to the shock wave problem with Sod's initial conditions [7]:

$$(\rho_L, u_L, p_L) = (1.0, 0.0, 1.0), \quad (\rho_R, u_R, p_R) = (0.125, 0.0, 0.1).$$

Fig. 1 gives the results at time step 30. The numerical results show that the LB model gives the correct position and the resolution of the shock wave and contact discontinuity are also good.

The second model is used to the Benard convection problem which comes from a thermal convection experiment of a fluid that Benard did in 1900.

A container is filled up with a fluid. The distance between the upper and the lower plates of the container is much smaller than the width and the length of the plates. Heating the lower plate continuously, the temperature gradient arises in the fluid. If the temperature difference between those two plates, $\Delta T = T_l - T_u$, is not large, the fluid is macroscopically still in the

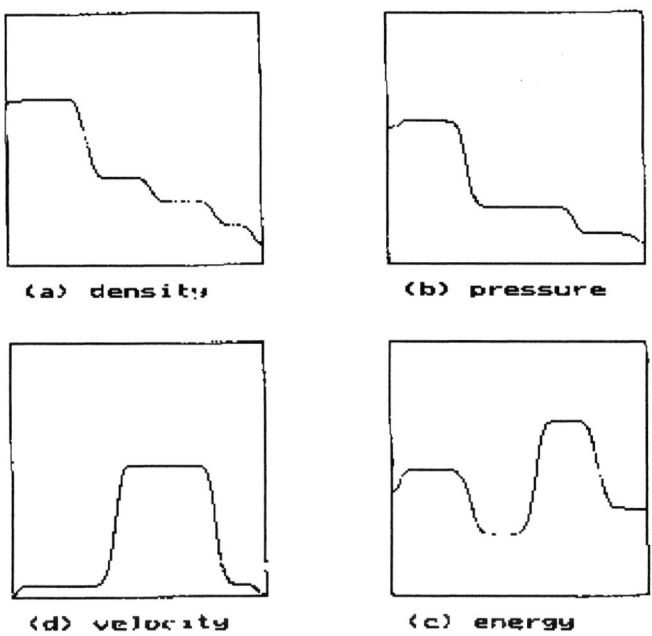

(a) density   (b) pressure

(d) velocity   (c) energy

Figure 1: The shock wave tube problem.

stationary state. Eventually, when ΔT exceeds a critical value, the stationary state of the fluid is broken down suddenly and is replaced by the convection state. The convection circuits, i.e., the Benard pattern, can be seen in the container.

The mathematical description about the Benard convection problem is the Boussineq equations [2]

$$\nabla \cdot \mathbf{u} = 0,$$
$$\frac{1}{P}\left(\frac{\partial \mathbf{u}}{\partial t} + (\mathbf{u} \cdot \nabla)\mathbf{u}\right) = -\nabla p + \Delta \mathbf{u} + R_a \alpha \theta \lambda,$$
$$\frac{\partial \theta}{\partial t} + \mathbf{u} \cdot \nabla \theta = \mathbf{u} \cdot \lambda + \Delta \theta.$$

The boundary conditions are

$$u = 0, \quad \theta = 0, \qquad x_3 = 0, 1,$$

where $\lambda = (0, 0, 1)$ is a upper vertical unit vector, θ denotes the difference of temperature distribution between a certain initial state of the fluid and its perturbed state. $P = \nu/\kappa$ is called the Prandtl number and $R_a = g\alpha d^3 \frac{\Delta T}{\nu \kappa}$

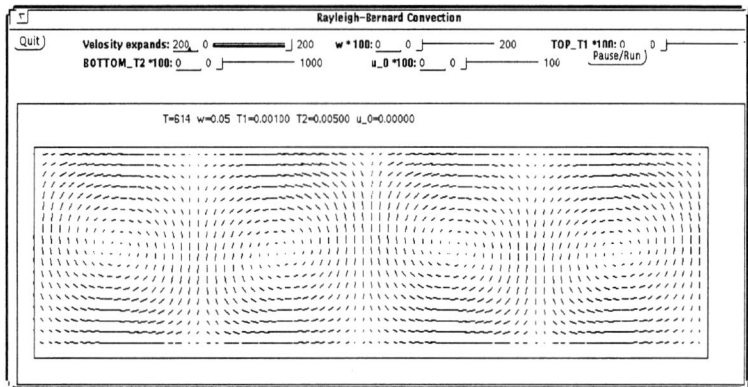

Figure 2: The Benard pattern.

the Rayleigh number in which ν, κ, g, α, and d are the shear viscosity, the heat conductivity, the gravity acceleration, the coefficient of heat expansion, and the thickness of the fluid layer, respectively. This is a bifurcation problem. As the parameter R_a changes, the solution of the problem also changes. When the Rayleigh number $R_a > 1708$, the heat conduction of the fluid layer changes into the Benard convection.

In our simulation, a rectangle container is chosen and it is divided into a square lattice of 80×20. The initial state of fluid in the container is given randomly and the boundary conditions are treated as in [8]. The result at time step 763 is shown in Fig. 2 and a very good convection pattern is obtained which is in a good agreement with the result of experiment.

Acknowledgments. This work was supported in part by National Natural Science Foundation of China.

References

[1] Alxander, F. J., Chen, H., Chen, S., and Doolen, G. R., Lattice Boltzmann model for compressible fluids, *Phys. rev. A* **46** (1992), 1967–1970.

[2] Chandrasekher, S., *Hydrodynamic and Hydromagnetic Stability*, Oxford Press, Clarendon, 1961.

[3] Chapman, S., and Cowling, T. G., *The Mathematical Theory of Non-Uniform Gases*, 3rd ed., Cambridge University Press, London, 1970.

[4] Chen, Y., Ohashi, H., and Akiyama, M., Heat transfer in lattice BGK modeled fluid, *J. Stat. Phys.* **81** (1995), 71–85.

[5] Qian, Y. H., Simulating thermohydrodynamics with lattice BGK models, *J. Sci. Comp.* **8** (1993), 231–242.

[6] Qian, Y. H., d'Humieres, D., and Lallmand, P., Lattice BGK models for Navier-Stokes equation, *Europhys. Lett.* **17** (1992), 479–484.

[7] Sod, G. A., A survey of several finite difference methods for systems of non-linear hyperbolic conservation laws, *J. Comp. Phys.* **27** (1978), 1–8.

[8] Zou, Q. and He, X., On pressure and velocity boundary conditioned for the lattice Boltzmann BGK model, *Phys. Fluids* **9**(1997), 1591–1598.

Block Implicit Computation of Flow Field in Solid Rocket Ramjets

ZHIBO MA   JIANSHI ZHU

Abstract

To compute the flow field in solid rocket ramjet (SRR) in which the chamber has a complex boundary, a block implicit algorithm (BIA) had been developed. The boundary conditions of three-dimensional steady-state Navier-Stokes (NS) equations were treated by modifying the discrete equations and the grids were generated through an algebraic way. These methods have been put into practice and proved to be valid and efficient in the computation of flow field in the chamber. The technique developed here applies to similar problems in porous medium flows.

KEYWORDS: rocket ramjet, numerical simulation, block implicit algorithm

1 Introduction

For the purpose of numerical study on the reaction flow field in SRR, the authors of this paper have developed a computer program in which the BIA has been adopted [2-4]. In this program all the control equations are in a cylindrical coordinate system and accordingly the domain of flow field is a simple column.

In the real SRR or so-called ducted rocket, the dome plate always presents the shape of a ellipsoid and the fuel nozzles are inserted into the forepart of the chamber. Such designs of structure make the field boundary complicated. For the sake of accommodating these situations, one may think that the curvilinear coordinate system would be the first choice, but this may not be the best idea because it increases the computational cost. In fact, the most part of the long chamber is in the shape of a column and it is still advantageous to use the cylindrical coordinate system if the boundary conditions could be processed properly. In this paper the coordinates of grids are calculated with an algebraic method and the treatment of boundary conditions of NS equations is discussed corresponding to these specific grids.

2   Generation of Grids

A new program has been developed to generate automatically grids. In the program the grids are generated in such a manner that the resulting computational domain can cover the actual domain. The boundary grids at the forepart of the chamber present the profile of ladders. This treatment can lead to a high accuracy in numerical simulation.

To eliminate the false numerical waves of pressure, a stagger-mesh has been adopted. In the stagger-mesh system, scalars are stored on the main grids and velocities are stored on the interface of the main grids. For three-dimensional problems, each control volume (or a main grid) has six interfaces. These interfaces can be divided into three kinds, each of which is perpendicular to the axes of x, r, and θ, respectively. During the process of calculations, each interface is labeled by a predefined integral number, which is stored in arrays named *judx(i,j,k), judr(i,j,k) and judth(i,j,k) corresponding to the* three kinds of interfaces. The integral numbers representing the characteristics of interfaces are defined as follows:

"0" .... means the interfaces are on the wall;

"1" .... means the interfaces are in the inner field;

"11" ....means the interfaces are at the front air inlets;

"12" ....means the interfaces are at the back air inlets;

"21" ....means the interfaces are at the first fuel nozzle;

"22" ....means the interfaces are at the second fuel nozzle;

"23" ....means the interfaces are at the third fuel nozzle.

When calculating flow fields, we must follow the conservation of mass of air and fuel, so in the process of discretization, the areas of interfaces must meet the conditions

$$\sum_{judx(i,j,k)=11} A_x(i,j,k)cosA_1 + \sum_{judr(i,j,k)=11} A_r(i,j,k)sinA_1 = AREA_1,$$

$$\sum_{judx(i,j,k)=12} A_x(i,j,k)cosA_2 + \sum_{judr(i,j,k)=12} A_r(i,j,k)sinA_2 = AREA_2,$$

where $AREA_1$ and $AREA_2$ denote the cross-section areas of front and back air inlets, respectively. The grids on the fuel nozzles occur in the same way.

3   Outline of Block Implicit Algorithms

The steady control equation on a finite volume can be generally described in an integral form

$$\oint_{\Delta \vec{S}} (\rho \vec{V} \phi - \Gamma_\phi \nabla \phi) \cdot d\vec{S} = \int_{\Delta V} S_\phi dV,$$

where ϕ denotes velocities or scalars.

When using a block implicit algorithm to solve the steady-state NS equations, the symmetrically coupled Gauss-Seidel (SCGS) technique is adopted in sweeping through the grids to smooth u, v, w, and p. For the grid with indices (i, j, k), the variables to be smoothed can be described as a column vector [1]

$$\vec{X} = [x_1 \ x_2 \ x_3 \ x_4 \ x_5 \ x_6 \ x_7],$$

and $\vec{X}$ is seperated into two parts as

$$\vec{X}_{new} = \vec{X}_{old} + \Delta \vec{X},$$

when it is updated. The corresponding discrete equations of $\Delta \vec{X}$ are

$$A\Delta \vec{X} = B. \tag{3.1}$$

It can be easily solved through a direct method.

4   Treatment of Boundary Conditions

To solve the NS equations of compressible flow in the chamber of SRR, computational conditions must be given by: (a) velocities and stagnation temperatures of air at the exit of inlets and those of fuel at the exit of fuel nozzles and (b) pressure at the exit of the chamber.

Generally, the mass flows of air and fuel are design parameters, so their velocities are unknown and must be recomputed continually to meet the given flows according to the densities at the air inlets and fuel nozzles. The density is obtained from the ideal gas state equation

$$\rho = p/RT,$$

where the temperature is computed from the thermal enthalpy H and special heat C_p such as

$$T = H/C_p.$$

Like the velocities on the entrance of the chamber, the pressure at the exit of the chamber (or the so-called p_{out}) is also an unknown quantity. The p_{out} must be continually updated along with the iteration cycles on u, v, w, and p to meet the real boundary conditions. In other words, p_{out} is regarded as a constant in each cycle mentioned above and finally determined by other predefined parameters and physical states such as the stagnation pressures of air and fuel, the throat area of the chamber nozzle, and the combustion efficiency. To avoid divergence in solving the NS equations using BIA, care must be taken to treat the pressure, density, and velocity at the boundaries. With regard to the problem about the SRR chamber, five kinds of boundaries are encountered. Except at the exit of the chamber, a zero-derivative condition is used on the variables p and ρ at the boundary faces. At the exit of the chamber, pressure is given by p_{out} as

$$p(imax - 1, j, k) = p_{out},$$

and the density is obtained from the ideal gas state equation.

The velocities can be computed through their discrete equations described as follows:

(a) **At the entrance of air and fuel**

At the entrance the velocities can be regarded as constants within each iteration cycle on u, v, w, and p, but they can not be allowed to be equal to the constants directly or it leads to divergence. In fact, the velocities on the boundaries are allowed to fluctuate, but finally they converge to the practical values. For example, at the entrance of air, if the three velocity components are U_{air}, V_{air}, and W_{air}, then their discrete equations corresponding to equation (3.1) are revised as

$$a_{11}\Delta u(i, j, k) + a_{17}\Delta p(i, j, k) = a_{11}[U_{air} - u(i, j, k)_{old}],$$
$$a_{33}\Delta v(i, j, k) + a_{37}\Delta p(i, j, k) = a_{33}[V_{air} - v(i, j, k)_{old}],$$
$$a_{55}\Delta w(i, j, k) + a_{57}\Delta p(i, j, k) = a_{55}[W_{air} - w(i, j, k)_{old}].$$

(b) **At the solid wall face**

As the non-slide condition is adopted, the speed of fluid is zero on the wall, so the discrete equations of $u(i, j, k)$, $v(i, j, k)$, and $w(i, j, k)$ are

$$a_{11}\Delta u(i, j, k) + a_{17}\Delta p(i, j, k) = -a_{11}u(i, j, k)_{old},$$
$$a_{33}\Delta v(i, j, k) + a_{37}\Delta p(i, j, k) = -a_{33}v(i, j, k)_{old},$$
$$a_{55}\Delta w(i, j, k) + a_{57}\Delta p(i, j, k) = -a_{55}w(i, j, k)_{old}.$$

(c) **At the symmetric plane**

In this situation, $w(i, j, k)$ or $w(i, j, k + 1)$ become zero, so their discrete equations are

$$a_{55}\Delta w(i, j, k) + a_{57}\Delta p(i, j, k) = -a_{55}w(i, j, k)_{old},$$
$$a_{66}\Delta w(i, j, k + 1) + a_{67}\Delta p(i, j, k) = -a_{66}w(i, j, k + 1)_{old}.$$

(d) **At the axis of the chamber**

At the first layer of grids near the axis of the chamber, the SCGS operator can not be applied, so the velocities are approximated by using those of the second layer of grids:

$$u(i, 1, k) = u(i, 2, k),$$
$$v(i, 1, k) = cos(0.5(\theta(i, j, k) + \theta(i, j, k + 1)))v_0(i),$$
$$w(i, 1, k) = sin(\theta(i, j, k))v_0(i),$$

where

$$v_0(i) = |\vec{v}_0(i)|.$$

In the above formula, $\vec{v}_0(i)$ is the velocity component projected in the (r, θ) plane of the fluid, the position of which is on the axis of chamber.

To keep the conservation of mass flow, the variables of $v(i, 2, k)$ must be obtained from the mass continuity equation.

(e) **At the exit of chamber**

The flow field can be regarded as fully developed at the exit of the chamber; therefore, the derivatives of velocities along with the chamber axis are considered to be zero. They are described as

$$u(imax, j, k) = u(imax - 1, j, k), \quad v(imax, j, k) = v(imax - 1, j, k),$$
$$w(imax, j, k) = w(imax - 1, j, k).$$

5 Numerical Example

To verify the quality of the treatment of boundary conditions, we have calculated a problem of a model ramjet that has two front air inlets and two back fuel nozzles. The structure is symmetrical in (r, θ) plane. The computation domain of θ is $0° \sim 90°$. The mesh contains $34 \times 17 \times 11$ grid nodes. To solve the compressible problem (neglecting the chemical reaction), it spends PC586 about 3.0 minutes when the norm of residuals converges to a level less than 10^{-5}. The normalized error of flow at the exit of chamber is less than 0.3%.

6 Conclusions

The numerical simulation on reaction flow in SRR is a burdensome task that spends much computation time, so it is important to develop a numerical method to solve efficiently this practical engineering problem. Through the approach described above, the grids are generated rapidly by an algebraic way. With the special treatment of boundary conditions, the NS equations can be solved quickly in the cylindrical coordinate system and the conservation of mass flow can be well observed even for the chamber with a large L/D ratio. The computer program developed under the guidance of these ideas plays a useful role in the research and development on SRR.

References

[1] Ma, Z., The numerical computation of after burner flows in solid rocket ramjets, Doctor Dissertation, Beijing University of Aeronautics and Astronautics, 1998.

[2] Ma, Z., Cai, X., and Zhang, Z., The development about computation method on flow field in the after burner of solid rocket ramjet, *J. Aviation Weapons* **5** (1997).

[3] Ma, Z., Zhang, Z., and Cai, X., Numerical study of mixing flows in a ducted rocket combustor, *Journal of Propulsion Technology* **4** (1998).

[4] Vanka, S. P., Fast numerical computation of viscous flow in a cube, *Numerical Heat Transfer* **20** (1991), 255–261.

Stable Conforming and Nonconforming Finite Element Methods for the Non-Newtonian Flow

PINGBING MING   ZHONG-CI SHI

Abstract

Some mixed finite element methods for a non-Newtonian flow problem are discussed in this pape. A new variational formulation is proposed to handle the Newtonian-viscosity dependent problem arising
from a three-field version of the White-Metzner model. It allows us
to derive Newtonian-viscosity independent error bounds and to facili-
tate the choice of stable finite element spaces. Moreover, a continuous
stress approximation can be obtained without any extra effort. The
enhanced-strain-oriented quadrilateral Wilson-P_1 element is used for
approximating the two-field version with or without a modification on
the variational formulation. Stability is justified and quasi-optimal
error bounds are presented.

KEYWORDS: mixed finite element, Wilson element enhanced strain method,
Korn inequality

1   Introduction

Non-Newtonian flows are extensively involved in a number of problems such
as the production of oil and gas from underground reservoirs. Such kind
of problem has an intrinsic, two-scale coupling which poses some challenges
in the numerical simulation, and the difficulty emanates from the nonlinear
viscosity. Numerical simulation of many problems in this field is still open.
In this paper we consider a simplified model, which possesses some intrinsic
characters of the original problem. It is believed that an effective algorithm
for the original problem should be effective for the simplified model considered
herein.

Our assumptions for the simplification are the following: only steady-state
flows are take into account, inertial effects are ignored, the Deborah number
is small, and, consequently, we only consider quasi-Newtonian fluids.

The line of the paper follows: in §2, the White-Metzner model is presented and a new variational principle is proposed to deal with the Newtonian-viscosity-dependent problem and in §3, the enhanced-strain-oriented quadrilateral Wilson-P_1 element is presented and analyzed.

2 Model of White-Metzener and Newtonian-Viscosity Dependent Problem

We give a brief description of a White-Metzner model for the viscoelastic flow. We refer to [7] for a more general description.

Let Ω be a bounded convex polygonal in $\mathcal{R}^N (N = 2, 3)$ with the Lipschitz boundary Γ. $\mathcal{R}^N$ is equipped with Cartesian coordinates $x_i, i = 1, \dots, N$. For a function u, $\frac{\partial u}{\partial x_i}$ is written as $u_{,i}$, and the Einstein convention for summation is used.

For a scalar function p, the gradient of p is a vector ∇p, and $(\nabla p)_i = p_{,i}$. If q is another scalar function, we denote $(p, q) = \int_\Omega pq$. For a vector function $\mathbf{u}$, the gradient of $\mathbf{u}$ is a tensor $\nabla \mathbf{u}$, $(\nabla \mathbf{u})_{ij} = \mathbf{u}_{i,j}$, div$\mathbf{u} = \mathbf{u}_{i,i}$, $\mathbf{u} \cdot \nabla = \mathbf{u}_i \frac{\partial}{\partial x_i}$. For a tensor function σ, $\nabla \cdot \sigma$ is a vector function, $(\nabla \cdot \sigma)_i = \sigma_{ij,j}$, $\sigma : \tau = \sigma_{i,j}\tau_{i,j}$, $|\sigma|^2 = \sigma : \sigma$, and $(\sigma, \tau) = \int_\Omega \sigma : \tau$.

To describe the flow, we use the pressure p (scalar), the velocity vector $\mathbf{u}$, and the total stress tensor σ_{tot}. $\mathcal{E}u = \frac{1}{2}(\nabla \mathbf{u} + \nabla^T \mathbf{u})$ is the rate of strain tensor, and $\mathcal{E}_{\mathrm{II}}(\mathbf{u}) = \frac{1}{2}\mathcal{E}_{ij}(\mathbf{u})\mathcal{E}_{ij}\mathbf{u}$ is the second invariant of $\mathcal{E}\mathbf{u}$.

A White-Metzner type model is described by the constitutive equations:

$$\sigma_{tot} = \sigma + \sigma_N - p\mathbf{I}, \ (\mathbf{I})_{ij} = \delta_{ij}, \ \sigma_N = 2\mu\mathcal{E}u, \ \sigma = 2\eta(\mathcal{E}_{\mathrm{II}}(\mathbf{u}))\mathcal{E}\mathbf{u},$$

where μ is the Newtonian part of the viscosity and η is the viscosity function for the viscoelastic part. The velocity $\mathbf{u}$ must satisfy the incompressible condition

$$\mathrm{div}\mathbf{u} = 0.$$

In this paper we consider the stationary creeping flow [11]. The fluid is subject to a density of force $\mathbf{f}$. Then the momentum equation is

$$-\mathrm{div}\sigma_{tot} = \mathbf{f}.$$

Two classical laws for η are the Power law and Carreau's law [11]:

Power law: $\eta_p(z) = \frac{1}{2}gz^{\frac{(r-2)}{2}}, \quad r > 1, \ g \geq 0;$

Carreau's law: $\eta_c(z) = (\eta_0 - \eta_\infty)(1 + \lambda z)^{\frac{r-2}{2}}, \quad 0 \leq \eta_\infty < \eta_0, r > 1.$

Sobolev spaces are needed. $\mathbf{T} = [L^{r'}(\Omega)]^{N(N+1)/2} = \{\tau = (\tau_{ij}) \mid \tau_{ij} = \tau_{ji}, \tau_{ij} \in L^{r'}(\Omega), i, j = 1, \ldots, N\}$ with the norm $\|\tau\|_{\mathbf{T}} = (\int_\Omega |\tau|^{r'})^{\frac{1}{r'}}$. $(\cdot, \cdot)$ denotes the inner product of $\mathbf{X} = [W_0^{1,r}(\Omega)]^N, M = L_0^r(\Omega) = \{q \in L^{r'}(\Omega) \mid \int_\Omega q = 0\}$. $\mathbf{X}$ and M are equipped with the norm $\|\mathbf{v}\|_{\mathbf{X}} = (\int_\Omega |\mathcal{E}\mathbf{v}|^r)^{\frac{1}{r}}$ and $\|q\|_M = (\int_\Omega |q|^{r'})^{\frac{1}{r'}}$, respectively. It is easy to see $\|\cdot\|_{\mathbf{X}}$ is an equivalent norm over X. We also denote $\mathbf{T}'$ and $\mathbf{X}'$ the dual space of $\mathbf{T}$ and $\mathbf{X}$, respectively, and $\langle , \rangle$ the dual multiple between $\mathbf{X}$ and $\mathbf{X}'$. For $1 < r < 2$, we must modify the definition of $\mathbf{X}$ and define $\mathbf{X}_1 = \mathbf{X} \cap [H_0^1(\Omega)]^N$, but we still denote it as $\mathbf{X}$ for simplicity .

Some progress has been achieved in the finite element approximation of this problem. In 1992, Baranger, et al. [2] gave the first finite element approximation based upon a special case of the following variational formulation with $\alpha = 1$. However, abstract error bounds they obtained are μ-dependent; i.e, the error bounds are deteriorated as the Newtonian viscosity approaches zero. Furthermore, no finite element space pair is available. Meanwhile, the continuous approximation for the extra stress is widely used in engineer literatures; however, a fairly large finite element space is needed to achieve this goal ([9, 20]) that would cause an extra cost and lose the accuracy. Recently, the so-called *EVSS* and its modification [8] are proposed to attack this problem, but it needs an extra variable that would increase the cost and even seriously it would lead to unsymmetric algebraic systems.

Below a new variational formulation ([1, 13]) is proposed to solve the above problems. Some operators are needed to introduce our method:

$$B : \mathbf{X} \to \mathbf{X}', \quad B(\mathbf{u}) = 2\eta(\mathcal{E}_{\mathrm{II}}(\mathbf{u}))\mathcal{E}\mathbf{u},$$
$$\mathcal{H}_\alpha : \mathbf{T} \times \mathbf{X} \times M \to \mathbf{T}' \times \mathbf{X}' \times M',$$
$$l : \mathbf{T} \times \mathbf{X} \times M \to \mathbf{T}' \times \mathbf{X}' \times M',$$
$$\mathbf{x} = (\sigma, \mathbf{u}, p), \quad \mathbf{y} = (\tau, \mathbf{v}, q),$$
$$(\mathcal{H}_\alpha(\mathbf{x}), \mathbf{y}) = \alpha(A(\sigma), \tau) - \alpha(\tau, \mathcal{E}\mathbf{u}) + \alpha(\sigma, \mathcal{E}\mathbf{u}) + (1 - \alpha)(B(\mathbf{u}), \mathcal{E}\mathbf{v})$$
$$+ 2\mu(\mathcal{E}\mathbf{u}, \mathcal{E}\mathbf{v}) - (p, \mathrm{div}\mathbf{v}) + (q, \mathrm{div}\mathbf{u}), \quad \alpha \in [0, 1],$$
$$< l, \mathbf{y} > = < \mathbf{f}, \mathbf{v} > .$$

We define the problem: *Problem $\mathcal{H}$*. Find $\mathbf{x} \in \mathbf{T} \times \mathbf{X} \times M$ such that

$$(H_\alpha(\mathbf{x}), \mathbf{y}) = < l, \mathbf{y} > \quad \forall \mathbf{y} \in \mathbf{T} \times \mathbf{X} \times M, \forall \alpha \in [0, 1].$$

It can be cast into the saddle point problem: *Problem $\mathcal{H}$*. Find $\mathbf{x} \in$

$\mathbf{T} \times \mathbf{X} \times M$ such that

$$\alpha(A(\sigma), \tau) - \alpha(\tau, \mathcal{E}\mathbf{u}) = 0 \quad \forall \tau \in \mathbf{T},$$
$$\alpha(\sigma, \mathcal{E}\mathbf{v}) + (1 - \alpha)(B(\mathbf{u}), \mathcal{E}\mathbf{v}) + 2\mu(\mathcal{E}\mathbf{u}, \mathcal{E}\mathbf{v})$$
$$-(p, \mathrm{div}\mathbf{v}) = < \mathbf{f}, \mathbf{v} > \quad \forall \mathbf{v} \in \mathbf{X},$$
$$(\mathrm{div}\mathbf{u}, q) = 0 \quad \forall q \in M, \quad \alpha \in [0, 1].$$

When α equals 1 or 0, *Problem* $\mathcal{H}$ degenerates to the problem which has been discussed in [5, 6], or the model discussed in the next section, respectively. We only consider the case $\alpha \in (0, 1)$ in this section.

Based on the classic theory of a mixed variational problem (see [3] for the linear case and [19] the for nonlinear case), the basic vehicle for the analysis of *Problem* $\mathcal{H}$ is the following two inequalities, which concern the monotony and continuity properties of $A(\sigma)$.

Though the explicit formula for $A(\sigma)$ is not available when $\eta = \eta_c$, we are still able to prove the monotony and continuity properties of $A(\sigma)$ in this case.

Theorem 2.1 *([14]) There exists a constant C independent of σ and τ such that, for any $\delta \geq 0$,*

$$(A(\sigma) - A(\tau), \sigma - \tau) \geq C|\sigma - \tau|^{2+\delta}(1 + |\sigma| + |\tau|)^{r'-2-\delta},$$
$$|A(\sigma) - A(\tau)| \leq C|\sigma - \tau|^{1-\delta}(1 + |\sigma| + |\tau|)^{r'-2+\delta}.$$

The proof of these inequalities is tremendously long since in this case the explicit form of $A(\sigma)$ is unknown. The baisc idea in our proof is the prolongation trick and the application of the known results for $A(\sigma)$ when $\eta = \eta_p$.

We assume that the triangulation $\mathcal{C}_h$ is a regular partition of Ω; the quasi-uniformity of $\mathcal{C}_h$ is not necessary unless we state it clearly. Let $\mathbf{T}_h$, $\mathbf{X}_h$, and M_h be the finite element space: $\mathbf{T}_h \in \mathbf{T}$, $\mathbf{X}_h \in \mathbf{X}$, and $M_h \in M$.

Problem $\mathcal{H}_h$. Find $\mathbf{x}_h = (\sigma_h, \mathbf{u}_h, p_h) \in \mathbf{T}_h \times \mathbf{X}_h \times M_h$ such that

$$(\mathcal{H}_\alpha(\mathbf{x}_h), \mathbf{y}) = < \mathbf{l}, \mathbf{y} > \quad \forall \mathbf{y} \in \mathbf{T}_h \times \mathbf{X}_h \times M, \quad \forall \alpha \in (0, 1).$$

The well-posedness of *Problem* $\mathcal{H}_h$ depends heavily on the *B-B inequality*

$$\exists \beta_h(r) > 0, \quad \inf_{q \in M_h} \sup_{\mathbf{v}_h \in \mathbf{X}_h} \frac{(\mathrm{div}\mathbf{v}_h, q)}{\|\mathbf{v}_h\|_{\mathbf{x}}\|q\|_M} \geq \beta_h(r).$$

Choosing $\mathbf{X}_h$ and M_h satisfy the above *B-B inequality* is a delicate task, though there are many examples for $r = 2$ (see [10] and [3] for a review).

Nevertheless, there exist far fewer examples for the case $r \neq 2$. In [16], a general criterion is given and exploited to prove that almost all the stable pairs for the case $r = 2$ is also stable for the case $r \neq 2$.

Remark 2.1 *Note that the B-B inequality for (σ, u) and $(\sigma_h, \mathbf{u}_h)$ is no longer needed either for Problem $\mathcal{H}$ or for Problem $\mathcal{H}_h$. An enhanced K-ellipticity is introduced in a natural, reasonable and unifying way. It is known that there are many finite element space pairs satisfying the B-B inequality either for (σ, u) or for $(\mathbf{u}, p)$, but they do not satisfy the two B-B inequalities simultaneously [18].*

To derive error bounds, we adopt the abstract quasi-norm introduced in [6], which is very useful. Let $(\sigma, \mathbf{u}) \in \mathbf{T} \times \mathbf{X}$ be the solution of *Problem (H)*. Then for $(\tau, \mathbf{v}) \in \mathbf{T} \times \mathbf{X}$, we define the quasi-norms

$$|\tau|_{r'}^{\rho'} = \int_\Omega |\tau|^2 (\theta + |\sigma| + |\tau|)^{r'-2}, \quad \rho' = \max(2, r'),$$
$$|\mathcal{E}\mathbf{v}|_r^\rho = \int_\Omega |\mathcal{E}\mathbf{v}|^2 (\theta + |\mathcal{E}\mathbf{u}| + |\mathcal{E}\mathbf{v}|)^{r'-2}, \quad \rho = \max(2, r).$$

If the usual approximation properties for the finite element spaces $\mathbf{T}_h$, $\mathbf{X}_h$, and M_h is assumed and the *B-B inequality* hlods, then we have the following error bounds.

Define $\kappa \in [r, r + (2-r)\theta]$ for $r \in (1, 2]$ and $\kappa \in [r, r + (r-2)\theta]$ for $r \in (1, 2]$ below.

Theorem 2.2 *Let $\alpha \in (0, 1)$ and $(\sigma_h, u_h) \in T_h \times V_h$ be the unique solution of Problem $\mathcal{H}_h$.*

(i) If $r \in (1, 2]$, we have

$$\|\sigma - \sigma_h\|_{\mathbf{T}} \leq C h^{k \frac{\kappa}{r'}}, \quad \|\mathbf{u} - \mathbf{u}_h\|_{1,r} + \|\mathbf{u} - \mathbf{u}_h\|_{1,2} \leq C h^{\frac{k\kappa}{2}}.$$

In particular, when $\theta = 1$, we even have $\|\sigma - \sigma_h\|_{L^2} \leq C_1 h^k$.

(ii) If $r \in [2, \infty)$, we have

$$\|\sigma - \sigma_h\|_{\mathbf{T}} \leq C h^{\frac{k\kappa'}{2}}, \quad \|\mathbf{u} - \mathbf{u}_h\|_{1,r} \leq C h^{\frac{k\kappa'}{r}}.$$

When $\theta = 1$, we even have

$$\|\mathbf{u} - \mathbf{u}_h\|_{1,2} \leq C h^k, \quad \|\sigma - \sigma_h\|_{L^2} \leq C h^{\frac{k\kappa'}{2}}.$$

(iii) Furthermore, if the mesh is quasi-uniformity, then

$$\|p - p_h\|_{L^{\nu'}} \leq C\zeta, \quad r \in (1, 2],$$
$$\zeta = \min\left(\mu^{\frac{1}{2}} h^{\frac{k\kappa}{2} + N(\frac{1}{2} - \frac{1}{\kappa})}, h^{k(\kappa-1)}\right), \tag{2.1}$$
$$\|p - p_h\|_{L^{\nu'}} \leq C h^{\frac{k\kappa'}{2}}, \quad r \in [2, \infty).$$

Remark 2.2 *When* $r \in (1,2]$, *the error bound* (2.1) *for the pressure is deteriorated. We find that the deterioration happens only in the limiting cases:* (1) $N = 2$, $k = 1$, (2) $N = 3$, $k = 1$, (3)$N = 3$, $k = 2(r \in (1, \frac{3}{2}))$. *If we assume* $\mu = h^{(\frac{N}{r}-k)(2-r)}$ *in these three cases, the accuracy can be recovered. In fact, this kind of assumption on* μ *is realistic when* μ *is very small. Recalling that in this case the proper Sobolev space for the pressure is* $L_0^2(\Omega)$, *we only need to derive error bounds in* $L_0^2(\Omega)$. *Since the norm on* $L_0^2(\Omega)$ *is weaker than that in* $L_0^{r'}(\Omega)$ *in the present situation, we can expect to get* μ-*independent error bounds; it is just the case.*

Corollary 2.1 *With the same assumptions as in case* (iii) *of Theorem 2.2, we have*

$$\|p - p_h\|_{L^2} \le Ch^{k(r-1)}, \quad \theta = 0, \quad \|p - p_h\|_{L^2} \le Ch^k, \quad \theta = 1.$$

Exploiting the special structure of our quasi-norm, we can get some error bounds with respect to stronger norm (see [14]).

3   Wilson Element for Non-Newtonian Flow

In 1995, Reddy and Simo [17] investigated the stability property of the enhanced strain method, which is identical to the Wilson element method under uniform square meshes. More recently, Zhang [24] has established the stability of the Wilson element and its variants [23] for the incompressible elastic problem over an arbitrary quadrilateral mesh with a minor modification of the original variational formulation. In [17], the enhanced strain oriented element has been subject to a severe numerical test, which indicated that such kind of element performs very well even on relatively coarse mesh. Therefore, it is worthwhile to study the Wilson element in the present context.

In this section, we study the convergence behavior of the enhanced strain oriented Wilson element for *Problem* $\mathcal{H}$ with $\alpha = 0$, with or without a modification on the variational formulation. We show that the same accuracy can be achieved compared to the conforming bilinear element for the variational formulation with a minor modification, and under the *bi-section condition* [21], the same assertion holds for the variational formulation without a modification.

The Wilson nonconforming finite element space [23] is

$$\{\mathbf{v} \in [L^2(\Omega)]^2 \mid \hat{\mathbf{v}} = \mathbf{v} \circ F_K \in \mathbf{P}(\hat{K}) \;\; \forall K \in \mathcal{C}_h\},$$

where $[Q_1(\hat{K})]^2 \subset \mathbf{P}(\hat{K}) \subset [P_2(\hat{K})]^2$. We write $\mathbf{P}(\hat{K}) = [Q_1(\hat{K})]^2 + \mathbf{B}(\hat{K})$, where $\mathbf{B}(\hat{K})$ contains the nonconforming part:

$$\mathbf{B}(\hat{K}) = \mathrm{Span}(\xi^2 - 1, \eta^2 - 1)^2.$$

We denote the Wilson nonconforming element space as $\mathbf{X}_h$ and the P_1 conforming element space as M_h, which is used for approximating the velocity and pressure, respectively. $\mathbf{X}_h$ can be decomposed into the conforming part $\mathbf{V}_h$ and the nonconforming part $\mathbf{B}_h$:

$$\mathbf{V}_h = \{\mathbf{v}^c \in X \mid \hat{\mathbf{v}}^c = \mathbf{v}^c \circ F_K \in [Q_1(\hat{K})]^2 \ \forall K \in \mathcal{C}_h\},$$
$$\mathbf{B}_h = \{\mathbf{v}^b \in [L^r(\Omega)]^2 \mid \hat{\mathbf{v}}^b = \mathbf{v}^b \circ F_K \in \mathbf{B}_K \ \forall K \in \mathcal{C}_h\}.$$

Define the semi-norm

$$\|\mathbf{v}\|_h = \Big(\sum_{K \in \mathcal{C}_h} |\mathbf{v}|_{1,r,K}^r \Big)^{\frac{1}{r}} = \Big(\sum_{K \in \mathcal{C}_h} |v_1|_{1,r,K}^r + \sum_{K \in \mathcal{C}_h} |v_2|_{1,r,K}^r \Big)^{\frac{1}{r}}.$$

We list some geometric properties of an arbitrary quadrilateral mesh [10]:

$$x^k = a_0 + a_1\xi + a_2\eta + a_{12}\xi\eta, \quad y^k = b_0 + b_1\xi + b_2\eta + b_{12}\xi\eta.$$

$$DF_K(\xi, \eta) = \begin{pmatrix} a_1 + a_{12}\eta & a_2 + a_{12}\xi \\ b_1 + b_{12}\eta & b_2 + b_{12}\xi \end{pmatrix},$$

and the Jacobian of F_K is $J_K(\xi, \eta) = det(DF_K) = J_0^K + J_1^K\xi + J_2^K\eta$, where $J_0^K = a_1b_2 - a_2b_1$, $J_1^K = a_1b_{12} - a_{12}b_1$, $J_2^K = a_{12}b_1 - a_2b_{12}$. Denote the inverse of F_K by F_K^{-1}. Then

$$(DF_K)^{-1}(\xi, \eta) = \frac{1}{J_K(\xi, \eta)} \begin{pmatrix} b_2 + b_{12}\xi & -a_2 - a_{12}\xi \\ -b_1 - b_{12}\eta & a_1 + a_{12}\eta \end{pmatrix}.$$

With these notation, for any $v \in W^{1,r}(\Omega)$ we define $\frac{\tilde{\partial}v}{\partial x}, \frac{\tilde{\partial}v}{\partial y}$ as follows:

$$J_K\frac{\tilde{\partial}v}{\partial x} = \frac{\partial y^K}{\partial \eta}(0,0)\frac{\partial \hat{v}}{\partial \xi} - \frac{\partial y^K}{\partial \xi}(0,0)\frac{\partial \hat{v}}{\partial \eta} = b_2\frac{\partial \hat{v}}{\partial \xi} - b_1\frac{\partial \hat{v}}{\partial \eta},$$
$$J_K\frac{\tilde{\partial}v}{\partial y} = -\frac{\partial x^K}{\partial \eta}(0,0)\frac{\partial \hat{v}}{\partial \xi} + \frac{\partial y^K}{\partial \xi}(0,0)\frac{\partial \hat{v}}{\partial \eta} = -a_2\frac{\partial \hat{v}}{\partial \xi} + a_1\frac{\partial \hat{v}}{\partial \eta}.$$

We define the modified divergence operator by

$$\widetilde{\mathrm{div}}\,\mathbf{v} = \frac{\tilde{\partial}v_1}{\partial x} + \frac{\tilde{\partial}v_2}{\partial y},$$

and

$$[B(\mathbf{u}), \mathcal{E}\mathbf{v}]_h = (A(\mathbf{u}^c), \mathcal{E}\mathbf{v}^c)_h + (A(\mathbf{u}^b), \mathcal{E}\mathbf{v}^b))_h,$$
$$[\mathrm{div}\,\mathbf{u}, p]_h = (\mathrm{div}\,\mathbf{u}^c, p) + (\widetilde{\mathrm{div}}\,\mathbf{u}^b, p)_h, \quad [\mathbf{f}, \mathbf{v}]_h = (\mathbf{f}, \mathbf{v}^c).$$

Problem $\mathcal{H}_h$. Find $(\mathbf{u}_h, p_h) \in \mathbf{X}_h \times M_h$ such that

$$[A(\mathbf{u}_h), \mathcal{E}\mathbf{v}]_h - [\text{div}\mathbf{v}, p_h]_h = [\mathbf{f}, \mathbf{v}]_h \quad \forall \mathbf{v} \in \mathbf{X}_h,$$
$$[\text{div}\mathbf{u}_h, q]_h = 0 \quad \forall q \in M_h.$$

To get the well-posedness of *Problem $\mathcal{H}_h$*, we need the following lemmas.

Lemma 3.1 ([15])

$$C_1(|\mathbf{v}^c|_{1,r} + \|\mathbf{v}^b\|_h) \le \|\mathbf{v}\|_h \le C_2(|\mathbf{v}^c|_{1,r} + \|\mathbf{v}^b\|_h) \quad \forall \mathbf{v} \in \mathbf{X}_h,$$
$$\|\mathbf{v}^b\|_{0,r} \le Ch\|\mathbf{v}^b\|_h \quad \forall \mathbf{v} \in \mathbf{X}_h.$$

This lemma can be proved by a combination of scaling trick and the equivalent norm theorem in finite dimensional spaces.

Lemma 3.2 ([15]) *The discrete Korn inequality*

$$\|\mathbf{v}\|_h \le C\|\mathcal{E}\mathbf{v}\|_{0,r} \quad \forall \mathbf{v} \in \mathbf{X}_h,$$

and the B-B inequality hold

$$\exists \beta_h(r) > 0, \quad \inf_{q \in M_h} \sup_{\mathbf{v} \in \mathbf{X}_h} \frac{[\text{div}\mathbf{v}, q]}{\|\mathbf{v}\|_h \|q\|_M} \ge \beta_h(r).$$

The proof of this *B-B inequality* is long and technical; we refer to [15] for more details.

Theorem 3.1 *If $(\mathbf{u}_h, p_h) \in \mathbf{X}_h \times M_h$ is the unique solution of (H_h), then*
i) for $r \in (1, 2]$, we have

$$\|\mathbf{u} - \mathbf{u}_h\|_h \le Ch^{\frac{\kappa}{2}}, \quad \|p - p_h\|_M \le Ch^{\kappa-1},$$

ii) for $r \in [2, \infty)$, we have

$$\|\mathbf{u} - \mathbf{u}_h\|_h \le Ch^{\frac{1}{\kappa-1}}, \quad \|p - p_h\|_M \le Ch^{\frac{\kappa'}{2}}.$$

When $\theta = 1$, we even have

$$\|\mathbf{u} - \mathbf{u}_h\|_{1,2} \le Ch^{\frac{\kappa}{2}}.$$

Corollary 3.1 *The nonconforming part of the approximate solution $\mathbf{u}_h \in \mathbf{X}_h$ tends to zero with the estimates*

$$\|\mathbf{u}_h^b\|_{0,r} + h\|\mathbf{u}_h^b\|_h \le Ch^{1+\frac{\kappa}{2}} \quad \forall r \in (1, 2],$$

$$\|\mathbf{u}_h^b\|_{0,r} + h\|\mathbf{u}_h^b\|_h \le Ch^{1+\frac{1}{\kappa-1}} \quad \forall r \in (2, \infty).$$

Furthermore,

$$\|\mathbf{u} - \mathbf{u}_h^c\|_h \le Ch^{\frac{\kappa}{2}} \quad \forall r \in (1, 2], \quad \|\mathbf{u} - \mathbf{u}_h^c\|_h \le Ch^{\frac{1}{\kappa-1}} \quad \forall r \in (2, \infty).$$

As for the variational formulation without a modification, we only need the *bi-section condition* (see [21]):

Bisection condition. The distance d_K between the midpoints of two diagonals of $K \in C_h$ is of order $O(h_K^2)$ uniformly for all elements K as $h \to 0$.

References

[1] Arnold, D. N. and Brezzi, F., Some new elements for the Reissner-Mindlin plate model, Boundary Value Problems for PDEs (Baiocchi C. and Lions J. L., eds.), Masson, Paris, 1992, 287–292.

[2] Baranger, J. and Sandri, D., A formulation of Stokes' problem and the linear elasticity equations suggested by the Oldroyd for viscoelastic flow, *RAIRO M^2AN* **26** (1992), 331–345.

[3] Brezzi, F. and Fortin, M., *Mixed and Hybrid Methods*, Springer-Verlag, New York, 1991.

[4] Brezzi, F., Fortin, M., and Marini. D., Mixed finite element methods with continuous stresses, *Math Model Meths Appl Sci.* **3** (1993), 275–287.

[5] Baranger, J., Najib, K., and Sandri, D., Numerical analysis of a three-field model for a Quasi-Newtonian flow, *Comput Meths Appl Mech Engrg.* **109** (1993), 281–192.

[6] Barrett, J. W. and Liu, W. B., Quasi-norm error bounds for the finite element approximation of a non-Newtonian flow, *Numer Math* **61** (1994), 437–456.

[7] Crochet, J., Davis, A. R., and Walters, K., *Numerical Simulations of Non-Newtonian Flow*, Elsevier, Amsterdam, Rheology Series Vol. 1, 1984.

[8] Fortin, M., Guénette, R., and Pierre, R., Numerical analysis of the modified EVSS method, *Comput Meths Appl Mech Engrg.* **143** (1997), 79–95.

[9] Fortin, M. and Pierre, R., On the convergence of the mixed method of Crochet and Marchal for viscoelastic flows, *Comput Meths Appl Mech Engrg.* **73** (1989), 341–350.

[10] Girault, V. and Raviart, R. A., *Finite Element Methods for Navier-Stokes Equations: Theory and Algorithms*, Springer-Verlag, Berlin-New York, 1986.

[11] Hornung, U., *Homogenization and Porous Media*, Vol. 6, Interdisciplinary Applied Mathematics, Springer-Verlag, New York, 1997.

[12] Lions, J. L. and Magenes, E., *Non-homogeneous Boundary Value Problems and Applications* I, Springer-Verlag, New York, 1972.

[13] Ming, P. and Shi, Z.-C., Dual combined finite element methods for a non-Newtonian flow (I): Nonlinear stabilized methods, 1998 Preprint.

[14] Ming, P. and Shi, Z.-C., Dual combined finite element methods for a non-Newtonian flow (I): Parameter-dependent problem, 1998 Preprint.

[15] Ming, P. and Shi, Z.-C., Analysis of Wilson finite element for a non-Newtonian flow, 1999, Preprint.

[16] Ming, P. and Shi, Z.-C., A technique for the analysis of B-B inequality for non-Newtonian flow, 1999, Preprint.

[17] Reddy, B. D. and Simo, J. C., Stability and convergence of a class of enhanced strain finite element methods, *SIAM J Numer Anal*, **32** (1995), 1705–1728.

[18] Sandri, D, Analysis d'une formulation á trois champs du problème de Stokes, *RAIRO M²AN* **27** (1993), 817–841.

[19] Scheurer, B., Existence et approximation de points selles pour certains problems nonlinéaires, *RAIRO* **4** (1992), 369–400.

[20] Schwab, C. and Suri, M., Mixed $h - p$ finite element methods for Stokes and Non-Newtonian Flow, Research report No. 97-19, Seminar für Angewandte Mathematik, ETH Zürich, 1997.

[21] Shi, Z.-C., A convergence condition for the quadrilateral Wilson element, *Numer Math* **44** (1984), 349–361.

[22] Simo, J. C. and Refai, S. C., A class of mixed assumed strain methods and the method of incompressible modes, *Int J Numer Meths Eng.* **29** (1990), 1595–1638.

[23] Wilson, E. L, Taylor, R. L, Doherty, W. and Ghaboussi, J., Incompressible displacement modes, *Numerical and Computer Models in Structural Mechanics*, Fenves, S. et al, eds. Academic Press, New York, 1973.

[24] Zhang, Z., Analysis of some quadrilateral nonconforming elements for incompressible elasticity, *SIAM J Numer Anal.* **34** (1997), 640–663.

Numerical Simulation of Compositional Fluid Flow in Porous Media

GUAN QIN   HONG WANG   RICHARD E. EWING
MAGNE S. ESPEDAL

Abstract

A new sequential solution method with selectively chosen primary variables is developed for modeling the enhanced oil recovery processes. A mixed finite element method (MFEM) is used to solve one phase pressure and pseudo total-volumetric velocity simultaneously. An Eulerian-Lagrangian localized adjoint method (ELLAM) is used to solve each transport equations. Computational results for two- and three-phase multi-component fluid flow occurring in enhanced oil recovery processes are presented, which show the strength of the method.

KEYWORDS: compositional flow, mixed finite element method, Eulerian-Lagrangian localized adjoint method

1 Introduction

The objective of reservoir simulation is to understand the complex chemical, physical, and fluid flow processes occurring in a petroleum reservoir sufficiently well to be able to optimize hydrocarbon recovery. To do this, one must be able to predict the reservoir performance under various exploitation schemes. In many enhanced oil recovery processes, some chemicals are injected into the reservoir to mobilize the residual oil trapped in the reservoir rocks. Thus, two or more fluids can flow in an immiscible fashion at certain times and in a miscible manner at other times. Different compositional models have been developed to model accurately these processes by characterizing the composition of reservoir fluid using a finite number of components [1, 2, 14], leading to strongly coupled systems of nonlinear partial differential equations (PDEs) and constraining equations. The governing PDEs are basically of an advection/diffusion type with advection being the dominant process. Although diffusion or dispersion is a small phenomenon relative to advection, it is important for miscible flow regimes and may at times describe important capillary pressure effects in the immiscible flows. The constraining equations provide further information on the distribution of mass for each individual component in different phases. In general, they are strongly nonlinear, implicit functions of pressure and mole fractions for each component in different phases. Consequently, extremely large number of PDEs and unknown variables are involved.

The combination of strong nonlinearities and close couplings between the equations and constraints cause severe numerical difficulties in the solution of the systems. Due to the enormous size of many field-scale applications, these equations cannot be solved in a fully-coupled, fully-implicit fashion, and some linearization techniques must be used to obtain the numerical solution. However, a blind linearization with little regard to the properties of the equations or the solutions can result in extremely large, ill-conditioned, nonlinear systems; the accurate solution of these equations can be extremely difficult and expensive. Choices of implicitness and decoupling of the equations must be analyzed and treated with great care for these difficult problems. Secondly, quite large grid-spacings must be used in the discretization of PDEs due to the enormous size and complexities of the problems. The use of large grid-spacings in space-centered finite difference methods (FDMs) and finite element methods (FEMs) often yield numerical solutions with excessive oscillations. On the other hand, upwind FDM and Petrov-Galerkin FEM can produce numerical solutions with severe numerical dispersion, and spurious effects related to the orientation of the grid [8, 15].

The ELLAM was originally introduced by Celia, Russell, Herrera, and Ewing for the solution of (one-dimensional constant-coefficient) advection-diffusion PDEs in a mass conservative manner [3]. It overcomes the principal shortcomings of many previous characteristic methods while maintaining their numerical advantages. Subsequently, ELLAM schemes have been successfully applied to linear transport PDEs, multi-component fluid flow problems in compressible media with wells, and many other problems (see [15, 16] and the references therein). In this paper we develop a numerical method for compositional flow by utilizing the basic properties of the flow more effectively. By choosing the primary variables properly, we ease the coupling and nonlinearity of the system. We then apply a sequential procedure to linearize and decouple the coupled system, and use an ELLAM scheme to solve each linearized transport equation and an MFEM to solve one phase pressure and

pseudo total-volumetric velocity simultaneously. Numerical results for two-
and three-phase multi-component flow problems are presented, which show
the strength of the method.

2   A Compositional Model

A compositional model is presented under the following assumptions [1, 2, 14]:

(i) The reservoir temperature is constant.

(ii) There is no mass exchange between the water phase and the oil and
gas phases.

(iii) The water and rock are incompressible, which implies that the density
ρ_w of the water phase is constant and that the porosity ϕ of the rock
is only a function of spatial variables.

(iv) The viscosity μ_w of the water phase is constant.

Since there exists mass exchange between phases, mass is not conserved
within each phase. Nevertheless, the total mass of each component is conserved, and the corresponding mass conservation can be described by

$$\phi\frac{\partial m^w}{\partial t} + \nabla \cdot \left(\frac{u_w}{v_w}m^w\right) = q^w, \tag{2.1}$$

$$\phi\frac{\partial m^i}{\partial t} + \nabla \cdot \left(\frac{u_o}{v_o}m_o^i + \frac{u_g}{v_g}m_g^i\right) = q^i, \quad i = 1,\ldots,N. \tag{2.2}$$

Here N is the number of hydrocarbon components. The superscript and the
subscript refer to the component and phase indices, respectively. In Eqs. (2.1)
and (2.2), the nomenclature is such that m^w (or m_o^i or m_g^i) represents the
moles of the corresponding component per pore volume; $m^i = m_o^i + m_g^i$ is the
overall moles of component i; $\underline{u}_w$, $\underline{u}_o$ and $\underline{u}_g$ are the phase velocities given by
Eqs. (2.4) below; v_w, v_o and v_g are the ratios of the water, oil and gas phase
volume to the pore volume, respectively; q^w and q^i are the mass flow rate of
water and the ith component. Under the assumption that the pore volume
of porous media is fully filled with fluids, the following volumetric constraint
holds [14]

$$v_T = v_w + v_o + v_g = 1. \tag{2.3}$$

The following multi-phase Darcy's law gives the relations between the phase velocities $\underline{u}_w$, $\underline{u}_o$ and $\underline{u}_g$ and the phase pressures p_w, p_o and p_g:

$$\underline{u}_\alpha = -\mathbf{K}\frac{k_{r\alpha}}{\mu_\alpha}\left(\nabla p_\alpha - \rho_\alpha g_c \nabla z\right), \quad \alpha = w, o, g. \tag{2.4}$$

In Eq. (2.4), $\mathbf{K}$ is the absolute permeability of rock which may be a full tensor for anisotropic and heterogeneous porous media. k_{rw}, k_{ro} and k_{rg} are the relative permeabilities of rock that are functions of saturations. μ_w, μ_o and μ_g (or ρ_w, ρ_o and ρ_g) are the viscosities (densities) of the fluid where μ_w (or ρ_w) is constant and μ_o and μ_g (or ρ_o and ρ_g) are the functions of pressure and composition of hydrocarbon. $\lambda_w = k_{rw}/\mu_w$, $\lambda_o = k_{ro}/\mu_o$ and $\lambda_g = k_{rg}/\mu_g$ are the phase mobilities. g_c is gravity acceleration.

In addition, the nonlinear PDEs presented above are coupled with a complex phase package that describes the relative phase equilibrium for a given pressure-volume-temperature through the equation of state (e.g., the Peng-Robinson equation of state).

Eqs. (2.1)–(2.4) form a strongly coupled system of nonlinear PDEs and constraints. By the Gibbs phase rule one concludes that the system is uniquely determined by $N + 2$ extensive variables, which are called primary variables [1]. Other variables can be expressed as functions of these variables. To solve this system efficiently, one should choose the primary variables carefully based on all the inherent physical and computational properties. We choose the first $N + 1$ primary variables as in many papers [1, 2, 4, 14], which are the pressure of oil phase p, the overall mass per pore volume of the hydrocarbon components m^i ($i = 1, \ldots, N$), and the water saturation s_w.

3 A Sequential Solution Method

3.1 An MFEM for the pressure and total velocity

Notice that $v_T = v_T(p, s_w, m^1, \ldots, m^N)$ is a function of p, s_w, and $m^1, \ldots, m^N$. If one differentiates the constraint equation (2.3) with respect to t and replaces $\partial s_w/\partial t$ and $\partial m^i/\partial t$ by Eqs. (2.1) and (2.2) incorporated with the Darcy's law (2.4), one obtains the following pressure equation [1, 14]:

$$\beta_T\frac{\partial p}{\partial t} - \mathbf{K}\left[\sum_{i=1}^N \frac{\partial v_T}{\partial m^i}\nabla \cdot \left(\frac{m_o^i k_{ro}}{v_o\mu_o} + \frac{m_g^i k_{rg}}{v_g\mu_g}\right)\nabla + \frac{\partial v_T}{\partial s_w}\nabla \cdot \left(\frac{k_{rw}}{\mu_w}\right)\nabla\right]p = r_p, \tag{3.1}$$

where p is the pressure of the oil phase, $\beta_T = -\phi(\partial v_T/\partial p)$ is the total fluid compressibility that is positive because $\partial v_T/\partial p$ is negative, and r_p is the volumetric discrepancy error.

Eq. (3.1) is a parabolic PDE with respect to the pressure p and can be solved by FDMs, FEMs, or finite volume methods (FVMs). After the numerical solution p^h is obtained, one computes the numerical phase velocities $\underline{u}_w^h$, $\underline{u}_o^h$ and $\underline{u}_g^h$ by substituting p^h into Eq. (2.4) and using the capillary pressure curves. However, when the subsurface geology is strongly heterogeneous, the absolute permeability $\mathbf{K}$ of the rock can be very rough. In this case the exact solution p of Eq. (3.1) is not necessarily smooth and so the numerical solution p^h might not be accurate. As a result, the numerical Darcy's velocities $\underline{u}_w^h$, $\underline{u}_o^h$, and $\underline{u}_g^h$ obtained from Eq. (2.4) by numerically differentiating p^h and multiplying p^h by a rough coefficient k/μ are even less accurate. This, in turn, affects the accuracy of the numerical approximations to other primary variables when $\underline{u}_w^h$, $\underline{u}_o^h$ and $\underline{u}_g^h$ are put into Eqs. (2.1) and (2.2).

While the pressure p may be rough, the total Darcy's velocity $\underline{u} = \underline{u}_w + \underline{u}_o + \underline{u}_g$ is usually smooth. Moreover, it is the phase velocities instead of the pressure that are needed in Eqs. (2.1) and (2.2). Hence, we adopt an MFEM to solve the following system of first-order PDEs for the total Darcy's velocity u and the pressure p simultaneously [12, 14]:

$$\frac{dp}{dt} + \nabla \cdot \underline{u} = R_p, \quad \underline{u} + \lambda_T \mathbf{K} \nabla p = R_{\underline{u}}, \tag{3.2}$$

where R_p and $R_{\underline{u}}$ involve the temporal and spatial derivatives of other primary unknowns, but not $\underline{u}$ and p. $\lambda_T = \lambda_w + \lambda_o + \lambda_g$ is the total mobility. The operator d/dt is defined by

$$\frac{d}{dt} = \beta_T \frac{\partial}{\partial t} + \sum_{i=1}^{N} \nabla \left(\frac{\partial v_T}{\partial m^i} \right) \cdot \left(\frac{m_o^i k_{ro}}{v_o \mu_o} + \frac{m_g^i k_{rg}}{v_g \mu_g} \right) \nabla \cdot . \tag{3.3}$$

After the total Darcy's velocity $\underline{u}$ is obtained from equation (3.2), the phase velocities $\underline{u}_w$, $\underline{u}_o$ and $\underline{u}_g$ can be computed by (with $\alpha =$ water, oil, and gas)

$$\underline{u}_\alpha = f_\alpha \underline{u} + f_\alpha \mathbf{K} \sum_{j \neq \alpha} \lambda_j \left[\nabla(p_{cjo} - p_{cao}) - (\rho_j - \rho_\alpha) g_c \nabla z \right], \tag{3.4}$$

where $f_\alpha = \lambda_\alpha / \lambda_T$ is the fractional flow function and p_{cjo} (or p_{cao}) is the capillary pressure between phase j (or phase α) and the oil phase.

An MFEM for Eqs. (3.2) can be written as follows: find $(\underline{u}, p) \in \mathcal{V}_h \times \mathcal{W}_h$, for any $t \in [0, T]$, such that

$$
\begin{aligned}
\left(\frac{dp}{dt}, \varsigma\right) + \left(\nabla \cdot \underline{u}, \varsigma\right) &= \left(R_p, \varsigma\right) \quad \forall \varsigma \in \mathcal{W}_h, \quad t \in [0, T], \\
\left((\lambda_T \mathbf{K})^{-1} \underline{u}, \underline{\eta}\right) - \left(p, \nabla \cdot \underline{\eta}\right) &= \left(R_{uv}, \underline{\eta}\right) \quad \forall \underline{\eta} \in \mathcal{V}_h, \quad t \in [0, T].
\end{aligned}
\tag{3.5}
$$

Here $\mathcal{V}_h$ and $\mathcal{W}_h$ are the well-known Raviart-Thomas spaces [5, 13].

3.2   An ELLAM for the transport equations

Substituting the phase velocities $\underline{u}_w$, $\underline{u}_o$, and $\underline{u}_g$ obtained from Eq. (3.4) into Eq. (2.1) and using the fact that the water phase is incompressible, we rewrite Eq. (2.1) as follows:

$$
\phi \frac{\partial s_w}{\partial t} + \nabla \cdot \left(\underline{u} f_w(s_w)\right) - \nabla \cdot \left(\mathbf{D}^w \nabla s_w\right) = R^w,
\tag{3.6}
$$

with R^w given by

$$
R^w = \nabla \cdot \left(\sum_{i=1}^{N} \mathbf{D}^i \nabla m^i\right) + q^w.
\tag{3.7}
$$

Likewise, we can rewrite Eq. (2.2) as

$$
\phi \frac{\partial m^i}{\partial t} + \nabla \cdot \left(\underline{V}^i m^i\right) - \nabla \cdot \left(\mathbf{D}^i \nabla m^i\right) = R_m^i, \quad i = 1, \dots, N,
\tag{3.8}
$$

where $\underline{V}^i$ and R_m^i are as follows:

$$
\underline{V}^i = \left[\left(\frac{m_o^i}{m^i}\right)\left(\frac{f_o}{v_o}\right) + \left(\frac{m_g^i}{m^i}\right)\left(\frac{f_g}{v_g}\right)\right] \underline{u},
\tag{3.9}
$$

$$
R_m^i = q_i + \nabla \cdot \left(\sum_{j=1; j \neq i}^{N} \mathbf{D}^j \nabla m^j\right).
\tag{3.10}
$$

The analysis in [6, 12] shows that the advection part of Eq. (3.6) is least coupled with system (3.8) and the coupling among the equations in system (3.8) is relatively weak. Therefore, in Eqs. (3.6) and (3.8) we move the off-diagonal diffusion terms to the right-hand side to decouple these equations. In the next subsection, we present a sequential solution method to solve the nonlinear advection-diffusion equation (3.6) for the water saturation s_w, and to solve the decoupled and linearized advection-diffusion equations (3.8) for the total moles m^i of each component.

Notice that $\underline{V}^i$ is in fact the barycentric velocity of the ith component in the multi-phase and multicomponent mixture. It depends on both the

process of fluid flow and the phase equilibrium. The physical interpretation
of Eqs. (3.8) is that a particle of component i in a multi-phase fluid is traveling
with the velocity $\underline{V}^i$ while simultaneously diffusing at the rate controlled by
$\mathbf{D}^i$. The coefficients of diffusion terms, $\mathbf{D}^w$ and $\mathbf{D}^i$, come from the capillary
pressure effects. In many cases, capillary pressure effects are small and the
advection terms dominate these equations. Consequently, the exact solutions
of the equations are usually smooth outside some small regions within which
the solutions have sharp fronts that need to be resolved accurately without
oscillations or numerical dispersions. Because of the numerical advantages
of ELLAM approach, we solve each linearized transport equation in (3.8) by
using an ELLAM scheme [3, 16]

$$
\begin{aligned}
\int_\Omega \phi m^i(\cdot, t_j) w(\cdot, t_j) d\underline{x} &+ \Delta t \int_\Omega \nabla w(\cdot, t_j) \mathbf{D}^i \nabla m^i(\cdot, t_j) d\underline{x} \\
&= \int_\Omega \phi m^i(\cdot, t_{j-1}) w(\cdot, t_{j-1}^+) d\underline{x} + \Delta t \int_\Omega R_m^i w(\cdot, t_j) d\underline{x},
\end{aligned}
\tag{3.11}
$$

where $w(\underline{x}, t_{j-1}^+) = \lim_{t \to t_{j-1}^+} w(\underline{x}, t)$. We refer readers to [15, 16] for discussions on implementational issues.

3.3   A sequential solution procedure

The system of Eqs. (3.2), (3.6), and (3.8) is a large, time-dependent, strongly
coupled system of nonlinear PDEs coupled with the nonlinear phase equilibrium constraints. We present a time-marching, sequential-implicit algorithm
to decouple the computations of phase equilibrium from the solutions of the
system at each time step. The algorithm can be outlined as follows:

1. At the time $t = 0$, the pressure p, water saturation s_w, and moles
   $m^1, \ldots, m^N$ of the hydrocarbon components are known from the given
   initial conditions for the model.

2. Apply the phase package to determine the phase distributions $m_o^1, \ldots, m_o^N$
   and $m_g^1, \ldots, m_g^N$ based on $m^1, \ldots, m^N$.

3. Compute the phase viscosities μ_o and μ_g (recall that μ_w is assumed
   to be constant) from the empirical correlation [10] and compute the
   densities ρ_o and ρ_g from the equation of state (e.g. Peng-Robinson).

4. Calculate the ratios v_o and v_g of phase volume to pore volume $v_o =$
   m_o/ρ_o and $v_g = m_g/\rho_g$, where $m_o = m_o^1 + \ldots + m_o^N$ and $m_g = m_g^1 +$

... $+ m_g^N$. Then, calculate the saturations s_o and s_g by $s_o = v_o/v_T$ and $s_g = v_g/v_T$ [14].

5. Compute the relative permeabilities after the saturations are obtained. Steps 2–5 yield the coefficients for equations (3.2), (3.6) and (3.8). One then proceeds to the next time step.

6. Apply an ELLAM [9] combined with the operator-splitting techniques to solve the nonlinear advection-diffusion PDE (3.6) for s_w, and the linearized transport PDEs (3.8) for $m^1, \ldots, m^N$. Here, in the coefficients of Eqs. (3.6) and (3.8), the values of the corresponding variables are computed based on their values at the previous time step.

7. Substitute the values of s_w and $m^1, \ldots, m^N$ at the current time step into equation (3.2), and apply an MFEM to solve Eq. (3.2) for the pressure p and the total velocity $\underline{u}$ at the current time step.

8. If necessary, perform some iterations between Eqs. (3.2), (3.6), (3.8) and the nonlinear constraints at the current time step.

9. Go back to Step 2 to update the coefficients at the current time step and repeat the above processes until one finally comes to time $t = T$.

4   Numerical Simulations

We use the method to simulate a 1D, three-phase, multi-component flow with complex *in situ* phase behavior. The reservoir is 250 feet in length with the sectional area of $50 \times 100 (\text{feet}^2)$. The porosity of the reservoir is 20%. The initial reservoir pressure is 2000 psi and the reservoir temperature is 160 F. The permeability of the reservoir is 2 Darcy, and the analytical relative permeabilities are given by the modified Corey's model

$$k_{rw} = 0.1\bar{s}_w^3, \quad k_{ro} = \bar{s}_o^{2.5}[1 - (1 - \bar{s}_o)^3], \quad k_{rg} = \bar{s}_g^{2.5}[1 - (1 - \bar{s}_g)^2], \quad (4.1)$$

where the normalized phase saturations are defined as follows

$$\bar{s}_w = \frac{s_w - s_{wr}}{1 - s_{wr} - s_{or}}, \quad \bar{s}_o = \frac{s_o - s_{or}}{1 - s_{wr} - s_{or}}, \quad \bar{s}_g = \frac{s_g - s_{gr}}{1 - s_{wr} - s_{or} - s_{gr}}. \quad (4.2)$$

The residual saturations of each phases are

$$s_{wr} = 0.35, \quad s_{or} = 0.25, \quad s_{gr} = 0. \quad (4.3)$$

Initially, the reservoir is filled with 20% water saturation and 80% undersaturated oil saturation. Water phase is incompressible. The molar density of water phase is 3.467 lb $-$ mole/ft^3 and the viscosity is 0.5 cp (centipoise). The oil phase is composed of 20% methane (light hydrocarbon component), 20% butane (medium hydrocarbon component), and 60% decane (heavy hydrocarbon component). The Peng-Robinson equation of state [11] is used in the flash calculations. The critical properties of these hydrocarbon components are listed in Table 1. The molar densities and viscosities of the hydrocarbon fluid are computed by the equation of state and Lohrenz rule [10], respectively.

	$p_c(psi)$	T_c (R^0)	V_c	w_t	ω
Methane	673.1	343.3	1.59	16.04	0.014
Buthane	550.7	765.4	4.08	58.12	0.193
Decane	306.0	1115	9.66	142.3	0.489

Table 1.   Critical Properties

We simulate an injection of 95% water and 5% hydrocarbon mixture (45% methane, 45% butane, and 10% decane) into the reservoir and the total injection rate is 1000 lb-mole per day. The water phase is mobilized by the injecting water, leading to a three-phase fluid flow process. 50 cells were used in space. The size of time steps is selected according to the time-step restrictions that are the maximum changes in (1) pressure, (2) overall composition of the hydrocarbon components, and (3) the phase saturations tolerated at each element. Since the initial guess for the flash calculations is given following the characteristics by the ELLAM scheme (3.10), restriction (2) is relaxed. The left and right figures in Fig. 1 show the computed water and gas saturations vs at 50 and 150 days, respectively. Since water is the main stream of the injecting fluid, a water saturation front is formed and propagated as time evolves (at about 70 feet at 50 days and at about 170 feet at 150 days). Because the reservoir pressure is increased due to the injection, the resident hydrocarbon fluid is vaporized to form a gas zone. For more numerical results, see [6].

5   Summary

The mathematical model for compositional flow involves a strongly coupled system of a large number of nonlinear PDEs and algebraic equations. Due

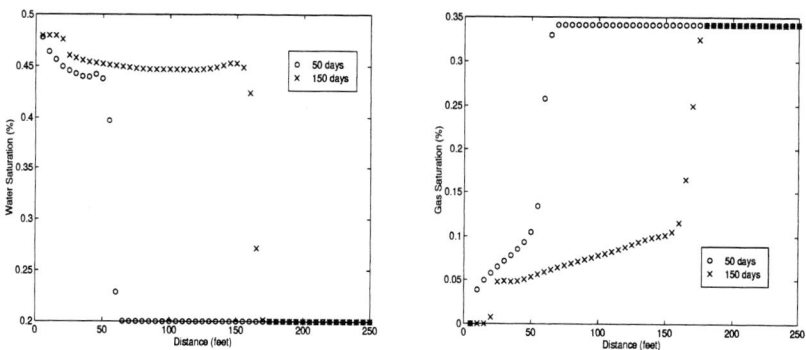

FIGURE 1: The profiles of the simulation at 50 and 150 days.

to the enormous size of the petroleum field applications, it is very expensive to solve the system in a fully-coupled and fully-implicit fashion in general. To derive a stable algorithm for solving the compositional system, we need to derive a proper form for the governing equations to alleviate the nonlinearities and couplings. Choices of primary variables, the proper forms of the equations to be solved, and implicitness and decoupling of the equations are of essential importance numerical simulations.

With a proper choice of primary variables, we develop a sequential solution procedure for a compositional model: (1) We use an MFEM to solve the oil phase pressure and the pseudo-total volumetric velocity. Instead of calculating the phase velocities, we compute the barycentric velocities and obtain fairly accurate and smooth approximations. (2) We decouple and linearize the coupled nonlinear transport PDEs by using the barycentric velocities. (3) We use an ELLAM scheme to solve each decoupled and linearized transport PDEs, which yields accurate numerical solutions that are free of oscillations and numerical dispersion and permits the use of large time steps. Moreover, the utilization of the ELLAM scheme generates an accurate initial guess and considerably improves the efficiency for the flash calculations, which in turn alleviates the restriction on time step size bounded by the maximum relative changes in the overall composition of the hydrocarbon components tolerated at each element. These results show the strong potential of the method.

References

[1] Acs, G., Doleschall, S., and Farkas, E., General purpose compositional model, *Soc. Pet. Eng. J.* **25** (1985), 543–553.

[2] Aziz, K. and Settari, A., *Petroleum Reservoir Simulation*, Applied Science Publisher Ltd, 1979.

[3] Celia, M. A., Russell, T. F., Herrera, I., Ewing, R. E., An Eulerian-Lagrangian localized adjoint method for the advection-diffusion equation, *Advances in Water Resources* **13** (1990), 187–206.

[4] Chavent, G. and Jaffre, J., *Mathematical Models and Finite Elements for Reservoir Simulation*, North Holland, Amsterdam, 1978.

[5] Chen Z., Ewing, R. E., and Espedal, M., Multiphase flow simulation with various boundary conditions. Peters, A. *et al.* (eds) *Computational Methods In Water Resources X*, Vol 2. Kluwer Academic Publishers, Dordrecht, Netherlands, 1994, pp. 925–932.

[6] Chen, Z., Qin, G., and Ewing, R. E., Analysis of a compositional model for fluid flow in porous media, *SIAM J. Appl. Math.*, to appear.

[7] Douglas, J., Jr. and Russell, T. F., Numerical methods for convection-dominated diffusion problems based on combining the method of characteristics with finite element or finite difference procedures, *SIAM J. Numer. Anal.* **19** (1982), 871–885.

[8] Ewing, R. E. (ed.), *The Mathematics of Reservoir Simulation*, Frontiers in Applied Mathematics, Vol. 1., SIAM, Philadelphia, 1984.

[9] Ewing, R. E., Operator splitting and Eulerian-Lagrangian localized adjoint methods for multiphase flow. Whitman ed., The Math. of Finite Elements and Applications VII, San Diego, CA, 1991, pp. 215–232.

[10] Lohrenz, J., Bray, B. and Clark, C., Calculating viscosities of reservoir fluids from their composition, *J. Pet. Tech.* **16** (1964), 1171–1176.

[11] Peng, D. Y. and Robinson, D. B., A new two-constant equation of state, *Ind. Eng. Chem. Fundam.* **15** (1976), 59–64.

[12] Qin, G., Numerical solution techniques for compositional model. Ph.D. Dissertation, Department of Chemical and Petroleum Engineering, University of Wyoming, 1995.

[13] Raviart, P. and Thomas, J., A mixed finite element method for 2nd order elliptic problems. Lecture Notes in Math., Vol. 606, 1977, pp. 292–315.

[14] Trangenstein, J. and Bell, J., Mathematical structure of compositional reservoir simulation, *SIAM J. Sci. Stat. Comput.* **10** (1989), 817–845.

[15] Wang, H., Ewing, R. E., Qin, G., Lyons, S. L., Al-Lawatia, M, and Man, S., A family of Eulerian-Lagrangian localized adjoint methods for multi-dimensional advection-reaction equations, *J. Comp. Physics* **152** (1999), 120–163.

[16] Wang, H., Liang, D., Ewing, R. E., Lyons, S. L., and Qin, G., An accurate approximation to compressible flow in porous media with wells, this volume.

Parallelization of a Compositional Reservoir Simulator

Hilde Reme   Geir Åge Øye   Magne S. Espedal
Gunnar E. Fladmark

Abstract

A finite volume dicretization has been used to solve compositional
flow in porous media. Secondary migration in fractured rocks has been
the main motivation for the work. Multipoint flux approximation has
·been implemented and adaptive local grid refinement, based on domain decomposition, is used at fractures and faults. The parallelization method, which is described in this paper, strongly promotes code
reuse and gives a very high level of parallelization despite low implementation costs. The programming framework is also portable to other
platforms or other applications. We have presented computer experiments to examine the parallel efficiency of the implemented parallel
simulator with respect to scalability and speedup.

KEYWORDS: porous media, multipoint flux approximation, domain decomposition, parallelization

1 Introduction

In this paper we are going to discuss some computational issues conected
to flow in fractured porous media. The simulation of secondary migration
[14, 16] has been the main motivation for the work [12, 18]. Secondary
migration is the movement of hydrocarbons through a carrier bed/reservoir
rock into a trap. In order to be able to simulate such a large and complex
process as secondary oil migration processes, the model has to be simplified.

Our model does not allow dynamical changes of the discontinuities. This
means that the fractures and faults may not open and close frequently. The
dynamical geometry variation, which exists in the model, is in the z-direction.
Due to the difference in the overburden pressure and the pore pressure we
may have a compaction of the rock or of the control volumes in the numerical
discretization. To represent discontinuities, local grid refinement is used.

For simplicity, only regular grid cells are used. The mathematical model describes compositional three-phase flow. Phase exchange between oil and gas and visa versa is allowed. In addition, the model includes thermodynamic equations, assuming equilibrium. The numerical model uses a block-centered finite difference discretization technique. With the use of this technique all the unknowns are located in the center of the control volume. Most of the existing models use a simple two-point flux approximation (TPFA) for the flux calculation. This may be incorrect if there exists a discontinuity in the permeability or if the principal permeability directions are not along any of the grid directions. The two-point flux approximation method is also used in the model, but in addition we have the possibility to use a multi-point flux approximation (MPFA) method if necessary. The computer code is highly object oriented and it is written in C++. Normally, the C++ programming language is computationally slower than the FORTRAN 90 programming language. However, it is often much easier to parallelize an object oriented C++ program than a FORTRAN 90 program. Another advantage when using C++ is that it is easy to add/delete new/old objects and classes.

2 Compositional Flow Model

The mathematical model includes mass conservation, energy conservation, a generalization of the Darcy low, equations of state, and equilibrium calculations. The components in the model are the hydrocarbons $(c_1, c_2, ..., c_n)$ and water. Totally the system includes $n_c = 1 + n$ components.

Even though our general simulator (SOM) [12, 18] includes this complex mathematical model, we have mostly used a simplified version of the model, named the symmetric Black Oil (SBO) model [11]. The phase calculation in this model is related to tables for bubble and dew points rather than fugacity calculations. In addition to allow some of the gas components to be in the oil phase, the SBO also allows some of the oil components to be in the gas phase. This means that the SBO model is a generalization of the black oil model [3]. The SBO model may also handle condensate systems.

The primary variables in both models are the temperature T, the water pressure p_w, and all the molar masses N_ν, $\nu = 1, 2, ..., n_c$.

The mass conservation of component ν of one chemical species which flows through the porous media is given by:

$$-\int_{CS} \dot{\vec{m}}_\nu \cdot d\vec{S} = \frac{\partial}{\partial t} \int_{CV} m_\nu dV + \int_{CV} q_\nu dV \,, \qquad \nu = 1, 2, ..., n_c \,. \quad (2.1)$$

Here $\dot{\vec{m}}_\nu$, m_ν, q_ν, CV, and CS denote the mass flux of component ν, mass density of component ν, source/sink density of component ν, control volume, and interface of the control volume. Let $C_{\nu m}$ denote the mass fraction of a component ν in phase m, $m = o, g, w$. Then the mass flux and the mass density of the component ν are respectively given by

$$\dot{\vec{m}}_\nu = C_{\nu g} \rho_g \vec{v}_g + C_{\nu o} \rho_o \vec{v}_o + C_{\nu w} \rho_w \vec{v}_w \,, \qquad \nu = 1, 2, ..., n_c \,,$$
$$m_\nu = \phi_p (C_{\nu g} \rho_g S_g + C_{\nu o} \rho_o S_o + C_{\nu w} \rho_w S_w) \,, \qquad \nu = 1, 2, ..., n_c \,.$$

where ρ_m, S_m, and ϕ denote mass density of the phase m, saturation of the phase m, and porosity. The Darcy law for multiphase flow is given by

$$\vec{v}_m = -\underline{K} \sum_{k=g,o,w} \frac{k_{r_{mk}}}{\mu_k} (\nabla p_k - \gamma_k \nabla d) \,, \qquad m = g, o, w \,,$$

where $\underline{K}$, $k_{r_{mk}}$, μ_k, p_k, γ_k, and d denote absolute permeability tensor, generalized relative permeability for coupled multi-phase flow, viscosity of the fluid phase k, fluid pressure of phase k, specific gravity of fluid phase k, and depth (positive in the gravity direction). Equation (2.1) gives n_c equations (one for each component), while we have $3n_c + 22$ unknowns. The first $3n_c + 21$ unknowns are $C_{\nu m}$, $\nu = 1, 2, ..., n_c$, ρ_m, S_m, p_m, μ_m and $k_{r_{mk}}$, $m, k = g, o, w$. The temperature T is the last unknown. To close the model we need to have $2n_c + 22$ more independent equations or relations.

We assume that the three phases fill the available pore space, which implies

$$S_g + S_o + S_w = 1 \,. \quad (2.2)$$

We also know that the sum of the mass fraction, for each phase, has to be one

$$\sum_{\nu=1}^{n_c} C_{\nu m} = 1 \,, \qquad m = g, o, w \,. \quad (2.3)$$

The density and viscosity are assumed to be functions of the temperature, the phase pressure and the mass fraction, while the generalized relative permeabilities and the capillary pressures are functions of the temperature and the saturations

$$
\begin{aligned}
\rho_m &= \rho_m(T, p_m, C_{1m}, ..., C_{n_cm}) , & m &= g, o, w , \\
\mu_m &= \mu_m(T, p_m, C_{1m}, ..., C_{n_cm}) , & m &= g, o, w , \\
k_{r_{mk}} &= k_{r_{mk}}(T, S_g, S_o, S_w) , & m, k &= g, o, w , & (2.4) \\
p_g - p_o &= p_{cgo}(T, S_g, S_o, S_w) , \\
p_o - p_w &= p_{cwo}(T, S_g, S_o, S_w) .
\end{aligned}
$$

As the temperature may vary in time we must include the heat flow equation. To ensure conservation of energy, the model uses the integral expression

$$
\frac{\partial}{\partial t} \int_{CV(t^n)} (\rho u) dV - \int_{CS(t^n)} (\underline{k}\nabla T) \cdot d\vec{S} = - \int_{CS(t^n)} h\rho\vec{u} \cdot d\vec{S} + \int_{CV(t^n)} q dV ,
\tag{2.5}
$$

where

$$
\rho u = \sum_{m=g,o,w} \phi_p S_m u_m \rho_m + u_r \rho_r (1 - \phi_p) ,
$$

$$
h\rho\vec{u} = \sum_{m=g,o,w} h_m \rho_m \vec{v}_m .
$$

where ρ_m, ρ_r, u_m, u_r, $\underline{k}$, T, and h denote mass density of phase m, mass density of rock, internal energy of phase m, internal energy of rock, bulk heat conductivity, temperature, and enthalpy, respectively. Since the movements in the system are very slow, all the terms including velocity may be neglected. This is the reason why we have neglected the kinetic energy, the potential energy, the viscous dissipation in the fluid flow, and the deformation energy in the solid phase and in the rock. The only mechanical work rate included is the term related to normal stresses in fluid flow.

Equations (2.2)-(2.5) give 22 new equations, we still need another $2n_c$ equations to close the model. These equations are given by the assumption of thermal equilibrium for each time-step. It is convenient to use the Gibbs function in the thermodynamics of phase equilibrium. The Gibbs function is defined by

$$
G = U + pV - TS ,
$$

where U, p, V, T, and S denote internal energy, pressure, volume, temperature, and entropy.

2.1   Pressure equation

The equations given above represent a closed set of equations to determine all the primary and secondary variables at a new time step. To calculate the primary variables we have used a sequential IMPEC (implicit pressure and explicit concentration). The secondary variables are determined by using experimentally based relationships and "flash calculations", which is discussed in this work. A pressure equation can be derived in several ways. In this work, however, we have used the volume balance method (VBM) [12, 17, 18] to obtain the water pressure equation. We define the residual volume R as

$$R = V_p - \sum_{k=g,o,w} V_k \ ,$$

where V_p and V_k are the pore volume and the volume of phase k, respectively. They both depend on the primary variables, stating

$$V_p = V_p(p_w, W) \ , \quad V_k = V_k(T, p_w, N_1^k, ..., N_{n_c}^k) \ , \qquad k = g, o, w \ ,$$

where

$$N_\nu^k = N_\nu^k(T, p_w, N_1, ..., N_{n_c}) \ , \qquad \nu = 1, 2, ..., n_c \ \text{ and } \ k = g, o, w \ .$$

The overburden pressure is given by $W = \sigma + p$, where σ and p are the effective stress and the pore pressure, respectively.

The residual volume is a function depending on time t. To satisfy the saturation condition (2.2) $\forall \, t$, the residual volume has to vanish $\forall \, t$. From this we obtain the volume balance

$$R(t) = 0 \qquad \forall \, t \ .$$

By assuming that we know the residual volume at some time t, denoted $R(t)$, we are able to find the value for the residual volume at time $t + \Delta t$. This can be done by constructing the truncated Taylor expansion

$$R(t + \Delta t) \approx R(t) + \frac{dR}{dt} \Delta t = 0 \ .$$

If we use the chain rule for partial differentiation on $\frac{dR}{dt}$, we get

$$\frac{\partial R}{\partial p_w} \frac{\partial p_w}{\partial t} + \sum_{\nu=1}^{n_c} \frac{\partial R}{\partial N_\nu} \frac{\partial N_\nu}{\partial t} = -\frac{R}{\Delta t} - \frac{\partial R}{\partial W} \frac{\partial W}{\partial t} \ . \qquad (2.6)$$

The partial derivatives of the residual volume with respect to the primary variables may be computed from equations of state and thermo-dynamical equilibrium conditions. Since we assume that the temperature is constant for this calculation, we do not have to include the term $\frac{\partial R}{\partial T} \frac{\partial T}{\partial t}$.

2.2 Boundary conditions

Normally, the equations governing the flow of fluids through porous media are second order partial differential equations. A Robin boundary condition is the most general type of boundary condition which is used for our problem and is given by

$$\alpha u + \beta \frac{\partial u}{\partial n} = \gamma .$$

Here $\vec{n}$ denotes the outward normal direction at some point on the boundary and α, β, and γ are constants which may be selected. The boundary conditons can be given explicitly or implicitly [12, 18].

3 Flux Discretization

For the numerical approximation we use a control volume finite difference block-centered disretization. Each control volume $CV_i, i = 1, 2, ..., N$ where N is the total number of grid cells, represents a fixed cell in the grid. The total interface for CV_i is denoted CS_i. We denote the set of grid cells connected to CV_i through a given flux molecule, including the grid itself, as $\mathcal{M}_i$. The common subset of the interface between CV_i and CV_j, $i \neq j$, is denoted $CS_{(i,j)}$. This means that

$$CS_{(i,j)} = CS_i \cap CS_j \quad \text{and} \quad CS_i = \sum_{l=1}^{n_i} CS_{(i,j_l)} ,$$

where n_i is the total number of sub-interfaces on CS_i and j_l is the global neighbor CV_j to CV_i which share the local sub-interface l. We then define $\mathcal{M}^{is}_{(i,j)}$ as the set of cells which influence the flux molecule for the interface $CS_{(i,j)}$. The set $\mathcal{M}^{is}_{(i,j)}$ depends (like the set $\mathcal{M}_i$) on the flux molecule method. If we use a two-point flux molecule approximation $\mathcal{M}^{is}_{(i,j)} = \{i, j\}$ [9], $\mathcal{M}^{is}_{(i,j)} = \{i, j, k, l, ...\}$ for other flux molecule methods. For a multi-point flux approximation the set $\mathcal{M}^{is}_{(i,j)}$ may contains 18 different cells for an inner sub-interface [1, 7, 8]. Between $\mathcal{M}_i$ and $\mathcal{M}^{is}_{(i,j)}$ we have the relationship

$$\mathcal{M}_i = \cap_{l=1}^{n_i} \mathcal{M}^{is}_{(i,j_l)} .$$

By use of these definitions we may write a general flux approximation as

$$f_{(i,j)} = -\int_{CS_{(i,j)}} (\lambda \underline{K} \nabla u) \cdot d\vec{S} \simeq \sum_{q \in \mathcal{M}^{is}_{(i,j)}} t_{iq} u_q ,$$

where t_{iq} are named the transmissibilities. Note that these transmissibilities include both a solution-dependent part and a geometry-dependent part, which may be calculated independently. In this work we have consentrate on the calculation of the geometry-dependent part. For the calculation of the solution-dependent part we refer to [12, 18].

3.1 The multi-point flux approximations (MPFA)

The easiest way to calculate the geometry-dependent part of the transmissibilities is to use the two-point flux approximation (TPFA) for each surface. With use of this method $\mathcal{M}^{is}_{(i,j)}$ only contains two cells and is given by $\mathcal{M}^{is}_{(i,j)} = \{i, j\}$. The TPFA method has been known for a long time and frequently used in the reservoir simulation literature; see, for instance, [9].

A multi-point flux approximations method has been developed lately [1, 7, 8]. As the name of this method indicates we use more than two points (two cell centers) to calculate the flux across the interface $CS_{(i,j)}$ to CV_i. In two dimensions $\mathcal{M}^{is}_{(i,j)}$ includes values from six different cells while the number of cells in $\mathcal{M}^{is}_{(i,j)}$ increases to eighteen in three dimensions.

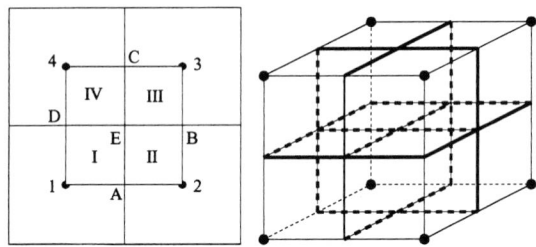

Figure 1: Grid cells with interaction regions in two- and three-dimensions.

When using this method the definition of the interaction region is essential. Figure 1 shows a sketch of this region in both two and three dimensions. The MPFA method covers the domain of the differential equations by such non-overlapping interaction regions. Instead of calculating the transmissibilities for $CS_{(i,j)}$ directly, this method calculates the transmissibilities across each sub-surface in the interaction region. To calculate the transmissibilities for $CS_{(i,j)}$, the method adds the contributions from the two interaction regions involved.

For simplicity, we have chosen to present the flux discretization techniques in two-dimensions. The implementation of these techniques in the SOM simulator are three-dimensional.

To calculate the flux $f^{AE}_{(1,2)}$ across the interface segment AE (see Figure 1), we may use the approximation

$$f^{AE}_{(1,2)} = - \int_{CS^{AE}_{(1,2)}} (\lambda \underline{K} \nabla u) \cdot d\vec{S} \simeq \sum_{q \in \mathcal{M}^{AE}_{(1,2)}} t_{Aq} u_q = \sum_{i=1}^{4} t_{Ai} u_i \ .$$

Since the flux must be zero when u_i is a constant vector, it follows that $\sum_{i=1}^{4} h_{Ai} = 0$. There exist several MPFA methods which determine t_{Ai} [1, 7, 8, 12, 18]. The main difference between these methods is where they require continuity in flux and potential. The O-method requiers continuity in both flux and potential at the points A, B, C and D, given in Figure 1. In [12, 18] they propose an improvement of this method, which ensures full continuity across the interfaces AE, BE, CE, and DE.

3.2 Non-matching grid lines and the OSCI-method

As far as we know, the MPFA method has not yet been given for non-matching grids in three dimensions. In [2] some two dimensional cases are shown. In this section we use slave cells to handle this problem in both two and three dimensions. Figure 2 shows how we implement slave cells

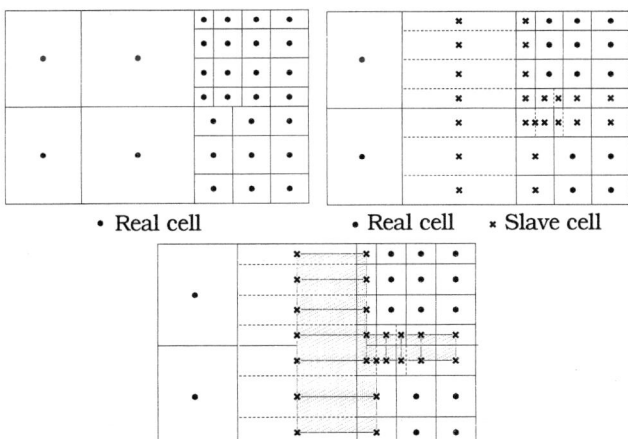

Figure 2: Figures show the grid with only real cells, with both real and slave cells and with the non-overlapping interaction regions between slave cells.

and create non-overlapping interaction regions between these cells. We refer to this MPFA method as the OSCI-method (O-method with slave cells and interpolation).

To create the new slave grid, all the lines from one refinement which touch another refinement or a coarse cell are extended to the next crossing line (see Figure 2). This means that one real coarse cell is divided in the number of fine cells touching the coarse interface. If lines from different refinements match, slave cells are identical to real cells. Thus, if the real cells create a non-matching grid, the OSCI-method creates non-overlapping interaction regions between slave cells. Then this method calculates the transmissibilities on this non-overlapping slave interaction regions in the same way as the O-method does across interfaces for matching grids. Note that this technique does not require regular grid density, but otherwise it is a general technique that fits all kinds of adaptive LGR.

After we have calculated the slave transmissibilities, an interpolation is needed. Either we may interpolate these contributions to the real cells or we may interpolate the quantities in the real cells out to the slave cells. We have chosen the latter alternative. So far only a linear interpolation has been carried out, but it should not be a problem to implement higher order and more accurate interpolation techniques. Since this problem is beyond the scope of this work, it is left for future work.

For a uniform grid in two dimensions with a homogeneous permeability tensor $\underline{K}$, it has been shown that the OSCI-method creates a second order method. This is identical to the O-method across interfaces for matching grids. Another advantage with this technique is that it is quite simple to implement. The drawback with this method is that it creates non-overlapping interaction regions between slave cells and not between real cells. Figure 3 shows how we cover a domain by non-overlapping real interaction regions and non-overlapping slave interaction regions. The O-method and OSCI-method

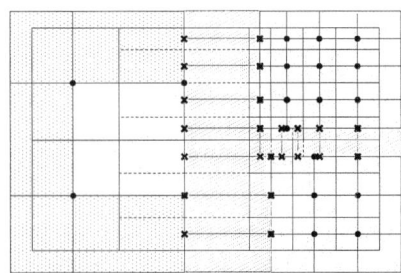

Figure 3: The domain covered by non-overlapping real interaction regions (grey regions) and non-overlapping slave interaction regions (red regions).

have been compared [12, 18] and it turns out that the OSCI method works well for nonmatching grids.

3.3   Adaptive use of the TPFA and the MPFA methods

Based on the differences which exist between a TPFA method and a MPFA method, it is possible to decide which flux approximation method to use in different cases. The MPFA method should be used for the cases [7]

- Whenever the medium is anisotropic and non-aligned with the local frame of reference.

- Whenever non-orthogonal and/or unstructured grids are employed.

- Whenever fine-scale cross-flow upscaling is performed particularly for cross-bedding.

The OSCI-method may also be used to match different solutions from different domains. Since the convergence of a domain decomposition method depends strongly on the accuracy of the exchange of information between different domains, one should expect that a MPFA method improves the convergence. Thus an improved Mortar method [4] may be developed.

It is only due to the large computational cost that we should avoid to use am MPFA method in the total domain. Normally, when using a LGR technique, an MPFA method should be used in the refined regions where the solution includes large gradients, while a TPFA method should be used in the coarse domain where no large changes exist. The SOM has the possibility to combine the two flux approximation methods like this. In an input file the user has to specify the flux approximation method for each sub-regions. The only restriction related to this specification is that TPFA method has to be used between coarse grid cells in a composite domain.

Trying to reduce the large computation time caused by MPFA, a splitting in the matrix level is suggested in [7]. This paper shows how we may decompose a general full tensor flux into a diagonal tensor flux together with cross terms. Then, time-split semi-implicit, stable, full tensor flux approximations are introduced within a general finite volume formalism. The standard diagonal tensor Jacobi matrix structure may now be retained while ensuring spatial consistency of the discretization. With the use of this M-matrix flux splitting method for a general full tensor discretization operator, the computation time is reduced by over 50% for quadrilateral grids.

4   Local Grid Refinements (LGR)

One of the main objectives in this work has been to implement a LGR in the SOM framework. In this section we try to describe how we have solved this problem. Even though the main interest is to use the LGR for fractured and faulted problems, we have selected a general LGR which also may be used in conjunction with resolving wells or larger gradients.

In this work we have used a fixed coarse grid and defined refined sub-domains from a pre-selected set of coarse grid blocks. In many situations, some care must be taken when constructing the coarse grid. If a large gradient is located near a coarse grid boundary, the numerical solution is affected. A dynamic adaptive coarse grid can be favorable in cases where we have large moving gradients (i.e., a moving front). In these cases, it is important to use a refined grid in a region containing large gradients. This has been used and studied in [6]. A fixed coarse grid is natural to use in cases where we try to resolve the dynamics at larger fractures and faults in addition to wells. If we know where these objects are located, this information must be used when constructing the coarse grid. As far as possible, regions with large gradients must be located inside each coarse block and not on or near the interface between them.

Based on the discussion above, we could also have used an adaptive LGR strategy. First, we create the fine grid, including large fractures and faults. Then the coarse grid could be constructed adaptively. The coarse grid must depend on the solution and the numerical behavior. The adjustment of the coarse grid could depend on the information from a first iteration. This option must be included in further development of the simulator.

There are two main difficulties related to LGR:

- to create the composite grid,

- to store all the data in an efficient and suitable manner.

We have chosen to use the existing coarse grid generator as a starting point. After we have generated a coarse grid, we determine all the coarse cells which have to be refined. Then we use a grid generator for each of prescribed refined subdomains. The gridding can be different in each sub-domain and the sub-domains can be arbitrary placed in the coarse grid. This means that we are able to create arbitrary refined domains; see Figure 4.

There is no global numbering of the geometry variables in this implementation. The grid generator also produces input files for the lithology mapping and boundary values mapping for each refined sub-domain. This means that we can prescribe an independent set of lithologies and boundary values for each sub-domain and the coarse domain. After we have created the geometry input files for each refinement and the coarse region, SOM has to be modified for achieving the use of these files. Since SOM is written in C++, this is done in an object-oriented framework. By using local numbering on each refined sub-domain, we can preserve the original regular data-structure for each refined patch. This means that we can use existing functions for discretization, linearization and fast linear solver for each refined sub-domain, even if the composite grid is not regular. This strategy may keep the cost related to the implementation of a LGR in the SOM framework at a minimum.

It is also possible to use dynamical allocation of new elements in the different class arrays in C++. This means that it is straightforward to add or delete refined sub-domains adaptively during the computations. Further details about the implementation is given in [12, 18].

5 A Composite Grid Solver

Local adaption introduces several design challenges including: grid structures, error control, efficient linear solvers and parallelism. The convergence rate and the complexity of iterative methods usually deteriorate with a decreasing size of the smallest mesh size. These small scales may even hinder direct methods since they determine the condition-number of the discretization matrix.

If we can use existing fast solution methods developed on regular grid, on non-regular composite grids, we have made a large step towards developing efficient parallel algorithms for solving the same composite problem.

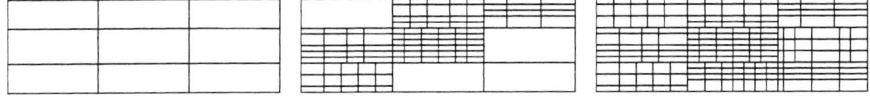

Figure 4: This Figure show how we divide the coarse grid into many refined no-regular non-matching sub-domains.

We assume that equations (2.5) and (2.6) have been discretized on a

composite grid as shown in Figure 4. Then we have the linear system to
solve

$$A\vec{x} = \vec{y}, \quad \vec{x}, \vec{y} \in \mathcal{R}^N. \tag{5.1}$$

Here A is the composite grid Jacobi matrix, $\vec{x}$ and $\vec{y}$ are the composite grid
solution vector and the right hand side vector, respectively. N is the total
number of grid cells in the composite grid.

A global numbering of the nodes for this composite problem would de-
stroy the nice regular banded structure in the matrix. This section describes
an iteration technique that allows to decouple the refined patches and the
underlying coarse grid. This means that the coupling is through the internal
boundary between the coarse and the refined regions. We can then use fast
regular solvers for each refined sub-domain. To attain this, we divide the
composite grid Ω into different disjoint sets; Ω_c consist of grid points in the
non-refined region and $\Omega_f = \bigcup_{i=1}^{p} \Omega_{f_i}$ is the union of all grid points in the
refined sub-domains. Here p is the number of refined sub-domains. After this
partitioning equation (5.1) can be written

$$\begin{bmatrix} A_{cc}^n & A_{cf}^n \\ A_{fc}^n & A_{ff}^n \end{bmatrix} \begin{bmatrix} \vec{x}_c \\ \vec{x}_f \end{bmatrix} = \begin{bmatrix} \vec{y}_c \\ \vec{y}_f \end{bmatrix}.$$

Our solution technique is based upon a two-step method, with a splitting
based on a Galerkin technique. This is a two level domain-decomposition
method, where we have one regular coarse grid and many regular refined subgrids. This means that we split the linear equation (5.1) into different regular
sub-problems, corresponding to each sub-domain. A coarse grid operator is
used to handle the communication between the sub-domains. The Galerkin
technique was first introduced in [15]. This technique has later been studied
in [10, 13, 12, 18]

5.1   A Galerkin composite grid solver

The algebraic algorithm, proposed in [10] for solving the linear equation (5.1),
can be formulated as follows. Let $\hat{\Omega}$ denote the grid without any refinement,
i.e. the regular coarse grid Ω_0 in Figure 4. Further, let $\hat{N}$ be the number
of grid blocks in $\hat{\Omega}$ and let $M_{\hat{i}} = \{$fine cells within coarse cell number $\hat{i}\}$,
$\hat{i} = 1, 2, ..., \hat{N}$. Defining the local refined sub-domains in this way, means
that we only have one coarse grid block under each refined patch. Let N
be the total number of cells in the composite grid, shown in Figure 4. We
define a basis vector $\vec{\Psi}_{\hat{i}} \in \mathcal{R}^N$, for the coarse cell $\hat{i}$. It consists of only zero

element, except for unit element at all $\hat{i} \in M_{\hat{i}}$, $i = 1, 2, ..., \hat{N}$. Let $\vec{x}^{(s)}$ be the (s) iterate of the composite solution vector $\vec{x}$. This gives us the function

$$\vec{z} = \vec{x}^{(s)} + \sum_{\hat{i}=1}^{\hat{N}} \hat{d}_{\hat{i}}^{(s)} \vec{\Psi}_{\hat{i}} = \begin{bmatrix} \vec{z}_c \\ \vec{z}_f \end{bmatrix}, \quad \vec{z} \in \mathcal{R}^N. \tag{5.2}$$

Here $\hat{d}_{\hat{i}}^{(s)}$, is the $\hat{i}$ th component of $\vec{\hat{d}}^{(s)}$ on $\hat{\Omega}$. Note that this function would describe an updated composite solution vector $\vec{z}$, from the error on the coarse grid $\hat{\Omega}$. Using $\vec{z}$ as a test-function in a Galerkin formulation, then for each coarse cell $\hat{j}$ the equation should be satisfied:

$$\langle \vec{\Psi}_{\hat{j}}, A\vec{z} \rangle = \langle \vec{\Psi}_{\hat{j}}, \vec{y} \rangle, \quad \hat{j} = 1, 2, ..., \hat{N},$$

here $\langle \cdot, \cdot \rangle$ is the inner product. This would give us a coarse linear system to solve for $\vec{\hat{d}}^{(s)}$

$$\hat{A}\vec{\hat{d}}^{(s)} = \vec{\hat{y}}, \quad \vec{\hat{d}}^{(s)}, \vec{\hat{y}} \in \mathcal{R}^{\hat{N}}, \tag{5.3}$$

where

$$\hat{a}_{\hat{k}\hat{l}} = \sum_{i \in M_{\hat{k}}} \sum_{j \in M_{\hat{l}}} a_{\hat{k}_i \hat{l}_j} \quad \text{and} \quad \hat{y}_{\hat{k}} = \sum_{i \in M_{\hat{k}}} \left(y_{\hat{k}_i} - \sum_{r=1}^{N} a_{\hat{k}_i r} x_r^{(s)} \right).$$

Here $\hat{k}_i$ and $\hat{l}_i$ indicate the local cell numbers i and j in the coarse cells $\hat{k}$ and $\hat{j}$, respectively. Note that if $\hat{k}$ is not refined, $M_{\hat{k}}$ consists of only one coarse cell. After solving equation (5.3) for $\vec{\hat{d}}^{(s)}$, we use the definition for $\vec{z}$ from equation (5.2) to get the new estimate for the composite solution vector

$$\vec{x}^{(s+1)} = \vec{z}.$$

This means that we choose $\vec{x}_c^{(s+1)} = \vec{z}_c^{(s+1)}$. When we solve for $\vec{x}_{f_q}$, on each of the refined sub-domains Ω_{f_q}, $q = 1, 2, ..., p$, we use the last updated solutions $\vec{x}_c^{(s+1)}$ and $\vec{x}_{f_q}^{(s)}$ as boundary conditions. This would be a kind of a two level block Jacobi formulation, and all the sub-domains Ω_{f_q} can be solved in parallel. We can also use a coloring of the sub-domains, or use a typical two level Gauss-Seidel formulation. By using a block Jacobi formulation, each sub-domain Ω_{f_q}, $q = 1, 2, ..., p$, can be given as

$$A_{f_q f_q}^n \vec{x}_{f_q}^{(s+1)} = \vec{y}_{f_q}, \quad \vec{x}_{f_q}, \vec{y}_{f_q} \in \mathcal{R}^{N_q},$$

where

$$\vec{y}_{f_q} \leftarrow \vec{y}_{f_q} - \sum_{l \neq q} A_{f_q f_l}^n \vec{x}_{f_l}^{(s)} - A_{f_q c}^n \vec{x}_c^{(s+1)}.$$

Note that these problem can be solved in parallel, by using the latest updated boundary conditions. Then the updated composite solution vector is given by

$$\vec{x}^{(s+1)} = \left[\begin{array}{c} \vec{x}_c^{(s+1)} \\ \vec{x}_f^{(s+1)} \end{array} \right].$$

The iteration goes back to the definition in equation (5.2) and it proceeds until the composite solution vector satisfy

$$\frac{||\vec{x}^{(s+1)} - \vec{x}^{(s)}||}{||\vec{x}^{(s)}||} < \epsilon.$$

Here ϵ is a prescribed tolerance.

6   Parallel Implementation

The Galerkin algorithm described earlier can be formulated as an iterative process, where in each iteration we solve updated boundary value problem for each sub-domain. The work on each sub-domain, consists mainly of building the corresponding Jacobian and the right hand side. Then solving the corresponding linear system restricted to one sub-domain, using values from its neighboring sub-domains as boundary conditions. The sub-problems can be solved in parallel because neighboring sub-problems are only coupled through previously computed values in the neighboring sub-domain. The basic building block for the Galerkin domain-decomposition method is the sub-domain solver. The sub-domains can be carried out completely independently, allowing the code to run on different processors of a parallel computer.

Traditionally, domain-decomposed parallelization of PDE is done at the level of local matrix/vector operations and linear solvers. However, we are interested in a parallelization strategy at the level of sub-domain simulators. In [5, 18], they have proposed a Simulator Parallel Model (SPM) for parallelizing existing sequential PDE simulators, which is a Single Program Multiple Data (SPMD) model. Basically, the SPM based on domain-decomposition for developing parallel PDE software is: *Assign one processor with one or several sequential simulators, each responsible for one sub-domain.* The coordination of the computation among the processors is left to a global administrator, implemented at a high abstraction level, close to the mathematical formulation of the Galerkin method.

The SPM strongly promotes code reuse, because most of the global administration and the related communications between processors can be extracted from specific applications. This means that we minimize the development costs and we have efficient control of performance parameters such as load balance, communication topology and domain-decomposition.

The generic framework object-oriented framework for implementing a parallel version of SOM, based on the simulator-parallel programming model, consists of three main parts: The sequential sub-domain simulator **Sub-domain_Simulator**, a communication part **Communicator** and a global administrator **SOM_Manager**. Figure 5 shows the simplified framework.

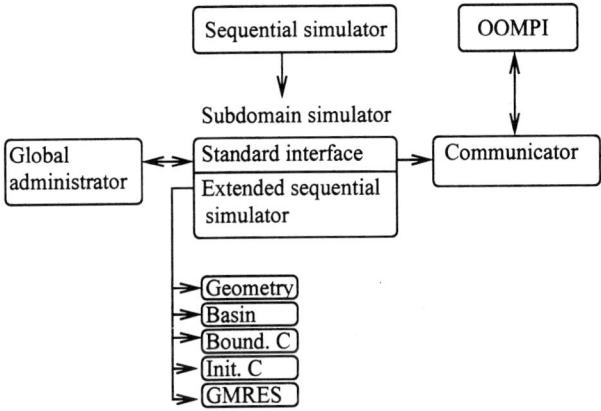

Figure 5: The figure shows the object-oriented framework for the simulator parallel model. The three main parts are: the sub-domain simulators with local data, the global administrator and a communicator class.

The sub-domain simulator is the most important building block in the model. It consists mainly of the code from the sequential simulator, but also of some ad-on functions to handle the parallel computing, synchronization and to be able to use communicated boundary values. We assign one sub-domain simulator, controlling one refined sub-domain, to each processor. In addition we let one processor controlling the coarse grid solver.

During the parallel simulation, the concrete communication between processors occurs in form of exchanging messages and it is all handled by objects of **Communicator**. On each processor there is one such object connected to the local sub-domain simulator. This class uses member functions, given by each of the sub-domain simulators, to retrieve and modify its local data. The only concern of this class is sending and receiving data between each processor, and no actual computing is done here, just assembling and storing data. To solve these two problems, this class have a close connection to the OOMPI library. We have implemented a manager class **SOM_Manager**, to administrate each of the sub-domain solvers. The class is controlled from the main program, and it has the global view if the concurrent solution process.

7   Some Computer Experiments

In this section we report parallel simulation results of some numerical experiments. The purpose is as follows: Examine the parallel efficiency of the implemented parallel simulator, due to scalability and speedup. We have carried out the experiments on a SGI Origin 2000, with R10000 processors, to explore the behavior of the method.

7.1   Computer experiment 1

In the first computer experiment, we are mainly interested in exploring the parallel implementation due to scalability and parallel efficiency. The scaling factors are dependent of two parts. First the independent variables, which are the problem size and the number of processors. Then we have the dependent variables, which are the variables inside the respectively parallel application. We concentrate on the first group. By varying the number of grid blocks and the number of sub-domains used, we want to explore the scaling properties and parallel performance compared to the sequential program.

The number of grid blocks, or the problem size, could have a large influence on the parallel efficiency and the scalability of the parallel program. Larger problem size means larger memory requirements, which again leads to how the actual computer handles a larger memory in the cache. A parallel program could have a great performance on small problems, but would be very slow on very large problems. Communication time depends on the number of grid blocks on each sub-domain and on the time for constructing and sending boundary data to neighboring sub-domain processor.

Also, we want to vary, is the number of processors used. When increasing the number of processors, we expect that the used wall clock time would decrease. Ideally, it would decrease in a linear way.

The partition of the domain Ω is as follows. We divide the domain in equally sized sub-domains M and connect one processor to one sub-domain. The decomposition of the grid Ω is done in 3D. In addition, we use one processor for the coarse grid. Thus, the total number of processors used becomes $P = M + 1$. With this partition of the domain we would have a

# P	Speedup	# N	NX × NY × NZ	CPU (s)
9	9.7	$2.88 \cdot 10^5$	40×30×30	214
9	9.5	$5.76 \cdot 10^5$	80×30×30	534
9	10.6	$11.52 \cdot 10^5$	80×60×30	1085
17	16.5	$2.88 \cdot 10^5$	40×30×30	125
17	18.7	$5.76 \cdot 10^5$	80×30×30	262
17	17.8	$11.52 \cdot 10^5$	80×60×30	598
33	32.6	$2.88 \cdot 10^5$	60×60×10	60
33	33.4	$5.76 \cdot 10^5$	60×60×20	132
33	33.7	$11.52 \cdot 10^5$	60×60×40	306
49	48.8	$11.52 \cdot 10^5$	80×60×30	220
49	48.3	$23.04 \cdot 10^5$	80×60×60	433

Table 1: The used CPU time (in seconds) and the parallel performance for different number of processors P and different problem-sizes N. $NX \times NY \times NZ$ are the number of grid blocks in each direction. In each grid cell we have 8 unknowns N: 6 hydro-carbon components, the water pressure and the temperature. $P = M + 1$, where M is the number of sub-domains. The speedup numbers are calculated by comparing with the sequential CPU time.

perfectly load balanced system. Since we use uniform lithology in Ω, the computational complexities would be the same for all sub-domains.

From Table 1, we can see that the parallel efficiency of the parallel version of the simulator is very good. When increasing the number of processors by a factor of two, we then approximately decrease the CPU time by a factor of two. This means that the parallel code is very efficient due to the computing/communication relation. The scalability is given by the expression $T(P, N) = T(kP, kN)$. Here T is the CPU time, P is the number of processors, N is the total number of unknowns and k is the scaling-factor. From Table 1, we see that this is almost given. The load on the Origin 2000 computer varies from time to time, which can explain some of the jumps in the given CPU time in Table 1.

Table 1 has also a number for speedup, which is the result by comparing the parallel version against the sequential version of the code. Table 1 shows that we have a super-linear speedup in this experiment. It is caused by the fact thata CPU working in parallel may have a faster access to the cache memory than a single processor has.

7.2   Computer experiment 2

The previous computer experiment was relatively simple. The rectangular domain-decompositions give a rectangular communication topology, which could improve the parallel performance. Many applications require a non-rectangular domain-decomposition/communication topology, which could affect the communication time.

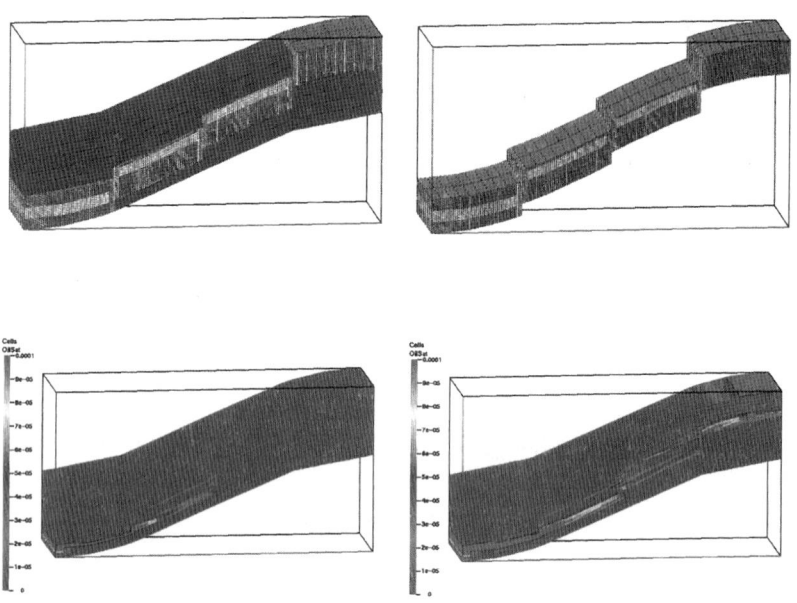

Figure 6: These figures show the oil which is migrating through a faulted carrier bed. The oil tends to follow the permeable layers connected by the very high permeability crushed zones (red color). The coarse grid blocks (green color), represents almost non-permeable lithology.

In the next problem we use a non-regular geometry with locally refined patches. In this implementation, we only use a very simple compositional model, the SBO [11]. This mainly interpolates from some dew-point tables and bubble-point tables. This simplification implies that, doing the flash calculation would not take much part of the time.

We want to simulate the migration of oil through a faulted area. Parallel to the faults there are crushed zones with high permeability and high porosity. These high permeability zones may create some computational complexities. The computing domain and the domain decomposition are shown in Figure 6.

The coarse grid blocks are representing a very low permeable lithololology. In Figure 6 the coarse grid blocks are removed and we can clearly detect the faulted area. The permeability varies from almost impermeable, to 500 mD in the crushed fault zone. The layered sediments have varying permeability, but all below 500 mD. The aspect ratio, which is the relation between the horizontal and vertical permeability, is 10 in this experiment.

The extent of the domain is 1000 m × 300 m × 140 m and the carrier bed are lifted 200 m above the horizon. As boundary condition we use a relatively large water flux $q_w = 1.0 \cdot 10^{-4}$ mol/m²s, from left to right. This is injected only in the permeable sediments. An oil flux $q_o = 1.0 \cdot 10^{-9}$ mol/m²s, is also injected in the lowest portion on the left side. Only water can go out of the sediments on the right-hand side. The simulation period is 300000 days, or about 822 years. It should be noticed that all the lithologies data and the boundary conditions are realistic.

Each refined subdomain follow a faulted block, containing 200 grid blocks, which means that we have about 2500 grid blocks in total. In Figure 6, we have given two snapshots of the oil saturation during the simulations after 50000 days and after 250000 days. Here we can clearly see the oil migrating through the higher permeability zones. These are connected via the crushed faulted zone with a very high permeability. We can also see that some of the oil is trapped, i.e. on the top of two of the faults in Figure 6. Table 2, shows the CPU measurements for different parts of the IMPEC solution strategy in the code. The CPU measurements are after 10 days of simulation. The system would get stabilized during the simulation, so a longer time period would not be relevant for these results.

We notice is that we have a great loss of parallel performance in the "Solve Pressure" function. This loss is mainly due to the bad speedup of the Galerkin solver, only 9.8. This experiment demonstrates that the geometrical complexities would strongly affect the time solving the pressure equation. In this experiment, we have located the high conductivity faults on the interface between two coarse grid blocks. We know that this may be a very bad choice. Instead, the faults should be located inside the coarse grid blocks, which probably would given a better numerical results. The overall speedup is still quite good 12.1, despite the poor performance in the pressure solver.

Function	13 CPU			Sequential	
	CPU (s)	% t. loop	speedup	CPU (s)	% t. loop
Update rock data	0.2	0.26	13.0	2.6	0.02
Calc coupl. coeffi.	0.5	0.67	13.0	6.5	0.7
Solve temperature	5.9	7.9	11.3	66.7	7.5
Solve Pressure	23.8	32.0	10.6	255.0	28.4
*Eq. setup	5.0	6.7	14.6	73.3	8.1
*Eq. solve	18.3	24.0	9.8	178.4	19.8
Calc molar masses	40.0	53.3	13.1	523.4	58.3
Do flash calc.	0.4	0.5	11.8	4.7	0.48
Update physical var.	0.8	1.0	11.3	9.0	0.88
Total time loop	74.4	100	12.1	897.2	100
Total simulation	74.8	*	12.1	899.0	*

Table 2: This table gives the CPU measurements from the IMPEC solution strategy. The CPU time is a result of 10 days of simulation.

8 Conclusions

In this work we have used a fixed adaptive LGR technique. In simulation of reservoirs with fixed wells, faults and large fractures, certain fixed LGR techniques have proven to be very effective. Our fixed adaptive LGR technique creates non-matching interfaces, as nearly all LGR techniques do.

The most important issue when working with fluid flow is to conserve the mass. To ensure mass conservation, the flux calculation across any of the interfaces (both matching and non-matching) have to be treated with great care. In most application, a two point flux approximation has been used. In this paper a muli point flux approximation (MPFA) is also implemented. As far as we now, a MPFA method has not yet been applied to cases with non-matching interfaces in three dimensions. In this work we propose to add slave cells close to the non-matching interfaces. Since these slave cells create a locally matching grid an existing MPFA method may be used. The Simulator Parallel Mode (SPM) which is described in this paper, strongly promotes code reuse and gives a very high level of parallelization despite low implementation costs. This programming framework is also portable to other platforms or other applications. With this model we have good control with performance parameters as load balance, synchronization and communication topology.

We have presented two computer experiments to examine the parallel efficiency of the implemented parallel simulator with respect to scalability and speedup. In the first experiment we examine the scalability for a large homogeneous one-phase problem. In the next experiment, we go a step further and study multi-phase heterogeneous problems. This problem requires a much larger portion of communication and synchronization, than in the first experiment. However, the experiments still show nice results for parallel efficiency and speedup. In many industrial groups, clusters of PCs or workstations are used. This means that for effective use of the computers, a parallel implementation based on explicit message passing inside the network should be considered. Our parallel implementation framework puts no restriction for running the code on such networks/clusters with PCs or workstations. This option should be implemented in the future. Asynchronous MPI implementation should be considered, due to the heterogeneous computers in a typical network.

Acknowledgments. The authors thank The Norwegian reserarch council and Norsk Hydro for support.

References

[1] Aavatsmark, I., Barkve, T., Bøe, Ø., and Mannseth, T., Discretization on unstructured grids for inhomogeneous, anisotropic media. Part I: Derivation of the methods, *SIAM J. Scient. Comput.* **19** (1998), 1700-1717.

[2] Aavatsmark, I., Barkve, T., Bøe, Ø., Reiso, E., and Teigland, R., Multipoint flux approximation for faults and grid refinement, submitted, 1999.

[3] Bear, J. and Bachmat, Y., *Introduction to Modeling of Transport Phenomena in Porous Media*, Kluwer Academic Publishers, Dordrecht, Netherlands, 1990.

[4] Bernardi, C., Maday, Y., and Patera, A. T., A new nonconforming approach to domain decomposition: the mortar element method, *Nonlin. PDEs and their Applications*, College de France Seminar, **XI**, 1992.

[5] Bruaset, A. M., Cai, X., Langtangen, H. P., and Tveito, A., Numerical solution of PDEs on parallel computers utilising sequential simulators. Technical Report, University of Oslo, Depart. of Inform., Oslo, Norway, 1998.

[6] Dahle, H. K., Espedal, M. S., and Sævareid, O. Characteristic, Local grid refinement techniques for reservoir flow problems, *Numerical Methods for Partial Differential Equations* **6** (1992), 279-309.

[7] Edwards, M. G., M-matrix flux splitting for general full tensor discretization operators, *J. Comput. Physics*, 1999, to appear.

[8] Edwards, M. G., Cross flow tensors and finite volume approximation with deferred correction *Comput. Methods Appl. Mech. Eng.* **151** (1998), 143-161.

[9] Ewing, R. E., Lazarov, R. D., and Vassilevski, P. S., Local refinement technique for elliptic problems on cell-center grids, I; Error analysis *Math. Comp.* **56** (1991), 437–461.

[10] Fladmark, G.E., A numerical method for local refinement of grid blocks applied to reservoir simulation, Internal Report, 1982, Norsk Hydro, Bergen, Norway.

[11] Fladmark, G.E., Secondary oil migration: Mathematical and numerical model, Internal Report, 1995, Norsk Hydro, Bergen, Norway.

[12] Reme, H., The preconditioning and multi-point flux approximation methods for solving secondary oil migration and upscaling problems, Dr. Scient. Thesis, Math. Dept., University of Bergen, Norway, 1999.

[13] Teigland, R., On multilevel methods for numerical reservoir simulation, Dr. Scient. Thesis, Math. Dept., University of Bergen, Norway, 1991.

[14] Verweij, J. M., *Hydrocarbon migration systems analysis*, Elsevier, Amsterdam, 1993.

[15] Wachspress, E. L., *Iterative Solution of Elliptic Systems and Applications to the Neutron Diffusion Equations of Reactor Physics*, Prentice-Hall Inc, Englewood Cliffs, 1966.

[16] Welte, D. H., Horsfield, B., and Baker, D. R., *Petroleum and Basin Evolution*, Springer-Verlag, Berlin, 1997.

[17] Wong, T. W., Firoozabadi, A., and Aziz, K., The relationship of the volume balance method of compositional simulation to the Newton-Raphson method, *SPE Reservoir Structure Symp.*, **5** (1990), 414–422.

[18] Øye, G. Å., An object-oriented parallel implementation of local grid refinement and Domain decomposition in a simulator for secondary oil migration, Dr. Scient. Thesis, Math. Dept., University of Bergen, Norway, 1999.

Relationships Among Some Conservative Discretization Methods

Thomas F. Russell

Abstract

Relationships among various mass-conservative discretization techniques for equations of the type $-\nabla \cdot \mathbf{K}\nabla p = q$ on distorted logically rectangular meshes are discussed. The case of heterogeneous, anisotropic $\mathbf{K}$ is important for applications to subsurface porous media, in particular the groundwater flow equation and the pressure equation of petroleum reservoir simulation. Some methods are based on $\mathbf{K}$ itself, others on $\mathbf{K}^{-1}$. Within one of these groups, mass lumping and quadrature can be keys to understanding connections between methods; incomplete inversion of the mass matrix is useful in relating one group to the other.

KEYWORDS: conservation, distorted grid, finite volume, mixed method

1 Introduction

The purpose of this paper is to introduce some concepts that could be useful in understanding relationships among various types of conservative discretization methods. We concern ourselves particularly with finite volume, flux-based finite difference, and mixed finite element methods designed to calculate accurate fluxes on distorted meshes for problems with heterogeneous, anisotropic conductivity. Such procedures can be important in a variety of applications, particularly subsurface flows in porous media.

This study is merely introductory, and is far from exhaustive or definitive. The methods discussed here are sufficiently varied and complex that a complete analysis of their relationships would constitute a much longer paper. It is not our goal here to compare the merits of these approaches; other forums will be more appropriate for that. We also do not address the nontrivial issues of solving the discrete equations arising from these methods. Rather, we seek some simple tools that can assist efforts to comprehend a "big picture" of discretizations that can fairly be described as confusing.

For simplicity of exposition, the spatial domain will be limited to 2-D, with 3-D mentioned only in passing. This is a significant simplification, because, for example, the edges of a quadrilateral (bilinear image of a square) are straight, but the faces of a hexahedron (trilinear image of a cube) need not be planar, and while the Jacobian of a bilinear mapping is linear, in general for a trilinear mapping it is nonlinear. The 2-D methods discussed here extend in one way or another to 3-D, and are in various stages of development for 3-D. However, for our introductory purposes here, we find it best to emphasize 2-D. For similar reasons, we consider only a linear pressure equation, and our focus is on logically rectangular meshes, though more complicated PDEs and grids of more general connectivity could be addressed.

Section 2 describes the lowest-order Raviart-Thomas (RT$_0$) [11, 12] mixed finite element (MFE) method, as a starting point to which other methods can be related. The analogous control-volume mixed (CVMFE) method [5] is introduced in Section 3, along with some modifications by Garanzha and Konshin [7] and some observations in [7] that relate the CVMFE to the support-operators (SO) method of Shashkov and co-workers [8, 10]. These methods are related in a different way in Section 4 to the multi-point flux approximation (MPFA) methods of Aavatsmark and co-workers [1, 2]. Section 4 also briefly discusses the expanded mixed (EM) method of Arbogast *et al.* [3, 4] and the mixed finite volume (MFV) method of Thomas and Trujillo [13], and Section 5 summarizes the paper.

2   RT$_0$ Mixed Method

A MFE represents a partial differential equation as a system of lower-order equations, solving these for multiple variables of physical interest. In the context of porous media, we assume incompressible flow, neglecting gravitational effects, so that the pressure equation (with no-flow boundary condition for simplicity) takes the form

$$-\nabla \cdot (\mathbf{K}\nabla p) \;\; = \;\; q, \qquad \mathbf{x} \in \Omega, \tag{2.1}$$

$$-\mathbf{K}\nabla p \cdot \mathbf{n} \;\; = \;\; 0, \qquad \mathbf{x} \in \partial\Omega, \tag{2.2}$$

where $\mathbf{K}$ (scalar or anisotropic tensor) is the mobility or hydraulic conductivity, p the pressure, q a source/sink (e.g., well) term, and Ω is the reservoir or aquifer with boundary $\partial\Omega$. Let $\mathbf{v}$ be the velocity vector, and express (2.1) as a system representing Darcy's law and conservation of mass, respectively:

$$\mathbf{v} = -\mathbf{K}\nabla p, \tag{2.3}$$

$$\nabla \cdot \mathbf{v} = q. \tag{2.4}$$

A primary goal is to obtain a more accurate $\mathbf{v}$, especially when $\mathbf{K}$ is heterogeneous, by solving the system (2.3)–(2.4) for $\mathbf{v}$ and p, instead of solving (2.1) for p and applying (2.3) to obtain $\mathbf{v}$. Piecewise-constant test functions in (2.7) below will also yield local mass conservation.

As in [11], let $\mathbf{V} = \{\mathbf{w} \in H(\mathrm{div}, \Omega) : \mathbf{w} \cdot \mathbf{n} = 0 \text{ on } \partial\Omega\}$, $P = L^2(\Omega)$, write (2.3) as

$$\mathbf{K}^{-1}\mathbf{v} + \nabla p = 0, \tag{2.5}$$

and arrive at the weak form of (2.3)–(2.4), which is to find $\mathbf{v} \in \mathbf{V}$ and $p \in P$ such that

$$\int_\Omega \mathbf{K}^{-1}\mathbf{v} \cdot \mathbf{w}\, dx - \int_\Omega \nabla \cdot \mathbf{w}\, p\, dx = 0, \qquad \mathbf{w} \in \mathbf{V}, \tag{2.6}$$

$$\int_\Omega \nabla \cdot \mathbf{v}\, z\, dx = \int_\Omega q z\, dx, \qquad z \in P. \tag{2.7}$$

The rectangular RT_0 elements define discrete subspaces $\mathbf{V}_h$ and P_h with respect to a cartesian grid. P_h consists of the piecewise-constant functions. $\mathbf{V}_h$ can best be viewed by associating a degree of freedom with the flux (constant normal component times edge length $|E|$) on each inter-block edge E. A typical basis function has flux 1 on one edge (D in Fig. 1) and 0 on all others, with the flux varying linearly in the direction of the velocity. For arbitrary quadrilateral grids, appropriate pressure and velocity spaces are defined via the Piola mapping [11, 12] (see Section 3.2).

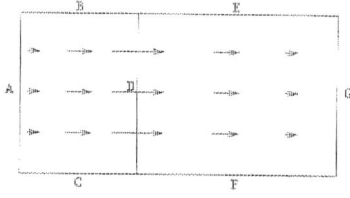

Figure 1. RT_0 velocity basis function on rectangles.

3   CVMFE and K^{-1} Methods

In addition to the local mass conservation provided by (2.7), the CVMFE seeks a local discrete Darcy law, which an engineer could view as applying to a cell-sized "tank" with pressures imposed at the ends. In the MFE context, this is analogous to the formulation of control-volume finite element (CVFE) methods from Galerkin FE, hence the name. We describe the formulation first on rectangles, then on distorted quadrilaterals.

3.1   Rectangular grid

Starting from the system (2.4)–(2.5), consider the pressure cells $Q_{i,j}$ and the control volumes $Q_{i+1/2,j}$ and $Q_{i,j+1/2}$ associated with the edge fluxes $(f_x)_{i+1/2,j}$ and $(f_y)_{i,j+1/2}$, as depicted in Fig. 2. For this case, we take $\mathbf{K} = k$ to be scalar. The unknowns are associated with cells and edges as in the MFE, and the trial functions for $\mathbf{v}$ and p are again the RT_0 spaces. In the MFE, the test functions $\mathbf{w}$ and z are from the same spaces; in the CVMFE, (2.4) is still treated in this way, but the x- and y-components of the vector Darcy law (2.5) are instead integrated over "tanks" $Q_{i+1/2,j}$ and $Q_{i,j+1/2}$, respectively. This is equivalent to taking the scalar product of (2.5) with $\mathbf{w} = (1,0)$ on $Q_{i+1/2,j}$ and $(0,0)$ elsewhere, and with $\mathbf{w} = (0,1)$ on $Q_{i,j+1/2}$ and $(0,0)$ elsewhere, respectively, and integrating. The resulting partial derivatives of p can be integrated out, leaving

$$\int_{x_i}^{x_{i+1}} \int_{y_{j-1/2}}^{y_{j+1/2}} k^{-1} v_x(x,y)\, dy\, dx + \int_{y_{j-1/2}}^{y_{j+1/2}} (p(x_{i+1},y) - p(x_i,y))\, dy = 0, \quad (3.1)$$

and similarly for the y-component. Using the RT_0 trial functions, the integrals such as (3.1) are expressed in terms of the unknowns p, f_x, and f_y. We then obtain the discrete Darcy equations on the "tanks": in the x-direction on $Q_{i+1/2,j}$,

$$a_{i+1/2,j;A}(f_x)_{i-1/2,j} + a_{i+1/2,j;D}(f_x)_{i+1/2,j}$$
$$+ a_{i+1/2,j;G}(f_x)_{i+3/2,j} + p_{i+1,j} - p_{i,j} \;\; = \;\; 0, \quad (3.2)$$

where A, D, G refer to edges as in Fig. 1, and

$$a_{i+1/2,j;A} \;\; = \;\; \frac{1}{8}\frac{k_{i,j}^{-1}}{|Q_{i,j}|}(\Delta x_i)^2, \quad (3.3)$$

$$a_{i+1/2,j;D} \;\; = \;\; \frac{3}{8}\frac{k_{i,j}^{-1}}{|Q_{i,j}|}(\Delta x_i)^2 + \frac{3}{8}\frac{k_{i+1,j}^{-1}}{|Q_{i+1,j}|}(\Delta x_{i+1})^2, \quad (3.4)$$

$$a_{i+1/2,j;G} = \frac{1}{8} \frac{k_{i+1,j}^{-1}}{|Q_{i+1,j}|} (\Delta x_{i+1})^2, \tag{3.5}$$

where $\Delta x_i = x_{i+1/2} - x_{i-1/2}$, $\Delta x_{i+1} = x_{i+3/2} - x_{i+1/2}$, with an analogous equation in the y-direction on $Q_{i,j+1/2}$.

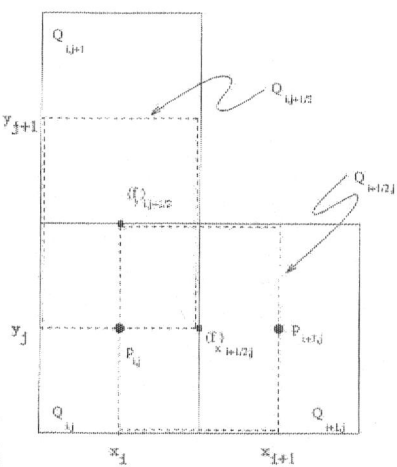

Figure 2. Cells, unknowns, and control volumes for rectangular grid.

Integration of (2.4) over the cell $Q_{i,j}$ (equivalent to multiplying by a test function $z = 1$ on $Q_{i,j}$ and 0 elsewhere), together with the Gauss divergence theorem, yields the discrete mass conservation:

$$(f_x)_{i-1/2,j} - (f_x)_{i+1/2,j} + (f_y)_{i,j-1/2} - (f_y)_{i,j+1/2} = -|Q_{i,j}|q_{i,j}. \tag{3.6}$$

Equation (3.2), its y-analogue, and (3.6) give rise to a symmetric system of linear equations that is solved for the pressures at block centers and the fluxes across edges.

3.2   Distorted quadrilateral grid

If Q is a convex quadrilateral, then there is a unique bilinear mapping of a reference square $\hat{Q} = [0,1]^2$ onto Q that sets up coordinates on Q. The pressure in Q is associated with the "center" of Q, meaning the image of the center of $\hat{Q}$: Note that this is not generally the centroid of Q.

The extension of the CVMFE method to general quadrilaterals requires that continuity of flux be maintained, so that the normal component of a velocity function must be constant on each edge. Then we can associate degrees of freedom with fluxes on edges, as in the rectangular case. In Fig. 3 we show two adjacent quadrilaterals with coordinates determined by their local bilinear mappings. The velocity vector function $\mathbf{v}_{i+1/2,j}$ that has flux 1 (hence normal component $1/|E_{i+1/2,j}|$) on the common edge $E_{i+1/2,j}$ (labeled in Fig. 4) and 0 on the other edges is pictured. It is oriented along x-coordinate lines, and has constant normal component on each complementary y-line, with the flux varying linearly in the x-direction.

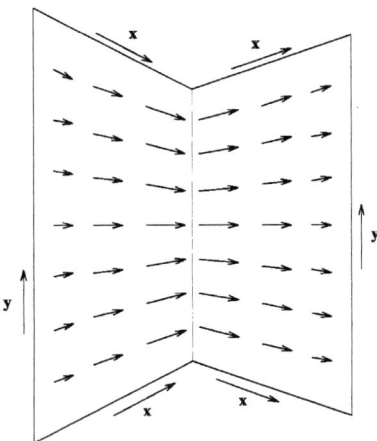

Figure 3. Velocity basis function $\mathbf{v}_{i+1/2,j}$ on quadrilaterals.

Let $\hat{x}, \hat{y} \in [0, 1]$ be the reference coordinates for a quadrilateral Q. Set

$$\mathbf{X}(\hat{x}, \hat{y}) = \left(\frac{\partial x}{\partial \hat{x}}, \frac{\partial y}{\partial \hat{x}} \right) \tag{3.7}$$

$$\mathbf{Y}(\hat{x}, \hat{y}) = \left(\frac{\partial x}{\partial \hat{y}}, \frac{\partial y}{\partial \hat{y}} \right) \tag{3.8}$$

to be the columns of the Jacobian matrix of the bilinear mapping. These can be viewed as the images of the vectors $(1, 0)$ and $(0, 1)$, respectively, under the mapping from $\hat{Q}$ to Q. For example, the $\mathbf{v}_{i+1/2,j}$ in Fig. 3 is parallel to $\mathbf{X}$, and after some manipulation, one can show that it is given on the left-hand quadrilateral $Q_{i,j}$ by

$$\mathbf{v}_{i+1/2,j}(x, y) = \frac{\hat{x}\mathbf{X}}{J_{i,j}(\hat{x}, \hat{y})}, \tag{3.9}$$

where

$$J_{i,j}(\hat{x}, \hat{y}) = \frac{\partial x}{\partial \hat{x}} \frac{\partial y}{\partial \hat{y}} - \frac{\partial x}{\partial \hat{y}} \frac{\partial y}{\partial \hat{x}} \tag{3.10}$$

is the Jacobian of the mapping from $\hat{Q}$ to $Q_{i,j}$. In the right-hand quadrilateral $Q_{i+1,j}$, everything is the same except that $1 - \hat{x}$ and $J_{i+1,j}$ replace $\hat{x}$ and $J_{i,j}$, respectively. These velocity trial functions and unknowns (fluxes across edges) can also be obtained from those on rectangles by a Piola transformation [12]. Define also the corresponding unit normal vectors, $\mathbf{n}_x \perp \mathbf{Y}$, $\mathbf{n}_y \perp \mathbf{X}$, pictured in Fig. 4:

$$\mathbf{n}_x \;=\; \frac{(\partial y/\partial \hat{y}, -\partial x/\partial \hat{y})}{[(\partial x/\partial \hat{y})^2 + (\partial y/\partial \hat{y})^2]^{1/2}}, \tag{3.11}$$

$$\mathbf{n}_y \;=\; \frac{(-\partial y/\partial \hat{x}, \partial x/\partial \hat{x})}{[(\partial x/\partial \hat{x})^2 + (\partial y/\partial \hat{x})^2]^{1/2}}. \tag{3.12}$$

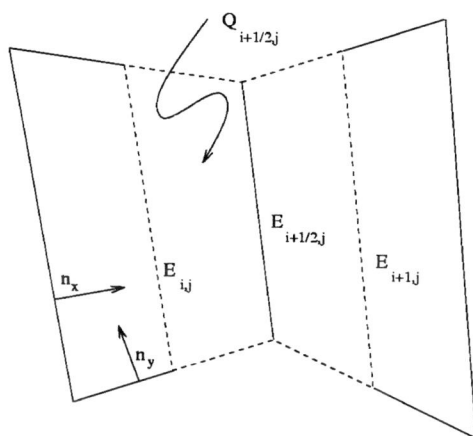

Figure 4. Control-volume mixed finite elements on quadrilaterals.

It remains to choose control volumes and test functions. For the integrations of (2.4), the control volumes are the quadrilateral blocks $Q_{i,j}$, and the test functions are scalar characteristic functions of the control volumes, i.e., functions that are 1 on one volume and zero elsewhere. The Gauss divergence theorem then yields (3.6) for quadrilaterals as well as rectangles.

For the integrations of (2.5), to mimic the steps leading to (3.1), we use images of rectangular control volumes $Q_{i+1/2,j}$ and $Q_{i,j+1/2}$ under the bilinear mapping, as seen in Fig. 4. We again denote such control volumes by $Q_{i+1/2,j}$ and $Q_{i,j+1/2}$. $Q_{i+1/2,j}$ in Fig. 4 will be the "tank" with pressures $p_{i,j}$ and $p_{i+1,j}$ at the two ends. We seek test functions that allow the gradient of p to be integrated out, leaving differences of p.

Let $Q_{i+1/4,j}$ and $Q_{i+3/4,j}$ denote the "left-hand half" and "right-hand half," respectively, of $Q_{i+1/2,j}$. Then $Q_{i+1/4,j}$ is the image of the right-hand

half, $(1/2, 1) \times (0, 1)$, of $\hat{Q}$ under the mapping to $Q_{i,j}$. The original CVMFE
[5] used $\mathbf{X}$ as the test function, which led to integrals that could be calculated analytically, with a factor of $1/J$ outside the integral; the modification
suggested by Garanzha and Konshin [7] instead uses $\mathbf{X}/J$, which requires numerical integration but gains accuracy on highly distorted meshes. By their
approach, the left-half p integral analogous to the one in (3.1) is

$$\int_{Q_{i+1/4,j}} \nabla p \cdot (\mathbf{X}_{i,j}/J_{i,j}) \, d\mathbf{x} = \int_0^1 \int_{1/2}^1 \frac{\partial p}{\partial \hat{x}} \, d\hat{x} \, d\hat{y}$$

$$= \int_0^1 p(1, \hat{y}) \, d\hat{y} - \int_0^1 p(1/2, \hat{y}) \, d\hat{y}$$

$$\approx p_{i+1/2,j} - p_{i,j}, \tag{3.13}$$

where the last step involves p at edge and cell centers and is exact for bilinear p. In defining the CVMFE equations, treat (3.13) as exact, i.e., ignore
truncation error. Similarly, the right-half integral yields

$$\int_{Q_{i+3/4,j}} \nabla p \cdot (\mathbf{X}_{i+1,j}/J_{i+1,j}) \, d\mathbf{x} = p_{i+1,j} - p_{i+1/2,j}. \tag{3.14}$$

Hence, by choosing the test vector field

$$\mathbf{w}_{i+1/2,j} = \begin{cases} \mathbf{X}_{i,j}/J_{i,j} & \text{on } Q_{i+1/4,j}, \\ \mathbf{X}_{i+1,j}/J_{i+1,j} & \text{on } Q_{i+3/4,j}, \\ 0 & \text{elsewhere}, \end{cases} \tag{3.15}$$

we combine (3.13)–(3.14) into

$$\int_{Q_{i+1/2,j}} \nabla p \cdot \mathbf{w}_{i+1/2,j} \, d\mathbf{x} = p_{i+1,j} - p_{i,j}, \tag{3.16}$$

and the edge value $p_{i+1/2,j}$ is eliminated. If desired, it can be recovered later
in a postprocessing step.

With the test function from (3.15), the $\mathbf{v}$ term of (2.5) can be calculated
as in [5], leading to the discrete Darcy equation analogous to (3.2), with edges
denoted as in Fig. 1:

$$a_{i+1/2,j;A}(f_x)_{i-1/2,j} + a_{i+1/2,j;D}(f_x)_{i+1/2,j} + a_{i+1/2,j;G}(f_x)_{i+3/2,j}$$

$$+ a_{i+1/2,j;B}(f_y)_{i,j+1/2} + a_{i+1/2,j;C}(f_y)_{i,j-1/2}$$

$$+ a_{i+1/2,j;E}(f_y)_{i+1,j+1/2} + a_{i+1/2,j;G}(f_y)_{i+1,j-1/2}$$

$$+ p_{i+1,j} - p_{i,j} = 0, \tag{3.17}$$

where the coefficients in (3.17) are given by

$$a_{i+1/2,j;D} = \int_0^1 \int_{1/2}^1 \hat{x}(\mathbf{K}_{i,j}^{-1}\mathbf{X}_{i,j}) \cdot \mathbf{X}_{i,j}/J_{i,j} \, d\hat{x} \, d\hat{y} \qquad (3.18)$$

$$\int_0^1 \int_0^{1/2} (1-\hat{x})(\mathbf{K}_{i+1,j}^{-1}\mathbf{X}_{i+1,j}) \cdot \mathbf{X}_{i+1,j}/J_{i+1,j} \, d\hat{x} \, d\hat{y},$$

$$a_{i+1/2,j;A} = \int_0^1 \int_{1/2}^1 (1-\hat{x})(\mathbf{K}_{i,j}^{-1}\mathbf{X}_{i,j}) \cdot \mathbf{X}_{i,j}/J_{i,j} \, d\hat{x} \, d\hat{y}, \qquad (3.19)$$

$$a_{i+1/2,j;G} = \int_0^1 \int_0^{1/2} \hat{x}(\mathbf{K}_{i+1,j}^{-1}\mathbf{X}_{i+1,j}) \cdot \mathbf{X}_{i+1,j}/J_{i+1,j} \, d\hat{x} \, d\hat{y}, \qquad (3.20)$$

$$a_{i+1/2,j;B} = \int_0^1 \int_{1/2}^1 \hat{y}(\mathbf{K}_{i,j}^{-1}\mathbf{Y}_{i,j}) \cdot \mathbf{X}_{i,j}/J_{i,j} \, d\hat{x} \, d\hat{y}, \qquad (3.21)$$

$$a_{i+1/2,j;C} = \int_0^1 \int_{1/2}^1 (1-\hat{y})(\mathbf{K}_{i,j}^{-1}\mathbf{Y}_{i,j}) \cdot \mathbf{X}_{i,j}/J_{i,j} \, d\hat{x} \, d\hat{y}, \qquad (3.22)$$

$$a_{i+1/2,j;E} = \int_0^1 \int_0^{1/2} \hat{y}(\mathbf{K}_{i+1,j}^{-1}\mathbf{Y}_{i+1,j}) \cdot \mathbf{X}_{i+1,j}/J_{i+1,j} \, d\hat{x} \, d\hat{y}, \qquad (3.23)$$

$$a_{i+1/2,j;F} = \int_0^1 \int_0^{1/2} (1-\hat{y})(\mathbf{K}_{i+1,j}^{-1}\mathbf{Y}_{i+1,j}) \cdot \mathbf{X}_{i+1,j}/J_{i+1,j} \, d\hat{x} \, d\hat{y}. \quad (3.24)$$

The Darcy equation for the horizontal edge $E_{i,j+1/2}$ is analogous, with the roles of the x- and y-directions reversed. That equation, with (3.6) and (3.17), forms the discrete system for CVMFE.

As the trial and test functions are different for CVMFE, the discrete equations may not be symmetric. A representative example is (3.21), in which test function $\mathbf{w}_{i+1/2,j}$ interacts with trial function $\mathbf{v}_{i,j+1/2}$. Symmetry would require that (3.21) yield the same result as

$$a_{i,j+1/2;B} = \int_{1/2}^1 \int_0^1 \hat{x}(\mathbf{K}_{i,j}^{-1}\mathbf{X}_{i,j}) \cdot \mathbf{Y}_{i,j}/J_{i,j} \, d\hat{x} \, d\hat{y}. \qquad (3.25)$$

Since (3.21) and (3.25) integrate over different (but overlapping) half-cells, symmetry cannot hold in general with variable J (with constant J, $\mathbf{X}$, $\mathbf{Y}$, as on a grid of parallelograms, symmetry does hold). However, with a quadrature rule that receives no contribution from the nonoverlapping parts of the half-cells, i.e., one whose nonoverlapping points have $\hat{y} = 0$ in (3.21) and $\hat{x} = 0$ in (3.25), symmetry is possible. As noted by Garanzha and Konshin

[7], for the half-cell in (3.21), a rule of the form

$$\int_0^1 \int_{1/2}^1 g \, d\hat{x} \, d\hat{y} \;\approx\; r(g(1/2,0) + 2g(1,1/2) + g(1/2,1))$$
$$+ \;\; (1/8 - r)(g(1,0) + 2g(1/2,1/2) + g(1,1)), \;\; (3.26)$$

where $0 \leq r \leq 1/8$, preserves symmetry and also integrates bilinear g exactly. With such a quadrature, the entire discrete system is symmetric, and for parallelograms all integrals are exact; in particular, the weights $1/8, 6/8, 1/8$ in (3.3)–(3.5) are preserved. We note here that the corresponding weights for MFE are $1/6, 4/6, 1/6$, and that MFE is symmetric with any consistent quadrature because the trial and test functions coincide. Because the MFE integrands that produce these weights are quadratic, it is possible within the MFE framework, with a rule that is not exact for quadratics, to produce the CVMFE weights and the CVMFE formulas resulting from quadrature rule (3.26) [7]. In general, it is not possible to produce the exact CVMFE formulas in this way, as they are nonsymmetric.

Quadrature provides a convenient framework within which to consider the technique of mass lumping, which is often used to simplify finite element methods, producing a diagonal mass matrix and reducibility to finite differences. In the context of CVMFE equations (3.18)–(3.24), this is equivalent to setting $\hat{x}$, $1 - \hat{x}$, $\hat{y}$, and $1 - \hat{y}$ equal to 1 whenever they are greater than $1/2$, and 0 otherwise, then integrating. In (3.26), this would correspond to a quadrature with $\hat{x} = 1$ and $\hat{y} = 0$ or 1, i.e.,

$$\int_0^1 \int_{1/2}^1 g \, d\hat{x} \, d\hat{y} \approx 1/4 \, (g(1,0) + g(1,1)). \tag{3.27}$$

This quadrature preserves symmetry, just as (3.26) does, but it does not integrate bilinear g exactly; indeed, it must not, because it changes the weights $1/8, 6/8, 1/8$ in (3.3)–(3.5) to $0, 1, 0$. Garanzha and Konshin [7] (p. 24) note that this lumping produces the SO method [8, 10]. Thus, CVMFE and SO are related in much the same way as CVFE and conventional point-centered finite differences.

4 Relationships to K Methods

In the petroleum industry, there is considerable recent work [1, 2, 6, 9] on discretizations that take a view dual to that of formulas such as (3.17). We call the methods of Section 3 "$\mathbf{K}^{-1}$ methods" because, when complicated details are ignored, they reduce to a relationship of the form

$$-\mathbf{K}^{-1}\mathbf{f} = \Delta p, \qquad (4.1)$$

where $\mathbf{f}$ is the flux vector and Δp the pressure difference. $\mathbf{K}^{-1}$ appears discretely as a mass matrix, so that a combination of fluxes equals a pressure drop. "$\mathbf{K}$ methods," on the other hand, adopt the perspective that

$$\mathbf{f} = -\mathbf{K}\,\Delta p, \qquad (4.2)$$

expressing an individual discrete flux as a combination of pressure drops. As a prototype of these methods, we consider the MPFA scheme of Aavatsmark *et al.* [1, 2] and seek to relate $\mathbf{K}^{-1}$ methods to it. It seems natural to attempt to pass from (4.1) to (4.2) by inverting $\mathbf{K}^{-1}$. To avoid a full stencil in (4.2) even for orthogonal grids, it is necessary to lump the principal weights in $\mathbf{K}^{-1}$, such as the $1/8, 6/8, 1/8$ for CVMFE. For the contributions to $\mathbf{K}^{-1}$ caused by non-orthogonality and/or anisotropy, some form of incomplete inversion is suggested.

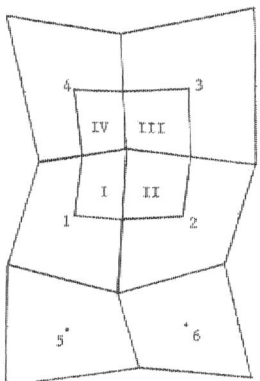

Figure 5. Cells and interaction region for MPFA method.

To describe the MPFA method, we refer to Fig. 5. The pressure p is to be determined at the cell centers and is assumed to be linear (not bilinear) on each piece I,II,III,IV of the interaction region. Thus, there are 12 degrees of freedom on that region, and 12 constraints are imposed: 4 cell-center values, continuity of p at the 4 points where the interaction-region boundary crosses a cell edge, and continuity of the flux across the 4 interfaces in the region.

To close this system, the relation

$$f_E = -\int_E \mathbf{K}\nabla p \cdot \mathbf{n}\, dS \qquad (4.3)$$

between the pressure and the fluxes across edges is needed. With a bilinear reference mapping to a distorted quadrilateral, as in Section 3, this can be written

$$f_x = a\frac{\partial p}{\partial \hat{x}} + c\frac{\partial p}{\partial \hat{y}},$$

$$f_y = c\frac{\partial p}{\partial \hat{x}} + b\frac{\partial p}{\partial \hat{y}}, \qquad (4.4)$$

where

$$a = -J\left(\mathbf{K}\nabla\hat{x}\right)\cdot\nabla\hat{x} = -(\mathbf{Y}\cdot\mathbf{Y}/J)\left(\mathbf{Kn}_x\right)\cdot\mathbf{n}_x,$$

$$b = -J\left(\mathbf{K}\nabla\hat{y}\right)\cdot\nabla\hat{y} = -(\mathbf{X}\cdot\mathbf{X}/J)\left(\mathbf{Kn}_y\right)\cdot\mathbf{n}_y,$$

$$c = -J\left(\mathbf{K}\nabla\hat{x}\right)\cdot\nabla\hat{y} = -(|\mathbf{X}||\mathbf{Y}|/J)\left(\mathbf{Kn}_x\right)\cdot\mathbf{n}_y. \qquad (4.5)$$

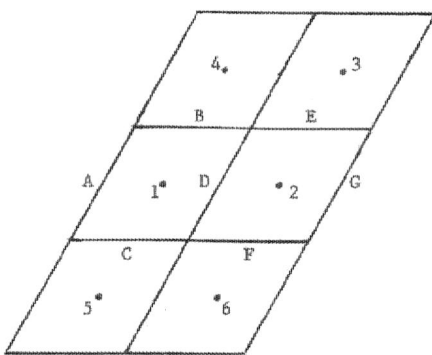

Figure 6. Uniform grid of parallograms.

This is best understood in a simple case, where a, b, c are constants. In Fig. 6 we show a grid of parallelograms that meets this criterion, since J, $\mathbf{X}$, $\mathbf{Y}$ are constants, and assume also that $\mathbf{K} = k$ is a scalar constant. Referring to Fig. 6, solution of the 12 equations leads to [1]

$$f_D = \left(a - \frac{c^2}{2b}\right)(p_2 - p_1) + \frac{c}{4}\left(1 + \frac{c}{b}\right)(p_3 - p_5) + \frac{c}{4}\left(1 - \frac{c}{b}\right)(p_4 - p_6). \quad (4.6)$$

(For no distortion, $c = 0$, and (4.6) reduces to $f_D = a(p_2 - p_1)$, as expected for finite differences.) If θ is the angle of distortion ($\theta = 0$ for rectangles,

$\mathbf{X} \cdot \mathbf{Y} = |\mathbf{X}||\mathbf{Y}| \sin \theta$, then $J = |\mathbf{X}||\mathbf{Y}| \cos \theta$, $\mathbf{n}_x \cdot \mathbf{n}_y = -\sin \theta$, and

$$a = \frac{-Jk}{\mathbf{X} \cdot \mathbf{X} \cos^2 \theta}, \qquad b = \frac{-Jk}{\mathbf{Y} \cdot \mathbf{Y} \cos^2 \theta}, \qquad c = k \tan \theta = k \frac{\mathbf{X} \cdot \mathbf{Y}}{J}, \quad (4.7)$$

whence

$$a - \frac{c^2}{2b} = -\frac{Jk}{\mathbf{X} \cdot \mathbf{X}}, \qquad \frac{c}{4} = \frac{1}{4} k \frac{\mathbf{X} \cdot \mathbf{Y}}{J}. \tag{4.8}$$

Next, we evaluate the coefficients in the $\mathbf{K}^{-1}$ methods. For the configuration in Fig. 6, we write the analogue of (3.17),

$$a_A f_A + a_B f_B + a_C f_C + a_D f_D + a_E f_E + a_F f_F + a_G f_G + p_2 - p_1 = 0, \quad (4.9)$$

where, by (3.18)–(3.24),

$$a_D = \frac{6}{8} \frac{k^{-1} \mathbf{X} \cdot \mathbf{X}}{J}, \qquad a_A = a_G = \frac{1}{8} \frac{k^{-1} \mathbf{X} \cdot \mathbf{X}}{J},$$

$$a_B = a_C = a_E = a_F = \frac{1}{4} \frac{k^{-1} \mathbf{Y} \cdot \mathbf{X}}{J}. \tag{4.10}$$

For MFE, the coefficients $6/8, 1/8$ in (4.10) become $4/6, 1/6$; for lumping (SO), they are $1, 0$; for all of the $\mathbf{K}^{-1}$ methods, the $1/4$ holds. As noted above, before inverting $\mathbf{K}^{-1}$, it is best to lump; doing so, and using (4.8),

$$p_2 - p_1 = -\frac{k^{-1} \mathbf{X} \cdot \mathbf{X}}{J}(f_D + \epsilon(f_B + f_C + f_E + f_F)),$$

$$= \left(a - \frac{c^2}{2b}\right)^{-1} ((\mathbf{I} + \epsilon \mathbf{N}) f)_D, \tag{4.11}$$

where

$$\epsilon = \frac{1}{4} \frac{\mathbf{X} \cdot \mathbf{Y}}{\mathbf{X} \cdot \mathbf{X}} \tag{4.12}$$

and $\mathbf{N}$ is a matrix with 1's on 4 off-diagonals. Now, because

$$(\mathbf{I} + \epsilon \mathbf{N})^{-1} = \mathbf{I} - \epsilon \mathbf{N} + O(\epsilon^2), \tag{4.13}$$

we can perform an incomplete inversion and write

$$f_D = \left(a - \frac{c^2}{2b}\right)(p_2 - p_1 - \epsilon[(p_4 - p_1) + (p_3 - p_2) + (p_1 - p_5) + (p_2 - p_6)]). \quad (4.14)$$

The leading terms of (4.6) and (4.14) match. The terms of first order in c would match if

$$c = -4\epsilon \left(a - \frac{c^2}{2b}\right). \tag{4.15}$$

A calculation from (4.8) and (4.12) shows that actually

$$c = -4\epsilon \left(a - \frac{c^2}{2b} \right) \left(\frac{\mathbf{X} \cdot \mathbf{X}}{J} \right)^2. \tag{4.16}$$

Thus, it could be of interest to study how these methods relate on significantly distorted grids with large aspect ratios.

Arbogast *et al.* [3, 4] and Thomas and Trujillo [13] have developed other distorted-grid $\mathbf{K}$ methods derived from MFE formulations. The complexities are such that we can only give sketchy qualitative descriptions here. The expanded mixed (EM) method circumvents the difficulty of inverting $\mathbf{K}^{-1}$ by introducing an additional auxiliary variable, $\tilde{\mathbf{v}} = -\nabla p$, and then incorporating the relation $\mathbf{v} = \mathbf{K}\tilde{\mathbf{v}}$ into the weak form. The usual MFE for div $(\mathbf{K}\,\mathrm{grad})$, if reduced to a single equation for the pressure, leads to a discrete matrix $\mathbf{N}^T\mathbf{M}^{-1}\mathbf{N}$, where the three factors essentially discretize div, $\mathbf{K}$, and grad, respectively; $\mathbf{M}$ is sparse, but $\mathbf{M}^{-1}$ is full. The EM puts $\mathbf{K}$ on the right-hand side of the system, where its connectivity is sparse, in place of $\mathbf{K}^{-1}$ on the left. Low-order integration (midpoint and trapezoidal rules) leads to lumping and a compact stencil (9-point in 2-D, 19-point in 3-D) for use in finite difference codes. It seems likely that this scheme is related to lumped incompletely inverted $\mathbf{K}^{-1}$ methods, because such methods can obtain the same stencils in 2-D and 3-D, but this is speculative at this point. The method also uses Lagrange multipliers, representing edge pressures, where the grid is not smooth; another speculation is that these are related to the CVMFE edge pressure values that were eliminated in (3.16). The aim of Thomas and Trujillo is a MFV formulation that admits both a discrete pressure and a discrete velocity that have the regularity of their physical counterparts, i.e., $p \in H^1(\Omega)$ and $\mathbf{v} \in H(\mathrm{div}, \Omega)$. In the usual MFE approaches, $\mathbf{v}$ satisfies this criterion, but p does not. The discrete MFV pressure is continuous piecewise bilinear on a primal grid, with test functions that are constant on dual cells centered around the vertices of the primal cells. The trial and test functions for $\mathbf{v}$ are similar to those for the CVMFE method, but on a finer mesh. This appears to be substantially different from all of the other methods discussed here.

5 Summary

The challenge of constructing accurate discretization methods for heterogeneous, anisotropic problems on distorted grids has given rise to a wide variety of approaches. The ones considered here (MFE, CVMFE, SO, MPFA, EM, MFV) share the property of conserving mass locally on discrete cells. Thus, in some sense they are all finite-volume methods. Broadly, they can be categorized as $\mathbf{K}$ or $\mathbf{K}^{-1}$ methods, depending on which way they formulate the conductivity coefficient. Typically a $\mathbf{K}$ method relates a flux to a combination of pressure differences, while a $\mathbf{K}^{-1}$ method relates a pressure difference to a combination of fluxes. Cell-centered finite difference and mixed finite element methods are standard representatives of these respective groups. Quadrature, mass lumping, and incomplete inversion of the mass matrix appear to be the concepts that can illuminate the interconnections between various schemes. Further study of these issues, particularly in 3-D, should be of significant benefit for practical modeling of flow in porous media.

Acknowledgments. This research was supported in part by National Science Foundation Grant No. DMS-9706866 and Army Research Office Grant No. 37119-GS-AAS.

References

[1] Aavatsmark, I., Barkve, T., Bøe, Ø., and Mannseth, T., Discretization on non-orthogonal, quadrilateral grids for inhomogeneous, anisotropic media, *J. Comp. Phys.* **127** (1996), 2–14.

[2] Aavatsmark, I., Barkve, T., and Mannseth, T., Control-volume discretization methods for 3D quadrilateral grids in inhomogeneous, anisotropic reservoirs, *Soc. Pet. Eng. J.* **3** (1998), 146–154.

[3] Arbogast, T., Keenan, P., Wheeler, M., and Yotov, I., Logically rectangular mixed methods for Darcy flow on general geometry, *Proc. 13th SPE Symposium on Reservoir Simulation*, Society of Petroleum Engineers, Dallas, 1995, pp. 51–59.

[4] Arbogast, T., Wheeler, M. F., and Yotov, I., Mixed finite elements for elliptic problems with tensor coefficients as cell-centered finite differences, *SIAM J. Numer. Anal.* **34** (1997), 828–852.

[5] Cai, Z., Jones, J. E., McCormick, S. F., and Russell, T. F., Control-volume mixed finite element methods, *Computational Geosciences* **1** (1997), 289–315.

[6] Edwards, M. G., Cross-flow, tensors and finite volume approximation with deferred correction, *Comp. Meth. Appl. Mech. Engrg.* **151** (1998), 143–161.

[7] Garanzha, V. A., and Konshin, V. N., Approximation schemes and discrete well models for the numerical simulation of the 2-D non-Darcy fluid flows in porous media, *Comm. on Appl. Math.*, Computer Centre, Russian Academy of Sciences, Moscow, 1999.

[8] Hyman, J., Shashkov, M., and Steinberg, S., The numerical solution of diffusion problems in strongly heterogeneous non-isotropic materials, *J. Comp. Phys.* **132** (1997), 130–148.

[9] Lee, S. H., Tchelepi, H., and DeChant, L. F., Implementation of a flux-continuous finite difference method for stratigraphic, hexahedron grids, *Proc. 15th SPE Symposium on Reservoir Simulation*, Society of Petroleum Engineers, Dallas, 1999, pp. 231–241.

[10] Morel, J. E., Hall, M. L., and Shashkov, M. J., A local support-operators diffusion discretization scheme for hexahedral meshes, Report LA-UR-99-4358, Los Alamos National Laboratory, 1999.

[11] Raviart, P. A., and Thomas, J.-M., A mixed finite element method for 2nd order elliptic problems, *Mathematical Aspects of Finite Element Methods*, I. Galligani and E. Magenes, ed., Springer-Verlag, 1977, pp. 292–315.

[12] Thomas, J.-M., *Sur l'analyse numérique des méthodes d'éléments finis hybrides et mixtes*, Ph.D. Thesis, Université Pierre et Marie Curie, 1977.

[13] Thomas, J.-M., and Trujillo, D., Analysis of finite volume methods, *Mathematical Modelling of Flow Through Porous Media*, A. Bourgeat *et al.*, ed., World Scientific, Singapore, 1995, pp. 318–336.

Parallel Methods for Solving Time-Dependent Problems Using the Fourier-Laplace Transformation

DONGWOO SHEEN

Abstract

In this paper we summarize recent progresses on the parallel method for solving time-dependent problems using the Fourier-Laplace transformation. These problems arise in the study of elastic wave equations with absorbing boundary conditions, for example. Instead of solving the time-dependent problems in the space-time domain, we solve them as follows. First, take the Fourier-Laplace transformation of given problems originally set in the space-time domain, and consider the corresponding problems in the space-frequency domain which form a set of indefinite, complex-valued elliptic problems. Such problems are solved in a natural parallel manner since each problem is independent of others. The Fourier-Laplace inversion formula will then recover the solution in the space-time domain.

KEYWORDS: parallel method, Fourier-Laplace transform, parabolic and hyperbolic problems

1 Introduction

A direction of solving in parallel a certain class of time-dependent problems will be discussed in this paper. Time-dependent problems are usually solved by using efficient time-marching algorithms, such as backward Euler, Crank-Nicolson, or any higher order methods. However, the nature of time marching essentially blocks the usual attempt to parallelize such algorithms along the time axis, since they require the knowledge of the solutions at previous time steps in order to advance to the next time step. The spirit of time marching is in the sequential computing rather than in the parallel computing. Therefore, when time-marching algorithms are to apply for time-dependent problems, parallelization is naturally sought in solving space problems for each time step using the solutions solved at previous time steps. In this direction, domain decomposition methods have been used in the past decade by decomposing the space domain into several subdomains and solving resulting elliptic problems restricted to subdomains for each time step. For such a direction, we refer [1, 2, 7, 8, 9, 10, 17] and so on, as well as recent publications in major numerical analysis journals. However, these methods require heavy communication cost among processors in order to pass informations between neighboring subdomains.

In 1993 and the subsequent year Douglas et al. [4, 3, 19] analyzed and solved acoustic and elastic wave equations in the space-frequency domain after taking in time the Fourier transforms of the original space-time problems, and then the solutions in the time-space domain are obtained by the inversion formula. Since the frequency has no hierarchy, Fourier-transformed elliptic problems can be solved in parallel in arbitrary order without any data communication. This provides a basis for natural parallelism. After these works, with the aid of several coworkers the author has been developing the theory and extending its applications to viscoelasticity, parabolic problems, and linearized Navier-Stokes equations. See [5, 11, 16, 20, 15, 14, 12, 20]. This paper essentially surveys in brief such approaches for solving time-dependent linear initial-boundary value problems via Fourier-Laplace transformation. However, we give a general setting for the description of problems and a unified approach to solve them.

2   The Algorithm

2.1   The model problem

Let Ω be a Lipschitz domain in $\mathcal{R}^d, d = 2, 3$ and $\partial\Omega$ its boundary. Set $J = (0, \infty)$. Consider the following problem: given $f = f(x, t)$, $g = g(x, t)$ $u_0 = u_0(x)$, and $u_1 = u_1(x)$, find $u = u(x, t)$ such that

$$a\frac{\partial^2 u}{\partial t^2} + b\frac{\partial u}{\partial t} + \mathcal{A}u = f(x, t), \quad (x, t) \in \Omega \times J,$$

$$ac\frac{\partial u}{\partial t} + \mathcal{B}u = g(x, t), \quad (x, t) \in \partial\Omega \times J, \tag{2.1}$$

$$u(x, 0) = u_0(x), \quad a(\frac{\partial u}{\partial t}(x, 0) - u_1(x)) = 0, \quad x \in \Omega.$$

Here, and in what follows, we assume that all the coefficients a, b, and c depend only on the space variable and are nonnegative, $\mathcal{A}$ is a time-independent, symmetric, elliptic operator, and $\mathcal{B}$ is another time-independent, symmetric operator. Prototypes for $\mathcal{A}$ and $\mathcal{B}$ are the $-\Delta$ and $\frac{\partial}{\partial \nu}$, where ν denotes the unit outward normal to $\partial\Omega$. Also it will be assumed that $f(x,t) = g(x,t) = 0$, $t < 0$. (2.1) covers parabolic and hyperbolic initial-boundary value problems in the cases of $a = 0$ and $a > 0$, respectively. If $a = 1, b = 0, c = 1, \mathcal{A} = -\Delta$, and $\mathcal{B} = \frac{\partial}{\partial \nu}$, problem (2.1) describes the wave equation with the first-order absorbing boundary condition.

In the sequel, problem (2.1) is assumed to be well-posed and numerical methods for solving it will be our main interests.

2.2 The transformed problem

Recall first that the Fourier-Laplace transform in time $\widehat{v}(\cdot, \varsigma), \varsigma = \sigma + i\omega = \sigma(\omega) + i\omega$, of a real-valued function $v(\cdot, t)$ vanishing for $t < 0$ is defined by

$$\widehat{v}(\cdot, \varsigma) = \int_0^\infty e^{-\varsigma t} v(\cdot, t) dt, \tag{2.2}$$

and if v is square integrable in t, that is $\int_0^\infty |v(\cdot, t)|^2 dt < \infty$, the inversion formula

$$\frac{1}{2\pi i} \int_\Gamma e^{\varsigma t} \widehat{v}(\cdot, \varsigma) d\varsigma = \begin{cases} v(\cdot, t), & t \geq 0, \\ 0, & t < 0, \end{cases} \tag{2.3}$$

holds, where Γ is a contour in the complex half-plane $\mathcal{C}$ with $\mathrm{Re}\,(\varsigma) \geq 0$, and the integration is evaluated as $\mathrm{Re}\,\varsigma$ on the contour is increasing. If the integral in (2.2) converges for $\varsigma_0 = \sigma_0 + i\omega_0$, it converges absolutely and uniformly for all ς with $\mathrm{Re}\,(\varsigma) > \mathrm{Re}\,(\varsigma_0)$. The contour Γ will retain as a straght line (i.e., $\sigma(\omega)$ is a constant function) if equation (2.1) is hyperbolic, and it can be deformed into the left half plane so that the spectrum of $-\mathcal{A}$ lies to the left side of the contour since the semigroup related with the parabolic problem will be analytic [20]. Let us restrict the contour Γ to be symmetric with respect to the real axis, and denote by Γ_+ the upper half part of Γ lying above the real axis. Then, using $\overline{\widehat{v}(\cdot, \varsigma)} = \widehat{v}(\cdot, \overline{\varsigma})$, the inversion formula (2.3) takes the simpler form:

$$\frac{1}{\pi} \mathrm{Im} \int_{\Gamma_+} e^{\varsigma t} \widehat{v}(\cdot, \varsigma) d\varsigma = \begin{cases} v(\cdot, t), & t \geq 0, \\ 0, & t < 0. \end{cases} \tag{2.4}$$

Assume that $u(x, \cdot), u_t(x, \cdot)$, and $au_{tt}(x, \cdot)$ are square integrable in the second variable. Then, utilizing

$$\widehat{u_t}(\cdot, \varsigma) = \varsigma \widehat{u}(\cdot, \varsigma) - u(\cdot, 0), \quad \widehat{u_{tt}}(\cdot, \varsigma) = \varsigma^2 \widehat{u}(\cdot, \varsigma) + \varsigma u(\cdot, 0) - u_t(\cdot, 0),$$

problem (2.1) can be transformed into the following set of complex-valued elliptic problems for ζ on Γ_+: find $\widehat{u} = \widehat{u}(\cdot, \zeta)$ such that

$$(a\zeta^2 + b\zeta + A)\widehat{u}(\cdot, \zeta) = \widehat{f}(\cdot, \zeta) - a(\zeta u_0(\cdot) - u_1(\cdot)) + bu_0(\cdot) \text{ in } \Omega,$$
$$(ac\zeta + B)\widehat{u}(\cdot, \zeta) = \widehat{g}(\cdot, \zeta) - acu_0(\cdot) \text{ on } \partial\Omega. \qquad (2.5)$$

Some remarks should be made on problem (2.5).

Remark 2.1. In the hyperbolic case, that is $a > 0$, the above problem is of Helmholtz type. Thus if, in addition, $acB = 0$ and $b = 0$, it turns out to be an eigenvalue problem and existence is not guaranteed for arbitrary $\zeta = i\omega$ on the imaginary axis; otherwise, existence and uniquenss follow immediately. However, if $acB \neq 0$ for all x, solvability follows from the unique continuation principle and the Fredholm alternative. See [4, 3] for the acoustic wave problem and [18, 19] for the elastic wave case with absorbing boundary conditions.

In the parabolic case, solvability is gauranteed with a Dirichlet, Neumann, or Robin boundary condition [16, 15, 20]

Remark 2.2. The Fourier transformation (i.e, $\zeta = i\omega$) has been considered in [4, 3, 19, 11, 13, 16, 15, 14, 12], while Fourier-Laplace transformation has been applied in [20].

2.3   The fully-discretization

For $\zeta \in \Gamma$, the solution to problem (2.5) can be approximated by any of the finite element, finite difference, or finite volume method, which is denoted by $\widehat{u}_h(\cdot, \zeta)$ so that it is a solution to the discrete problem:

$$(a\zeta^2 + b\zeta + A_h)\widehat{u}(\cdot, \zeta) = \widehat{f}(\cdot, \zeta) - a(\zeta u_0(\cdot) - u_1(\cdot)) + bu_0(\cdot) \text{ in } \Omega,$$
$$(ac\zeta + B_h)\widehat{u}(\cdot, \zeta) = \widehat{g}(\cdot, \zeta) - acu_0(\cdot) \text{ on } \partial\Omega, \qquad (2.6)$$

with A_h and B_h being suitable discrete approximations to A and B, respectively.

Then the *semidiscrete approximation* $u_h(\cdot, t)$ to the solution of (2.1) can be obtained by using the inversion formula (2.4). We trun to the numerical evaluation of the indefinite integral (2.4). For this, it may be useful to transform Γ_+ into a finite contour by a smooth monotone function $\psi : (0, \infty) \to [0, 1]$ such that $\overline{\psi(0, \infty)} = [0, 1]$. In [20] $\psi(\omega) = e^{-\omega\tau/q}$ is used with parameters τ and q, which will be explained later. Also in [21], $\psi(\omega) = \tanh(\tau\omega)$ is used. To evaluate the resulting integral on the transformed compact contour, we then apply standard composite quadrature rules of order q based on an N

uniform subdivision of the compact contour. Let $y_j, w_j, j = 0, \cdots, N$ be suitable quadrature points and weights on $[0,1]$ of order q accuracy. Then, for the *fully-discrete approximation* $u_{h,N}$ to the solution $u(\cdot, t)$ of (2.1), we have

$$u_{h,N}(\cdot, t) = \frac{1}{\pi} \text{Im} \sum_{j=0}^{N} e^{\{\sigma(\psi^{-1}(y_j)) + i\psi^{-1}(y_j)\}t} \widehat{u}_h\left(\cdot, \sigma(\psi^{-1}(y_j)) + i\psi^{-1}(y_j)\right)$$

$$\{\sigma'(\psi^{-1}(y_j)) + i\} \frac{d\psi^{-1}}{dy}(y_j) w_j,$$

since

$$u_h(\cdot, t) = \frac{1}{2\pi i} \int_{\Gamma} e^{\zeta t} \widehat{u}_h(\cdot, \zeta)\, d\zeta$$

$$= \frac{1}{\pi} \text{Im} \int_{\Gamma+} e^{\zeta t} \widehat{u}_h(\cdot, z)\, d\zeta$$

$$= \frac{1}{\pi} \text{Im} \int_{0}^{\infty} e^{\{\sigma(\omega) + i\omega\}t} \widehat{u}_h(\cdot, \sigma(\omega) + i\omega) \{\sigma'(\omega) + i\} d\omega$$

$$= \frac{1}{\pi} \text{Im} \int_{0}^{1} e^{\{\sigma(\psi^{-1}(y)) + i\psi^{-1}(y)\}t} \widehat{u}_h\left(\cdot, \sigma(\psi^{-1}(y)) + i\psi^{-1}(y)\right)$$

$$\{\sigma'(\psi^{-1}(y)) + i\} \frac{d\psi^{-1}}{dy}(y)\, dy.$$

Remark 2.3. Observe that $\widehat{u}_h\left(\cdot, \sigma(\psi^{-1}(y_j)) + i\psi^{-1}(y_j)\right)$ is independent of t. This is a crucial fact to our parallel method to be effective as for each j, $\widehat{u}_h\left(\cdot, \sigma(\psi^{-1}(y_j)) + i\psi^{-1}(y_j)\right)$ can be computed as solutions to Problem (2.6) simultaneously.

Remark 2.4. For efficient calculation of the approximation solution with the transformation $\psi(\omega) = e^{-\omega\tau/q}$ it is recommended to choose the parameter τ such that $t \in (\tau, q\tau]$ where t is the time for the solution to be calculated.

2.4 The algorithm

We summarize the algorithm in the following form.

Let q be the accuracy of the quadrature rule on $[0, 1]$, and choose appropriate contour Γ_+ and the transformation function ψ.

Algorithm

Step 1. Take the Fourier-Laplace transformation of the given problem (2.1), and obtain the transformed problem (2.5).

Step 2. Apply any spatial discreteization to (2.5) and solve the resulting problem (2.6) *in parallel*.

Step 3. By the inversion formula (2.4), obtain the solution $u_{h,N}$.

3   Applications

3.1   Parabolic problems: initial value problems

We will abstract some results from [20] which will appear somewhere else.

Consider the initial value problem:

$$u_t - \Delta u = 0, \text{ for } t > 0, \text{ with } u(0) = u_0. \tag{3.1}$$

Then solve the transformed complex-valued elliptic problems:

$$\zeta \widehat{u} - \Delta \widehat{u} = u_0, \quad \text{for } \operatorname{Re}\zeta \geq -\gamma,$$

for all ζ on a contour $\Gamma = \Gamma_\gamma = \{\zeta = -\gamma - \omega \pm i\omega; \omega \geq 0\}$ with $\operatorname{Im}\zeta$ increasing from $-\infty$ to ∞.

The solution $u(t)$ of (3.1) is then obtained by the inversion formula after the application of the transformation $\psi(\omega) = e^{-\omega\tau/q}$.

The following error analysis has been obtained in [20].

Theorem 3.1 *Let $u_N(t)$ denote the semidiscrete approximation to the solution $u(t)$ of the homogeneous inital value problem obtained using the above method. then there exists $C = C(\lambda_0 - \gamma) > 0$, such that*

$$\|u_N(t) - u(t)\| \leq C\|u_0\|e^{-\gamma t} \begin{cases} \dfrac{1}{N^q}\left(\dfrac{1+t^q}{\tau^q(1+t-\tau)} + \dfrac{t^q}{\tau^q}\log_+\dfrac{1}{t-\tau}\right), t > \tau, \\[2ex] \dfrac{1}{N^q}\left(\log\log N + \dfrac{1}{\tau^q} + \log_+\dfrac{1}{\tau}\right), N \geq 3, t = \tau, \\[2ex] \dfrac{1}{N^{qt/\tau}}\left(\dfrac{1+\tau^q}{\tau^q} + \log_+\dfrac{1}{\tau-t} + \log_+\dfrac{1}{t}\right), \end{cases}$$

where $0 < t < \tau$ in the last one.

The fully-discretized error $\|u_{h,N} - u_N\|$ is then the usual finite element, or finite difference error plus the error in the above theorem.

3.2   Parabolic problems: inhomogeneous problems

We consider the initial value problem:

$$u_t - \Delta u = f, \text{ for } t > 0, \text{ with } u(0) = 0.$$

Then solve the transformed complex-valued elliptic problems:

$$\zeta \widehat{u} - \Delta \widehat{u} = \widehat{f},$$

for all ζ on a contour $\Gamma = \Gamma_\gamma = \{\zeta = i\omega; \omega \geq 0\}$, which is the ususal Fourier transformation taken.

Again the solution $u(t)$ is obtained by the Fourier inversion formula.

$$u(t) = \frac{1}{2\pi i} \int_\Gamma e^{\zeta t} R(\zeta; -A)\widehat{f}\, d\zeta = \frac{1}{\pi} \text{Im} \int_{\Gamma+} e^{\zeta t}\widehat{u}(\zeta)\, d\zeta \quad \text{for } t > 0.$$

For more details of analysis concerning this approach, we refer to [16, 14], for instance.

3.3 Hyperbolic problems with absorbing boundary conditions

We consider the initial value problem:

$$u_{tt} - \Delta u = f, \qquad \Omega \times (0, \infty),$$

$$u_t + \frac{\partial u}{\partial \nu} = 0, \qquad \partial\Omega \times (0, \infty),$$

$$u(0) = u_t(0) = 0, \qquad \Omega.$$

Take the Fourier transform ($\zeta = i\omega$) to the above equation to get

$$-\omega^2 \widehat{u} + \Delta\widehat{u} = \widehat{f}, \qquad \Omega,$$

$$i\omega\widehat{u} + \frac{\partial u}{\partial \nu} = 0, \qquad \partial\Omega,$$

for all ω.

Again the solution $u(t)$ is obtained by the Fourier inversion formula

$$u(t) = \frac{1}{\pi}\text{Im} \int_{\Gamma+} e^{\zeta t}\widehat{u}(\zeta)\, d\zeta \quad \text{for } t > 0,$$

For this approach, see [4, 3].

3.4 Other applications

For the elastic wave equations with absorbing boundary conditions, the above approach with Fourier transformation has been applied [19]. The method of applying Fourier transformation has been used to treat viscoelastic problems in [11, 12], and linearized Navier-Stokes equations in [15]. Also, an application to solve certain class of semilinear parabolic problems is now in progress by Ganesh and the author [6].

Acknowledgments. The preparation of this paper was partially supported by ARC (Australian Research Council), GARC, KOSEF 97-0701-01-01-3, and BSRI-98-1417.

References

[1] Chan, T. F., Glowinski, R., Périaux, J., and Widlund, O. B. (eds.), *Proceedings of the Second International Symposium on Domain Decomposition Methods for Partial Differential Equations*, Philadelphia, SIAM, 1989.

[2] Chan, T. F., Glowinski, R., Périaux, J., and Widlund, O. B. (eds.), *Third International Symposium on Domain Decomposition Methods for Partial Differential Equations*, Philadelphia, SIAM, 1990.

[3] Douglas, J., Jr., Santos, J. E., and Sheen, D., Approximation of scalar waves in the space–frequency domain. *Math. Models Meth. Appl. Sci.* **4** (1994), 509–531.

[4] Douglas, J., Jr., Santos, J. E., Sheen, D., and Bennethum, L. S., Frequency domain treatment of one–dimensional scalar waves. *Math. Models Meth. Appl. Sci.* **3** (1993), 171–194.

[5] Feng, X and Sheen, D., An elliptic regularity estimate for a problem arising from the frequency domain treatment of waves. *Trans. Amer. Math. Soc* **346** (1994), 475–487.

[6] Ganesh, M. and Sheen D., A naturally parallelizable computational method for parabolic problems, in preparation.

[7] Glowinski, R., Golub, G. H., Meurant, G. A., and Périaux, J. (eds.), *First International Symposium on Domain Decomposition Methods for Partial Differential Equations*, Philadelphia, SIAM, 1988.

[8] Glowinski, R., Kuznetsov, Yu. A., Meurant, G. A., Périaux, J., and Widlund, O. B. (eds.), *Fourth International Symposium on Domain Decomposition Methods for Partial Differential Equations*, Philadelphia, SIAM, 1991.

[9] Keyes, D. E. Keyes, Chan, T. F., Meurant, G. A., Scroggs, J. S., and R. G. Voigt, (eds.), *Fifth International Symposium on Domain Decomposition Methods for Partial Differential Equations*, Philadelphia, SIAM, 1992.

[10] Keyes, D. E. and Xu, J. (eds.), *Domain Decomposition Methods in Scientific and Engineering Computing: Proceedings of the Seventh International Conference on Domain Decomposition*, volume 180 of *Contemporary Mathematics*, Providence, Rhode Island, American Mathematical Society, 1994.

[11] Kim, D., Kim, J., and Sheen, D., Absorbing boundary conditions for wave propagations in viscoelastic media, *J. Comput. Appl. Math.* **76** (1996), 301–314.

[12] Kim, D. and Sheen, D., Finite element methods for a viscoelastic system with absorbing boundary conditions in the space-frequency domain, to appear.

[13] Kim, D. and Sheen, D., An elliptic regularity of a Helmholtz-type problem with an absorbing boundary condition, *Bull. Korean Math. Soc.* **34** (1997), 135–146.

[14] Lee, C.-O., Lee, J., and Sheen, D., A frequency–domain method for finite element solutions of parabolic problems, to appear.

[15] Lee, C.-O., Lee, J., and Sheen, D., Frequency domain formulation of linearized Navier-Stokes equations. *Comput. Methods Appl. Mech. Engng.*, in press.

[16] Lee, C.-O., Lee, J., Sheen, D., and Yeom, Y., A frequency–domain parallel method for the numerical approximation of parabolic problems, *Comput. Methods Appl. Mech. Engng.* **169** (1999), 19–29.

[17] Quarteroni, A., Périaux, J., Kuznetsov, Yu. A., and Widlund, O. B. (eds.), *Domain Decomposition Methods in Science and Engineering: The Sixth International Conference on Domain Decomposition*, volume 157 of *Contemporary Mathematics*, Providence, Rhode Island, American Mathematical Society, 1994.

[18] Ravazzoli, C. L., Douglas, J., Jr., Santos, J. E., and Sheen, D., On the solution of the equations of motion for nearly elastic solids in the frequency domain, In *Anales de la 4ª. Reunion de Trabajo en Procesamiento de la Información y Control, RPIC '91*, pp. 231–235, Buenos Aires, November 18–22, 1991.

[19] Sheen, D., Douglas, J., Jr., and Santos, J. E., Frequency–domain parallel algorithms for the approximation of acoustic and elastic waves. In *Numerical Analysis (Finite Element Methods), Proceedings of Applied Mathematics Workshop*, pp. 243–288, KAIST, Taejon, February 16–18, 1993.

[20] Sheen, D., Sloan, I. H., and Thomée, V., A parallel method for time-discretization of parabolic problems based on contour integral representation quadrature, *Math. Comp.*, 1999, in press.

[21] Sheen, D., Sloan, I. H., and Thomée, V., A parallel method for time-discretization of parabolic problems based on contour integral representation quadrature II, *title subject to change*, in preparation.

Cascadic Multigrid Methods for Parabolic Pressure Problems

ZHONG-CI SHI   XUEJUN XU

Abstract

In this paper we develop the cascadic multigrid method for parabolic problems, which arise as the pressure equations for the flow of compressible fluids in porous media. The optimal convergence accuracy and computation complexity are obtained.

KEYWORDS: cascadic multigrid, finite element, parabolic problem

1 Introduction

Bornemann and Deuflhard [2, 3] have presented a new type of multigrid method, the so-called cascadic multigrid. Compared with usual multigrid methods, it requires no coarse grid corrections at all that may be viewed as a "one way" multigrid. Another distinctive feature is that it performs more iterations on coarser levels so as to obtain less iterations on finer levels. Numerical experiments show that this method is very effective for second order elliptic problems.

In this paper we consider the cascadic multigrid for parabolic problems, which arise as the pressure equations for the flow of compressible fluids in porous media. Here we must treat the effect of discrete time steps. As pointed out in [1], for a small time step $\tau \leq O(h^2)$, where h is the space mesh size, some standard iterative methods, like the Richardson iteration, can guarantee a good convergence for the discrete system. But for a relative large time step τ, [1] recommended multigrid methods; see [4] for details. Now, we consider to use the cascadic multigrid. Similar to second order elliptic problems, it is proved that the cascadic multigrid with the conjugate gradient (CG) iteration as a smoother is accurate with the optimal complexity in 2D and 3D and nearly optimal in 1D. As for other traditional iterative methods, like the Richardson iteration, the cascadic multigrid still yields the optimal accuracy and complexity in 2D and 3D and in a certain case of 1D. Notice that for the second order elliptic problem, the cascadic multigrid with these iterative methods gives the optimal accuracy and computation complexity only in 3D and nearly optimal in 2D. They cannot be used for 1D.

2 Model Problem and its Finite Element

Consider the parabolic problem: Find $u(x,t)$ such that

$$
\begin{aligned}
\frac{\partial u}{\partial t} + \mathcal{L}u &= f && \text{in} && \Omega \times [0,T], \\
u(x,t) &= 0 && \text{in} && \partial\Omega \times [0,T], \\
u(x,0) &= u_0(x),
\end{aligned}
\tag{2.1}
$$

where $\Omega \subset R^d$ $(d = 1,2,3)$ is a bounded domain, $f \in L^2(\Omega)$, and $\mathcal{L}$ is an elliptic operator

$$
\mathcal{L}u = - \sum_{i,j=1}^{d} \frac{\partial}{\partial x_i}\left(a_{ij}(x)\frac{\partial u}{\partial x_j}\right).
$$

Here $a_{ij}(x)$ satisfies

$$
c\xi^t\xi \le \sum_{i,j=1}^{d} a_{ij}\xi_i\xi_j \le C\xi^t\xi \quad \forall x \in \Omega, \xi \in R^d,
$$

where c and C are positive constants. The variational form of (2.1) is to find $u \in H_0^1(\Omega)$, $u(x,0) = u_0(x)$ such that

$$
\left(\frac{\partial u}{\partial t}, v\right) + B(u,v) = (f,v) \quad \forall v \in H_0^1(\Omega), \quad t \in [0,T],
$$

where

$$
B(u,v) = \int_\Omega \sum_{i,j=1}^{d} a_{ij}\frac{\partial u}{\partial x_j}\frac{\partial v}{\partial x_i}dx \quad \forall u,v \in H^1(\Omega),
$$
$$
(f,v) = \int_\Omega fvdx.
$$

We use the backward Euler scheme and Crank-Nicolson scheme for the time discretization [8]. Both schemes are absolutely stable [6]. Let Δt_n be the nth time step and M the number of steps. Then $\sum_{n=1}^{M} \Delta t_n = T$. Consider the problem: For a given function $g_{n-1} \in H^{-1}(\Omega)$, find $w \in H_0^1(\Omega)$ such that

$$
A_\tau(w,v) = \tau^{-1}(w,v) + B(w,v) = (g_{n-1},v) \quad \forall v \in H_0^1(\Omega), \tag{2.2}
$$

where τ is the time step parameter. For the backward Euler scheme, we have

$$
w = u^n - u^{n-1}, \quad \tau = \Delta t_n, \quad (g_{n-1},v) = (f,v) - B(u^{n-1},v),
$$

and for the Crank-Nicolson scheme,

$$w = u^n - u^{n-1}, \quad \tau = \Delta t_n/2, \quad (g_{n-1}, v) = 2((f, v) - B(u^{n-1}, v)).$$

Now we define the τ-norm by

$$\|v\|_\tau^2 = \tau^{-1}(v, v) + B(v, v) \quad \forall v \in H_0^1(\Omega).$$

Let Γ_l ($l \geq 0$) be a quasiuniform triangular partition of Ω with the mesh size $h_l = h_0 2^{-l}$. Γ_l is obtained by linking the midpoints of three edges of triangles on Γ_{l-1}. We assume that $\bar{\Omega} = \cup_{K \in \Gamma_l} \bar{K}$. Let V_l denote the P_1-conforming finite element space on Γ_l. Then we obtain the discrete form of (2.2): Find $u_l \in V_l$ such that

$$A_\tau(u_l, v_l) = (g, v_l) \quad \forall v_l \in V_l. \tag{2.3}$$

Define the operator $A_{l,\tau} : V_l \to V_l$ by

$$(A_{l,\tau} u_l, v_l) = A_\tau(u_l, v_l) \quad \forall u_l, v_l \in V_l.$$

Then (2.3) can be expressed by

$$A_{l,\tau} u_l = g_l, \tag{2.4}$$

where $g_l \in V_l$, $(g_l, v) = (g, v)$ $v \in V_l$.

3 Cascadic Multigrid Method

We use the cascadic multigrid to solve (2.4) at each time step. Define the cascadic algorithm for (2.4) as follows: (1) set $u_0^0 = u_0^* = u_0$ and let $u_l^0 = u_{l-1}^*$; (2) for $l = 1, \ldots, L$, set $u_l^{m_l} = C_{l,\tau}^{m_l} u_l^0$; (3) set $u_l^* = u_l^{m_l}$, where $C_{l,\tau}$ denotes the Richardson iteration procedure, i.e.,

$$u_l - C_{l,\tau}^{m_l} u_l^0 = T_{l,\tau}^{m_l}(u_l - u_l^0) = (I - R_{l,\tau} A_{l,\tau})(u_l - u_l^0).$$

Here $R_{l,\tau} = (\lambda_l + \tau^{-1})^{-1} I$ and $\lambda_l = O(h_l^{-2})$.

Following [2], we call a cascadic multigrid method optimal on the level L if we obtain both the accuracy

$$\|u_L - u_L^*\|_\tau \approx \|u - u_L\|_\tau,$$

which means that the iterative error is comparable to the approximation error, and the multigrid complexity

$$amount \ of \ work = O(n_L),$$

where $n_L = dim V_L$. Note that $h_l = h_L 2^{L-l}$.

Consider sequences $m_1, m_2, \ldots, m_L$ of the kind

$$m_l = [\beta^{L-l} m_L], \tag{3.1}$$

for some fixed $\beta > 0$, where $[\cdot]$ means the choosing integral function. If τ satisfies that $\tau \leq \lambda_L^{-1}$, based on the observation in [1], we know that some usual iterative methods, like the Richadson iteration, can already guarantee good convergence. Therefore, we only consider the case where $\tau \geq \lambda_L^{-1}$. In such case, for any fixed τ, we can find a positive constant $0 < \gamma_0 < 1$ which satisfies

$$\tau \leq \lambda_L^{-1}/\gamma_0, \tag{3.2}$$

where γ_0 is dependent of τ.

Theorem 3.1. *The accuracy of the cascadic multigrid with the Richardson iteration for the parabolic problem is*

$$\|u_L - u_L^*\|_\tau \leq C \frac{h_L}{m_L^{\frac{1}{2}}} \frac{1}{1 - \frac{2}{\beta^{\frac{1}{2}}(1+\gamma_0)}} \|g\|_0 \quad for \quad \beta > (\frac{2}{1+\gamma_0})^2,$$

where β and m_L are defined in (3.1) and τ is in (3.2).

According to Lemma 1.4 in [2], we have

Lemma 3.1. *The computational cost of the cascadic multigrid is proportional to*

$$\sum_{l=1}^{L} m_l n_l \leq C \frac{1}{1 - \frac{\beta}{2^d}} m_L n_L \quad for \quad \beta < 2^d.$$

Theorem 3.1 indicates that a large β can yield an optimal accuracy. Meanwhile, Lemma 3.1 shows that the optimal complexity of the method can be achieved only for a small β. Therefore, we have

Theorem 3.2. *If β in (3.1) satisfies*

$$(\frac{2}{1+\gamma_0})^2 < \beta < 2^d, \quad d = 1, 2, 3,$$

then both the optimal accuracy and complexity of the cascadic multigrid with the Ricchardson iteration can be obtained.

Remark 3.1. From Theorem 3.2, it is seen that the cascadic multigrid with the Richardson iteration gives the optimal accuracy and complexity for 2D and 3D parabolic problems. But for 1D problem, it requires that the parameter β must be chosen to satisfy

$$(2/(1+\gamma_0))^2 < \beta < 2,$$

which turns out that the value γ_0 in (3.2) should be greater than $2^{1/2} - 1$ that prevents choices of a relatively large time step parameter τ, say of order h in the Crank-Nicolson scheme.

Remark 3.2. Compared with the parabolic case, for 3D elliptic problems, the cascadic multigrid with the Richardson iteration gives the optimal accuracy and complexity. But for the problem in 2D, it gives only nearly optimal complexity. It cannot be used for 1D elliptic problems at all [2, 7].

4 Conjugate Gradient Method

Assume that u_l^0 is an initial value of the CG method on the level l. Let $C_{l,\tau}^{m_l} u_l^0$ be the m_l steps of the CG iteration. Then the error of the CG method can be expressed by

$$\|u_l - C_{l,\tau}^{m_l} u_l^0\|_\tau = \min_{p \in P_{m_l}, p(0)=1} \|p(A_{l,\tau})(u_l - u_l^0)\|_\tau,$$

where P_{m_l} denotes the set of polynomials p with degree $\leq m_l$ [3].

Using a same argument of Theorem 2.2 in [2], we have

Lemma 4.1. *There exists a linear operator* $T_{l,\tau} = \phi_{\lambda_l, m_l}(A_{l,\tau})$, *where* $\phi_{\lambda,m} \in P_m$ *and* $\phi_{\lambda,m}(0) = 1$, *such that*

$$\|T_{l,\tau}^{m_l} v_l\|_\tau \leq \frac{(\lambda_l + \tau^{-1})^{\frac{1}{2}}}{2m_l + 1} \|v_l\|_0, \quad \|T_l^{m_l} v_l\|_\tau \leq \|v_l\|_\tau \quad \forall v_l \in V_l.$$

Using Lemma 4.1 and following the same line of Lemma 1.3 as in [2], we have

Lemma 4.2. *Assume that the time step parameter* $\tau \geq O(h_L^2)$. *Then the accuracy of the cascadic multigrid with the CG method as smoother is*

$$\|u_L - u_L^*\|_\tau \leq \begin{cases} C\dfrac{1}{1 - (\frac{2}{\beta})}\dfrac{h_L}{m_L}\|g\|_0 & for \quad \beta > 2, \\[3mm] CL\dfrac{h_L}{m_L}\|g\|_0 & for \quad \beta = 2. \end{cases}$$

Remark 4.1. It should be noticed that the assumption on the time step parameter $\tau \geq O(h_L^2)$ in Lemma 4.2 is not a real restriction since we can always assume $\tau = O(h_L^2)$ for the backward Euler scheme and $\tau = O(h_L)$ for the Crank-Nicolson Scheme. Moreover, as pointed out in [1], for a small time step parameter $\tau \leq O(h_L^2)$, some standard iterative methods alone are efficient enough to guarantee a good convergence.

Combining Lemma 4.2 with Lemma 1.4 in [2], we get

Theorem 4.1. *(1) For 2D and 3D parabolic problems, the optimal accuracy and complexity can be obtained for the cascadic multigrid with the CG iteration. (2) For 1D parabolic problems, if we choose $\beta = 2$ and the number of iterations on the level L is*

$$m_L = [m_* L],$$

then the error of the cascadic multigrid method is

$$\|u_L - u_L^*\|_\tau \leq C \frac{h_L}{m_*} \|g\|_0$$

and the complexity of computation is

$$\sum_{l=1}^{L} m_l n_l \leq c m_* n_L (1 + \log n_L)^2.$$

Remark 4.2. Besides the P_1 conforming finite element, the above results are also valid for other conforming or nonconforming finite elements of the second order problem (see [7]).

Remark 4.3. In practical computation, the right hand term g_l in (2.4) is related to the cascadic multigrid solution of the last time step. According to [8], the backward Euler and Crank-Nicolson scheme are absolutely stable, so the small perturbation of right hand term in (2.4) still assure the efficiency of our algorithm.

Acknowledgments. The authors thank Dr Mo Mu and Dr Qiang Du of Hong Kong University of Science and Technology for useful discussions.

References

[1] Bank, R. E. and Dupont, T., An optimal order process for solving finite element equations, *Math. Comp.* **36** (1980), 35–51.

[2] Bornemann, F. A. and Deuflhard, P., The cascadic multigrid method for elliptic problems, *Numer. Math.* **75** (1996), 135–152.

[3] Bornemann, F. A. and Deuflhard, P., The cascadic multigrid method, *The Eighth International Conference on Domain Decomposition Methods for Partial Differential Equation*, R.Glowinski, J.Periaux, Z.Shi, O.Widlund, eds. Wiley & Sons, 1997.

[4] Bramble, J. H., *Multigrid Methods*, Pitman, 1993.

[5] Ciarlet, P. G., *The Finite Element Method for Elliptic Problems*, North-Holland, Amsterdam, 1978.

[6] Johnson, C., *Numerical Solution of Partial Differential Equations by the Finite Element Method*, Cambridge University Press, Cambridge, 1987.

[7] Shi, Z. and Xu, X., Cascadic multigrid method for the elliptic problem, *East-West J. Numer.Math.* **7** (1999), 209–218.

[8] Thomée, V., Galerkin Finite Element Methods for Parabolic Problems, *Lecture Notes in Mathematics 1054*, Springer-Verlag, 1984.

Estimation in the Presence of Outliers: The Capillary Pressure Case

SAM SUBBEY   JAN-ERIK NORDTVEDT

Abstract

The inversion of laboratory centrifuge data to obtain capillary pressure functions in petroleum science leads to a *Volterra* integral equation of the first kind with a right-hand side defined by a set of discrete data. The problem is ill-posed in the sense of Hadamard [4]. The discrete data lead to a discretized equation of the form

$$A\vec{c} = \vec{b} + \vec{\epsilon},$$

where $\vec{b}$ represents the observation vector, A is an ill-conditioned matrix derived from the forward problem, $\vec{c}$ is the coefficients in a representation of the inverse capillary function, i.e., parameters to be determined, and $\vec{\epsilon}$ is the error vector associated with $\vec{b}$. If $\vec{\epsilon} \sim N(0, \sigma^2)$, and satisfies the *Gauss-Markov* (G-M) conditions, then an estimate, $\vec{c}_\lambda$, of $\vec{c}$ is BLUE [9]. In the presence of outliers, the G-M conditions and/or the normality assumption can be violated.

In this paper we parameterize the capillary pressure function using B-splines and address the issue of ill-posedness by reformulating the problem as a constrained optimization task involving the determination of the spline coefficients. By the nature of the experimental procedure, we expect the G-M conditions to be satisfied. A systematic method of outlier elimination and a choice of knots is employed to ensure satisfaction of the normality assumption and thereby derive capillary pressure curves to a high degree of accuracy. A robust method for estimating the solution curve, which accommodates both outliers and influential points, namely the L_1-norm solution, is also presented. The method is demonstrated on synthetic data.

KEYWORDS: Volterra, outliers, ill-posed, regularization, capillary pressure, estimation

1   Introduction

The problem of determining capillary pressure functions from centrifuge data
leads to an integral equation of the form

$$\int_a^x K(x,t)f(t)dt = g(x), \quad x \in [a,b], \tag{1.1}$$

where the kernel K is known exactly and given by the underlying mathematical model, g is only known with a certain degree of accuracy in a finite set
of points $x_1, \ldots, x_M$, and the sought function $f(t)$ is however continuous. By
the nature of the right-hand side, $g(x)$, eq. (1.1) is a discrete inverse problem
which is ill-posed in the sense of Hadamard [4]. By a parameterization of the
sought function, eq. (1.1) reduces to a system of linear equations of the form

$$A\vec{c} = \vec{b} + \vec{\epsilon}, \tag{1.2}$$

where $\vec{b}$ is the observation vector, A arises from discretization of the forward
problem, $\vec{\epsilon}$ is the error vector associated with $\vec{b}$, and $\vec{c}$ contains the model
parameters. The matrix A is usually ill-conditioned. The ill-conditioning is
closely connected to the parameterization of the probelm. As the dimension
of the parameters increases, the spectra of A become increasingly dominated
by very low singular values. Thus the condition of A increases with the
increasing parameter dimension.

   The usual assumption is that $\vec{\epsilon}$ satisfies the following conditions:

$$\mathrm{E}(\epsilon_i) = 0, \cdot \quad \mathrm{E}(\epsilon_i^2) = \sigma^2, \quad \mathrm{E}(\epsilon_i, \epsilon_j) = 0 \quad \text{when } i \neq j. \tag{1.3}$$

Eqs. (1.3) above are the *Gauss-Markov* (G-M) conditions. To be able to
choose between the feasible vectors $\vec{c}$, we introduce a *merit number* to define
the goodness of fit associated with a feasible vector $\vec{c}$. Suppose we have M
observations; we form the vector $\vec{\delta} \in R^M$ by setting

$$\delta_i = b_i - a_r\vec{c} \quad i = 1, \ldots, M, \tag{1.4}$$

where a_r refers to the rth row of the matrix A and δ_i defines the residual
for data point i, i.e., the difference between the measured values b_i and the
calculated value.

   We first make the assumption that the ϵ_i's are normally distributed. If we
maintain this assumption and in addition assume that the G-M conditions
hold, then the residuals δ_i's are independently and normally distributed with

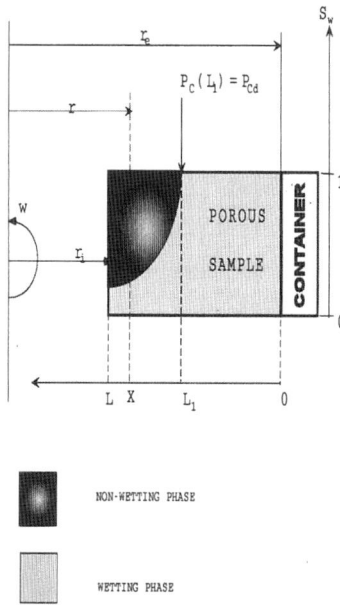

Figure 1: Schematic diagram of the centrifuge system.

zero mean and a variance of σ^2, i.e., $\delta_i \sim N(0, \sigma^2)$, [9]. If this is the case, then the least square (L_2) solution of eq. (1.2) is theoretically optimal, i.e., has the least variance. In fact, it is the best linear unbiased estimate (BLUE) [9]. In practice, it is frequently assumed that experimental errors are normally distributed. This and the fact that of all L_p-norms only the L_2-norm approach is linear account in part for the popular use of the L_2-norm solution approach.

A single point or a small number of points may violate either or both of our basic assumptions of normality and G-M conditions. A solution in L_2-norm will then *not* be theoretically optimal. These points which violate one or all of the G-M conditions and/or the normality assumption are termed *outliers*. In essence, we define an outlier as an observation inconsistent with the assumed model of the random process generating the observations.

In this paper, by the nature of the experimental process, we expect the G-M conditions to be satisfied. We concern ourselves with the case when the normality assumptions are violated. We demonstrate a method for testing small sample data for the normality assumption and show how a systematic outlier detection and elimination can be employed to obtain the satisfaction of the normality condition and thereby a solution in the L_2-norm. We demonstrate this on the ill-posed problem of determining capillary pressure functions from centrifuge data. We show how capillary pressure curves can be estimated by reformulation of the problem as a constrained optimization task where the constraints are dictated by the physics of the problem.

In a situation where the outlier accommodation rather than elimination is required, we present a robust method for estimating the capillary pressure functions namely the L_1-norm solution. Here we treat only drainage capillary pressure curves.

2   The Physical Problem

From FIG. 1, the pressure at a radius r, $r_i \leq r \leq r_e$, is given by

$$P_i(r, \omega) = \rho_i \omega^2 (r_e^2 - r^2)/2. \tag{2.1}$$

By assumption, the capillary pressure $P_c = P_{nw} - P_w = 0$ at $r = r_e$. Hence

$$P_c(r_i, \omega) = \frac{1}{2} \Delta \rho \omega^2 (r_e^2 - r_i^2). \tag{2.2}$$

The average saturation of the porous medium is defined as

$$\overline{S} = \frac{1}{r_e - r_i} \int_{r_i}^{r_e} S(r) dr. \tag{2.3}$$

We substitute eq. (2.2) into eq. (2.3), and by a change of variables arrive at eq. (2.4) that relates the average core saturation, $\overline{S}$, measured at the outlet, the capillary pressure, $P_c(\omega)$ and the sought point-saturation along the core, S. Thus, based on experimental data we need to invert:

$$\overline{S}(P_c(r_i, \omega)) = \int_0^{P_c(r_i, \omega)} K(P_c(r_i, \omega), P_c(\omega)) S(P_c(\omega)) dP_c(\omega), \tag{2.4}$$

where

$$K(P_c(r_i, \omega), P_c(\omega)) = \frac{1 + \frac{r_i}{r_e}}{2P_c(r_i, \omega)} \frac{1}{\sqrt{\left\{1 - \frac{P_c(\omega)\left\{1 - \left\{\frac{r_i}{r_e}\right\}^2\right\}}{P_c(r_i, \omega)}\right\}}}.$$

Eq. (2.4) above is a Volterra integral equation of the first kind. It is ill-posed and has numerical instabilities. Higher order numerical methods will therefore diverge. The equation may be reformulated into the Volterra equation of the second kind which is more stable but this is not recommended as this will involve the numerical differentiation of data. The equation must therefore be regularized. Among the methods of regularization which have proven be effective in solving eq. (2.4) are the Tikhonov regularization and the regularization by convexity constraints [10]. In this paper regularization is carried out in two stages. A pre-regularization by adequate paramterization to satisfy the normality condition is followed by regularization by imposition of convexity constraints.

3 Problem Formulation

We parameterize eq. (2.4) above by representing $S(P_c(\omega))$ by an mth order B-Spline as

$$S(P_c(\omega)) = \sum_{j=1}^{N} c_j N_j^m(P_c(\omega), \vec{y}). \tag{3.1}$$

Here N_j^m are the normalized B-spline basis functions, $\vec{y}$ is the spline partition, the c_j's are the spline coefficients, and N is the number of the basis functions, otherwise refered to as the spline dimension. Substitute eq. (3.1) into eq. (2.4) to arrive at

$$\overline{S}(P_c(r_i, \omega)) = \sum_{j=1}^{N} c_j \int_0^{P_c(r_i, \omega)} N_j^m(P_c(\omega), \vec{y}) K(P_c(r_i, \omega), P_c(\omega)) dP_c(\omega).$$

We split the integral above into two parts; from the zero capillary pressure to the displacement threshold pressure P_{cd} and from P_{cd} to the capillary pressure at the inner face of the core. We know that $S(P_c(\omega)) = 1$, for $0 \leq P_c(\omega) \leq P_{cd}$, and define

$$f(P_{cd}, \omega) = \int_0^{P_{cd}} K(P_c(r_i, \omega), P_c(\omega)) dP_c(\omega). \tag{3.2}$$

Thus we have

$$\overline{S}(P_c(r_i, \omega)) - f(P_{cd}, \omega)$$
$$= \sum_{j=1}^{N} c_j \int_{P_{cd}}^{P_c(r_i, \omega)} N_j^m(P_c(\omega), \vec{y}) K(P_c(r_i, \omega), P_c(\omega)) dP_c(\omega). \tag{3.3}$$

P_{cd} enters into eq. (3.3) nonlinearly through $f(P_{cd}, \omega)$ in eq. (3.3) and must be determined using the experimental data. A zero order search is performed within the range $0 \leq P_{cd} < P_c(\omega)|_{\overline{S}(P_c(r_i, \omega)) < 1}$ for the optimal value. $P_c(\omega)|_{\overline{S}(P_c(r_i, \omega)) < 1}$ stands for the first experimental capillary pressure for

which the average water saturation is less than unity. Here we for simplicity assume that P_{cd} is known. The P_{cd} problem can be easily handled as
outlined above; for a more complete analysis; see [10]. $\overline{S}(P_c(r_i, \omega))$ has been
measured at M points $(P_c(\omega_k), \ k = 1, \ldots, M\ )$ and the saturation values
$(\overline{S}(P_c(r_i, \omega_k)), \ k = 1, \ldots, M)$ have been obtained in the form of readings.
We therefore write the equation above as, $k = 1, \ldots, M$,

$$
\overline{S}(P_c(r_i, \omega_k)) - f(P_{cd}, \omega_k)
$$
$$
= \sum_{j=1}^{N} c_j \int_{P_{cd}}^{P_c(r_i, \omega_k)} N_j^m(P_c(\omega), \vec{y}) K(P_c(r_i, \omega_k), P_c(\omega)) dP_c(\omega). \tag{3.4}
$$

We note that eq. (3.4) constitutes a system of linear equations of the form

$$
\vec{b} = A\vec{c}, \tag{3.5}
$$

where $\vec{c}$ contains the spline coefficients and the elements in the matrix A are
given by the integral over the basis functions. We define

$$
b_i = \overline{S}(P_c(r_i, \omega_i)) - f(P_{cd}, \omega_i),
$$
$$
A_{i,j} = \int_{P_{cd}}^{P_c(r_i, \omega_i)} N_j^m(P_c(\omega), \vec{y}) K(P_c(r_i, \omega_i), P_c(\omega)) dP_c(\omega),
$$
$$
i = 1, \ldots, M, \ j = 1, \ldots, N.
$$

By the physics of the problem we require that the solution curve satisfy
monotonicity and convexity constraints [10]. Monotonicity translates into
demanding that the N-coefficients of our splines are ordered according to

$$
1 \geq c_1 \geq c_2 \geq c_3 \ldots \geq c_N \geq 0.
$$

For $m = 3$, the above is a necessary and sufficient condition for monotonicity
[7] and can be expressed in matrix form as

$$
E\vec{c} \geq \vec{F},
$$

$$
\begin{bmatrix}
-1 & 0 & 0 & \cdots & 0 & 0 & 0 \\
1 & -1 & 0 & \cdots & 0 & 0 & 0 \\
0 & 1 & -1 & 0 & \cdots & 0 & 0 \\
\vdots & & & & & & \\
0 & 0 & 0 & \cdots & 0 & 1 & -1 \\
0 & 0 & 0 & \cdots & 0 & 0 & 1
\end{bmatrix}
\begin{bmatrix}
c_1 \\ c_2 \\ \vdots \\ c_{N-1} \\ c_N
\end{bmatrix}
\geq
\begin{bmatrix}
-1 \\ 0 \\ \vdots \\ 0 \\ 0
\end{bmatrix}.
$$

By convexity constraints, we require that the second derivative of the capillary pressure function be positive. The second derivative of a B-spline expressed as a third order polynomial is

$$\frac{d^2}{dP_c(\omega)}(a_{l,j}P_c^2(\omega) + b_{l,j}P_c(\omega) + c_{l,j}) = 2a_{l,j}.$$

The coefficients $a_{l,j}$ are known since they have to be derived during the formation of matrix A. Our convexity constraint translates into a matrix equation of the form

$$G\vec{c} \geq \vec{H},$$

where H is an $N-2$ column vector with zero elements and G is a $(N-2 \times N)$ matrix defined as

$$G(l,j) = \begin{cases} 2a_{l,j}, & j = l(1)l + 2, \\ 0 & \text{Otherwise} \end{cases}$$

Based on our assumptions that the nature of our experimental procedure guarantees a satisfaction of the G-M conditions, if we further assume that the residuals resulting from the solution of eq. (3.5) are normally distributed, then a solution in the L_2-norm, expressed through eq. (3.6), is the BLUE of $\vec{c}$. However, due to the ill-conditioned nature of the system, we still need to regularize the system by introducing monotonicity and convexity constraints in the form of eq. (3.7). Hence the task is reduced to

$$\min_{\vec{c}} \|\vec{b} - A\vec{c}\|^2, \tag{3.6}$$

subject to

$$\begin{bmatrix} E \\ G \end{bmatrix} [\ \vec{c}\ ] \geq \begin{bmatrix} F \\ H \end{bmatrix}. \tag{3.7}$$

4 Checking the Normality Assumption

We adopt two methods of test, namely an approximate *rankit* plot and a Shapiro-Wilk normality statistic check. We discuss them briefly in this section.

4.1 Rankit plot

If $\delta_{(1)} < \delta_{(2)} < \ldots < \delta_{(M)}$ are ordered values of M independent and identically distributed $N(\mu, \sigma^2)$ random variables, then (see [2])

$$E[\delta_{(i)}] \approx \mu + \gamma_i, \quad \gamma_i = \sigma \cdot \Phi^{-1}[(i - 3/8)/(M + 1/4), \quad \Phi(x) = \frac{1}{\sqrt{2\pi}} \int_{-\infty}^{x} e^{-\frac{1}{2}t^2} dt.$$

From this equation a plot of $\delta_{(i)}$ against γ_i, where $\delta_{(i)}$ are the ordered values of δ_i, can give an approximately straight line. In that case the residuals could be taken to be approximately normally distributed and the errors to be approximately normal.

4.2   Shapiro-Wilk test

If we assume the same distribution for $\delta_{(i)}$ as above and set

$$s^2 = \frac{1}{(M-1)} \sum_{i=1}^{M} (\delta_i - \overline{\delta_i})^2, \quad \overline{\delta_i} = \frac{1}{M} \sum_{i=1}^{M} \delta_i,$$

then the Shapiro-Wilk test statistic is given by

$$W = \sum_{i=1}^{M} a_i \delta_{(i)}/s, \quad 0 < W < 1,$$

where $a_1, \ldots, a_M$ depend on the expected values of the order statistic from a standard normal distribution and are tabulated in Shapiro-Wilk [8]. We reject the null hypothesis of normality if $W \leq W_\alpha$, where W_α is a tabulated critical value, also given in [8]. Values of W close to 1 indicate near-normality. Based on the rankit-plot and the Shapiro-Wilk statistic, we establish the following algorithm, where W^j is the Shapiro-Wilk statistic for spline dimension j (the spline order is held constant).

Algorithm 1 *(Shapiro-Wilk and rankit plot: L_2-norm)*
Initialize: swap=0, dim=3
   1. **for** $\kappa = dim$ step 1 **do**

$$w \quad = \quad swap$$
$$\textbf{if } W^\kappa \quad < \quad w \textbf{ then goto } step\ 2$$
$$swap \quad = \quad W^\kappa$$

   2. **if** $W^{\kappa-1} \geq W_\alpha$ *such that* $P \geq 90\%$, *accept null hypothesis. Adopt model with* $W^{\kappa-1}$, **else**

   3. *Inspect rankit plot for model with* W^κ *and eliminate obvious outliers.* **Goto** *initialize.*

5   Validation-Synthetic Model

Verification of the algorithm on synthetic data is attractive as our estimation results can be compared to the true capillary pressure curve. We use a modified form of the parameterized capillary pressure function by Bentsen and Anli [1] in the following form:

$$P_c = P_{cd} \left(\frac{S_w - S_{wi}}{1 - S_{wi}} \right)^{-1/\theta}.$$

The integral in eq. (2.4) is then given by

$$\frac{\overline{S}_w(\omega) - S_{wi}}{1 - S_{wi}} = \frac{L_1}{L} + \frac{r_e}{L} \left(\frac{2P_{cd}}{\Delta\rho\omega^2 r_e^2} \right)^2 \int_{L/r_e}^{L_1/r_e} \frac{dx}{[(2-x)x]^\theta},$$

$$L_1 = r_e \left(1 - \sqrt{1 - \frac{2P_{cd}}{\Delta\rho\omega^2 r_e^2}} \right).$$

(5.1)

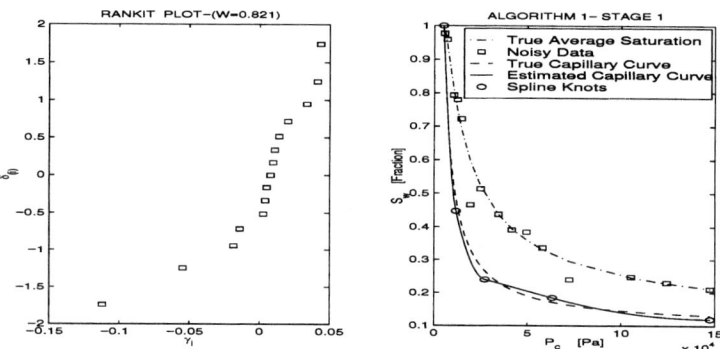

Figure 2: Rankit plot & cap.pressure curve-stage 1.

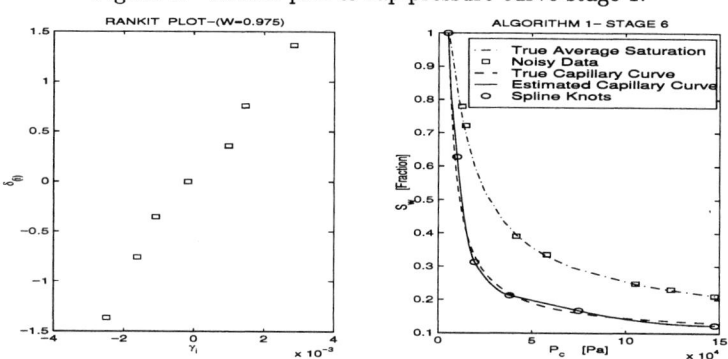

Figure 3: Rankit plot & cap.pressure curve-stage 6.

Eq. (5.1) can be solved numerically for $\overline{S}_w(\omega_i)$. In the particular cases shown here, we have used the parameters in this presentation as $P_{cd} = 5kPa$, $S_{wi} = 0.1$, $L = 5cm$, $r_e = 9.38cm$, $\Delta\rho = 1g/cm\,3$, and $\theta = 1$. For 15 different angular velocities, we calculate the average saturation using eq. (5.1). (In fact, for these parameter values, the integral is analytical). We generate data for two cases. To include outliers we perturb the synthetic data using the relation

$$\overline{S}_w^{perturbed}(\omega_j) = 1.2\overline{S}_w^{true}(\omega_j), \quad j = 5, 10,$$
$$\overline{S}_w^{perturbed}(\omega_j) = 0.8\overline{S}_w^{true}(\omega_j), \quad j = 6, 12. \tag{5.2}$$

All 15 data points including those perturbed above are refered to as partially perturbed data (pp-data). We add white noise using the relation

$$\overline{S}_w^{meas}(\omega_j) = \overline{S}_w^{pp-data}(\omega_j)(1 + 0.02\epsilon_{S,j}), \quad j = 1, \ldots, 15.$$

$\epsilon_{S,j}$ is drawn from a normal distribution with zero mean and unit standard deviation. This constitutes our first case, case 1. For case 2, we also perturb the last data point, 15, according to eq. (5.2).

6 Results and Discussions

Stage	N	max.W	P	min. W	M
1	6	0.877	$2\% < P < 1\%$	0.821	15
2	6	0.960	$50\% < P < 90\%$	0.935	13
3	6	0.961	$50\% < P < 90\%$	0.928	12
4	6	0.952	$50\% < P < 90\%$	0.838	10
5	6	0.966	$50\% < P < 90\%$	0.928	8
6	7	0.975	$P = 90\%$		7

Table 1: Algorithm 1-case 1.

In the first approach, the problem involing case 1 is solved by using a spline dimension of 6 to 7, which appears to give sufficient flexibility in the calculated capillary pressure curve. Regularization is achieved by the imposition of convexity constraints on the solution curve, and Algorithm 1 is employed to achieve compliance with the normality assumption.

Table 6 shows important data at the various stages of application of Algorithm 1. Tabulated are the spline dimension, N, the maximum Shapiro-Wilk statistic obtained with the corresponding percentage level, P, the minimum Shapiro-Wilk statistic, and the number of experimental data employed at the given stage, M.

FIG. 2 shows the rankit plots as well as the estimated capillary pressure curves for stage 1 of Algorithm 1. The plot order of the residuals at $W =$

0.821 is $\delta_6 < \delta_{12} < \delta_1 < \delta_3 < \delta_{13} < \delta_{15} < \delta_{14} < \delta_2 < \delta_{11} < \delta_9 < \delta_7 < \delta_8 <$ $\delta_{10} < \delta_4$. Inspection of the rankit plot at stage 1 shows that the 6th and 12th data points (δ_6, δ_{12}) deviate considerably from any straight line about which majority of the residuals have minimum deviation. These points could be flagged out as possible outliers.

FIG. 3 shows the rankit plot as well as the estimated capillary pressure curve at stage 6 of Algorithm 1. At this point, we can state, at 90% level, that our residuals and errors are normally distributed (refer to Table 6). We see that the estimated capillary pressure curve to a high degree reconciles the true one. From experience, a confidence level $\geq 90\%$ has been found to be adequate.

Case 2 presents a unique situation where the last data point (15) is an outlier. Algorithm 1 could still be applied to this case. Indeed, at stage 5 of Algorithm 1, experimental data number 15 could be flagged as a possible outlier. However, from a practical point of view, it will be unwise to delete data number 15 since we would then be unable to estimate the capillary pressure curve in the range between the 14th and 15th average saturation values. We therefore need a robust norm definition for the residuals which, when applied in combination with our monotonicity and convexity constraints, accommodates the outlier. This we do by the L_1-norm expressed as

$$\min_{\vec{c}} \sum |\vec{b} - A\vec{c}|. \tag{6.1}$$

Thus we use a low-dimensional spline $(N = 5)$ and solve eq. (6.1) subject to the constraints in eq. (3.7) to arrive at FIG. 2. Indeed, for this case, we adopted the spline dimension with the highest W in stage 1. Thus the solution curve represents that which is nearest to normality in stage 1. From FIG. 2, we realize that we are able to estimate the capillary presssure curve to a fairly high degree, accommodating all outliers.

7 Conclusions

1. We have developed an algorithm for detecting outliers based on violation of the normality assumption.

2. Outliers can be handled through Algorithm 1.

3. In the case where the last data point is an outlier, a solution in the L_1-norm is proposed, which accommodates the outlier.

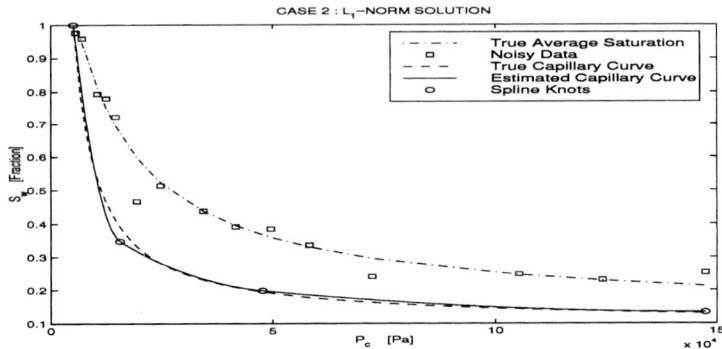

Figure 4: Case 2 cap.pressure curve-L_1-norm.

References

[1] Bentsen, R. G. and Anli, J., Using parameter sstimation technique to convert centrifuge data into a capillary pressure curve, *SPEJ*, Feb. 1977, 57–71.

[2] Blom, G., *Statistical Estimates and Transformed Beta Variates*, New York Wiley, 1958.

[3] Gustafson, S.-Å, *Regularizing a Volterra integral equation problem by means of convex programming*, Working Paper Nr. 138, 1991.

[4] Hadamard, J., *Lectures on the Cauchy Problem in Linear Partial Differential Equations*, Yale University Press, 1923.

[5] Hassler, G. L. and Brunner, E., Measurements of sapillary pressures in small core samples, *Trans. AIME* **160** (1945), 114–121.

[6] Madansky, A., *Prescription for Working Statisticians* , Springer-Verlag, 1988.

[7] Schumaker, L. L., *Spline Functions: Basic Theory*, J. Wiley & Sons, New York City, 1981.

[8] Shapiro, S. S. and Wilk, M. B., An analysis of variance test for normality (complete samples), *Biometrika* **52** (1965), 519–611.

[9] Srivasta, S. and Sen, A., *Regression Analysis-Theory, Methods, and Application*, Springer-Verlag, 1990.

[10] Subbey, S. and Nordtvedt, J.-E., Estimating sapillary pressure functions from centrifuge data, submitted to AIChe.

A Comparison of ELLAM with ENO/WENO Schemes for Linear Transport Equations

Hong Wang   Mohamed Al-Lawatia

Abstract

We present an Eulerian-Lagrangian localized adjoint method (EL-LAM) for linear advection-reaction partial differential equations in multiple space dimensions. We carry out numerical experiments to compare the performance of the ELLAM scheme with the essentially non-oscillatory (ENO) schemes and weighted essentially non-oscillatory (WENO) schemes, which shows that the ELLAM scheme outperforms ENO and WENO schemes in the context of linear transport PDEs.

KEYWORDS: advection-reaction equations, characteristic methods, comparison of numerical methods, essentially non-oscillatory schemes, Eulerian-Lagrangian methods, transport equations

1   Introduction

Advection-dominated partial differential equations (PDEs) describe the displacement of oil by injected fluid in petroleum recovery, the subsurface contaminant transport and remediation, and many other applications [3, 8, 19]. Because of the moving steep fronts present in their solutions, the numerical treatment of these PDEs often presents severe difficulties. Standard finite difference or finite element methods (FDMs, FEMs) tend to generate solutions with severe non-physical oscillations. While classical upwind FDM could eliminate these oscillations, they yield solutions with excessive smearing and potentially spurious effects related to the orientation of the grid. Two general classes of improved approximations can be identified from the literature: upwind methods that use fixed spatial grids with some form of upwinding and the standard temporal discretization, and the characteristic methods that carry out the temporal discretization by characteristic tracking.

The Eulerian-Lagrangian localized adjoint method (ELLAM) [5] was introduced by Celia, Russell, Herrera, and Ewing in solving (one-dimensional constant-coefficient) advection-diffusion PDEs. The ELLAM methodology provides a general characteristic solution procedure and a consistent framework for treating general boundary conditions and conserving mass. Thus, it overcomes the two principal shortcomings of the previous characteristic methods while maintaining their numerical advantages. We conducted numerical experiments to observe the performance of the ELLAM scheme with many widely used methods [1, 20], including the upwind FDM, Galerkin FEM, quadratic and cubic Petrov-Galerkin FEMs [2, 4, 6], the streamline diffusion FEM [13], the continuous and discontinuous Galerkin FEMs [14, 16], the monotone upstream-centered scheme for conservation laws (MUSCL) and the minmod scheme [7, 18]. These experiments show that the ELLAM scheme outperforms these methods in the context of linear transport PDEs. We also proved optimal-order error estimates for the ELLAM schemes in [9, 10].

In this paper we present an ELLAM scheme and numerically compare the performance of the ELLAM scheme with the essentially non-oscillatory (ENO) and weighted ENO (WENO) schemes [11, 12, 15].

2   An ELLAM Scheme

2.1   Definition of test functions

We consider a multi-dimensional linear advection-reaction PDE

$$c_t + \nabla \cdot (\mathbf{v}c(\mathbf{x}, t)) + K(\mathbf{x}, t)c = F(\mathbf{x}, t), \quad \mathbf{x} \in \Omega, \ t \in (0, T], \qquad (2.1)$$

where $\Omega \subset \mathbb{R}^d$ is a bounded domain with a Lipschitz continuous boundary $\Gamma = \partial\Omega$. A boundary condition

$$c(\mathbf{x}, t) = g(\mathbf{x}, t), \quad (\mathbf{x}, t) \in \Gamma^{(I)} \times [0, T] \qquad (2.2)$$

is specified only at the inflow boundary $\Gamma^{(I)}$ identified by $\Gamma^{(I)} = \{\mathbf{x} \mid \mathbf{x} \in \Gamma, \ \mathbf{v} \cdot \mathbf{n} < 0\}$. In addition, an initial condition $c(\mathbf{x}, 0) = c_0(\mathbf{x})$ is specified to close Eq. (2.1).

We define a quasi-uniform temporal partition on $[0, T]$ by $0 = t_0 < t_1 < t_2 < \ldots < t_{N-1} < t_N = T$. Multiplying Eq. (2.1) by space-time test functions $w(\mathbf{x}, t)$ that are continuous and piecewise smooth, vanish outside the space-time strip $\Omega \times [t_{n-1}, t_n]$, and are discontinuous in time at time t_{n-1}, we obtain a space-time weak formulation

$$\int_{\Omega} c(\mathbf{x}, t_n) w(\mathbf{x}, t_n) dx + \int_{t_{n-1}}^{t_n} \int_{\Gamma} \mathbf{v} \cdot \mathbf{n}\, c(\mathbf{x}, t) w(\mathbf{x}, t) ds dt$$

$$- \int_{t_{n-1}}^{t_n} \int_{\Omega} c(\mathbf{x}, t)(w_t + \mathbf{v} \cdot \nabla w - Kw)(\mathbf{x}, t) dx dt \qquad (2.3)$$

$$= \int_{\Omega} c(\mathbf{x}, t_{n-1}) w(\mathbf{x}, t_{n-1}^+) dx + \int_{t_{n-1}}^{t_n} \int_{\Omega} F(\mathbf{x}, t) w(\mathbf{x}, t) dx dt,$$

where $w(\mathbf{x}, t_{n-1}^+) = \lim_{t \to t_{n-1}^+} w(\mathbf{x}, t)$, which takes into account the fact that $w(\mathbf{x}, t)$ is discontinuous in time at time t_{n-1}.

In the ELLAM framework [5], the test functions w are chosen to satisfy the adjoint equation of Eq. (2.1)

$$w_t + \mathbf{v} \cdot \nabla w - K\, w = 0. \qquad (2.4)$$

Let $\mathbf{y} = \mathbf{r}(\theta; \bar{\mathbf{x}}, \bar{t})$, with $\bar{t} \in [t_{n-1}, t_n]$, be the characteristic determined by

$$\frac{d\mathbf{y}}{d\theta} = \mathbf{v}(\mathbf{y}, \theta), \quad \text{with} \quad \mathbf{y}|_{\theta = \bar{t}} = \bar{\mathbf{x}}. \qquad (2.5)$$

Eq. (2.4) is rewritten as

$$-\frac{d}{d\theta} w(\mathbf{r}(\theta; \bar{\mathbf{x}}, \bar{t}), \theta) + K(\mathbf{r}(\theta; \bar{\mathbf{x}}, \bar{t}), \theta) w(\mathbf{r}(\theta; \bar{\mathbf{x}}, \bar{t}), \theta) = 0,$$
$$w(\mathbf{r}(\theta; \bar{\mathbf{x}}, \bar{t}), \theta)|_{\theta = \bar{t}} = w(\bar{\mathbf{x}}, \bar{t}). \qquad (2.6)$$

Solving Eq. (2.6) yields the following expression for w

$$w(\mathbf{r}(\theta; \bar{\mathbf{x}}, \bar{t}), \theta) = w(\bar{\mathbf{x}}, \bar{t}) e^{-\int_{\theta}^{\bar{t}} K(\mathbf{r}(\gamma; \bar{\mathbf{x}}, \bar{t}), \gamma) d\gamma}. \qquad (2.7)$$

Therefore, the test functions w in Eq. (2.3) should vary exponentially along the characteristics $\mathbf{r}(\theta; \bar{\mathbf{x}}, \bar{t})$. Once $w(\bar{\mathbf{x}}, \bar{t})$ is specified, $w(\mathbf{r}(\theta; \bar{\mathbf{x}}, \bar{t}), \theta)$ is determined completely along the characteristic $\mathbf{r}(\theta; \bar{\mathbf{x}}, \bar{t})$. Thus, to define the test functions w in the space-time strip $\Omega \times [t_{n-1}, t_n]$, we only need to define w on $\bar{\Omega}$ at the time t_n and on the space-time outflow boundary $\Gamma^{(O)} \times [t_{n-1}, t_n]$ with $\Gamma^{(O)} = \{\mathbf{x} \in \Gamma \mid \mathbf{v} \cdot \mathbf{n} > 0\}$.

2.2 Derivation of a reference equation

To avoid confusion, we replace the dummy variables $\mathbf{x}$ and t in the second term on the right-hand side of Eq. (2.3) by $\mathbf{y}$ and θ and reserve $\mathbf{x}$ and t for the points in Ω at time t_n or on $\Gamma \times [t_{n-1}, t_n]$. Let $\Omega(\theta) \subset \Omega$ be the set of the points that will flow out of the domain Ω during the time period $[\theta, t_n]$. For any $\mathbf{y} \in \Omega \backslash \Omega(\theta)$, there exists an $\mathbf{x} \in \Omega$ such that $\mathbf{y} = \mathbf{r}(\theta; \mathbf{x}, t_n)$. Likewise, for any $(\mathbf{y}, \theta) \in \Omega(\theta)$, there exists a pair $(\mathbf{x}, t) \in \Gamma^{(O)} \times [t_{n-1}, t_n]$ such that $\mathbf{y} = \mathbf{r}(\theta; \mathbf{x}, t)$. Therefore,

$$
\begin{aligned}
\int_{t_{n-1}}^{t_n} & \int_\Omega F(\mathbf{y}, \theta) w(\mathbf{y}, \theta) \, d\mathbf{y} d\theta \\
&= \int_{t_{n-1}}^{t_n} \int_{\Omega \backslash \Omega(\theta)} F(\mathbf{r}(\theta; \mathbf{x}, t_n), \theta) \, w(\mathbf{r}(\theta; \mathbf{x}, t_n), \theta) \, d\mathbf{r} d\theta \\
&\quad + \int_{t_{n-1}}^{t_n} \int_{\Omega(\theta)} F(\mathbf{r}(\theta; \mathbf{x}, t), \theta) \, w(\mathbf{r}(\theta; \mathbf{x}, t), \theta) \, d\mathbf{r} d\theta.
\end{aligned}
\tag{2.8}
$$

The first term on the right-hand side of Eq. (2.8) is evaluated by applying the Euler formula at time t_n, leading to

$$
\begin{aligned}
\int_{t_{n-1}}^{t_n} & \int_{\Omega \backslash \Omega(\theta)} F(\mathbf{r}(\theta; \mathbf{x}, t_n), \theta) w(\mathbf{r}(\theta; \mathbf{x}, t_n), \theta) d\mathbf{r} d\theta \\
&= \int_\Omega \int_{t^*(\mathbf{x})}^{t_n} F(\mathbf{r}(\theta; \mathbf{x}, t_n), \theta) w(\mathbf{r}(\theta; \mathbf{x}, t_n), \theta) |\frac{\partial \mathbf{r}(\theta; \mathbf{x}, t_n)}{\partial \mathbf{x}} | d\theta d\mathbf{x} \\
&= \int_\Omega F(\mathbf{x}, t_n) w(\mathbf{x}, t_n) \left[\int_{t^*(\mathbf{x})}^{t_n} e^{-K(\mathbf{x}, t_n)(t_n - \theta)} d\theta \right] d\mathbf{x} + E_1(f, w) \\
&= \int_\Omega \phi^{(1)}(\mathbf{x}, t_n) F(\mathbf{x}, t_n) w(\mathbf{x}, t_n) d\mathbf{x} + E_1(F, w).
\end{aligned}
\tag{2.9}
$$

Here the space-dependent time step $\Delta t^{(I)}(\mathbf{x}) = t_n - t^*(\mathbf{x})$, where $t^*(\mathbf{x}) = t_{n-1}$ if the characteristic $\mathbf{r}(\theta; \mathbf{x}, t_n)$ does not backtrack to the boundary Γ during the time period $[t_{n-1}, t_n]$, or $t^*(\mathbf{x}) \in [t_{n-1}, t_n]$ is the time when $\mathbf{r}(\theta; \mathbf{x}, t_n)$ intersects the boundary Γ otherwise.

$$
\phi^{(1)}(x, t_n) = (1 - e^{-K(\mathbf{x}, t_n) \Delta t^{(I)}(\mathbf{x})}) / K(\mathbf{x}, t_n),
$$

if $K(\mathbf{x}, t_n) \neq 0$, or $\Delta t^{(I)}(\mathbf{x})$ otherwise. $E_1(F, w)$ is the local truncation error.

The second term on the right-hand side of Eq. (2.8) is treated similarly. We obtain

$$
\begin{aligned}
\int_{t_{n-1}}^{t_n} & \int_{\Omega(\theta)} F(\mathbf{r}(\theta; \mathbf{x}, t), \theta) \, w(\mathbf{r}(\theta; \mathbf{x}, t), \theta) \, d\mathbf{r} d\theta \\
&= \int_{t_{n-1}}^{t_n} \int_{\Gamma^{(O)}} \mathbf{v} \cdot \mathbf{n} \phi^{(2)}(\mathbf{x}, t) F(\mathbf{x}, t) w(\mathbf{x}, t) ds dt + E_2(F, w).
\end{aligned}
\tag{2.10}
$$

Here $\Delta t^{(O)}(\mathbf{x}, t) = t_n - t^*(\mathbf{x}, t)$ for $(\mathbf{x}, t) \in \Gamma^{(O)} \times [t_{n-1}, t_n]$, where $t^*(\mathbf{x}, t) = t_{n-1}$ if $\mathbf{r}(\theta; \mathbf{x}, t)$ does not backtrack to the boundary Γ during the time period $[t_{n-1}, t]$, or $t^*(\mathbf{x}, t) \in [t_{n-1}, t]$ is the time when $\mathbf{r}(\theta; \mathbf{x}, t)$ intersects the boundary Γ otherwise. $\phi^{(2)}(\mathbf{x}, t) = (1 - e^{-K(\mathbf{x}, t) \Delta t^{(O)}(\mathbf{x}, t)}) / K(\mathbf{x}, t)$ if $K(\mathbf{x}, t) \neq 0$ or $\Delta t^{(O)}(\mathbf{x}, t)$ otherwise. $E_2(F, w)$ is the local truncation error.

Incorporating Eqs. (2.8)–(2.10) and the inflow boundary condition (2.2) into Eq. (2.3), we obtain the following reference equation

$$
\begin{aligned}
\int_{\Omega} c(\mathbf{x}, t_n) w(\mathbf{x}, t_n) d\mathbf{x} &+ \int_{t_{n-1}}^{t_n} \int_{\Gamma^{(O)}} \mathbf{v}(\mathbf{x}, t) \cdot \mathbf{n}(\mathbf{x}) c(\mathbf{x}, t) w(\mathbf{x}, t) ds dt \\
&= \int_{\Omega} c(\mathbf{x}, t_{n-1}) w(\mathbf{x}, t_{n-1}^+) d\mathbf{x} + \int_{\Omega} \phi^{(1)}(\mathbf{x}, t_n) F(\mathbf{x}, t_n) w(\mathbf{x}, t_n) d\mathbf{x} \\
&+ \int_{t_{n-1}}^{t_n} \int_{\Gamma^{(O)}} \phi^{(2)}(\mathbf{x}, t) \mathbf{v}(\mathbf{x}, t) \cdot \mathbf{n}(\mathbf{x}) F(\mathbf{x}, t) \ w(\mathbf{x}, t) \ ds dt \\
&- \int_{t_{n-1}}^{t_n} \int_{\Gamma^{(I)}} \mathbf{v}(\mathbf{x}, t) \cdot \mathbf{n}(\mathbf{x}) \ g(\mathbf{x}, t) w(\mathbf{x}, t) ds dt + E(w),
\end{aligned}
\tag{2.11}
$$

where $E(w) = \int_{t_{n-1}}^{t_n} \int_{\Omega} c(\mathbf{x}, t)[w_t(\mathbf{x}, t) + \mathbf{v}(\mathbf{x}, t) \cdot \nabla w(\mathbf{x}, t) - K(\mathbf{x}, t) w(\mathbf{x}, t)] d\mathbf{x} dt + E_1(F, w) + E_2(F, w)$.

2.3   A numerical scheme

In the ELLAM scheme, the trial space $\mathcal{S}_h$ consists of piecewise linear (or d-linear) functions in $\overline{\Omega}$ at time t_n and on $\Gamma^{(O)} \times [t_{n-1}, t_n]$. Because of the boundary condition (2.2), no degrees of freedom should be introduced at $\Gamma^{(I)}$ at time t_n. Similarly, since the solutions are known at the time step t_{n-1}, no degrees of freedom should be introduced at $\Gamma^{(O)}$ at time t_{n-1}. However, to conserve mass, all the test functions should sum exactly to one [5]. Therefore, the test space $\overline{\mathcal{S}}_h$ is obtained by modifying the trial space $\mathcal{S}_h$: For a basis function associated with a node at $\Gamma^{(I)}$ at time t_n, we add it to the basis function associated with its adjacent node that is inside Ω. The basis functions associated with the nodes at $\Gamma^{(O)}$ at time t_{n-1} are treated similarly. An ELLAM scheme is formulated as follows: find $c(\mathbf{x}, t) \in \mathcal{S}_h$ which satisfies the boundary condition (2.2), such that $\forall w(\mathbf{x}, t) \in \overline{\mathcal{S}}_h$

$$
\begin{aligned}
\int_{\Omega} c(\mathbf{x}, t_n) w(\mathbf{x}, t_n) d\mathbf{x} &+ \int_{t_{n-1}}^{t_n} \int_{\Gamma^{(O)}} \mathbf{v}(\mathbf{x}, t) \cdot \mathbf{n}(\mathbf{x}) c(\mathbf{x}, t) w(\mathbf{x}, t) ds dt \\
&= \int_{\Omega} c(\mathbf{x}, t_{n-1}) w(\mathbf{x}, t_{n-1}^+) d\mathbf{x} + \int_{\Omega} \phi^{(1)}(\mathbf{x}, t_n) F(\mathbf{x}, t_n) w(\mathbf{x}, t_n) d\mathbf{x} \\
&+ \int_{t_{n-1}}^{t_n} \int_{\Gamma^{(O)}} \phi^{(2)}(\mathbf{x}, t) \mathbf{v}(\mathbf{x}, t) \cdot \mathbf{n}(\mathbf{x}) F(\mathbf{x}, t) \ w(\mathbf{x}, t) \ ds dt \\
&- \int_{t_{n-1}}^{t_n} \int_{\Gamma^{(I)}} \mathbf{v}(\mathbf{x}, t) \cdot \mathbf{n}(\mathbf{x}) \ g(\mathbf{x}, t) w(\mathbf{x}, t) ds dt.
\end{aligned}
\tag{2.12}
$$

By using a Lagrangian coordinate, the ELLAM scheme (2.12) significantly reduces the temporal truncation errors and generates accurate solutions even if very large time steps are used. Moreover, it symmetrizes the governing

PDE (2.1), and generates a well-conditioned, symmetric and positive-definite coefficient matrix. Thus, the discrete system can be solved efficiently by, for example, the conjugate gradient method in an optimal order without any preconditioning needed. Furthermore, this scheme naturally incorporates the boundary condition (2.2) into its formulation and conserves mass. Finally, most terms in the scheme are standard in the FEM and can be solved in a straightforward manner. In the first term on the right-hand side, the trial and test functions are actually defined at different time steps. Hence, its evaluation could be very challenging and raises serious numerical difficulties. We use a forward Euler or second-order Runge-Kutta tracking algorithm [17] to evaluate this term.

3   Description of ENO and WENO Schemes

The ENO scheme was first introduced by Harten *et al* [11, 12] as an improvement over traditional fixed-stencil high-order FD interpolations, which are known to be oscillatory in nature especially near discontinuities of the solutions. These oscillations do not decay as the mesh is refined and lead to further instabilities in the solution. By choosing the smoothest stencil among several candidates, ENO provides a uniformly high order approximation of the fluxes at cell boundaries while avoiding the spurious oscillations near steep fronts and shock discontinuities that are associated with traditional FDMs. As a further improvement, Liu *et al* [15] provide Weighted ENO (WENO) that is based on using a convex combination of the candidate stencils instead of just one. This treatment maintains the advantages of the original ENO and in addition obtains a higher-order accuracy.

In this section we describe the ENO and WENO schemes of order $k = 4$ for Eq. (2.1) defined on $\mathbb{R}^2$ with a uniform partition $x_i = i\Delta x$ and $y_j = j\Delta y$. If we assume that no reaction or source term is present and let $\mathbf{v} = (V_1, V_2)$, Eq. (2.1) can be rewritten as

$$c_t + f_x(c) + g_y(c) = 0, \quad \text{with} \quad f(c) = V_1 c, \quad g(c) = V_2 c, \quad (3.1)$$

We can write the spatially discretized ENO and WENO schemes as

$$\frac{dc_{ij}(t)}{dt} = L(c) \equiv -\frac{1}{\Delta x}(\hat{f}_{i+1/2,j} - \hat{f}_{i-1/2,j}) - \frac{1}{\Delta y}(\hat{g}_{i,j+1/2} - \hat{g}_{i,j-1/2}). \quad (3.2)$$

Below we define the numerical flux $\hat{f}_{i+1/2,j}$ for a fixed j, and will skip the subscript j when it is clear from the context. The flux $\hat{g}_{i+1/2,j}$ in the y-

direction is defined by symmetry. For the fixed j, we identify the cell averages $\bar{q}_i = f(c_{ij})$. Upwinding is guaranteed by computing the Roe speed given by

$$a_{i+1/2} = \frac{f(c_{i+1,j}) - f(c_{ij})}{c_{i+1,j} - c_{ij}}. \tag{3.3}$$

The flux $\hat{f}_{i+1/2,j}$ is then given by

$$\hat{f}_{i+1/2,j} = \begin{cases} q^-_{i+1/2}, & \text{if } a_{i+1/2} \geq 0, \\ q^+_{i+1/2}, & \text{otherwise.} \end{cases} \tag{3.4}$$

The ENO and WENO schemes differ in the reconstruction procedure that provides $q^+_{i+1/2}$ and $q^-_{i+1/2}$. The ENO scheme starts by computing the undivided differences for degrees 1 to $k = 4$

$$\begin{aligned} Q[x_{i-1/2}, x_{i+1/2}] &= \bar{q}_i, \\ Q[x_{i-1/2}, \ldots, x_{i+j+1/2}] &= Q[x_{i+1/2}, \ldots, x_{i+j+1/2}] \\ &\quad - Q[x_{i-1/2}, \ldots, x_{i+j-1/2}]. \end{aligned} \tag{3.5}$$

The selection of the smoothest 5-point stencil (corresponding to $k = 4$) starts with the two point stencil

$$S = \left\{ x_{i-1/2}, x_{i+1/2} \right\}. \tag{3.6}$$

If

$$|Q[x_{i-3/2}, x_{i-1/2}, x_{i+1/2}]| < |Q[x_{i-1/2}, x_{i+1/2}, x_{i+3/2}]|, \tag{3.7}$$

then $x_{i-3/2}$ is added to the stencil S. Otherwise, $x_{i+3/2}$ is added. We repeatedly add points in a similar manner until we come up with a 5-point stencil. Assume that the stencil S is composed of r cells to the left and s cells to the right of the current cell. We introduce

$$q^-_{i+1/2} = \sum_{j=0}^{3} c_{rj} \bar{q}_{i-r+j}, \quad q^+_{i-1/2} = \sum_{j=0}^{3} c_{r-1,j} \bar{q}_{i-r+j}, \tag{3.8}$$

where

$$c_{rj} = \sum_{m=j+1}^{k} \left(\frac{\sum_{l=0, l \neq m}^{k} \Pi_{p=0, p \neq m,l}^{k} (r - p + 1)}{\Pi_{l=0, l \neq m}^{k} (m - l)} \right). \tag{3.9}$$

The WENO reconstruction is based on a convex combination of four different reconstruction values

$$q^-_{i+1/2} = \sum_{r=0}^{3} w_r q^{(r)}_{i+1/2}, \quad q^+_{i-1/2} = \sum_{r=0}^{3} \tilde{w}_r q^{(r)}_{i-1/2}, \tag{3.10}$$

where the reconstructed values $q^{(r)}_{i+1/2}$ and $q^{(r)}_{i-1/2}$ $(r = 0, 1, 2, 3)$ are defined by (3.8). The weights are given by

$$\begin{aligned}
\omega_r &= \frac{\alpha_r}{\sum_{s=0}^{3} \alpha_s}, & \alpha_r &= \frac{d_r}{(\varepsilon + \beta_r)^2}, \\
\tilde{\omega}_r &= \frac{\tilde{\alpha}_r}{\sum_{s=0}^{3} \tilde{\alpha}_s}, & \tilde{\alpha}_r &= \frac{d_{3-r}}{(\varepsilon + \beta_r)^2}, & 0 \le r \le 3.
\end{aligned} \tag{3.11}$$

Here the constants are $d_0 = 1/35$, $d_1 = 12/35$, $d_2 = 18/35$, $d_3 = 4/35$, and $\epsilon = 10^{-6}$. The smoothness indicators are

$$\begin{aligned}
\beta_r &= \left[\bar{q}_{i+3-r} - 3\bar{q}_{i+2-r} + 3\bar{q}_{i+1-r} - \bar{q}_{i-r} \right]^2 \\
&\quad + \frac{1}{2} \left[(\bar{q}_{i+2-r} - 2\bar{q}_{i+1-r} + \bar{q}_{i-r})^2 + (\bar{q}_{i+3-r} - 2\bar{q}_{i+2-r} + \bar{q}_{i+1-r})^2 \right] \quad (3.12) \\
&\quad + \frac{1}{3} \left[(\bar{q}_{i+1-r} - \bar{q}_{i-r})^2 + (\bar{q}_{i+2-r} - \bar{q}_{i+1-r})^2 + (\bar{q}_{i+3-r} - \bar{q}_{i+2-r})^2 \right].
\end{aligned}$$

For $n = 1, 2, \dots, N$, a fully discrete ENO or WENO scheme with a fourth-order Runge-Kutta temporal discretization is defined by

$$\begin{aligned}
c^{(1)} &= c_{n-1} + \frac{\Delta t}{2} L(c_{n-1}), \quad c^{(2)} = c_{n-1} + \frac{\Delta t}{2} L(c^{(1)}), \\
c^{(3)} &= c_{n-1} + \Delta t L(c^{(2)}), \\
c_n &= \frac{1}{3} \left(-c_{n-1} + c^{(1)} + cu^{(2)} + c^{(3)} \right) + \frac{\Delta t}{6} L(c^{(3)}).
\end{aligned} \tag{3.13}$$

4 Numerical Experiments

The spatial domain is $\Omega = (-0.5, 0.5) \times (-0.5, 0.5)$, a rotating velocity field is imposed as $\mathbf{v}(\mathbf{x}) = \mathbf{v}(x, y) = (-4y, 4x)$. The time interval is $[0, T] = [0, \pi/2]$, which is the time period required for one complete rotation. The initial condition $c_0(\mathbf{x})$ is given by

$$c_0(\mathbf{x}) = \exp\left(-\frac{|\mathbf{x} - \mathbf{x}_c|^2}{2\sigma^2} \right), \tag{4.1}$$

where $\mathbf{x}_c = (x_c, y_c)$, and σ are the centered and standard deviations. The corresponding analytical solution for Eq. (2.1) with $f = 0$ is

$$c(\mathbf{x}, t) = \exp\left(-\frac{|\bar{\mathbf{x}} - \mathbf{x}_c|^2}{2\sigma^2} - \int_0^t K(\mathbf{r}(\theta; \bar{\mathbf{x}}, 0), \theta) d\theta \right), \tag{4.2}$$

where $\bar{\mathbf{x}} = (\bar{x}, \bar{y}) = (x \cos(4t) + y \sin(4t), -x \sin(4t) + x \cos(4t))$, and $\mathbf{r}(\theta; \bar{\mathbf{x}}, 0) = (\bar{x} \cos(4\theta) - \bar{y} \sin(4\theta), \bar{x} \sin(4\theta) + \bar{y} \cos(4\theta))$.

This example can be viewed as an incompressible flow in a two-dimensional homogeneous medium with a known analytical solution, and has been widely used to test for numerical artifacts of different schemes, such as numerical stability, numerical dispersion, spurious oscillations, deformation, and phase errors, etc. For the performance of the ELLAM scheme for problems with discontinuities, please see [1, 20].

Scheme	h	Δt	Max.	Min.	CPU	CFL #	Fig. #
anal	1/64	N/A	N/A	1	0	N/A	
ELLAM	1/64	$\pi/8$	0.9987	0	1m 5s	71.25	1
ENO	1/64	$\pi/580$	0.8757	0	1m 10s	0.98	
	1/64	$\pi/6000$	0.8756	0	12m 5s	0.09	
	1/96	$\pi/860$	0.9552	-0.0004	3m 45s	0.99	
	1/96	$\pi/4000$	0.9552	0	17m 26s	0.21	
	1/196	$\pi/1700$	∞	∞	30m 2s	1.0	
	1/196	$\pi/2400$	0.9443	-0.0040	42m 7s	0.71	
	1/256	$\pi/2400$	0.9967	-0.0311	1h 15m	0.95	
	1/256	$\pi/3000$	0.9836	-0.0013	1h 34m	0.76	2
	1/512	$\pi/6000$	0.9993	0	12h 52m	0.76	
WENO	1/64	$\pi/580$	0.9617	0	2m 17s	0.98	
	1/64	$\pi/6000$	0.9617	0	23m 37s	0.09	
	1/96	$\pi/860$	0.9832	0	7m 34s	0.99	
	1/96	$\pi/4000$	0.9832	0	34m 57s	0.21	
	1/196	$\pi/1700$	0.9964	0	1h 1m	1.0	
	1/196	$\pi/2400$	0.9964	0	1h 27m	0.71	
	1/256	$\pi/2400$	0.9979	0	2h 31m	0.95	3
	1/256	$\pi/3000$	0.9979	0	3h 9m	0.76	
	1/512	$\pi/6000$	0.9995	0	1day 1h	0.76	

Table 1. The Performance of the ELLAM, ENO, and WENO.

In the example runs, the data are chosen as follows: $K = f = 0$, $\mathbf{x}_c = (-0.25, 0)$, $\sigma = 0.0447$. A uniform spatial grid $h = \Delta x = \Delta y = \frac{1}{64}$ and a time step of $\Delta t = \pi/8$ are used in the ELLAM simulation. A second-order Runge-Kutta method with a micro-time step of $\Delta t_m = \Delta t/20$ is used to track the characteristics. For the given h, a largest admissible (satisfying the CFL condition) time step of $\Delta t = \frac{\pi}{580}$ is used in the ENO and WENO simulations. We then systematically reduce the sizes of h and Δt to examine the performance of ENO and WENO schemes until their solutions are roughly compatible with the ELLAM solution. The numerical results are presented in Table 1 with selected solutions are plotted in Figures 1–3.

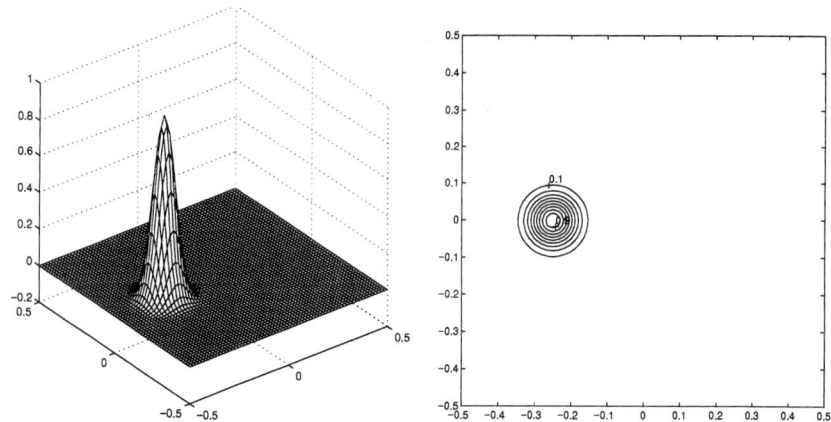

Figure 1: ELLAM, min = 0, max = 0.9987.

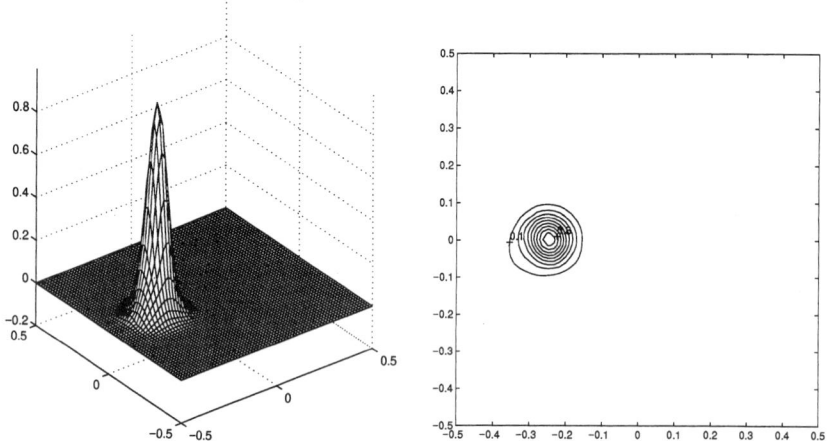

Figure 2: ENO, min = −0.0013, max = 0.9836.

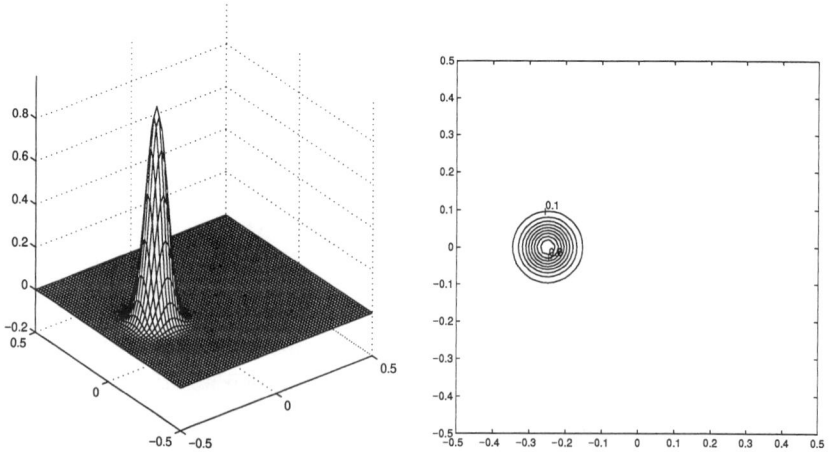

Figure 3: WENO, min = 0, max = 0.9979.

5   Summary

We present an ELLAM scheme for linear advection-reaction PDEs in multiple space dimensions and carry out numerical experiments to compare the performance of the ELLAM scheme with ENO and WENO schemes. We have the following observations: (i) Even though very large time steps and coarse spatial grids are used, the ELLAM scheme still generates much more accurate solutions than those obtained with the ENO and WENO schemes with much finer spatial and temporal grid sizes. Hence, the ELLAM scheme outperforms ENO and WENO schemes in the context of linear transport PDEs. (ii) The mathematical models used to describe multiphase and/or multicomponent fluid flow processes in porous medium are typically strongly coupled systems of nonlinear PDEs [3, 8, 20]. Nevertheless, most of these PDEs govern the transport of different components in these fluids and are advection-dominated PDEs that are linear in terms of the concerned concentrations. Thus, after some decoupling/linearization techniques are applied, these transport PDEs are reduced to linear advection-dominated transport PDEs at each time step. Therefore, ELLAM schemes are expected to perform very well for porous medium flow problems, which has in fact been confirmed in [19]. (iii) In this paper the ELLAM scheme is compared with the ENO/WENO schemes in terms of linear transport PDEs. The latter were originally developed for nonlinear hyperbolic conservation laws and have been successful in resolving shock discontinuities and complex solution structures. The investigation on the performance of the methods for nonlinear hyperbolic conservation laws are beyond the scope of this paper.

References

[1] Al-Lawatia, M., Sharpley, R. C., and Wang, H., Second-order characteristic methods for advection-diffusion equations and comparison to other schemes, *Advances in Water Resources* **22** (1999), 741–768.

[2] Barrett, J. W. and Morton, K. W., Approximate symmetrization and Petrov-Galerkin methods for diffusion-convection problems, *Comp. Meth. Appl. Mech. Engrg.* **45** (1984), 97–122.

[3] Bear, J., *Hydraulics of Groundwater*, McGraw-Hill, New York, 1979

[4] Bouloutas, E. T. and Celia, M. A., An improved cubic Petrov-Galerkin method for simulation of transient advection-diffusion processes in rect-

angularly decomposable domains, *Comp. Meth. Appl. Mech. Engrg.* **91** (1991), 289–308.

[5] Celia, M. A., Russell, T. F., Herrera, I., Ewing, R. E., An Eulerian-Lagrangian localized adjoint method for the advection-diffusion equation, *Adv. Wat. Res.* **13** (1990), 187–206

[6] Christie, I., Griffiths, D. F., Mitchell, A. R., and Zienkiewicz, O. C., Finite element methods for second order differential equations with significant first derivatives, *Int. J. Num. Engrg.* **10** (1976), 1389–1396.

[7] Colella, P., A direct Eulerian MUSCL scheme for gas dynamics, *SIAM J. Sci. Stat. Comp.* **6** (1985), 104–117.

[8] Ewing, R. E. (ed.), *The Mathematics of Reservoir Simulation*, Research Frontiers in Applied Mathematics. **1**, SIAM, Philadelphia, 1984

[9] Ewing, R. E. and Wang, H., Eulerian-Lagrangian localized adjoint methods for linear advection equations, *Computational Mechanics*, Springer International, 1991, pp. 245–250.

[10] Ewing, R. E. and Wang, H., An optimal-order error estimate to Eulerian-Lagrangian localized adjoint method for variable-coefficient advection-reaction problems, *SIAM Num. Anal.* **33** (1996), 318–348.

[11] Harten, A., Engquist, B., Osher, S., and Chakravarthy, S., Uniformly high order accurate essentially nonoscillatory schemes, III, *J. Comp. Phys.* **71** (1987), 231–241.

[12] Harten, A. and Osher, S., Uniformly high-order accurate non-oscillatory schemes, I, *SIAM J. Num. Anal.* **24** (1987), 279–309.

[13] Hughes, T. J. R. and Brooks, A. N., A multidimensional upwinding scheme with no crosswind diffusion, Hughes (ed.), *Finite Element Methods for Convection Dominated Flows* **34**, ASME, New York, 1979.

[14] Johnson, C. and Pitkäranta, J., An analysis of discontinuous Galerkin methods for a scalar hyperbolic equation, *Math. Comp.* **46** (1986), 1–26.

[15] Liu, X.-D., Osher, S., and Chan, T., Weighted essentially nonoscillatory schemes, *J. Comput. Phys.* **115** (1994), 200–212.

[16] Richter, G. R., An optimal-order error estimate for the discontinuous Galerkin method, *Math. Comp.* **50** (1988), 75–88.

[17] Russell, T. F. and Trujillo, R. V., Eulerian-Lagrangian localized adjoint methods with variable coefficients in multiple dimensions, Gambolati, *et al.* (ed.), Computational Methods in Surface Hydrology, Springer-Verlag, Berlin, 1990, 357–363.

[18] van Leer, B., On the relation between the upwind-differencing schemes of Godunov, Engquist-Osher, and Roe, *SIAM J. Sci. Stat. Comp.* **5** (1984), 1–20.

[19] Wang, H., Liang, D., Ewing, R. E., Lyons, S. L., and Qin, G., An accurate approximation to compressible flow in porous media with wells, this volume.

[20] Wang, H., Ewing, R. E., Qin, G., Lyons, S. L., Al-Lawatia, M, and Man, S, A family of Eulerian-Lagrangian localized adjoint methods for multi-dimensional advection-reaction equations, *J. Comp. Phys.* **152** (1999), 120–163.

An Accurate Approximation to Compressible Flow in Porous Media with Wells

Hong Wang   Dong Liang   Richard E. Ewing
Stephen L. Lyons   Guan Qin

Abstract

An Eulerian-Lagrangian localized adjoint method (ELLAM) is presented for compressible flow occurring in compressible porous media with wells. The ELLAM scheme symmetrizes the governing transport equation, greatly eliminates non-physical oscillation and/or excessive numerical dispersion present in many large-scale simulators widely used in industrial applications, and conserves mass. Computational experiments show that the ELLAM scheme can accurately simulate incompressible and compressible fluid flows in porous media with wells, even though coarse spatial grids and very large time steps, which are one or two orders of magnitude larger than those used in many numerical methods, are used. The ELLAM scheme can treat large mobility ratios, discontinuous permeabilities and porosities, anisotropic dispersion in tensor form, and wells.

KEYWORDS: characteristic method, compressible flow, Eulerian-Lagrangian methods, porous medium flow, wells

1   A Mathematical Model

The objective of subsurface fluid flow modeling is to simulate complex fluid flow processes occurring in subsurface porous media sufficiently well to optimize the recovery of hydrocarbon or to accurately predict and thoroughly remediate the contamination in groundwater transport processes. In order to do this, one must build mathematical models to describe the essential phenomena and the fundamental laws, and design numerical methods to discretize these models and to represent the basic features as well as possible without introducing serious nonphysical phenomena.

Let $p(\mathbf{x}, t)$ and $\mathbf{u}(\mathbf{x}, t)$ be the pressure and the Darcy velocity of a fluid mixture, and $c(\mathbf{x}, t)$ be the concentration of an invading fluid or concerned solute/solvent in the fluid mixture. The equation of mass conservation for the fluid mixture and Darcy's law lead to the following coupled system of partial differential equations (PDEs) that describes fluid flow processes in a porous medium reservoir Ω with injection and production wells [2, 5, 9]

$$
\begin{aligned}
\frac{\partial}{\partial t}(\phi\rho) + \nabla \cdot (\rho\mathbf{u}) &= q, & \mathbf{x} \in \Omega, \ t \in (0, T], \\
\mathbf{u} &= -\frac{\mathbf{K}}{\mu(c)}(\nabla p - \rho g \nabla d), & \mathbf{x} \in \Omega, \ t \in (0, T].
\end{aligned}
\tag{1.1}
$$

In many cases, the thickness of the medium is significantly smaller than its length and width. Hence, it is reasonable to average the medium properties vertically and to assume $\Omega \subset \mathbb{R}^2$ with a nonuniform local elevation. $\mathbf{K}(\mathbf{x})$ is the permeability tensor of the medium, $\mu(c)$ is the concentration-dependent viscosity of the fluid mixture, which is determined by some mixing rule

$$
\mu(c) = \mu_o[(1-c) + M^{\frac{1}{4}}c]^{-4}
\tag{1.2}
$$

where μ_o is the viscosity of the resident fluid and M is the mobility ratio. ρ is the density of the fluid mixture, g is the magnitude of gravitational acceleration, $d(\mathbf{x})$ is the reservoir depth, $q(\mathbf{x}, t)$ is a source and sink term that accounts for the effect of injection and production wells, ϕ is the porosity of the medium (proportion of volume available to porous medium flows).

For a fluid of constant compressibility c_ρ, the following equation of state

$$
\rho = \rho_r \exp(c_\rho(p - p_r))
\tag{1.3}
$$

holds, where ρ_r is the reference density at the reference pressure p_r. Eq. (1.3) and its simplified versions have been widely used in modeling subsurface contaminant transport and remediation in the hydro-science community. It can also be applied to compressible fluid flow processes in reservoir simulation, unless the fluids contain large quantities of dissolved gas [2, 9].

Due to the effect of large pressure changes involved in porous medium fluid flow processes and the type of the medium of the reservoir, the porous medium can deform. Let $c_\phi(\mathbf{x})$ be the compressibility of the medium. The following constitutive relation is often used to model the porosity ϕ [1]

$$
\phi = \phi_r(\mathbf{x}) \exp(c_\phi(\mathbf{x})(p - p_r)),
\tag{1.4}
$$

where ϕ_r is the porosity of the medium at the reference pressure p_r.

Incorporating Eqs. (1.3) and (1.4) into the system (1.1), and introducing the mass flow rate $\boldsymbol{\sigma} = \rho\mathbf{u}$ as a primary variable, we obtain the following system of PDEs for the pressure p and the mass flow rate $\boldsymbol{\sigma}$

$$S_p(\mathbf{x}, p)\frac{\partial p}{\partial t} + \nabla \cdot \boldsymbol{\sigma} = q, \qquad \mathbf{x} \in \Omega, \; t \in (0, T],$$

$$\boldsymbol{\sigma} = -\frac{\rho \mathbf{K}}{\mu(c)}(\nabla p - \rho g \nabla d), \qquad \mathbf{x} \in \Omega, \; t \in (0, T], \qquad (1.5)$$

where $S_p(\mathbf{x}, p)$ is the storage term defined by

$$S_p(\mathbf{x}, p) = \frac{\partial(\phi\rho)}{\partial p} = \rho\phi(\mathbf{x}, p)(c_\phi(\mathbf{x}) + c_\rho). \qquad (1.6)$$

The equation of mass conservation for the concerned component can be expressed in terms of the mass flow rate $\boldsymbol{\sigma}$ as

$$\frac{\partial(\phi\rho c)}{\partial t} + \nabla \cdot (\boldsymbol{\sigma} c - \mathbf{D}(\boldsymbol{\sigma}, p)\nabla c) = c^* q, \qquad \mathbf{x} \in \Omega, \;\; t \in (0, T], \qquad (1.7)$$

where c^* is a prescribed concentration at sources or is equal to c at sinks, $\mathbf{D}(\boldsymbol{\sigma}, p)$ is the diffusion-dispersion tensor that consists of molecular diffusion and (anisotropic velocity-dependent) mechanical dispersion

$$\mathbf{D}(\boldsymbol{\sigma}, p) = d_m \phi\rho \, \mathbf{I} + d_t|\boldsymbol{\sigma}| \, \mathbf{I} + \frac{d_l - d_t}{|\boldsymbol{\sigma}|} \begin{pmatrix} \sigma_x^2 & \sigma_x\sigma_y \\ \sigma_x\sigma_y & \sigma_y^2 \end{pmatrix}, \qquad (1.8)$$

$\boldsymbol{\sigma} = (\sigma_x, \sigma_y)$, d_m is the molecular diffusion coefficient, $\mathbf{I}$ is the identity tensor, and d_t and d_l are the transverse and longitudinal dispersivities, respectively.

System (1.5) and (1.7) needs to be closed by the initial and boundary conditions. In petroleum reservoir simulation the boundary $\partial\Omega$ is typically impermeable, leading to no-flow boundary conditions of the form [5, 9]

$$\boldsymbol{\sigma} \cdot \mathbf{n} = 0, \quad (\mathbf{D}(\boldsymbol{\sigma}, p)\nabla c) \cdot \mathbf{n} = 0, \quad (\mathbf{x}, t) \in \partial\Omega \times [0, T]. \qquad (1.9)$$

These conditions also arise in environmental modeling although other types of boundary conditions are possible [2]. For simplicity, we assume boundary conditions (1.9) and a rectangular domain $\Omega = (a_x, b_x) \times (a_y, b_y)$ [2, 9].

If the fluid and the medium are incompressible, $\rho = $ constant and $\phi = \phi(\mathbf{x})$. The system (1.5) and (1.7) is reduced to the mathematical model for incompressible fluid flow, which has been widely used previously [5, 4, 6].

Because diffusion or dispersion is often a small phenomenon relative to advection, Eq. (1.7) is an advection-diffusion equation with advection being the dominant phenomenon. Additional features of (1.5) and (1.7) include the singularities of the solutions at wells, discontinuous permeabilities and porosities, a large adverse mobility ratio in the flow processes that could cause viscous fingering phenomena, anisotropic dispersion in tensor form, as well as the enormous size of field-scale applications.

2   An ELLAM Scheme

We use a sequential decoupling and linearization technique for system (1.5) and (1.7), and a mixed finite element method to solve p and σ from (1.5) [5, 6]. For simplicity, we describe only an ELLAM scheme for Eq. (1.7) assuming that the pressure p and the mass flow rate σ in (1.7) are known.

The ELLAM was originally introduced by Celia, Russell, Herrera, and Ewing for the solution of (one-dimensional constant-coefficient) advection-diffusion PDEs [3, 8]. Let $0 = t_0 < t_1 < \ldots < t_n < \ldots < t_{N-1} < t_N = T$ be a partition of the time interval $[0, T]$ with $\Delta t_n = t_n - t_{n-1}$. In the ELLAM formulation, we multiply (1.7) by space-time test functions w that are continuous and piecewise smooth, vanish outside the space-time strip $\overline{\Omega} \times (t_{n-1}, t_n]$, and are discontinuous in time at time t_{n-1}. This yields a space-time weak formulation

$$
\int_\Omega \phi \rho c(\mathbf{x}, t_n) w(\mathbf{x}, t_n) d\mathbf{x} + \int_{t_{n-1}}^{t_n} \int_\Omega \nabla w(\mathbf{y}, \theta) \cdot \mathbf{D}(\sigma, p) \nabla c(\mathbf{y}, \theta) d\mathbf{y} d\theta
$$
$$
- \int_{t_{n-1}}^{t_n} \int_\Omega c(\mathbf{y}, \theta) \left[\phi \rho \frac{\partial w(\mathbf{y}, \theta)}{\partial \theta} + \sigma \cdot \nabla w(\mathbf{y}, \theta) \right] d\mathbf{y} d\theta \qquad (2.1)
$$
$$
= \int_\Omega \phi \rho\, c(\mathbf{x}, t_{n-1}) w(\mathbf{x}, t_{n-1}^+) d\mathbf{x} + \int_{t_{n-1}}^{t_n} \int_\Omega (c^* q w)(\mathbf{y}, \theta) d\mathbf{y} d\theta.
$$

Here $w(\mathbf{x}, t_{n-1}^+) = \lim\limits_{t \to t_{n-1}, t > t_{n-1}} w(\mathbf{x}, t)$ to take into the fact that $w(\mathbf{x}, t)$ is discontinuous in time at time t_{n-1}.

Careful analysis of various operator splittings in the ELLAM framework concludes that the test functions $w(\mathbf{y}, \theta)$ in (2.1) should be chosen to satisfy the hyperbolic part of the adjoint equation of (1.7) (e.g., see [3])

$$
\phi \rho \frac{\partial w}{\partial \theta}(\mathbf{y}, \theta) + \sigma \cdot \nabla w(\mathbf{y}, \theta) = 0, \quad \mathbf{y} \in \Omega, \quad \theta \in [t_{n-1}, t_n]. \qquad (2.2)
$$

Equation (2.2) implies that the test functions $w(\mathbf{y}, \theta)$ should be constant along the characteristics $\mathbf{y} = \mathbf{r}(\theta; \mathbf{x}, t_n)$, defined by the differential equation

$$
\frac{d\mathbf{r}}{d\theta} = \frac{\sigma}{\phi \rho}, \quad \theta \in [t_{n-1}, t_n],
$$
$$
\mathbf{r}(\theta; \mathbf{x}, t)\Big|_{\theta = t} = \mathbf{x}. \qquad (2.3)
$$

In the ELLAM scheme, we choose the test functions $w(\mathbf{x}, t_n)$ to be piecewise-bilinear functions for $\mathbf{x} \in \overline{\Omega}$ at time t_n and define them by constant extension along the characteristics $\mathbf{r}(\theta; \mathbf{x}, t_n)$ to the space-time strip $\overline{\Omega} \times (t_{n-1}, t_n]$.

We enforce the Euler quadrature at time t_n to evaluate the source and sink term in Eq. (2.1). Note that for any $(\mathbf{y}, \theta) \in \Omega \times [t_{n-1}, t_n]$, there exists an $\mathbf{x} \in \Omega$ such that $\mathbf{y} = \mathbf{r}(\theta; \mathbf{x}, t_n)$. We obtain

$$
\int_{t_{n-1}}^{t_n} \int_{\Omega} c^*(\mathbf{y}, \theta) q(\mathbf{y}, \theta) w(\mathbf{y}, \theta) \, d\mathbf{y} d\theta
$$
$$
= \int_{\Omega} \int_{t_{n-1}}^{t_n} c^*(\mathbf{r}(\theta; \mathbf{x}, t_n), \theta) q(\mathbf{r}(\theta; \mathbf{x}, t_n), \theta) w(\mathbf{x}, t_n) \left| \frac{\partial(\mathbf{r}, \theta)}{\partial(\mathbf{x}, t_n)} \right| d\theta d\mathbf{y} \quad (2.4)
$$
$$
= \Delta t_n \int_{\Omega} c^*(\mathbf{x}, t_n) q(\mathbf{x}, t_n) w(\mathbf{x}, t_n) d\mathbf{x} + E_q(c^*, w),
$$

where $\left| \frac{\partial(\mathbf{r}, \theta)}{\partial(\mathbf{x}, t_n)} \right| = 1 + \mathcal{O}(t_n - \theta)$ is the Jacobian of the transformation, $E_q(c^*, w)$ is the local truncation error.

We can evaluate the diffusion-dispersion term similarly and obtain

$$
\int_{t_{n-1}}^{t_n} \int_{\Omega} \nabla w(\mathbf{y}, \theta) \cdot \mathbf{D}(\sigma, p) \nabla c(\mathbf{y}, \theta) \, d\mathbf{y} d\theta
$$
$$
= \Delta t_n \int_{\Omega} \nabla w(\mathbf{x}, t_n) \cdot \mathbf{D}(\sigma, p)(\mathbf{x}, t_n) \nabla c(\mathbf{x}, t_n) d\mathbf{x} + E_{\mathbf{D}}(c, w), \quad (2.5)
$$

where $E_{\mathbf{D}}(c, w)$ is the local truncation error term.

Using Eq. (2.2), we see that the last term on the left-hand side of Eq. (2.1) vanishes along the characteristics $\mathbf{r}(\theta; \mathbf{x}, t_n)$ defined by Eq. (2.3). However, for a general mass flow rate field $\sigma(\mathbf{x}, t)$, porosity $\phi(\mathbf{x}, p)$ and density $\rho(p)$, one cannot analytically solve the problem (2.3) to track $\mathbf{r}(\theta; \mathbf{x}, t_n)$ exactly. Hence, numerical means have to be used to approximate the characteristics. For linear transport PDEs where the fluid velocity is assumed to be a known smooth function, we were able to utilize an Euler or a Runge-Kutta quadrature to track the characteristics and to obtain accurate numerical solutions [11]. For the coupled system (1.5) and (1.7), a semi-analytical tracking algorithm should be used [7]. When a numerical method is used to approximate the characteristics $\mathbf{r}(\theta; \mathbf{x}, t_n)$, the last term on the left-hand side of Eq. (2.1) does not vanish. Nevertheless, it has been proven that dropping this term does not affect the optimal order convergence rate of the ELLAM scheme [10].

In the ELLAM scheme, we substitute (2.4) and (2.5) into Eq. (2.1) and drop $E_q(c^*, w)$, $E_{\mathbf{D}}(c, w)$, and the last term on the left-hand side of Eq. (2.1). We define the trial functions $c(\mathbf{x}, t_n)$ to be piecewise bilinear functions on $\overline{\Omega}$ at time step t_n as in the standard finite element method. Note that the trial functions coincide with the test functions on $\overline{\Omega}$ at time level t_n. But the trial functions c are defined at time step t_n only while the test functions w are defined on the space-time strip $\overline{\Omega} \times (t_{n-1}, t_n]$ by constant extension along characteristics from $\overline{\Omega}$ at time t_n. This leads to the following ELLAM scheme

$$
\begin{aligned}
\int_\Omega &\phi(\mathbf{x}, p(\mathbf{x}, t_n))\rho(p(\mathbf{x}, t_n))c(\mathbf{x}, t_n)w(\mathbf{x}, t_n)d\mathbf{x} \\
&+\Delta t_n \int_\Omega \nabla w(\mathbf{x}, t_n) \cdot \mathbf{D}(\sigma(\mathbf{x}, t_n), p(\mathbf{x}, t_n))\nabla c(\mathbf{x}, t_n)d\mathbf{x} \\
&= \int_\Omega \phi(\mathbf{x}, p(\mathbf{x}, t_{n-1}))\rho(p(\mathbf{x}, t_{n-1}))c(\mathbf{x}, t_{n-1})w(\mathbf{x}, t_{n-1}^+)d\mathbf{x} \\
&+\Delta t_n \int_\Omega c^*(\mathbf{x}, t_n)q(\mathbf{x}, t_n)w(\mathbf{x}, t_n)d\mathbf{x}.
\end{aligned}
\tag{2.6}
$$

By using a characteristic tracking, the ELLAM scheme (2.6) significantly reduces the temporal truncation errors and generates accurate numerical solutions even if very large time steps are used. Moreover, the ELLAM scheme conserves mass [3], which is of essential importance in applications. Furthermore, the ELLAM scheme symmetrizes the transport PDE (1.7) and generates a 9-banded, symmetric and positive definite coefficient matrix. Finally, except for the first term on the right-hand side, all other terms in (2.6) are standard in the finite element method and can be computed in a straightforward manner. In this term, the trial and test functions are actually defined at different time steps. Hence, the evaluation of this term is very challenging and raises various numerical difficulties . We refer interested readers to [11] for detailed implementational issues.

3   Numerical Experiments

In this section we apply the ELLAM scheme (2.6) to solve the system (1.5) and (1.7). The test problems are standard five spot pattern displacements in reservoir simulation. Example 1 is the simulation of an incompressible fluid flow in a rigid homogeneous medium, while Example 2 is the simulation of a compressible fluid flow in a compressible, heterogeneous medium. The (common) data in these experiments is as follows: The spatial domain $\Omega = (0, 1000) \times (0, 1000)$ ft^2, the viscosity of oil $\mu_o = 1.0$ cp, the mobility ratio $M = 41$. The injection well is located at the upper-right corner $(1000, 1000)$ with an injection rate of $q = 30$ ft^2/day and $c^* = 1.0$. The production well is located at the bottom-left corner $(0,0)$ with a production rate of $q = 30$ ft^2/day. A spatial grid of $\Delta x = \Delta y = 50$ ft, and a time step of $\Delta t = 360$ days (one year) are used. Previously, the time steps used in these simulations

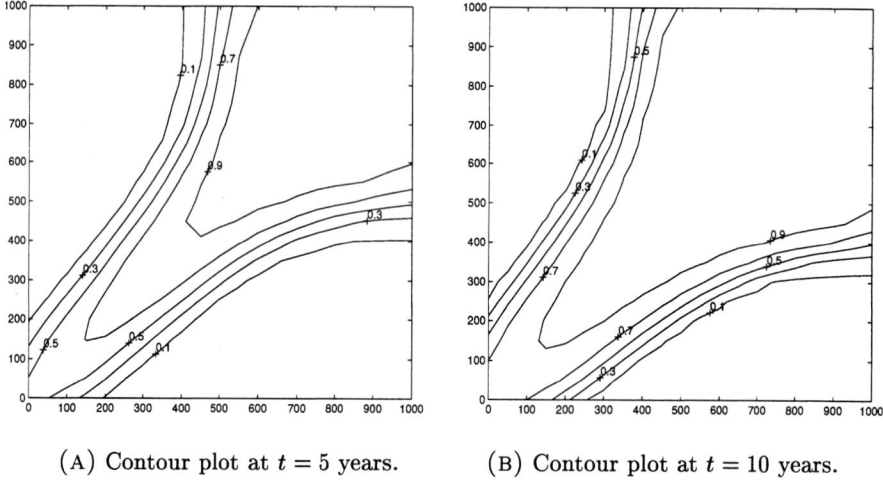

(A) Contour plot at $t = 5$ years.   (B) Contour plot at $t = 10$ years.

Figure 1: The concentration plots of invading component in Example 1.

range from a few days for the upwind finite difference method (UFDM) to one month for the modified method of characteristics (MMOC) [5, 6].

Example 1: Incompressible flow in a rigid homogeneous medium.
In this example run, we choose the remaining data as follows: The porosity $\phi = 0.1$, the permeability coefficients (diagonal entries) are $k_x = k_y = 80$ md, the molecular diffusion is $D_m = \phi d_m = 0$, the longitudinal and transverse dispersions are $D_l = \phi d_l = 4.0$ ft and $D_t = \phi d_t = 0.4$ ft, respectively. The initial concentration is $c_0(x, y) = 0$. The contour plots for the concentration of the invading fluid at 5 and 10 years are presented in Fig. 1(a) and 1(b).

Example 2: Compressible flow in a compressible, heterogeneous medium. In this example run, the compressibility of the medium $c_\phi = 0.000001$, the reference density $\rho_r = 0.8$ g/cm^3 $= 49.942$ lb/ft^3, the compressibility of the fluid $c_\rho = 0.0001$, the reference pressure $p_r = 1$ atm. $= 14.696$ psia, the initial pressure $p_0(x, y) = 3000$ psia, and the initial concentration $c_0(x, y) = 0.0$. The subdomain $\Omega^{(1)} = (150, 600) \times (150, 600)$ ft^2. The remaining parameters on this subdomain are as follows: The reference porosity $\phi_r = 0.09$, the permeability coefficients $k_x = k_y = 35$ md, the molecular diffusion coefficient $D_m = 0.0$ ft^2/day, the longitudinal and transverse dispersions are $D_l = 3.6$ ft and $D_t = 0.36$ ft, respectively. On the subdomain $\Omega^{(2)} = \Omega - \Omega^{(1)}$, the reference porosity $\phi_r = 0.1$, the permeability coefficients $k_x = k_y = 80$ md, the molecular diffusion coefficient $D_m = 0.0$ ft^2/day, the

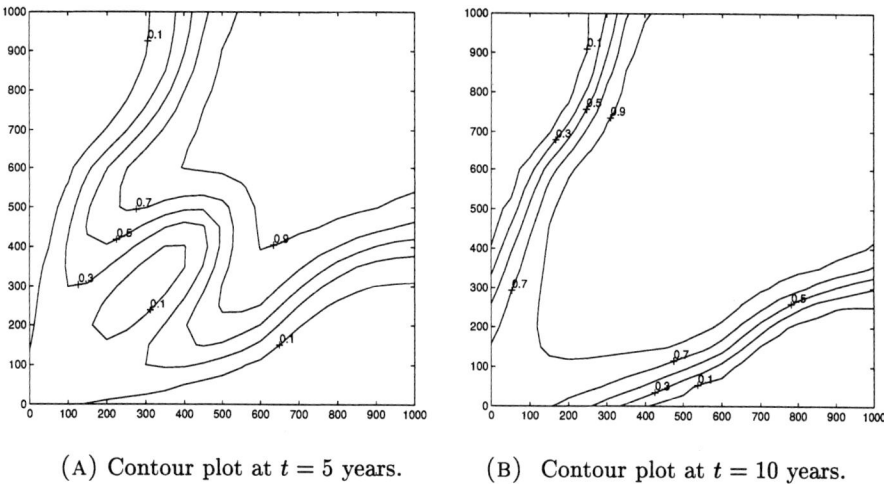

(A) Contour plot at $t = 5$ years. (B) Contour plot at $t = 10$ years.

Figure 2: The concentration plots of invading component in Example 2.

longitudinal and transverse dispersions are $D_l = 4$ ft and $D_t = 0.4$ ft, respectively. The contour plots for the concentration of the invading fluid at 5 and 10 years are presented in Fig. 2(a) and 2(b).

These results show that the ELLAM scheme can accurately simulate incompressible and compressible fluid flows in porous media with wells, even though coarse spatial grids and very large time steps, which are one or two orders of magnitude larger than those used in many numerical methods, are used, implying a significantly improved efficiency and accuracy. The EL-LAM scheme can treat large mobility ratios, discontinuous permeabilities and porosities, anisotropic dispersion in tensor form, and wells. The ELLAM scheme conserves mass.

References

[1] Aziz, H. and Settari, A., *Petroleum Reservoir Simulation*, Applied Science Publishers, 1979.

[2] Bear, J., *Hydraulics of Groundwater*, McGraw-Hill, New York, 1979.

[3] Celia, M. A., Russell, T. F., Herrera, I., and Ewing, R. E., An Eulerian-Lagrangian localized adjoint method for the advection-diffusion equation, *Advances in Water Resources* **13** (1990), 187–206.

[4] Douglas, J., Jr., Ewing, R. E., and Wheeler, M. F., A time-discretization procedure for a mixed finite element approximation of miscible displacement in porous media, *RARIO* **17** (1983), 249–265.

[5] Ewing, R. E. (ed.), *The Mathematics of Reservoir Simulation*, Research Frontiers in Applied Mathematics, **1**, SIAM, Philadelphia, 1984.

[6] Ewing, R. E., Russell, T. F., and Wheeler, M. F., Simulation of miscible displacement using mixed methods and a modified method of characteristics, SPE 12241 (1983), 71–81.

[7] Healy, R. W. and Russell, T. F., A finite-volume Eulerian-Lagrangian localized adjoint method for solution of the advection-dispersion equation, *Water Resources Research* **29** (1993), 2399–2413.

[8] Herrera, I., Ewing, R. E., Celia, M. A., and Russell, T. F., Eulerian-Lagrangian localized adjoint methods: the theoretical framework, *Numerical Methods for Partial Differential Equations* **9** (1993), 431–458.

[9] Peaceman, D. W., *Fundamentals of Numerical Reservoir Simulation*, Elsevier, Amsterdam, 1977.

[10] Wang, H., A family of ELLAM schemes for advection-diffusion-reaction equations and their convergence analyses, *Numerical Methods for Partial Differential Equations* **14** (1998), 739–780.

[11] Wang, H., Ewing, R. E., Qin, G., Lyons, S. L., Al-Lawatia, M, and Man, S, A family of Eulerian-Lagrangian localized adjoint methods for multi-dimensional advection-reaction equations, *J. Comput. Physics* **152** (1999), 120–163.

Fast Convergent Algorithms for Solving 2D Integral Equations of the First Kind

YAN-FEI WANG   TING-YAN XIAO

Abstract

Based on Tikhonov's regularization method, this paper applies two fast convergent algorithms developed by the authors to solving 2D integral equations of the first kind. The procedures of discretization and regularization are discussed. The numerical tests are presented to show high efficiency and numerical stability. The integral equations of the first kind can be seen in determination of capillary pressure functions.

KEYWORDS: regularization method, fast convergent algorithm, 2D integral equation of the first kind

1 Introduction

In many application areas, such as determination of capillary pressure functions in porous media, signal reconstruction, image restoration, and operator identification [1, 2, 5, 3], we encounter integral equaitons of the first kind. As a genaral framework of dealing with inverse problems, equations of this kind are inherently ill-posed and solving them is time consuming. Thus some kind of regularization methods must be employed. So far, many efforts have been put to the 1D case [14, 9, 4, 10, 8, 13, 16], while few successful results have been obtained for the 2D case. The main difficulty lies in the huge amount of computation time in the iterative process of choosing an appropriate regularization parameter. This results in need for the study of fast convergent algorithms.

Let us consider the general 2D integral equation of the first kind

$$Az = \int_{a_1}^{b_1} \int_{c_1}^{d_1} K(x,y;\xi,\eta)z(\xi,\eta)d\xi d\eta = u(x,y), \ (x,y) \in [a_2,b_2] \times [c_2,d_2].$$

(1.1)

The kernal $K(\cdot, \cdot)$ is continuous and $A : W_2^1[a_1, b_1; c_1, d_1] \rightarrow L^2[a_2, b_2; c_2, d_2]$ is a bounded linear and compact operator. As known, when the range of A, i.e., $R(A)$, is not closed, (1.1) is ill-posed.

Generally speaking, the right-side member u and the operator A of (1.1) are roughly known. Suppose that the approximate version of the input data (A, u) is (A_h, u_δ), with the error

$$\|A_h - A\| \leq h, \ \|u - u_\delta\| \leq \delta, \quad h, \delta \geq 0. \tag{1.2}$$

Due to the ill-posedness of this problem, we cannot take z_η given by

$$z_\eta = (A_h)^+ u_\delta, \quad \eta = (h, \delta) \tag{1.3}$$

as the stable approximation to the exact solution of the problem (1.1):

$$z_T = A^+ u, \tag{1.4}$$

where A^+ and A_h^+ denote the generalized inverses of the related operators. To get a stable approximation of the generalized solution of $A^+ u_\delta$ (or $A_h^+ u_\delta$), we employ Tikhonov's regularization method, which consists of the steps

1. Constructing the regularized operator $R(u, \alpha)$;
2. Selecting the regularization parameter $\alpha = \alpha(\delta)$ (or $\alpha = \alpha(\delta, h)$) matching the error $\eta = (\delta, h)$ of the input data.

The regularized solution $R(u_\delta, \alpha)$ can be obtained by minimizing the smoothing functional

$$M^\alpha[z, u_\delta] = \|Az - u_\delta\|^2 + \alpha\Omega[z], \quad \alpha > 0, \tag{1.5}$$

or

$$M^\alpha[z, u_\delta] = \|A_h z - u_\delta\|^2 + \alpha\Omega[z], \quad \alpha > 0. \tag{1.6}$$

Obviously, the minimizer z_δ^α or z_η^α must satisfy the Euler equation

$$(A^*A + \alpha\Omega')z = A^*u_\delta, \tag{1.7}$$

or

$$(A_h{}^*A_h + \alpha\Omega')z = A_h{}^*u_\delta. \tag{1.8}$$

As a posterior strategy, the regularization parameter should satisfy the so-called discrepancy equation [13, 12]

$$\phi(\alpha) = \|Az_\delta^\alpha - u_\delta\|^2 - \delta^2 = 0, \tag{1.9}$$

or the generalized discrepancy equation [15]

$$\phi(\alpha) = \|A_h z_\eta^\alpha - u_\delta\|^2 - (\delta + h\|z_\eta^\alpha\|)^2 = 0 \tag{1.10}$$

Here A_h^* (or A^*) is the Hilbert-adjoint operator of A_h (or A), $\Omega[\cdot]$ denotes the identity operator or the positively defined, linearly differentiable operator, and $\Omega'[\cdot]$ denotes the Frechet derivative.

Denote the root of (1.9) or (1.10) as $\alpha = \alpha(\delta)$ or $\alpha = \alpha(\eta)$. Then the regularized solution $R(u_\delta, \alpha(\delta))$ or $R(u_\delta, \alpha(\eta))$ is the stable approximation to the exact solution z_T.

The paper is organized as follows. §2 discusses the discretization and regularization of (1.1). §3 presents two fast convergent algorithms for solving (1.9) or (1.10) and gives the related convergence theorem of the algorithms. The numerical tests are presented in §4 to compare the efficiency of the new algorithms with Newton's method.

2 Discretization and Regularization

We use a finite-difference method to discretize (1.1). Let the step sizes in the direction of ξ, η, x, and y be h_1, h_2, h_3, and h_4, respectively. The difference scheme of (1.1) is

$$A z^{h_1 h_2} = u^{h_3 h_4}, \tag{2.1}$$

where

$$h_1 = \frac{b_1 - a_1}{K}, \quad h_2 = \frac{d_1 - c_1}{J}, \quad h_3 = \frac{b_2 - a_2}{L}, \quad h_4 = \frac{d_2 - c_2}{M},$$

K, J, L, and M are given positive integers, and

$$u^{h_3 h_4} \in L^{h_3 h_4} = \{u^{h_3 h_4} | u^{h_3 h_4} = (u_{11}, u_{21}, \ldots, u_{L1}, u_{12}, \ldots, u_{LM})^T\},$$
$$u_{lm} = u(x_l, x_m), \quad l = 1, 2, \ldots, L, \quad m = 1, 2, \ldots, M,$$
$$z^{h_1 h_2} \in W^{h_1 h_2} = \{z^{h_1 h_2} | z^{h_1 h_2} = (z_{11}, z_{21}, \ldots, z_{K1}, z_{12}, \ldots, z_{KJ})^T\},$$
$$z_{kj} = z(\xi_k, \eta_j), \quad k = 1, 2, \ldots, K; \ j = 1, 2, \ldots, J,$$
$$A = (A^{lm})_{L \times M}, \ A^{lm} = (A_{kj}^{lm})_{K \times J},$$
$$A_{kj}^{lm} = \int_{\xi_k}^{\xi_{k+1}} \int_{\eta_j}^{\eta_{j+1}} K(x_l, y_m; \xi, \eta) d\xi d\eta.$$

Define the norms $\| \cdot \|_W^{h_1 h_2}$ and $\| \cdot \|_L^{h_3 h_4}$ by

$$\|z^{h_1 h_2}\|_{W^{h_1 h_2}}^2 = (z^{h_1 h_2}, z^{h_1 h_2})_{W^{h_1 h_2}},$$

$$(z^{h_1 h_2}, r^{h_1 h_2})_{W^{h_1 h_2}} = \sum_{k=1}^{K} \sum_{j=1}^{J} h_1 h_2 z_{kj} \bar{r}_{kj} + \sum_{k=1}^{K} \sum_{j=1}^{J} h_1 h_2 \triangle z_{kj} \triangle \bar{r}_{kj},$$

$$\triangle z_{kj} = \frac{z_{k+1,j} - z_{kj}}{h_1} + \frac{z_{k,j+1} - z_{kj}}{h_2},$$

$$k = 1, 2, \ldots, K, \quad j = 1, 2, \ldots, J, \quad z_{kj}, r_{kj} \in W^{h_1 h_2},$$

$$\|u^{h_3 h_4}\|_{L^{h_3 h_4}}^2 = (u^{h_3 h_4}, u^{h_3 h_4})_{L^{h_3 h_4}},$$

$$(u^{h_3 h_4}, v^{h_3 h_4})_{L^{h_3 h_4}} = \sum_{l=1}^{L} \sum_{m=1}^{M} h_3 h_4 u_{lm} \bar{v}_{lm}, \quad u^{h_3 h_4}, v^{h_3 h_4} \in L^{h_3 h_4},$$

where $\bar{r}_{kj}$ and $\bar{v}_{lm}$ are the conjugates of r_{kj} and v_{lm}, respectively.

We can show that $\|z^{h_1 h_2}\|_{W_2^1}^2$ is the difference expression of $\|z(\xi, \eta)\|_{W_2^1}^2$ given by

$$\|z(\xi, \eta)\|_{W_2^1}^2 = \int_{a_1}^{d_1} \int_{c_1}^{d_1} (|z(\xi, \eta)|^2 + |\frac{\partial z}{\partial \xi} + \frac{\partial z}{\partial \eta}|^2) d\xi d\eta.$$

Then the discrete regularized solution of (1.1) can be obtained by minimizing the discreted Tikhonov functional

$$M^\alpha[z^{h_1 h_2}] = \|Az^{h_1 h_2} - u^{h_3 h_4}\|_{L^{h_3 h_4}}^2 + \alpha \|z^{h_1 h_2}\|_{W^{h_1 h_2}}^2. \tag{2.2}$$

The following theorem is classical.

Theorem 2.1 *For any $u^{h_3 h_4} \in L^{h_3 h_4}$ and any positive number α, there exists a unique element $z_\alpha^{h_1 h_2} \in W^{h_1 h_2}$ such that*

$$M^\alpha[z_\alpha^{h_1 h_2}] = \inf M^\alpha[z^{h_1 h_2}], \tag{2.3}$$

and $z_\alpha^{h_1 h_2}$ satisfy the Euler equation

$$(A^* A + \alpha B) z^{h_1 h_2} = A^* u^{h_3 h_4}, \tag{2.4}$$

where A^* is the adjoint matrix of A and B is in the form

$$B = \frac{h_3 h_4}{h_2} \begin{bmatrix} C & -H & & & & \\ -H & C & -H & & & \\ & -H & C & -H & & \\ & & \ddots & \ddots & \ddots & \\ & & & -H & C & -H \\ & & & & -H & C \end{bmatrix}_{L \times M},$$

$$C = \frac{h_3 h_4}{h_1} \begin{bmatrix} \frac{2(h_1^2+h_2^2)}{h_1^2 h_2^2}+2 & \frac{-1}{h_2^2} & & & & \\ \frac{-1}{h_1^2} & \frac{2(h_1^2+h_2^2)}{h_1^2 h_2^2}+2 & \frac{-1}{h_1^2} & & & \\ \ddots & & \ddots & & \ddots & \\ & & \frac{-1}{h_1^2} & \frac{2(h_1^2+h_2^2)}{h_1^2 h_2^2}+2 & \frac{-1}{h_1^2} & \\ & & & \frac{-1}{h_1^2} & \frac{2(h_1^2+h_2^2)}{h_1^2 h_2^2}+2 \end{bmatrix}_{K \times J},$$

$$H = \begin{bmatrix} \frac{1}{h_2^2} & & & & \\ & \frac{1}{h_2^2} & & & \\ & & \ddots & & \\ & & & \frac{1}{h_2^2} & \\ & & & & \frac{1}{h_2^2} \end{bmatrix}_{K \times J}.$$

For the proof, the first part is mainly based on the completeness of the space $W^{h_1 h_2}$, while the remaining part is a common business of variational calculas.

We can see that B is a 5-diagonal matrix and is of the diagonal dominance, so B is non-singular. Thus (17) can be rewritten as

$$(B^{-1}A^*A + \alpha I)z^{h_1 h_2} = B^{-1}A^* u^{h_3 h_4}. \tag{2.5}$$

It is more convenient to solve (2.5). By the way, (2.4) or (2.5) is a regularization of (2.1). Considering the huge amount of computation, solving it along with the discrepancy equation (1.9) or (1.10) is not easy. This urges us to find efficient methods to solve this problem.

3   Fast Convergent Algorithms

Without loss of generality, we only consider the method for solving equation (1.9). To do this, we first quote some results for the existence of the root and the convergence of the corresponding regularization solution z_δ^α. We define

$$\beta(\alpha; u_\delta) = \|Az_\delta^\alpha - u_\delta\|, \tag{3.1}$$

the discrepancy or discrepancy function of the approximate solution z_δ^α [7], where z_δ^α is the solution of (1.7).

The following theorems are useful in our further considerations.

Theorem 3.1 *Assume that $u \in R(A)$ (the range of A) and the signal-to-noise ratio of the input data u_δ is $\|u_\delta\|/\delta > 1$, i.e.,*

$$\|u - u_\delta\| \leq \delta < \|u_\delta\|. \tag{3.2}$$

Then the function $\alpha \rightarrow \beta(\alpha; u_\delta)$ is continuous and non-decreasing, and δ belongs to the range of the function.

From the above theorem, there exists a unique $\alpha^* = \alpha^*(\delta)$ satisfying the discrepancy equation

$$\beta(\alpha^*(\delta); u_\delta) = \delta, \tag{3.3}$$

or

$$\phi(\alpha^*(\delta)) = 0. \tag{3.4}$$

Theorem 3.2 *Under the same conditions of Theorem 3.1, if the regularization parameter $\alpha = \alpha(\delta)$ is determined by (3.3) or (3.4), then we have*

$$\lim_{\delta \to 0} z_\delta^{\alpha(\delta)} = z_T. \tag{3.5}$$

The proof of the above theorems is analogous to that of the theorems in [7]. In addition, by differentiating the two sides of (1.7) with regard to α directly, we can easily obtain the following theorem.

Theorem 3.3 *For any regularization parameter $\alpha > 0$, the solution z_δ^α of (1.7), along with the function $\phi(\alpha)$, is infinitely differentiable with respect to α and we have*

$$(A^*A + \alpha\Omega')(z_\delta^\alpha)' = -\Omega' z_\delta^\alpha, \tag{3.6}$$

$$(A^*A + \alpha\Omega')(z_\delta^\alpha)^{(k)} = -k\Omega'(z_\delta^\alpha)^{(k-1)}, \quad k = 2, 3, \ldots. \tag{3.7}$$

In principle, there are two ways to improve the efficiency and decrease the amount of computation in dealing with the discrepancy equation (1.9).

1. Try one's best to improve the convergence rate while solving the equation (1.9) by iterative methods. By making full use of the good property of the function $\phi(\alpha)$ (it is infinitely differentiable about α) and the higher-order Taylor's expansion, we may construct algorithms which possess a higher-order convergence rate.

2. Associated with the above tactics, we should decrease the total amount of computation for evaluating the functions $\phi(\alpha)$, $\phi'(\alpha)$, and $\phi''(\alpha)$. This can be accomplished by Cholesky's decomposition, along with three times of the back substitution.

Let us state the iterative formulas of two algorithms.

Given, by truncating the Taylor expantion of $\phi(\alpha)$ after the $(\alpha - \alpha_n)^3$ term at $\alpha = \alpha_n$, we obtain the two iterative formulas.

Iterative formula 1:

$$\alpha_{n+1} = \alpha_n - \frac{2\phi(\alpha_n)}{\phi'(\alpha_n) + sign(\phi'(\alpha_n))\sqrt{\phi'(\alpha_n)^2 - 2\phi(\alpha_n)\phi''(\alpha_n)}}. \tag{3.8}$$

Iterative formula 2:

$$\alpha_{n+1} = \alpha_n - \frac{\phi(\alpha_n)}{\phi'(\alpha_n)} - \frac{\phi''(\alpha_n)}{2\phi'(\alpha_n)}\left(\frac{\phi(\alpha_n)}{\phi'(\alpha_n)}\right)^2. \tag{3.9}$$

Below we see that the iterative formula (3.8) or (3.9) possesses the property of a third-order convergence rate. It should be pointed out that $\phi(\alpha_n)$, $\phi'(\alpha_n)$, and $\phi''(\alpha_n)$ can be computed as follows:

$$\phi'(\alpha) = -2\alpha\left(\frac{dz_\alpha}{d\alpha}, z_\alpha\right), \tag{3.10}$$

$$\phi''(\alpha) = -2\left(\frac{dz_\alpha}{d\alpha}, z_\alpha\right) - 2\alpha\left[\left(\frac{dz_\alpha}{d\alpha}, \frac{dz_\alpha}{d\alpha}\right) + \left(z_\alpha, \frac{d^2z_\alpha}{d\alpha^2}\right)\right], \tag{3.11}$$

where z_α, $dz_\alpha/d\alpha$, and $d^2z_\alpha/d\alpha^2$ can be obtained by solving a series of equations

$$(A^*A + \alpha_k\Omega')z_{\alpha_k} = A^*u_\delta, \tag{3.12}$$

$$(A^*A + \alpha_k\Omega')z'_{\alpha_k} = -\Omega'z_{\alpha_k}, \tag{3.13}$$

$$(A^*A + \alpha_k\Omega')z''_{\alpha_k} = -2\Omega'z'_{\alpha_k}. \tag{3.14}$$

From (3.12)–(3.14) we can see that their differences only lie in the right-hand side. This hints that, in one cycle of the iterations, by using Cholesky's decomposition only once with three times of back substitutions, the vectors z_{α_k}, z'_{α_k}, and z''_{α_k} are obtained. In this way, the computation time is greatly saved.

Assume that the operator equation of the first kind (1.1) has been discretized and regularized. Then the iterative scheme can be given as follows.

Algorithm 1.

step 1: input α_0, δ, ϵ, A, and u, and set $k = 0$;

step 2: Give a decomposition: $A^TA + \alpha_k\Omega' = L_kL_k^T$;

step 3: Solve the equation: $L_ky = A^Tu_\delta$, $L_k^Tz^{\alpha_k}_\delta = y$;

step 4: Compute $\phi(\alpha_k)$, if $|\phi(\alpha_k)| < \epsilon$; then go to step 10;

step 5: Solve the equation: $L_ky = -z^{\alpha_k}_\delta$, $L_k^T(z^{\alpha_k}_\delta)' = y$;

step 6: Compute $\phi'(\alpha_k)$, set $d_k = \frac{\phi(\alpha_k)}{\phi'(\alpha_k)}$, and do the Newton step iteration

$$\alpha_k := \alpha_k - d_k; \tag{3.15}$$

step 7: Solve the equation: $L_k y = -2(z_\delta^{\alpha_k})'$, $L_k^T (z_\delta^{\alpha_k})'' = y$;

step 8: Compute $\phi''(\alpha_k)$ and set $\rho_k = \phi'(\alpha_k)^2 - 2\phi(\alpha_k)\phi''(\alpha_k)$; if $\rho_k < 0$, then go to step 9; else, proceed with the another iteration

$$\alpha_k := \alpha_k - \frac{2\phi(\alpha_k)}{\phi'(\alpha_k) + sgn(\phi'(\alpha_k))\rho_k^{\frac{1}{2}}}; \tag{3.16}$$

step 9: Set $k:=k+1$ and go to step 2;

step 10: Set $\alpha^* = \alpha_k$ and $z_\delta^{\alpha^*} = z_\delta^{\alpha_k}$; stop.

Algorithm 2.

steps 1–7, 9, and 10 are the same as in Algorithm 1; step 8 is changed as

step 8:

$$\alpha_k := \alpha_k - d_k^2 \frac{\phi''(\alpha_k)}{2\phi'(\alpha_k)}. \tag{3.17}$$

For the above two algorithms, we have the following convergence theorem.

Theorem 3.4 *Assume that $\phi(\alpha)$ has a root at $\alpha = \alpha^*$: $\phi(\alpha^*) = 0$. Then $\exists \epsilon > 0$ such that, $\forall \alpha_0 \in U(\alpha^*, \epsilon)$, the sequence $\{\alpha_k\}_{k=1}^\infty$ generated by Algorithms 1 and 2 converges at a third-order rate to α^*.*

For the proof of the theorem, see [16].

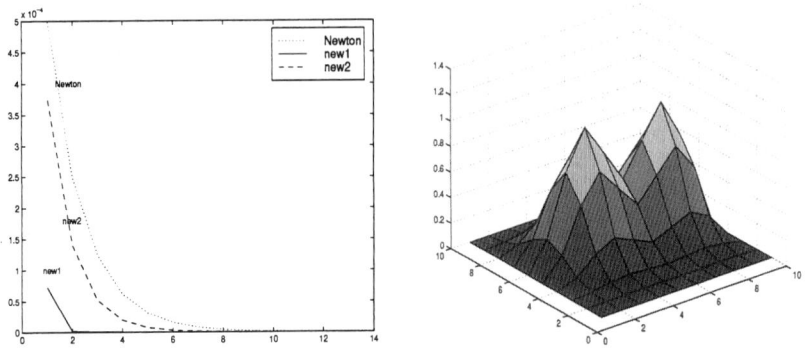

Figure 1:   Convergent rate comparison and exact solution.

Notice that after the decomposition is obtained, the amount of computing the kth order derivative $\phi^{(k)}(\alpha)$ is $O(n)$. Thus, in solving equation $Cx = b$ and computing $\phi(\alpha)$, $\phi'(\alpha)$, $\phi''(\alpha)$, their total amount of computation is about $n^3/6$. Thus, by employing our new algorithms, the convergence rate is greatly improved and the CPU time is also greatly saved. Moreover, if we let N denote the iterative steps of Newton's method and M denote the iterative steps of the new algorithms, then as the order number n is sufficiently large, the ratio of the amounts of computation (so is the CPU time) is near to N/M.

For (1.10), our algorithms are also applicable. It needs only some modifications of the expression $\phi'(\alpha)$ and $\phi''(\alpha)$.

4 Numerical Experiments and Conclusions

To compare the efficiency, we implement numerical experiments. Let us denote the Newton method as Newton and let new1 and new2 represent Algorithm 1 and Algorithm 2 based on (3.8) and (3.9), respectively. The numerical experiments are done with MATLAB on an IBM-PC machine.

Consider the integral equation (see [15])

$$Az = \int_0^1 \int_0^1 exp\{-80[(x-s-0.5)^2 + (y-t-0.5)^2]\}z(s,t)dadt = g(x,y),$$
(4.1)

where the kernal is $K(x,y;s,t) = exp\{-80[(x-s-0.5)^2 + (y-t-0.5)^2]\}$, the interval is $[0,1] \times [0,1]$, and $x,y \in (0,2)$. The exact solution is

$$\bar{z}(s,t) = \left(\frac{e^{-\frac{(s-0.3)^2}{0.03}} + e^{-\frac{s-0.7)^2}{0.03}}}{0.955040800} - 0.052130913 \right) e^{-\frac{(t-0.3)^2}{0.03}}.$$

We use the compound trapezoid formula to discretize (4.1). Suppose that $D = \{(s,t)|0 \le s \le 1, 0 \le t \le 1\}$ and the related points are equally spaced points:

$$s_i = ih_1, \ h_1 = 1/K, \ i = 0,1,\ldots,K, t_j = jh_2, \ h_2 = 1/J, \ j = 0,1,\ldots,J.$$

Then a series of small rectangle are

$$D_{ij} = \{(s,t)|s_i \le s_{i+1}, t_j \le t_{j+1}\},$$

where $i = 0,1,\ldots,K$ and $j = 0,1,\cdots,J$.

Applying the trapezoid integral formula on each sub-rectangle, then the compound trapezoid formula of 2D is found

$$\sum_{l=0}^L \sum_{m=0}^M A_{KJ}z(s_l,t_m) = g(x_l,y_m),$$
(4.2)

where

$$A_{KJ} = \frac{1}{4KJ} \sum_{i=0}^K \sum_{j=0}^J p_{ij}K(x_l,y_m;s_i,t_j),$$

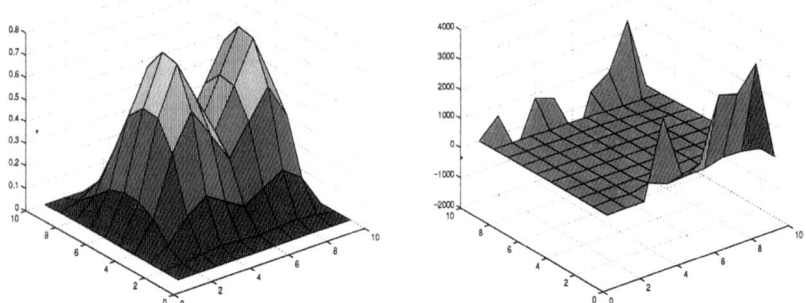

Figure 2.   Computed solutions.

and p_{ij}, $i = 0, 1, 2, \ldots, K$, $j = 0, 1, 2, \ldots, J$, are the elements of matrix P:
$P = (p_{ij})_{(K+1) \times (J+1)}$, with

$$
P = \begin{bmatrix}
1 & 2 & 2 & \ldots & 2 & 2 & 1 \\
2 & 4 & 4 & \ldots & 4 & 4 & 2 \\
2 & 4 & 4 & \ldots & 4 & 4 & 2 \\
\vdots & \vdots & \vdots & \ldots & \vdots & \vdots & \vdots \\
2 & 4 & 4 & \ldots & 4 & 4 & 2 \\
2 & 4 & 4 & \ldots & 4 & 4 & 2 \\
1 & 2 & 2 & \ldots & 2 & 2 & 1
\end{bmatrix}.
$$

So the discrete matrix is

$$
(A_{KJ})_{L \times M} z^{h_1 h_2} = g^{h_3 h_4}. \tag{4.3}
$$

Due to the ill-posedness of (4.1), $(A_{KJ})_{L \times M}$ is ill-conditioned. Thus we should employ a regularization method to get its stable numerical solution. In addition, to compare the stability of the above three algorithms, we give a strong perturbation to the right-hand member of (4.3):

$$
g_{\delta_k}^{h_3 h_4} = g_k^{h_3 h_4} + g_k^{h_3 h_4} \delta sin(\omega t_k),
$$

where $\omega = 500$. Let $\delta = 5.9161e - 6$ and $\alpha_0 = 0.001$. The results are shown in Figs. 1 and 2 and Table 1.

The left figure in Fig. 1 is the convergent rate comparison of three algorithms and illustrates that new1 and new2 converge to the regularized solution more rapidly than Newton. Meanwhile new1 is the fastest one. Under the case that $K = J = L = M = 35$, the concrete computed results of the three algorithms are shown in Table 1.

The numerical test also indicates that the three algorithms all converge well. They are not sensitive to the perturbation of the right-hand member. This can be seen from the right figure in Fig. 1 (the exact solution of the integral equation with disturbing right-hand members of new1 method) and the left figure in Fig. 2 (the computed solution of the integral equation with disturbing right-hand members of new1 method). The right figure in Fig. 2 (the computed solution of the integral equation with disturbing right-hand members without using the regularization method) illustrates that without using the regularization method, the computed solution cannot be accepted.

Algorithms	iterative steps	time consuming	α^*	relative error
Newton	13	1.04	1.2502e-6	2.6440e-7
*new*1	4	0.38	1.2502e-6	2.6438e-7
*new*2	10	0.99	1.2502e-6	2.6438e-7

Table 1. Comparison of three algorithms at $K = J = L = M = 35$.

In summary, the new algorithms for solving 2D integral equations of the first kind in this paper are stable and effective and new1 may be used as a good solver. In the future, we will consider the 3-D case and the nonlinear case.

Acknowledgements . The work was completed with financial support from the Natural Science Foundation of Hebei Province.

References

[1] Chen, W. D., A new extrapolation algorithm for band-limited signals using the regularization method, *IEEE Transactions on Signal Processing*, **41** (1993), 1048–1060.

[2] Chen, Y. M. and Liu, J. Q., A numerical algorithm for remote sensing of thermal conductivity, *J. Comput. Phys.* **43** (1981), 315–328.

[3] Colton, D., Piana, M., and Potthast, R., A simple method using Morozov's discrepancy principle for solving inverse scattering problems, *Inverse Problems* **13** (1997), 1477–1493.

[4] Conference, P., *Ill-Posed Problems in Natural Sciences*, Proceedings of the International Conference Held in Moscow, August 19–25, 1991.

[5] Dou, L. and Hodgson, R., Application of the regularization method to the inverse black radiation problems, *IEEE Transactions on Antennas and Propagation* **40** (1992), 1249–1253.

[6] Eldén, A., An efficient algorithm for the regularization of ill-conditioned least squares problems with triangular Toeplitz matrix, *Siam J. Sci. Static. Comput.* **5** (1984), 220–236.

[7] Groetsch, C. W., *The Theory of Tikhonov Regularization for Fredholm Equations of the First Kind*, Pitman, London, 1984.

[8] Groetsch, C. W., *Inverse Problems in the Mathematical Sciences*, Braunschweig, Wiesbaden, Vieweg, 1993.

[9] Groetsch, C. W. and Engl, H. W. (ed), *Inverse and Ill-Posed Problems*, New York, Academic, 1987.

[10] Hansen, P. C., Numerical tools for analysis and solution of Fredholm integral equations of the first kind, *Inverse Problems* **8** (1992), 849–872.

[11] Lam, P. K. and Eldén, A., Numerical solution of first kind Volterra equations by sequential Tikhonov regularization, *SIAM J. Numer. Anal.* **34** (1997), 1432–1450.

[12] Morozov, V. A., *Methods for Solving Incorrectly Posed Problems*, Springer-Verlag, New York, 1984.

[13] Tikhonov, A. N. and Arsenin, J., *Solutions of Ill-Posed Problems*, Wiley, New York, 1977.

[14] Tikhonov, A. N. and Goncharsky, A. V. (ed), *Ill-Posed Problems in the Natural Sciences*, Moscow, MIR, 1987.

[15] Tikhonov, A. N., Goncharsky, A., and Yagola, A., *Numerical Methods for the Solution of Ill-posed Problems*, Dordrecht, Kluwer, 1995.

[16] Wang, Y.-F. and Xiao, T.-Y., Third-order convergence algorithms for choosing regularization parameter based on discrepancy principle, to appear in *JCM*.

[17] Xiao, T.-Y., On the algorithms for solving the discrepancy equation in ill-posed problems, '96 Symposium on Computation Physics for Chinese Overseas and at Home, June 23-28, 1996, Beijing.

A Two-Grid Finite Difference Method for Nonlinear Parabolic Equations

ZITING WANG   XIANGGUI LI

Abstract

A two-level finite difference scheme is given for the approximation of nonlinear parabolic equations. The analysis of the scheme is given by assuming an implicit time discretization. In this two-level scheme the full nonlinear problem is solved on a coarse grid of size H. The nonlinear term is expanded about the coarse grid solution and an approximate interpolation operator is used to provide values of the coarse grid solution on the fine grid in terms of superconvergent node points. The nonlinear equations analyzed arise in the flow and transport processes of fluids in porous media and the analysis carried out applies to the numerical simulation of these processes.

KEYWORDS: finite difference, two-grid method, nonlinear parabolic equation, error estimate, superconvergence

1 Introduction

We consider a finite difference scheme for the nonlinear parabolic differential equation

$$
\begin{aligned}
\tfrac{\partial p}{\partial t} - \nabla \cdot (K(x,p)\nabla p) = f(t,x) \quad &\text{in} \quad (0,T] \times \Omega, \\
p(0,x) = p^0(x) \quad &\text{in} \quad \Omega, \\
-(K(p)\nabla p) \cdot \nu = g \quad &\text{on} \quad (0,T] \times \Gamma,
\end{aligned}
\tag{1.1}
$$

where Ω is a rectangular domain in $\mathcal{R}^d$ ($d = 1, 2$ or 3) with boundary Γ, ν is the outward unit normal vector on Γ, and $K : \Omega \times \mathcal{R} \to \mathcal{R}^{d \times d}$ is a symmetric, positive definite second-order diagonal tensor, i.e., $K = diag(k_{ll})$, $l = 1, \ldots, d$.

To avoid time-step constrains, it is often preferable to solve (1.1) implicitly in time. However, for fine meshes, the resulting large systems of nonlinear equations can be expensive to solve. To decrease the amount of work necessary to solve (1.1), we consider a two-level method where the nonlinear problem is solved only on a coarse grid of diameter H and a linear problem is solved on a fine grid of diameter $h \ll H$. On the fine grid, we approximate $K(p)$ by a first-order Taylor expansion about the solution from the coarse grid. Thus, instead of solving a large nonlinear problem on the fine grid, we solve a small nonlinear problem on the coarse grid and a large linearized problem on the fine grid.

2   A Coarse Grid Nonlinear Finite Difference

Let $H_s(\Omega)$, for a positive integer s, be the Sobolev space $W_2^s(\Omega)$. Denote the inner product in H_s by

$$(f,g)_s = \sum_{|\alpha| \leq s} \int_\Omega D^\alpha f \cdot D^\alpha g \, d\Omega,$$

where $f, g \in H_s(\Omega)$. We denote by $C^{p,1}(\Omega)$ the space of functions whose pth spatial derivative is Lipschitz continuous.

Let $V = H(\Omega, \mathrm{div}) = \{v \in (L^2(\Omega))^d : \nabla \cdot v \in L^2(\Omega)\}$ and $W = L^2(\Omega)$. We denote the subspaces of V containing functions with normal traces weakly equal to 0 and g^n as V^0 and V^n, respectively.

We consider two quasi-uniform triangulations of Ω, a coarse triangulation with mesh size H denoted by τ_H and a refinement of this triangulation with mesh size h denoted by τ_h. We consider the lowest-order RTN space on rectangles. Thus, on an element $E \in \tau_k$, $k = h$ or H, we have

$$V_k(E) = \{(\alpha_1 x_1 + \beta_1, \alpha_2 x_2 + \beta_2, \alpha_3 x_3 + \beta_3)^T : \alpha_i, \beta_i \in \mathcal{R}\},$$
$$W_k(E) = \{\alpha : \alpha \in \mathcal{R}\}.$$

Define $v_k^0 = V^0 \bigcap V_k$ and $V_k^n = V^n \bigcap V_k$. We use the standard nodal basis, where for V_k the nodes are at the midpoints of edges or faces of the elements, and for W_k the nodes are the centers of the elements. We define discrete products corresponding to the application of the midpoint (M), trapezoidal (T), and midpoint by trapezoidal (TM) quadrature rules, and denote the associated norms by $\| \cdot \|_r$ and the error in approximating an integral by the given rule by $E_T(q, r) = (q, r) - (q, r)_T$. The error in approximating an integral by either the trapezoidal or the trapezoidal by midpoint rule is

$$|E_Q(q,v)| \leq C \sum_{E \in \tau_h} \sum_{|\alpha|=2} \|\frac{\partial^\alpha}{\partial x^\alpha}(q,v)\|_{L^1(E)} h^2.$$

For any $\phi \in L^2(\Omega)$, let $\widehat{\phi}_k$ denote the L^2 projection of ϕ onto W_k; i.e.,

$$(\phi, w) = (\widehat{\phi}_k, w) \quad \forall w \in W_k.$$

This L^2 projection operator has the approximation property for $\phi \in H^j(\Omega)$

$$\|\widehat{\phi} - \phi\| \leq C\|\phi\|_j k^j, \quad 0 \leq j \leq 1, \quad k = h \text{ or } H.$$

Associated with the RTN mixed finite element spaces is the projection operator $\Pi : (H^1(\Omega))^d \to V_k$, defined by

$$(\nabla \cdot (\Pi q), w) = (\nabla \cdot q, w) \quad \forall w \in W_k,$$

with the approximation properties

$$\|q - \Pi q\| \leq C\|q\|_1 k, \quad \|\nabla \cdot (q - \Pi q)\| \leq C\|\nabla \cdot q\|_1 k.$$

Furthermore, by the definition of Πq and the midpoint rule of integration, we have that the error in the first component of the projection evaluated at the center of a grid block side is given by

$$|(\Pi q)^x - q^x|_{(x_{i+1/2}, y_j)} \leq Ch^2 \|q^x\|_{W_2^\infty(\Omega)}.$$

Using this estimate, we can bound the L^∞ norm of the projection by

$$\|\Pi q - q\|_{L^\infty(\Omega)} \leq Ch\|q\|_{W_2^\infty(\Omega)}.$$

In the expanded mixed formulation of (1.1), we define the variable $\tilde{u} = -\nabla p$ and $u = K(p)\tilde{u}$. The analysis uses the estimate

$$\|\Pi u^n - u^n\|_{TM} + \|\Pi \tilde{u}^n - \tilde{u}^n\|_{TM} \leq Ck^2 \|\tilde{u}\|_2.$$

For the lowest-order RTN spaces on rectangles, for any $q = (q^x, q^y) \in H^1(\Omega)$ and $E \in \tau_k$,

$$\left\|\frac{\partial}{\partial x}(\Pi q)^x\right\|_{L^2(E)} \leq \left\|\frac{\partial q^x}{\partial x}\right\|_{L^2(E)}, \quad \left\|\frac{\partial}{\partial y}(\Pi q)^y\right\|_{L^2(E)} \leq \left\|\frac{\partial q^y}{\partial y}\right\|_{L^2(E)}.$$

We uses the auxiliary variables, $\tilde{u}^n$ and u^n, defined by

$$\tilde{u}^n \equiv -\nabla p^n, \quad u^n \equiv K(p^n)\tilde{u}^n.$$

The problem is to find $(p^n, \tilde{u}^n, u^n) \in (W \times V \times V)$ satisfying

$$\begin{aligned}
(p_t^n, w) + (\nabla \cdot u^n, w) &= (f^n, w) && \forall w \in W, \\
(\tilde{u}^n, v) &= (p^n, \nabla \cdot v) && \forall v \in V_H^0, \\
(u^n, v) &= (K(p^n)\tilde{u}^n, v) && \forall v \in V.
\end{aligned} \tag{2.1}$$

We choose cell-centered finite difference approximations $P_H^n \in W_H$, $\tilde{U}_H^n \in V_H$, and $U_H^n \in V_H^n$ to the functions $p(t^n, \cdot)$, $\tilde{u}(t^n, \cdot)$, and $u(t^n, \cdot)$, respectively, for each $n = 1, \ldots, N$, satisfying

$$
\begin{aligned}
(d_t P_H^n, w) + (\nabla \cdot U_H^n, w) &= (f^n, w) \quad \forall w \in W_H, \\
(\tilde{U}_H^n, v)_{TM} &= (P_H^n, \nabla \cdot v) \quad \forall v \in V_H^0, \\
(U_H^n, v)_{TM} &= (K(\mathcal{P}_H(P_H^n))\tilde{U}_H^n, v)_T \quad \forall v \in V_H.
\end{aligned}
\tag{2.2}
$$

and we take $P_H^0 = \hat{p}_H(t^0, \cdot)$. We define $\mathcal{P}_H(p)$ from the values of p_{ij} for $i = 1, \ldots, \hat{N}_x$ and $j = 1, \ldots, \hat{N}_y$ by the bilinear interpolation operator. If p is twice differentiable in space, for $\mathcal{P}_h(p)$ we have

$$
\|\mathcal{P}_H(p) - p\|_\infty \le C H^2.
$$

Theorem 2.1 *Assume $f^n \in L^2(\Omega)$ for each n and K is continuously differentiable in its arguments. Then, for Δt sufficiently small, there exists a unique solution to problem (2.2).*

We make the smoothness assumptions

(H1) $f \in W_\infty^1(0, T; L^2(\Omega))$;

(H2) $K_{ll}(x, p) \in C^1(\overline{\Omega} \times \mathcal{R}) \cap W_\infty^2(\Omega \times \mathcal{R})$, $l = 1, \ldots, d$, and K_{ll} and $\frac{\partial K_{ll}}{\partial p}$ are uniformly Lipschitz continuous in p;

(H3)   There exist positive constants K_* and K^* such that, for $z \in \mathcal{R}^d$,

$$
K_* \|z\|^2 \le z^t K(x, p) z \le K^* \|z\|^2, \qquad x \in \Omega,\ p \in \mathcal{R};
$$

(H4) $p \in W_\infty^2(0, T; C^{3,1}(\Omega))$;

(H5) $u, \tilde{u} \in W_\infty^2(0, T; C^1(\overline{\Omega}))^d \cap W_\infty^2(0, T; W_\infty^2(\Omega))^d$.

Theorem 2.2 *For each $n = 1, \ldots, N$, let $(\underline{p}_H^n, \underline{\tilde{U}}_H^n, \underline{U}_H^n) \in (W_H \times V_H \times V_H^n)$ satisfy*

$$
\begin{aligned}
(\nabla \cdot \underline{U}_H^n, w) &= (b^n, w) \quad \forall w \in W_H, \\
(\underline{\tilde{U}}, v)_{TM} &= (\underline{P}_H^n, \nabla \cdot v) \quad \forall v \in V_H^0, \\
(\underline{U}_H^n, v)_{TM} &= (K(\mathcal{P}_H(p^n))\underline{\tilde{U}}_H^n, v)_T \quad \forall v \in V_H,
\end{aligned}
$$

with $b^n = f^n - p_t^n$ and $\underline{P}_H^0 = \hat{P}_H^0$. Then, under assumptions (H1)-(H5),

$$
\begin{aligned}
\|\underline{U}_H^n - u^n\|_{TM} + \|\underline{\tilde{U}}_H^n - \tilde{u}^n\|_{TM} + \|\underline{P}_H^n - p^n\|_M &\le C H^2, \\
\|d_t \underline{P}_H^n - d_t p^n\|_M &\le C(H^2 + \Delta t).
\end{aligned}
$$

We have the theorem about the convergence of the above difference scheme

Theorem 2.3 *Let* P_H^n, $\tilde{U}_H^n$, *and* $U_H^n, n = 1, \ldots, N$ *be defined by equation* (2.1) *with the initial value* $P_H^0 = \tilde{p}_H(t^0, \cdot)$. *Assume that* (H1)-(H5) *hold. Then there exists a positive constant* C, *independent of* H *and* Δt, *such that*

$$\|P_H^N - P^N\|_M + \{\Delta t \sum_{n=1}^{N} K_* \|\tilde{U}_H^n - \tilde{u}\|_T^2\}^{1/2} \leq C(H^2 + \Delta t).$$

3   Fine Grid Linear Scheme

We now consider a linear cell-centered finite difference scheme on the fine grid where we make use of the nonlinear solution on the coarse grid. We solve the problem for $P_h^n \in W_h, \tilde{U}_h^n \in V_h, U_h^n \in V_H^n$, $n = 1, \ldots, N$,

$$(d_t P_h^n, w) = -(\nabla \cdot U_h^n, w) + (f^n, w) \quad \forall w \in W_h,$$
$$(\tilde{U}_h^n, v)_{TM} = (P_h^n, \nabla \cdot v),$$
$$(U_h^n, v)_{TM} = (K(\mathcal{P}_H(P_H^n))\tilde{U}_H^n, v)_T,$$
$$+(K_p(\mathcal{P}_H(P_H^n))\mathcal{Q}_H(\tilde{U}_H^n)(\mathcal{P}_H(P_h^n) - \mathcal{P}_H(P_H^n)), v)_T \quad \forall v \in V_h.$$

We define $\mathcal{Q}_H(\tilde{u})$ as a vector quantity with entries $\mathcal{Q}_H^x(\tilde{u}^x)$ and $\mathcal{Q}_H^y(\tilde{u}^y)$. The entry $\mathcal{Q}_H^x(\tilde{u}^x)$ is defined from the values of $\tilde{u}_{i+1/2,j}^x$ for $i = 0, \ldots, \widehat{N}_x$ and $j = 1, \ldots, \widehat{N}_y$ as follows. For points (x, y) such that $x_{i-1/2} \leq x \leq x_{i+1/2}$, $i \in \{1, \ldots, \widehat{N}_x\}$ and $y_j \leq y \leq y_{j+1}$, $j \in \{1, \ldots, \widehat{N}_y\}$, we take $\mathcal{Q}_H^x(\tilde{u}^x)$ to be the bilinear interpolate of $\tilde{u}_{i-1/2,j}^x$, $\tilde{u}_{i+1/2,j}^x$, $\tilde{u}_{i-1/2,j+1}^x$, and $\tilde{u}_{i+1/2,j+1}^x$. This leaves a strip half a cell in height along the top and bottom of the domain. We consider the bottom strip. For $i = 0, \ldots \widehat{N}_x$, set

$$\mathcal{Q}_H^x(\tilde{u}^x)(x_{i+1/2}, y_{1/2}) = (2H_1^y + H_2^y)\tilde{u}_{i+1/2,1}^x - H_1^y \tilde{u}_{i+1/2,2}^x/(H_1^y + H_2^y).$$

For other points, an analogous interpolation definition is defined. We have the theorem about the convergence of the above linear finite difference scheme.

Theorem 3.1 *Let* P_h^n, $\tilde{U}_h^n$, *and* U_h^n, $n = 1, \ldots, N$, *be defined by equation* (2.2) *with the initial value* $P_h^0 = \hat{p}_h(t^0, \cdot)$. *Assume that* (H1)-(H5) *hold and that* H *and* $\Delta t H^{-d/2}$ *are sufficiently small. Then there exists a positive constant* C, *independent of* h, H, *and* Δt *such that*

$$\|P_h^N - p^N\|_M + \left\{\Delta t \sum_{n=1}^{N} K_* \|\tilde{U}_h^n - \tilde{u}^n\|_T^2\right\}^{1/2} \leq C(H^{4-d/2} + h^2 + \Delta t).$$

4   Conclusions

We have presented and derived error estimates for a two-level finite difference scheme for nonlinear parabolic equations and have shown and optimal-order convergence in both H^1 and L^2 for the coarse and fine grids. We remark that we have only considered the case of the Neumann boundary condition and a diagonal tensor K.

The estimates derived in this paper use the inverse estimate to bound the L^∞-norm in terms of the L^2-norm. As a result, our estimates may not be as sharp as possible. However, no better L^∞ estimates exist at this time for the expanded mixed finite element method.

The two-level scheme described above could be extended by adding levels and expanding about the next coarse solution in the nonlinear term at each new level. This corresponds to adding more Newton-like iterations with each iteration taking place on the next finer grid. This possibility is under investigation. We are currently implementing these two-level methods for equations of interest to flow in porous media.

References

[1] Xu, J., A novel two-grid method for semilinear elliptic equations, *SIAM J. Sci. Comput.* **15** (1994), 231–237.

[2] Xu, J., Two-grid discretization techniques for linear and nonlinear PDEs, *SIAM J. Numer. Anal.* **33** (1996), 1759–1777.

A Compact Operator Method for the Omega Equation

FRANCISCO R. VILLATORO   JESÚS GARCÍA-LAFUENTE

Abstract

The ageostrophic vertical velocity field for the mesoscale dynamics of the Alboran Sea is determined by means of synoptic data and numerical solution of the omega equation. For the stationary, horizontal geostrophic velocity field, a fourth-order, compact operator differentiation is used. The vertical ageostrophic flow is determined by means of a numerical solution of the omega equation with a Q vector formulation, solved by means of a fourth-order accurate, compact operator method. The numerical results confirm that on a macroscale, upward motion occurs upstream of the anticyclonic rotation while downward motion mesoscale takes place downstream.

KEYWORDS: oceanography, ageostrophic flow, omega equation, compact operator method

1 Introduction

The geographical location of the Alboran Sea between the South of Spain and the North of Africa, receiving water from the Atlantic Ocean (AO) through the Strait of Gibraltar, makes its study of fundamental importance for the research of the superficial circulation on the Western basin of the Mediterranean Sea. An expedition made in July 1993 shows the picture: the Atlantic jet from the AO proceeds as a wave to the East generating two anticiclonic gyres where the water from the AO accumulates near the African coast, the so called West (WE) and East (EE) gyres, which are separated by Cape Tres Focas and the Alboran Island [1]. The WE, which has also been found by other expeditions, is known to be permanent but there is some controversy related to the EE, which may be seasonal.

The data for the temperature and salinity distribution found in the Alboran Sea are synoptic, showing the general situation, but incapable to track the temporal evolution, and staggered, so these data have been filtered and interpolated on a regular grid as shown in [1], to study the stationary state of the fluid. From the experimental data the pressure/density ratio $\Delta\phi$ $[m^2/s^2]$ and the Brunt-Väisäla frequency N $[s^{-1}]$ are obtained every 5 meters from 2 to 202 m, in a regular grid of 10×10 km. In this paper these data have been used to obtain the stationary, horizontal geostrohpic velocity field in the Alboran Sea by means of a compact operator differentiation of fourth-order accuracy.

The estimation of the vertical velocity yields information on the exchanges of heat and salt near the surface of the ocean and is responsible for the vertical transport and distribution of nutrients and plackton [4]. As direct measures of vertical velocities are difficult, the solution of the omega equation can be used. In this paper the quasigeostrophic omega equation [2, 3], valid in regions of low dynamics and Rossby numbers, is solved by means of a fourth-order accurate, compact operator method [6]. It is standard to close this numerical method using homogeneous Dirichlet boundary conditions [8, 7].

When the horizontal velocities are smaller in one direction than in the other, the three-dimensional omega equation can be simplified to a two-dimensional omega equation for the stream function. A compact operator method has also been developed for this equation. Dirichlet boundary conditions have been used except for upstream and downstream, for which Roache, the null second-order derivative of the stream function, and Neumann, respectively, boundary conditions are used due to their better accuracy [5].

In §2 , the equations relevant to the problem studied in this paper are presented. §3 is devoted to the presentation of the numerical, fourth-order, compact operator methods developed in this paper. §4 shows the main results obtained in this paper. Finally, the last section is devoted to the presentation of conclusions.

2   Presentation of the Problem

When the Coriolis forces approximately balance with pressure gradients, the flow field $\mathbf{U} = (U, V, W)$ can be separated in a geostrophic and quasi-geostrophic flow as

$$U = U_g + u_a, \qquad V = V_g + v_a, \qquad W = w \equiv w_a,$$

where $|W| \ll |U|, |V|$, $|u_a| \ll |U_g|$, and $|v_a| \ll |V_g|$.

The geostrophic field can be calculated from the dynamic height using the expressions [3]:

$$U_g = -\frac{1}{f}\frac{\partial \Delta \phi}{\partial y}, \qquad V_g = \frac{1}{f}\frac{\partial \Delta \phi}{\partial x}, \tag{2.1}$$

where $f = 2\Omega \sin \psi$ is the local frequency of rotation of the Earth at latitude $\psi \approx 36°$ at the center of Alboran Sea and $f = 8.55 \times 10^{-5}\,[s^{-1}]$.

The steady-state omega equations for the quasi-geostrophic flow are as follows [2, 3]:

$$\frac{\partial(N^2 w)}{\partial x} - f^2 \frac{\partial u_a}{\partial z} = Q^x, \qquad \frac{\partial(N^2 w)}{\partial y} - f^2 \frac{\partial v_a}{\partial z} = Q^y, \qquad \frac{\partial u_a}{\partial x} + \frac{\partial v_a}{\partial y} + \frac{\partial w}{\partial z} = 0, \tag{2.2}$$

where the potential vorticity Q is obtained from the geostrophic flow as

$$Q^x = 2f\left(\frac{\partial V_g}{\partial x}\frac{\partial U_g}{\partial z} + \frac{\partial V_g}{\partial y}\frac{\partial V_g}{\partial z}\right), \qquad Q^y = -2f\left(\frac{\partial U_g}{\partial x}\frac{\partial U_g}{\partial z} + \frac{\partial U_g}{\partial y}\frac{\partial V_g}{\partial z}\right).$$

A simple omega equation for the vertical velocity flow can be easily obtained using the continuity equation in (2.2), yielding

$$f^2 \frac{\partial^2 w}{\partial z^2} + \left(\frac{\partial^2}{\partial x^2} + \frac{\partial^2}{\partial y^2}\right)(N^2 w) = \frac{\partial Q^x}{\partial x} + \frac{\partial Q^y}{\partial y} \equiv F. \tag{2.3}$$

The boundary conditions for this equation are difficult to determine since the synoptic experimental data have been interpolated in a regular grid. The bottom and top of the grid are fictitious boundaries at 202 and 2 m., respectively, of depth. The south and north boundaries correspond to water and coast points. The west and east boundaries correspond to water incoming from the Atlantic ocean and water going out from the Alboran Sea. Since the detailed boundary conditions are very difficult to obtain, Dirichlet boundary conditions have been used for all the boundaries of the computational grid since Viudez et al. [8] have shown they are enough accurate for this problem.

Eq. (2.3) can be further simplified if the variation of the horizontal velocity in the x-direction is smaller than that in the y-direction:

$$\left|\frac{\partial u_a}{\partial x}\right| \ll \left|\frac{\partial v_a}{\partial y}\right|.$$

By introducing the stream function

$$v_a = -\frac{\partial \psi}{\partial z}, \qquad w = \frac{\partial \psi}{\partial y},$$

we have the simplified omega equation

$$\frac{\partial}{\partial y}\left(N^2\frac{\partial\psi}{\partial y}\right) + f^2\frac{\partial^2\psi}{\partial z^2} = Q^y. \tag{2.4}$$

Similarly, when

$$\left|\frac{\partial v_a}{\partial y}\right| \ll \left|\frac{\partial u_a}{\partial x}\right|,$$

another stream function can be introduced

$$u_a = -\frac{\partial\psi'}{\partial z}, \qquad w = \frac{\partial\psi'}{\partial y},$$

yielding the simplified omega equation

$$\frac{\partial}{\partial x}\left(N^2\frac{\partial\psi'}{\partial x}\right) + f^2\frac{\partial^2\psi'}{\partial z^2} = Q^x. \tag{2.5}$$

For the boundary conditions for the stream function equation, homogeneous Dirichlet ones have been used for the north, south, bottom, and top boundaries, Neumann ones for the east, as considered as outflow, and Roache ones for the west inflow. When solving the stream function equation, Roache boundary conditions, i.e., the null second-order derivative, are more accurate than Neumann ones [5].

3   Fourth-Order Compact Operator Method

The calculation of the geostrophic field using eq. (2.1) requires the use of a differentiation formula. In this paper a compact operator method of fourth-order accuracy is used. This implicit method uses the difference expression

$$\dot{U}_i \equiv \frac{dU_i}{dx} = \frac{1}{2\,\Delta x}\frac{\nabla+\Delta}{1+\delta^2/6}U_i, \qquad i = 0, 1, \ldots, N, \tag{3.1}$$

where the finite difference operators are defined as

$$E\,U_i = U_{i+1}, \qquad \nabla = E - 1, \quad \Delta = 1 - E^{-1}, \quad \delta^2 = \nabla - \Delta.$$

Eq. (3.1) yields the linear, tridiagonal system of equations

$$\dot{U}_{i-1} + 4\dot{U}_i + \dot{U}_{i+1} = \frac{3}{\Delta x}(U_{i+1} - U_{i-1}),$$

whose truncation error expansion is

$$\dot{U}_i = \frac{dU_i}{dx} - \frac{\Delta x^4}{180}\frac{d^5U_i}{dx^5} + O(\Delta x^6), \tag{3.2}$$

showing its fourth-order accuracy. For the boundary conditions, asymmetrical formulas are used

$$5\dot{U}_0 + 16\dot{U}_1 - 3\dot{U}_2 = -\frac{1}{3\Delta x}\left(44U_0 - 26U_1 - 27U_2 + 10U_3 - U_4\right),$$

$$5\dot{U}_N + 16\dot{U}_{N-1} - 3\dot{U}_{N-2} = \frac{1}{3\,\Delta x}\,(44U_N - 26U_{N-1} - 27U_{N-2} \tag{3.3}$$
$$+10U_{N-3} - U_{N-4}),$$

for the left and right, respectively, boundaries. These expressions have been selected since they yield exactly the same truncation error terms, up to the fifth-order included, as the inner formula, i.e., eq. (3.2). The solution of the corresponding tridiagonal systems is obtained by using the Thomas method.

The second-order derivatives appearing in the elliptic equations (2.3)–(2.5) are also approximated using a compact operator method, e.g.,

$$\ddot{U}_i \equiv \frac{d^2U_i}{dx^2} = \frac{1}{\Delta x^2}\,\frac{\delta^2}{1 + \delta^2/12}\,U_i, \tag{3.4}$$

which yields the tridiagonal system of linear equations

$$\ddot{U}_{i-1} + 10\,\ddot{U}_i + \ddot{U}_{i+1} = \frac{12}{\Delta x^2}\,(U_{i+1} - 2\,U_i + U_{i-1}),$$

whose truncation error term is

$$\ddot{U}_i = \frac{d^2U_i}{dx^2} - \frac{\Delta x^4}{240}\,\frac{d^6U_i}{dx^6} + O\left(\Delta x^6\right). \tag{3.5}$$

When the compact operator approximation (3.4) is used for the inner points in the grid, different boundary conditions can be used. In this paper Dirichlet homogeneous boundary conditions are used for eq. (2.3) and Dirichlet, Neumann, and Roache homogeneous boundary conditions for both eqs. (2.4) and (2.5). For the Dirichlet boundary conditions, $U_0 = U_N = 0$, for the Neumann ones, eqs. (3.3) are used and for the Roache boundary conditions, the stencil is used at the left boundary:

$$16\ddot{U}_0 + 151\ddot{U}_1 - 83\ddot{U}_2$$
$$= \frac{1}{\Delta x^2}\left(\frac{771}{4}U_0 - \frac{2019}{4}U_1 + \frac{885}{2}U_2 - \frac{285}{2}U_3 + \frac{51}{4}U_4 - \frac{3}{4}U_5\right),$$

and the same expression with U_{N-i} replacing U_i for the right one. This expression has truncation error terms equal to eq. (3.5) up to the fifth-order (included). The substitution of eq. (3.4) into eq. (2.3) yields

$$f^2(12 + \delta_x^2)(12 + \delta_y^2)\frac{\delta_z^2}{\Delta z^2}w_{ijk} + (12 + \delta_z^2)(12 + \delta_y^2)\frac{\delta_x^2}{\Delta x^2}\left(N_{ijk}^2 w_{ijk}\right)$$

$$+(12 + \delta_z^2)(12 + \delta_x^2)\frac{\delta_y^2}{\Delta y^2}\left(N_{ijk}^2 w_{ijk}\right) = (12 + \delta_z^2)(12 + \delta_x^2)(12 + \delta_y^2)F_{ijk},$$

where $w_{ijk} \approx w(i\,\Delta x, j\,\Delta y, k\,\Delta z)$ and homogeneous Dirichlet boundary conditions are used. For the calculation of the horizontal ageostrophic flow, eqs. (2.2), an explicit, fourth-order, Runge-Kutta method is used with homogeneous Dirichlet boundary conditions, e.g., $u_a(0) = 0$ and $v_a(0) = 0$.

Eq. (3.4) can be easily substituted into eq. (2.5) or into eq. (2.4), separating this equation as

$$\frac{\partial \psi}{\partial y} = \phi, \quad \frac{\partial}{\partial y}\left(N^2\,\phi\right) + f^2\frac{\partial^2 \psi}{\partial z^2} = Q^y,$$

and yielding

$$\frac{\nabla_y + \Delta_y}{2\Delta y}\psi_{jk} = (6 + \delta_y^2)\phi_{jk},$$

$$(12 + \delta_z^2)\frac{\nabla_y + \Delta_y}{2\Delta y}\phi_{jk} + f^2(6 + \delta_y^2)\frac{\delta_z^2}{\Delta z^2}\psi_{jk} = (12 + \delta_z^2)(6 + \delta_y^2)Q_{jk}^y,$$

where $\phi_{jk} \approx \phi(j\,\Delta y, k\,\Delta z)$, a Neumann boundary condition is used in the left (inflow) boundary, a Roache one in the right (outflow), and Dirichlet ones in the other four ones. The v_a and w components of the ageostrophic velocity are calculated by differentiation of the stream function ψ by means of the compact operator eq. (3.1); the remaining component u_a is obtained solving the first equation of (2.2), with $u_a(0) = 0$, by means of an explicit, fourth-order, Runge-Kutta method.

The algebraic equations to be solved when the compact operator method is applied to eq. (2.3) have 15 non-null diagonals and 9 non-null ones for eqs. (2.4) or (2.5) and are well-conditioned. These linear systems have been solved by means of the stabilized bi-conjugate gradient method using incomplete LU factorization as a preconditioner. The mean of required iterations is approximately 10 for a residual error treshold of 10^{-10}.

4   Presentation of Results

Fig. 1 shows the geostrophic flow field resulting from the experimental data available. These figures show the west gyre, which continue being visible for all depths, and the smaller east gyre, which apparently disappear as depth is increased. A zoom of the bottom right plot shows that the east gyre remains despite with very small velocities. The code has been verified checking that the conservation law for the momentum approximately conserves.

Fig. 2 shows the ageostrophic flow field calculated using the simplified omega equation (2.5), based on the stream function formulation. Similar results have been obtained for the omega equation (2.4). Although the conservation of momentum approximately also holds for the ageostrophic velocity field (u_a, v_a, w), the results are not expected by physical principles applied to this problem; i.e., upward motion must occur upstream of the anticyclonic gyres while downward motion must take place downstream. The reason for this behaviour is that the hypothesis required to derive eq. (2.5) or eq (2.4) does not hold since the numerical results yield

$$\left|\frac{\partial u_a}{\partial x}\right| \approx \left|\frac{\partial v_a}{\partial y}\right|.$$

However, the results are encouraging since the velocities we have obtained have the correct orders of magnitude

$$|w| \ll |v_a| \approx |u_a| \ll |U_g| \approx |V_g|,$$

required for the validity of the complete omega equation (2.3).

Fig. 3 shows the vertical velocity field for the solution of the complete omega equation (2.3). The conservation law also approximately holds for this numerical method assuring the improved accuracy of the fourth-order, compact operator method over a standard second-order one. The resulting velocity field has similar orders of magnitude than the one obtained with the simplified omega equation, e.g.,

$$|w| \approx 10^{-7}\, m/s, \qquad |v_a| \approx |u_a| \approx 10^{-2}\, m/s, \qquad |U_g| \approx |V_g| \approx 1\, m/s.$$

Fig. 3 shows the vertical velocity contours for a yz-plane at x-positions corresponding to the two sides of both the West (top plots) and East (bottom plots) gyres. This figure and other figures not shown here show that upward and downward motion occur at the south and north directions, respectively, near the west side of the West gyre and the reverse, i.e., downward and upward motion at the south and north directions, in the east side of this gyre. Similar behaviour, but less remarked, is found in the East gyre. These results are not sufficiently confident so as to confirm the East gyre as a stationary phenomena but confirm that the numerical method yield physically realistic results.

5 Conclusions

A fourth-order, compact operator method has been developed for the calculation of the geostrophic velocity field and the solution of both the simplified

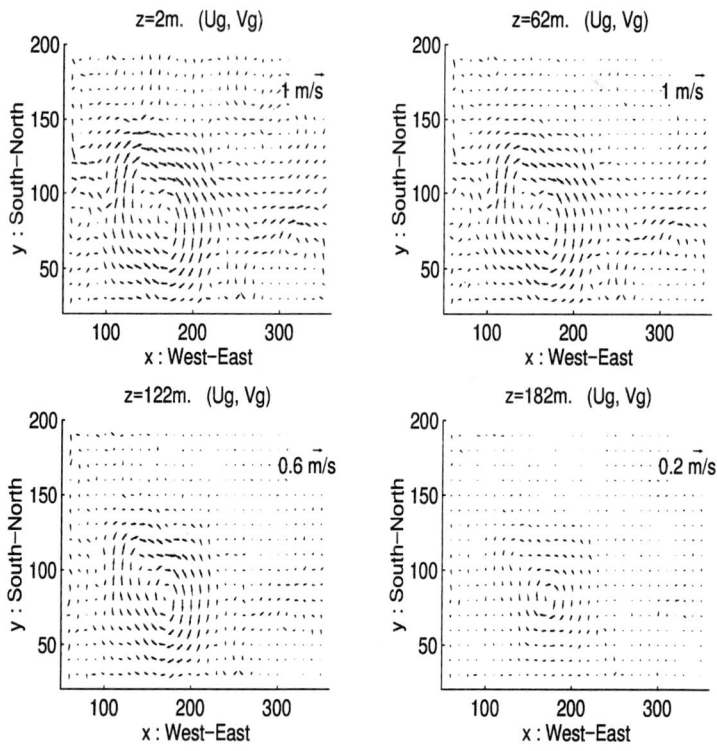

Figure 1: Geostrophic flow field for depths 2, 62, 122, and 182 meters from top left to bottom right.

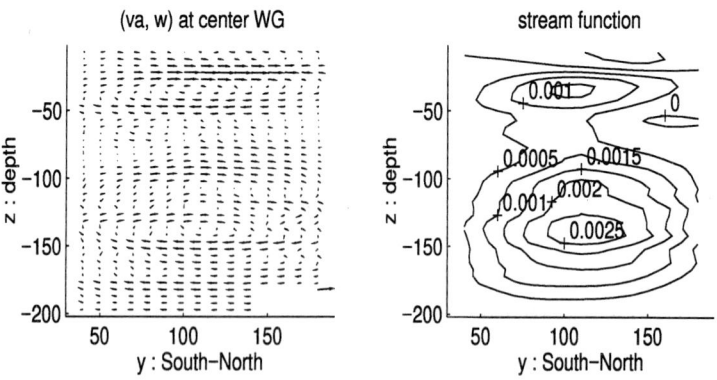

Figure 2: Vertical velocity flow both (v_a, w) (left) and stream function isocontours (bottom)) for a y-plane (South-North direction) trough the center of the West gyre in the Alboran Sea.

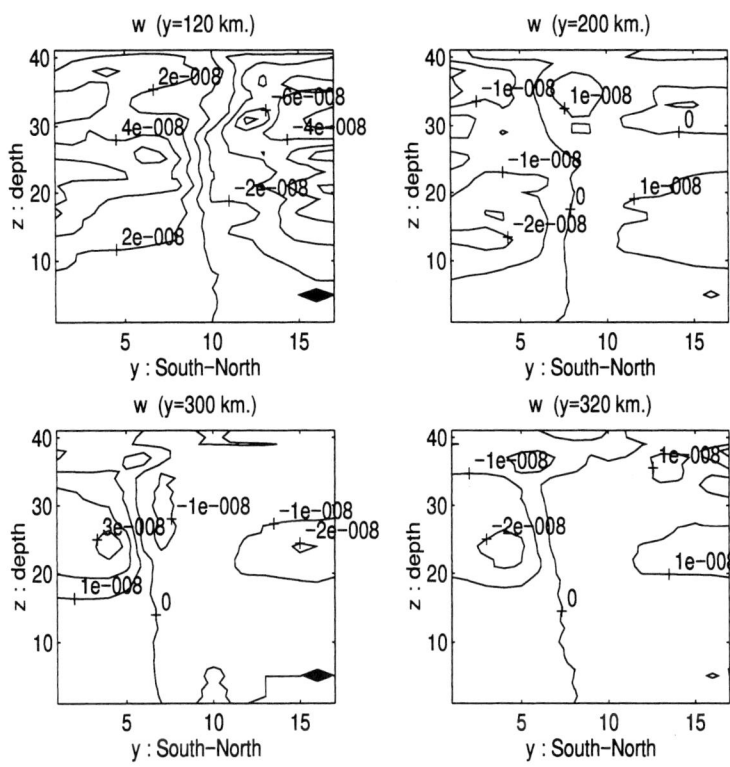

Figure 3: Isocontours of the vertical velocity flow at yz-planes at the west (left plots) and east (right plots) of both the West (top plots) and East (bottom plots) gyres found in Fig. 1.

omega equation for the stream function and the three-dimensional complete omega equation. The resulting methods are highly accurate and allow an easy treatment of Dirichlet, Neumann, and Roache boundary conditions. The system of linear algebraic equations to be solved is well-conditioned and has been solved by means of the stabilized bi-conjugate gradient method using the incomplete LU factorization as a preconditioner.

The numerical method developed in this paper has been applied to a set of experimental data which have been previously interpolated and smoothed in a regular grid. The geostrophic velocity field presents clearly both the West and East gyres as expected from previous studies. The ageostrophic vertical velocity field does not satisfy the hypothesis required for the use of the simplified omega equation based on the stream function. However, the results obtained in that case have the correct order of magnitude for the ageostrophic velocities.

The solution of the three-dimensional omega equation for the calculation of the vertical ageostrophic flow field confirms the presence of the West
gyre, since upward motion occur upstream of this anticyclonic gyre while
downward motion takes place downstream, but is not confident enough for
the East one. The only reason we can point out is that the interpolation
and smoothing procedures used to obtain a regular grid from the staggered
original experimental data are not correct when the velocities are small.

Our results indicate that further experimental data are required to model
accurately the vertical flow for the East gyre appearing in the geostrophic
flow field. Also new techniques for the interpolation and smoothing of the
experimental data must be developed. Finally, new numerical methods based
on compact operator methods using staggered grids instead of regular ones
must also be developed.

References

[1] García-Lafuente, J., Cano, N., Vargas, M., Rubín, J. P., and Hernández-
Guerra, A., Evolution of the Alboran Sea hydrographic structures durgin
July 1993, *Deep-Sea Res. Part I* **45** (1998), 39–65.

[2] Hoskins, B. J., Draghici, I., and Davies, H., A new look at the omega
equation, *Quart. J. Roy. Meteor. Soc.* **104** (1978), 31–38.

[3] Pedlosky, J., *Geophysical Fluid Dynamics*, 2nd ed., Springer-Verlag, New
York, 1987.

[4] Pinot, J. M., Tintoré, J., and Wang, D. P., A study of the omega equation
for diagnosing vertical motions at ocean fronts, *J. Mar. Res.* **54** (1996),
239–259.

[5] Ramos, J. I., Upstream and downstream boundary conditions in computational fluid mechanics, in *Proceedings of the Fourth IMACS International Symposium on Computer Methods for Partial Differential Equations*, edited by R. Vichnevetsky and R.S. Stepleman, Lehigh University,
Pennsylvania, USA, 1981.

[6] Smith, G. D., *Numerical Solution of Partial Differential equations*, 3rd
ed., Clarendon Press, Oxford, 1985.

[7] Tintoré, J. Gomis, D., and Alonso, S., Mesoscale dynamics and vertical
motion in the Alborán Sea, *J. Phys. Ocean.* **21** (1991), 811–823.

[8] Viudez, A., Tintoré, J., and Haney, R., Circulation in the Alboran Sea as determined by quasi-synoptic hydrographic observations. Part I: Three-dimensional structure of the two anticyclonic gyres, *J. Phys. Ocean.* **26** (1996), 684–705.

Domain Decomposition Algprithm for a New Characteristic Mixed Finite Element Method for Compressible Miscible Displacement

DANPING YANG

Abstract

A Schwarz type domain decomposition algorithm is formulated to solve an approximation for miscible displacement of compressible fluids in porous media and convergence rate of the algorithm is analyzed. First, a splitting positive definite mixed element procedure is used to treat the pressure equation of parabolic type. The coefficient matrix of the mixed element system is symmetric positive definite and the flux equation is separated from the pressure equation so that the approximate solution of the flux function can be independently obtained, and a characteristic finite element method is used to treat the convection-diffusion equations of the concentrations. Then a Schwarz type domain decomposition algorithm is introduced to solve the symmetric positive definite systems step by step. How many iterative cycles are needed at each time level? A convergence analysis is given.

KEYWORDS: compressible flow, miscible displacement, domain decomposition, characteristics, mixed element, convergence analysis

1 Introduction

The displacement of multi-phase multi-component compressible flow in porous media is governed by a nonlinear system consisting of a parabolic equation of pressure of the mixture and convection-diffusion equations of concentrations of fluids. To illustrate the method, we consider as our model the single phase and two component displacement of one compressible fluid by another in a porous medium. Let Ω be a convex bounded domain in R^2 with a boundary Γ. Under the assumptions that no volume change results from the mixing of the components and that a pressure-density relation exists for each component in a form that is independent of the mixing and that the components are of slight compressibility, Douglas and Roberts proposed a mathematical model [10] (with $x \in \Omega$, $0 < t \leq T$)

$$
\begin{aligned}
(a) \quad & d(c)p_t + \nabla \cdot \mathbf{u} = q, \quad \mathbf{u} = -a(c)(\nabla p + \gamma(c)\nabla H), \\
(b) \quad & \phi c_t + \mathbf{u} \cdot \nabla c - \nabla \cdot (\mathcal{D}(\mathbf{u})\nabla c) + b(c)p_t = (\tilde{c} - c)q,
\end{aligned}
\tag{1.1}
$$

where $c = c_1 = 1 - c_2$ denotes the (volumetric) concentration of the first component of the fluid mixture, q the volumetric rate of external flow, $\tilde{c}$ the concentration of the first component in the external flow, which must be specified at injection points ($q > 0$) and be assumed to equal c at production points ($q < 0$), and z_1 and z_2 the constant compressibility factors for the i-th component, $i = 1, 2$. $\kappa = \kappa(x)$ and $\phi = \phi(x)$ are the permeability and the porosity of the rock, respectively. $\mu = \mu(c)$ is the viscosity of the fluid, $\gamma = \gamma(c)$ is the gravitational coefficient and H is the depth function of oil reservoir. p is the pressure of fluid and $\mathbf{u}$ is Darcy's velocity. $\mathcal{D}(\mathbf{u}) = \phi[d_m\mathcal{I} + |\mathbf{u}|(d_l\mathcal{E}(\mathbf{u}) + d_t\mathcal{E}^{\perp}(\mathbf{u}))]$ denotes the diffusion matrix describing the effects of molecular diffusion and dispersion where $\mathcal{E}(\mathbf{u}) = (u_i u_j / |\mathbf{u}|^2)_{2\times 2}$ is the 2×2 matrix representing orthogonal projection along the velocity vector, $\mathcal{E}^{\perp}(\mathbf{u}) = \mathcal{I} - \mathcal{E}(\mathbf{u})$ is its orthogonal complement, and d_m, d_l, and d_t are the molecular diffusion and dispersion coefficients, respectively.

$$
a(c) = \frac{\kappa}{\mu(c)}, \quad d(c) = \phi \sum_{i=1}^{2} z_i c_i, \quad b(c) = \phi(z_1 - z_2)c(1 - c).
$$

We also assume that no flow occurs across the boundary:

$$
\mathbf{u} \cdot \nu = 0, \quad (\mathcal{D}(\mathbf{u})\nabla c) \cdot \nu = 0 \quad \text{on } \Gamma, \quad 0 \leq t \leq T, \tag{1.2}
$$

where ν is the unit vector outer normal to Γ. In addition, the initial conditions

$$
p = p_0, \quad c = c_0 \quad \text{in } \Omega, \quad t = 0, \tag{1.3}
$$

must be given.

Many works on approximations for the displacement of multiphase multi-component flows have been done; for instance, see Douglas [8, 9, 11], Douglas, Ewing and Wheeler [12, 13], Duran [14], Yuan [22], and Yang [20] for time-discretization procedures of miscible displacement of incompressible fluids in porous media by a modified method of characteristics combining with the classical mixed finite element methods, and see Douglas and Roberts [10] and Chou and Li [6] for numerical methods for a model for compressible miscible displacement in porous media.

In [21], we formulated a new approximation based on a modified method of characteristics combining with a splitting positive definite mixed element procedure. We use a splitting positive definite mixed element to treat the pressure equation of parabolic type. The coefficient matrix of the mixed element system is symmetric and positive definite and the flux equation is separated from the pressure equation so that we can independently obtain the approximate solution of the flux function and then, if required, obtain the approximation of the pressure almost explicitly. We use the characteristic finite element method to treat the convection-diffusion equations of the concentrations.

The purpose of this article is to formulate a Schwarz type domain decomposition algorithm to solve the approximation described in [21] and to analyze the convergent rate of the algorithm. It is clear that if iterative cycles are indefinitely implemented at each time level, we can obtain the solution of the fully discrete systems. But indefinite cycles are impossible and unnecessary. How many cycles are needed at each time level? We give a convergence analysis.

The article is organized as follows. In §2, we summarize the numerical method for the compressible displacement problem (1.1) based on a modified method of characteristics combining with a splitting positive definite mixed element procedure. In §3, we formulated a Schwarz type domain decomposition algorithm to solve the fully discrete scheme defined in §2. In §4 we give the error estimate of the approximate solution.

2   A Characteristic Mixed Element Procedure

It is easily seen that the equation (1.1(b)) is convection-dominated and directly depends on Darcy's velocity $\mathbf{u}$. A better numerical result can be obtained by using mixed element methods to approximate directly $\mathbf{u}$ and characteristic methods to treat the convection term. We assume that the coefficients $a(c)$ and $d(c)$ in system (1.1) are below bounded positively as in [10]. Introduce the space $\mathcal{H} = \{\mathbf{w} \in (L^2(\Omega))^2; \nabla \cdot \mathbf{w} \in L^2(\Omega), \mathbf{w} \cdot \nu = 0$ on $\Gamma\}$, $\mathcal{S} = H^1(\Omega)$, and $\mathcal{M} = L^2(\Omega)$. Let $\alpha(c) = 1/a(c)$ and $\beta(c) = 1/d(c)$. The system (1.1) is equivalent to the following characteristic mixed weak form that seeks $(c, \mathbf{u}, p) \in \mathcal{S} \times \mathcal{H} \times \mathcal{M}$ such that

$$
\begin{aligned}
(a) \quad & (\psi c_\sigma, z) + (\mathcal{D}(\mathbf{u})\nabla c, \nabla z) + (b(c)\beta(c)(q - \nabla \cdot \mathbf{u}), z) \\
& = ((\tilde{c} - c)q, z), \quad \forall \, z \in \mathcal{S}, \\
(b) \quad & ((\alpha(c)\mathbf{u})_t, \mathbf{w}) + (\beta(c)\nabla \cdot \mathbf{u}, \nabla \cdot \mathbf{w}) \\
& = (\beta(c)q, \nabla \cdot \mathbf{w}) - (\gamma(c)_t \nabla H, \mathbf{w}), \quad \forall \, \mathbf{w} \in \mathcal{H}, \\
(c) \quad & (p_t, v) = (\beta(c)(q - \nabla \cdot \mathbf{u}), v), \quad \forall \, v \in \mathcal{M},
\end{aligned}
\tag{2.1}
$$

where $\psi = \sqrt{\phi^2 + |\mathbf{u}|^2}$ and

$$
c_\sigma = \frac{\phi}{\psi} c_t + \frac{\mathbf{u}}{\psi} \cdot \nabla c
\tag{2.2}
$$

denotes the direction derivative along the characteristic line of the convection term $\phi c_t + \mathbf{u} \cdot \nabla c$.

Make a time partition $0 = t_0 < \ldots < t_n < \ldots < t_N = T$. Set $\tau_n = t_n - t_{n-1}$. Take $\tau = \max_{1 \leq n \leq N} \tau_n$ as a time step size. Let $\mathcal{T}_{h_c}$, $\mathcal{T}_{h_u}$, and $\mathcal{T}_{h_p}$ be three families of quasi-regular finite element partitions of the domain Ω, and h_c, h_u and h_p be the mesh parameters, which generally denote the largest of diameters of elements in the partitions, respectively. Let $\mathcal{H}_{h_u} \subset \mathcal{H}$ be each of the classical mixed elements defined on the partition $\mathcal{T}_{h_u}$, such as BDDM [2], BDFM [3], BDM [4], CD [5], and RT elements in [18]. Let $\mathcal{S}_{h_c} \subset \mathcal{S}$ and $\mathcal{M}_{h_p} \subset \mathcal{M}$ be usual finite element spaces defined on the partitions $\mathcal{T}_{h_c}$ and $\mathcal{T}_{h_p}$. Based on (2.1), we define a characteristic mixed element scheme.

Characteristic mixed element scheme. *Given an initial approximation* $(c_h^0, \mathbf{u}_h^0, p_h^0) \in \mathcal{S}_{h_c} \times \mathcal{H}_{h_u} \times \mathcal{M}_{h_p}$ *such that*

$$
\begin{aligned}
(a) \quad & (c_h^0, Z) = (c_0, Z) \quad \forall \, Z \in \mathcal{S}_{h_c}, \\
(b) \quad & (\alpha(c_0)\mathbf{u}_h^0, \mathbf{W}) = -(\nabla p_0 + \gamma(c_0)\nabla H, \mathbf{W}) \quad \forall \, \mathbf{W} \in \mathcal{H}_{h_u}, \\
(c) \quad & (p_h^0, V) = (p_0, V) \quad \forall \, V \in \mathcal{M}_{h_p},
\end{aligned}
\tag{2.3}
$$

seek $c_h^n \in \mathcal{S}_{h_c}$ *such that*

$$
\begin{aligned}
& (\phi c_h^n, Z) + \tau_n(\mathcal{D}(\mathbf{u}_h^{n-1})\nabla c_h^n, \nabla Z) + \tau_n(q_+^n c_h^n, Z) \\
& = (\phi \bar{c}_h^{n-1}, Z) + \tau_n(b(c_h^{n-1})\beta(c_h^{n-1})(\nabla \cdot \mathbf{u}_h^{n-1} - q^n), Z) \\
& \quad + \tau_n(q_+^n \tilde{c}^n, Z) \quad \forall \, Z \in \mathcal{S}_{h_c},
\end{aligned}
\tag{2.4}
$$

where $q_+^n = \max(0, q^n)$ *and* $\bar{c}_h^{n-1}$ *is given by*

$$
\bar{c}_h^{n-1} = c_h^{n-1}(\bar{x}), \quad \bar{x} = x - \tau_n \mathbf{u}_h^{n-1}(x)/\phi(x) \quad \forall \, x \in \Omega;
\tag{2.5}
$$

seek $\mathbf{u}_h^n \in \mathcal{H}_{h_u}$ *such that*

$$(\alpha(c_h^n)\mathbf{u}_h^n, \mathbf{W}) + \tau_n(\beta(c_h^n)\nabla \cdot \mathbf{u}_h^n, \nabla \cdot \mathbf{W})$$
$$= (\alpha(c_h^{n-1})\mathbf{u}_h^{n-1}, \mathbf{W}) + \tau_n(\beta(c_h^n)q^n, \nabla \cdot \mathbf{W}) \qquad (2.6)$$
$$-((\gamma(c_h^n) - \gamma(c_h^{n-1}))\nabla H, \mathbf{W}) \quad \forall \, \mathbf{W} \in \mathcal{H}_{h_u};$$

seek $p_h^n \in \mathcal{M}_{h_p}$ *such that*

$$(p_h^n, V) = (p_h^{n-1}, V) + \tau_n(\beta(c_h^n)(q^n - \nabla \cdot \mathbf{u}_h^n), V) \quad \forall \, V \in \mathcal{M}_{h_p}, \qquad (2.7)$$

for $n = 1, 2, \ldots, N$.

Because point $\bar{x}$ may lie outside domain Ω , we must continuously extend the value of function to the exterior of Ω. Define the operator $\mathcal{Q}$ as a projection from the exterior domain $\Omega^c = R^2 \backslash \bar{\Omega}$ to boundary Γ and the point $\mathcal{F}(x)$ as the symmetric point of point $x \in \Omega^c$ corresponding to the projection point $\mathcal{Q}(x)$. We say a domain Ω to satisfy the property A, if there exists a constant $\delta > 0$ such that for each point $x \in \Omega^c \bigcap \Omega_\delta$, where $\Omega_\delta = \{x; \inf_{y \in \Omega} |x - y| \leq \delta\}$, the projection point $\mathcal{Q}(x)$ is uniquely determined and $\mathcal{F}(x) \in \Omega$. In this case, we can define an extension by

$$\varphi(x) = \begin{cases} \varphi(x), & x \in \Omega, \\ \varphi(\mathcal{F}(x)), & x \in \Omega_\delta \backslash \Omega. \end{cases} \qquad (2.8)$$

(2.8) defines the linear bounded extending operator from $H^1(\Omega)$ to $H^1(\Omega_\delta)$. A domain satisfies the property A, if its boundary Γ is smooth. In the case of convex polygon, the domain Ω satisfies also the property A. After the function c_h^{n-1} is extended through use of above-mentioned method (2.8), $\bar{c}_h^{n-1}$ is defined for sufficiently small τ_n.

It is easily seen that the coefficient matrix of the mixed element system in schemes (2.6) and (2.7) are symmetric positive definite and the flux equations (2.6) on the flux variable $\mathbf{u}_h^n$ is separated from the pressure equation (2.7) on the pressure function p_h^n. The matching relation (i.e., the BBL-condition) between the mixed element space $\mathcal{M}_{h_p}$ and $\mathcal{H}_{h_u}$ required by the classical mixed element spaces in [2, 3, 4, 5, 18] now is not necessary. The flux function $\mathbf{u}_h^n$ can be solved independently and then, if required, the pressure function p_h^n can be obtained almost explicitly.

From the viewpoint of calculation, one can choose usual continuous finite element spaces as $\mathcal{H}_{h_u}$. But usual continuous finite element spaces, in general, cannot yield an approximate solution with optimal accuracy in L^2-norm. To get the optimal approximation, we still choose the classical mixed elements as

$\mathcal{H}_{h_u}$. The usual continuous finite element spaces may be used as the pressure function finite element spaces $\mathcal{M}_{h_p}$. It is clear that the number of unknowns in systems (2.6) and (2.7) is less than the number of the unknowns in the classical mixed element systems. Since systems (2.6) and (2.7) are symmetric positive definite systems, many fast effective algorithms can be used.

We assume that the finite element spaces $\mathcal{S}_{h_c}$, $\mathcal{H}_{h_u}$ and $\mathcal{M}_{h_p}$ have approximate properties that there exist integers $k_1 \geq k \geq 0$, $r \geq 0$ and $s \geq 1$ such that, for any $1 \leq q \leq \infty$

(a) $\displaystyle\inf_{\mathbf{W} \in \mathcal{H}_{h_u}} \|\mathbf{w} - \mathbf{W}\|_{(L^q(\Omega))^2} \leq K h_u^{k+1} \|\mathbf{w}\|_{(W^{k+1,q}(\Omega))^2}$

$\quad \forall\, \mathbf{w} \in \mathcal{H} \bigcap (W^{k+1,q}(\Omega))^2$,

$\displaystyle\inf_{\mathbf{W} \in \mathcal{H}_{h_u}} \|\nabla \cdot (\mathbf{w} - \mathbf{W})\|_{L^q(\Omega)} \leq K h_u^{k_1} \|\mathbf{w}\|_{(W^{k_1+1,q}(\Omega))^2}$

$\quad \forall\, \mathbf{w} \in \mathcal{H} \bigcap (W^{k_1+1,q}(\Omega))^2$,

(b) $\displaystyle\inf_{z \in \mathcal{S}_{h_c}} \|z - Z\|_{L^q(\Omega)} \leq K h_c^{s+1} \|z\|_{W^{s+1,q}(\Omega)} \,\, \forall\, z \in \mathcal{S} \bigcap W^{s+1,q}(\Omega)$,

(c) $\displaystyle\inf_{V \in \mathcal{M}_{h_p}} \|v - V\|_{L^q(\Omega)} \leq K h_p^{r+1} \|v\|_{W^{r+1,q}(\Omega)} \,\, \forall\, v \in \mathcal{M} \bigcap W^{r+1,q}(\Omega)$,

$$(2.9)$$

where $k_1 = k + 1$ in the cases that $\mathcal{H}_{h_u}$ is each of BDFM and RT mixed elements and $k_1 = k$ in the cases that $\mathcal{H}_{h_u}$ is each of BDDM and BDM mixed elements. Both cases are included in the CD mixed elements. In [21], we proved the convergence result.

Theorem 2.1 *Suppose that the finite element space $\mathcal{H}_{h_u}$ is each of the classical mixed elements described in [2, 3, 4, 5, 18] and that the coefficients in system (1.1) have the first and second order continuous derivatives and the solution of system (1.1) is smooth. If the mesh parameters h_c, h_u, and τ satisfy the relations*

(a) $K_* h_c^{s+2} \leq h_u^2, \quad h_u = o(h_c) \ (k = 0), \quad h_u = O(h_c) \ (k \geq 1)$,

(b) $\tau = o(h_c), \quad \tau = o(h_u)$,

$$(2.10)$$

then the a priori error estimates

(a) $\displaystyle\max_{1 \leq n \leq N} \|c^n - c_h^n\|_{L^2(\Omega)} + \max_{1 \leq n \leq N} \|\mathbf{u}^n - \mathbf{u}_h^n\|_{(L^2(\Omega))^2}$

$\quad \leq K\{h_c^{s+1} + h_u^{k+1} + \tau\}$,

(b) $\displaystyle\max_{1 \leq n \leq N} \|p^n - p_h^n\|_{L^2(\Omega)} \leq K\{h_p^{r+1} + h_c^{s+1} + h_u^{k_1} + \tau\}$

$$(2.11)$$

hold, where the constant K only depends upon some norms of the solution of system (1.1) and time T.

In the next section, we formulate a Schwarz alternating domain decomposition algorithm to solve the systems (2.4) and (2.6).

3   Domain Decomposition Algorithm

In this section, we focus on the study of the multiplicative Schwarz iterative algorithm for solving the systems (2.4) and (2.6). For sake of convenience, we assume that $\mathcal{T}_{h_c} = \mathcal{T}_{h_u} = \mathcal{T}_h$ and $k = s$. We assume that a set of overlapping sub-domains $\{\Omega_i\}_{i=1}^Q$ are given, whose boundaries are aligned with the mesh of partitions $\mathcal{T}_h$, $\Omega = \bigcup_{i=1}^Q \Omega_i$. Let $\mathcal{T}_{h,i}$ be the restriction of $\mathcal{T}_h$ in Ω_i, and $\mathcal{H}_h(\Omega_i)$ and $\mathcal{S}_h(\Omega_i)$ defined on $\mathcal{T}_{h,i}$ be the restriction of $\mathcal{H}_h$ and $\mathcal{S}_h$ in Ω_i, respectively. Define $\mathcal{H}_{h,0}(\Omega_i) = \{\mathbf{W} \in \mathcal{H}_h(\Omega); \mathbf{W} = 0 \text{ in } \Omega\backslash\Omega_i\}$ and $\mathcal{S}_{h,0}(\Omega_i) = \{Z \in \mathcal{S}_h; Z = 0 \text{ in } \Omega\backslash\Omega_i\}$. We assume that the domain decomposition satisfies the basic condition

CONDITION A. *For each $x \in \bar{\Omega}$, there exists an open domain D_x and $j \in \{1, 2, \ldots, Q\}$ such that $x \in D_x$ and $D_x \bigcap \Omega \subset \Omega_j$.*

We can define a Schwarz alternating iterative scheme to solve the system (2.4) and (2.6) step by step.

Alternating Iterative Scheme. *Given an initial approximation $(C^0, \mathbf{U}^0)$ $\in \mathcal{S}_{h_c} \times \mathcal{H}_{h_u}$ such that*

$$
\begin{aligned}
(a) \quad & (C^0, Z) = (c_0, Z) \ \forall \ Z \in \mathcal{S}_{h_c}, \\
(b) \quad & (\alpha(c_0)\mathbf{U}^0, \mathbf{W}) = -(\nabla p_0 + \gamma(c_0)\nabla H, \mathbf{W}) \ \forall \ \mathbf{W} \in \mathcal{H}_{h_u},
\end{aligned}
\tag{3.1}
$$

for $n = 1, 2, \ldots, N$, seek $(C^n, \mathbf{U}^n) \in \mathcal{S}_{h_c} \times \mathcal{H}_{h_u}$ by the following iterative method. Seek $C^n \in \mathcal{S}_{h_c}$ in the three steps: (1) Let $C_0^n = C^{n-1}$; (2) Find $C_{jQ+i}^n \in \mathcal{S}_h$, for $i = 1, 2, \ldots Q$, such that

$$
\begin{aligned}
(\phi C_{jQ+i}^n, Z) &+ \tau_n(\mathcal{D}(\mathbf{U}^{n-1})\nabla C_{jQ+i}^n, \nabla Z) + \tau_n(q_+^n C_{jQ+i}^n, Z) \\
&= (\phi \bar{C}^{n-1}, Z) + \tau_n(b(C^{n-1})\beta(C^{n-1})(\nabla \cdot \mathbf{U}^{n-1} - q^n), Z) \\
&+ \tau_n(q_+^n \tilde{c}^n, Z) \ \forall \ Z \in \mathcal{S}_{h,0}(\Omega_i),
\end{aligned}
\tag{3.2}
$$

$$
C_{jQ+i}^n = C_{jQ+i-1}^n \quad \text{in } \Omega\backslash\Omega_i,
\tag{3.3}
$$

where $\bar{C}^{n-1}$ is given by

$$
\bar{C}^{n-1} = C^{n-1}(\bar{x}), \quad \bar{x} = x - \tau_n \mathbf{U}^{n-1}(x)/\phi(x), \ \forall \ x \in \Omega,
\tag{3.4}
$$

for $j = 0, 1, \ldots, m - 1$; (3) Let $C^n = C_{mQ}^n$.

Seek $\mathbf{U}^n \in \mathcal{H}_{h_u}$ in the three steps: (1) Let $\mathbf{U}_0^n = \mathbf{U}^{n-1}$; (2) Find $\mathbf{U}_{jQ+i}^n \in \mathcal{H}_h$, for $i = 1, 2, \ldots Q$, such that

$$
\begin{aligned}
(\alpha(C^n)\mathbf{U}_{jQ+i}^n, \mathbf{W}) &+ \tau_n(\beta(C^n)\nabla \cdot \mathbf{U}_{jQ+i}^n, \nabla \cdot \mathbf{W}) \\
&= (\alpha(C^{n-1})\mathbf{U}^{n-1}, \mathbf{W}) + \tau_n(\beta(C^n)q^n, \nabla \cdot \mathbf{W}) \\
&- ((\gamma(C^n) - \gamma(C^{n-1})\nabla H, \mathbf{W}), \ \forall \ \mathbf{W} \in \mathcal{H}_{h_u,0}(\Omega_i),
\end{aligned}
\tag{3.5}
$$

$$\mathbf{U}_{jQ+i}^n = \mathbf{U}_{jQ+i-1}^n \quad \text{in } \Omega \backslash \Omega_i, \tag{3.6}$$

for $j = 0, 1, \ldots, m-1$; (3) Let $\mathbf{U}^n = \mathbf{U}_{mQ}^n$, where m denotes the iterative times at every time step.

According to the general theory of the Schwarz alternating algorithm (see [15, 16, 17]), the iterative sequence $\{(C_{mQ}^n, \mathbf{U}_{mQ}^n)\}_{m=1}^\infty$ converges to the solution $(c_h^n, \mathbf{u}_h^n)$ at each time step as iterative times m tends to infinity and under the condition A. But an infinite iteration is impossible. We want to know the convergent rate of the iterative sequence $\{(C_{mQ}^n, \mathbf{U}_{mQ}^n)\}_{m=1}^\infty$ so that we can stop the iteration at the certain accuracy. On the other hand, we hope that the iterative sequence $\{(C_{mQ}^n, \mathbf{U}_{mQ}^n)\}_{m=1}^\infty$ converges as fast as possible so that we can implement very small iterative cycles to obtain very good approximate solutions.

In next section, we state the convergence theorem.

4 Convergence Analysis

In this section, we analyze how the convergent rate of the iterative sequence $\{(C_{mQ}^n, \mathbf{U}_{mQ}^n)\}_{m=1}^\infty$ depends upon the mesh parameters h and τ under some stronger conditions than the condition A, which are practical and reasonable.

CONDITION B. *The sub-regions $\{\Omega_j\}_{j=1}^Q$ can be divided into four parts:*

$$D_j = \sum_{r_{j-1}+1 \leq i \leq r_j} \Omega_i, \quad j = 1, 2, 3, 4, \quad r_0 = 0, \quad r_4 = Q, \tag{4.1}$$

where subdomains in D_j are disjoint, $\{D_1, D_2\}$, $\{D_3, D_4\}$ and $\{D_1 \bigcup D_2, D_3 \bigcup D_4\}$ are the domain decomposition of $D_1 \bigcup D_2$, $D_3 \bigcup D_4$ and Ω, respectively, which satisfy condition A.

CONDITION C. *The sub-regions $\{\Omega_j\}_{j=1}^Q$ can be divided into κ parts:*

$$D_j = \sum_{r_{j-1}+1 \leq i \leq r_j} \Omega_i, \quad j = 1, 2, \ldots, \kappa, \quad r_0 = 0, \quad r_\kappa = Q, \tag{4.2}$$

such that

(1) $\{D_j\}_{j=1}^\kappa$ is a domain decomposition of Ω satisfying the condition A and $D_j \bigcap D_l = \emptyset$ for $l \neq j-1, j+1$.

(2) For each $1 \leq j \leq \kappa$, $\{\Omega_i\}_{i=r_{j-1}+1}^{r_j}$ is a domain decomposition of D_j satisfying the condition A and $\Omega_i \bigcap \Omega_l = \emptyset$ for $l \neq i-1, i+1$.

Conditions B and C can be satisfied easily.

Theorem 4.1 *Assume that the domain decomposition satisfies the condition B or condition C. Let* $(c^n, \mathbf{u}^n)$ *be the solution of the system* (1.1), $(C^n, \mathbf{U}^n)$ *be the solution of the alternating iterative scheme, and m be the iterative times at each time step. Then there exists the constant K, which is independent of h, τ and m, such that*

$$
\begin{aligned}
\max_{1 \leq n \leq N} \|c^n - C^n\|_{L^2(\Omega)} + \max_{1 \leq n \leq N} \|\mathbf{u}^n - \mathbf{U}^n\|_{(L^2(\Omega))^2} \\
\leq K\{h^{k+1} + \tau + \tau^{-\frac{1}{2}}(\tau + h)^{\frac{m}{4}}\}.
\end{aligned}
\tag{4.3}
$$

5   Conclusions

From the a priori error estimate (4.3), we see that the error results from two parts, one from the approximation error of the finite element function spaces and time discretization, which is bounded by $O(h^{k+1} + \tau)$, and other from alternating iteration, which is bounded by $O(\tau^{-1/2}(\tau + h)^{m/4})$. To get an approximate solution with an optimal global accuracy, infinitely iterative cycles are not needed. Generally, if $\tau = O(h^{k+1})$ and $m \geq 6(k+1)$, then the global error is bounded by $O(h^{k+1} + \tau)$. This means that only 6(k+1) cycles are required.

Acknowledgments.   This research was supported in part by China State Major Key Project for Basic Researches and by both The Research Fund for Doctoral Program of High Education and Trans-Century Training Program Foundation for the Talents by China State Education Commission.

References

[1] Adams, R. A., *Sobolev Spaces*, New York. Academic Press, 1975.

[2] Brezzi, F., Douglas, J., Jr., Duran, R., and Marini, L. D., Mixed finite elements for second order elliptic problems in three space variables, *Numer. Math.* **51** (1987), 237–250.

[3] Brezzi, F., Douglas, J., Jr., Fortin, M., and Marini, L. D., Efficient rectangular mixed finite elements in two and three space variables, *RAIRO Model Math. Anal. Numer.* **4(21)** (1987), 581–604.

[4] Brezzi, F., Douglas, J., Jr., and Marini, L. D., Two families of mixed finite elements for second order elliptic problems, *Numer. Math.* **47** (1985), 217–235.

[5] Chen, Z. and Douglas, J., Jr., Prismatic mixed finite elements for second order elliptic problems, *Calcolo* **26** (1989), 135–148.

[6] Chou, S. H. and Li, Q., Mixed finite element methods for compressible miscible displacement in porous media, *Math. Comp.* **57** (1991), 507–527.

[7] Ciarlet, P. G., *The Finite Element Methods for Elliptic Problems*, North-Holland, New York, 1978.

[8] Douglas, J., Jr., The numerical simulation of miscible displacement in porous media, *Comput. Methods Nonlin. Mech.*, North-Holland, Amsterdam, 1980.

[9] Douglas, J., Jr., Finite difference methods for two-phase incompressible flow in porous media, *SIAM J. Numer. Anal.* **20** (1983), 681–696.

[10] Douglas, J., Jr. and Roberts, J. E., Numerical methods for a model for compressible miscible Displacement in porous Media, *Math. Comp.* **41** (1983), 441–459.

[11] Douglas, J., Jr., Simulation of miscible displacement in porous media by a modified method of characteristics procedure, *Numerical Analysis. Lecture Notes in Math* 912, Springer-Verlag, Berlin, 1982.

[12] Douglas, J., Jr., Ewing, R. E., and Wheeler, M. F., The approximation of the pressure by a mixed method in the simulation of miscible displacement, *RAIRO, Anal. Numer.* **17** (1983), 17–33.

[13] Douglas, J., Jr., Ewing, R. E., and Wheeler, M. F., A time-discretization procedure for a mixed finite element approximation of miscible displacement in poro media, *RAIRO Anal. Numer.* **17** (1983), 249–256.

[14] Duran, R., On the approximation of miscible displacement in porous media by the modified methods of characteristics combined with a mixed method. *SIAM J. Numer. Anal.* **25** (1988), 989–1001.

[15] Dryja, M. and Widlund, O. B., Some domain decomposition algorithms for elliptic problems, *Proceedings of the Conference on Iterative Methods for Large Linear System held in Austin*, D. Young, Jr. ed., Academic Press, Orlando, 1989, 121–134.

[16] Lions, P. L., On Schwarz alternating method, Part I, *Proceedings of the first international symposium on domain decomposition methods for partial differential equations*, SIAM. Philadelphia, Glowinski, R., Golub, G. H., Meurant, G. A., and Périaux, J. (eds.), 1988.

[17] Lions, P. L., On Schwarz alternating method, Part II, *Proceedings of the second international symposium on domain decomposition methods for partial differential equations*, SIAM. Philadelphia, Chan, T., Glowinski, R., Périaux, J., and Widlund, O. B., eds., 1989.

[18] Raviart, P. A. and Thomas, J. M., A mixed finite element method for 2nd order elliptic problems, *Mathematical Aspects of Finite Element Methods, Lecture Notes in Math* 606, Springer-Verlag, Berlin and New York, 1977, 292–315.

[19] Wheeler, M. F., A priori error estimates for Galerkin approximations to parabolic partial differential equations, *SIAM J. Numer. Anal.* 4 (1973), 723–759.

[20] Yang, D. P., Approximation and its optimal error estimate of displacement of two-phase incompressible flow by mixed finite element and a modified method of characteristics, *Chinese Science Bulletin* 35 (1990), 1686–1689.

[21] Yang, D. P., A characteristic mixed finite element method for displacement problem of compressible flows in porous media, *Science in China, Series A* (1998).

[22] Yuan, Y. R., Characteristic-mixed finite element method for enhanced oil recovery simulation and optimal order error estimate, *Chinese Science Bulletin* 38 (1993), 1761–1766.

A Boundary Element Method for Viscous Flow on Multi-Connected Domains

DEQUAN YANG   TIGUI FAN   XINYU YANG

Abstract

A viscous flow problem in multi-connected domains is studied using a boundary element method. The results obtained are compared with those in [3] using a finite element method. These results show that the method used can treat the viscous flow with many different kinds of boundary conditions. This method is convenient and faster than others.

KEYWORDS: boundary element method, multi-connected domain, viscous flow

1 Introduction

In practical viscous flow problems in multi-connected domains such as porous springs, the piping and well drilling distribution in oil extraction is very important for engineering designs and is a very difficult problem to solve. Finite difference methods face many difficulties because of complicated boundary conditions. Finite element methods can treat many different boundary conditions, but they need lots of input messages, occupy many RAMs, and take a lot of time. Due to the nonlinearity of equations in fluid mechanics and non-selfadjointness of differential operators, relevant variational functionals hardly exist. For this reason, the finite element methods based on the method of weighted residuals such as Galerkin's approach, least squares, and collocation methods [2, 1] do not possess any physical meaning. They are nothing but a pure optimum scheme for numerical treatment. Using newly developed boundary element methods [4, 5], we study the above mentioned problems in this paper. We change multi-connected domain viscous flows into nonlinear boundary integration equations of the boundary velocity and boundary pressure, then discretize boundary integral equations, and finally change the velocity and pressure differential into the differential of the fundamental solution. This treatment that the differential of the unknown variables is changed into that of the known variable makes the problem easier to solve. The boundary variable is determined singly. After determining boundary points, any inner points can be calculated. Other methods are difficult to do that. In calculating nonlinear problems, creating grids is easier than finite element methods. Integral equations are discretized in algebraic equations. Using iterative methods, the unknown velocity and pressure are calculated. It is important that the velocity and pressure is calculated separately. Comparing with [3], this method is more effective.

2 A Fundamental Equation

Let the multi-connected domain be Ω with boundary $\Gamma = \sum_i \Gamma_{ti} + \sum_j \Gamma_{vj} (i = 1, 2, \ldots, t, j = 1, 2, \ldots, s)$. The known velocity and pressure on boundary Γ_{vj} and Γ_{ti} are shown in Fig. 1. In Ω, the dimensionless form of incompressible viscous fluid is

$$\nabla \cdot T(V) = \nabla \cdot (VV), \quad \nabla \cdot V = 0 \quad \text{in } \Omega, \tag{2.1}$$

with the boundary conditions

$$T \cdot \vec{n}|_{\Gamma_{ti}} = \vec{t}_i, \, j = 1, 2, \ldots, t, \quad V|_{\Gamma_{vj}} = V_j, \, j = 1, 2, \ldots, s,$$

where $T(V)$ is the stress tensor corresponding to V and $\vec{n}$ is the outward unit normal to Γ.

If W^k and q^k are fundamental solutions of Stokes's equation and $T_{ij}(W^k)$ is the stress tensor corresponding to W^k, we can obtain the integral equation

$$\begin{aligned}
C(X)v_k(X) = & \oint_\Gamma n_i(X_0)T_{ij}(W^k(X - X_0)_0 v_j(X_0)d\Gamma_0 \\
& + \oint_\Gamma n_i(X_0)v_i(X_0)v_j(X_0)w_j^k(X - X_0)d\Gamma_0 \\
& - \oint_\Gamma n_i(X_0)T_{ij}(V(X_0))_0 w_j^k(X - X_0)d\Gamma_0 \\
& + \int_\Omega v_i(X_0)w_{j,i}(X - X_0)v_j(X_0)d\Omega_0, \quad k = 1, 2, 3,
\end{aligned} \tag{2.2}$$

where

$$C(X) = 1 \quad \text{in } \Omega, \quad 1/2 \quad \text{on } \Gamma, \quad \text{and} \quad 0 \quad \text{not in } \overline{\Omega}.$$

Formulas determining pressures are

$$P(X) = \oint_\Gamma n_i(X_0)\{q^i(X - X_0)[v_i(X_0)v_j(X_0) - T_{ij}(V(X_0))_0]$$
$$+ \tfrac{2}{R_e} q_{i,j0}(X - X_0)v_j(X_0)\}d\Gamma_0 \qquad (2.3)$$
$$- \int_\Omega v_i(X_0)v_j(X_0)q_{i,j0}(X - X_0)d\Omega, \quad x \in \Omega.$$

(2.2) and (2.3) are the fundamental equation treating a multi-connected domain.

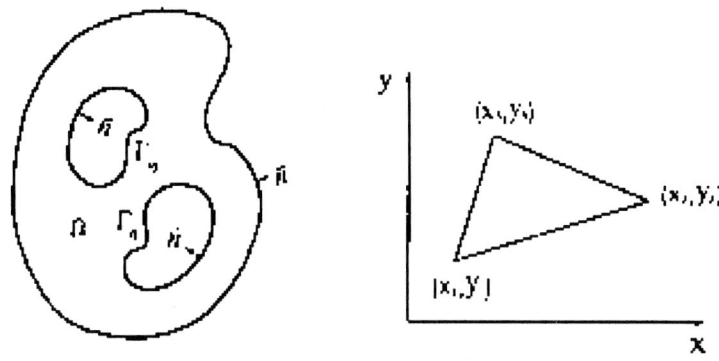

Fig 1. multi connected domain   Fig 2. triangle grid

3 Discretization of Boundary Integrals

Change Γ into N elements (for 2D) or N polygonal elements (for 3D) Γ_β, $\beta = 1, 2, \ldots, N$. Change the flow region Ω into E elements Ω_γ, $\gamma = 1, 2, \ldots, E$. The velocity and pressure on Γ_β are replaced by that at the center of the figure. The velocity of Ω_γ is also replaced by that at the center of the figure.

Formulas for discretization are given by

$$C_\alpha v_{k\alpha} = \sum_{\beta ij} L_{kij\alpha\beta} n_{i\beta} v_{j\beta} - \sum_{\beta j} M_{kj\alpha\beta} t_{j\beta}$$
$$+ \sum_{\beta ij} M_{kj\alpha\beta} n_{i\beta} v_{i\beta} v_{j\beta} - \sum_{\gamma ij} N_{kij\alpha\gamma} v_{ie_\gamma} v_{je_\gamma}, \qquad (3.1)$$

where $\alpha = \Gamma_1, \Gamma_2, \ldots, \Gamma_N$, $e_1, e_2, \ldots, e_E$, $\beta = 1, 2, \ldots, N$, $\gamma = 1, 2, \ldots, E$, $i, j, k = 1, 2, 3$,

$$L_{kij\alpha\beta} = \int_{\Gamma_\beta} T_{ij}(W^k(X_\alpha - X_0))_0 d\Gamma_0,$$

$$M_{kj\alpha\beta} = \int_{\Gamma_\beta} w_j^k (X_\alpha - X_0) d\Gamma_0,$$

$$N_{kij\alpha\gamma} = \int_{e_\gamma} w_{j,i_0}^k (X_\alpha - X_0) d\Omega_0.$$

When Γ_β is a small segment element, Ω_γ is a triangle element (shown in Fig. 2). Three sides of the triangle are as follows:

$$y = y_1 + k_1(x - x_1), \quad y = y_2 + k_2(x - x_2), \quad y = y_3 + k_3(x - x_3).$$

The integrals of an arbitrary function $f(x, y, x_\alpha, y_\alpha)$ on the triangle region are

$$\int_{\Delta_\gamma} f(x, y, x_\alpha, y_\alpha) d\Omega = -\sum_{l=1}^{3} \int_{x_l}^{x_m} dx \int_{0}^{y_l - k_l(x-x_1)} f(x, y, x_\alpha, y_\alpha) dy,$$

where $l = 1, 2, 3$ and $m = 2, 3, 1$. Analytic forms of geometric coefficients can be obtained.

4   Numerical Solutions

After calculating $L_{kij\alpha\beta}$, $M_{kj\alpha\beta}$, and $N_{kij\alpha\gamma}$, equation (3.1) can be solved by using an iterative method. The pressure $t_{j\beta}$ on Γ_v and velocity $v_{i\beta}$ on Γ_t can be calculated first. We have $N \times N$ linear equations

$$\frac{1}{2} v_{k\alpha}^{l+1} = \sum_{\beta in} L_{kij\alpha\beta} n_{i\beta} v_{i\beta}^{l+1} - \sum_{\beta j} M_{kj\alpha\beta} t_{j\beta}^{i+1}$$
$$+ \sum_{\beta ij} M_{kj\alpha\beta} n_{i\beta} v_{i\beta}^l v_{j\beta}^l - \sum_{\gamma ij} N_{kij\alpha\gamma} v_{ie_\gamma}^l v_{je_\gamma}^l,$$

where $k = 1, 2, 3$, $\alpha = \Gamma_1, \Gamma_2, \ldots, \Gamma_N$, and

$$v_{k\alpha}^l = \sum_{\beta ij} L_{kij\alpha\beta} n_{i\beta} v_{j\beta} - \sum_{\beta j} M_{kj\alpha\beta} t_{j\beta}^l$$
$$+ \sum_{\beta ij} M_{kj\alpha\beta} v_{i\beta}^l v_{j\beta}^l n_{i\beta} - \sum_{\gamma ij} N_{kij\alpha\gamma} v_{ie_\gamma}^{l-1} v_{je_r}^{l-1}.$$

The boundary stress and velocity are given by boundary conditions.

5   An Example and Conclusions

As an example, to compare with [3], we calculate the viscous flow in a rectangular cavity having a rectangular obstacle. The rectangular cavity is 10×12 and rectangular obstacle is 1.7×1.7. The boundary conditions are

$$u_{\Gamma_i} = \begin{cases} 1 - \left(\dfrac{y}{H}\right)^2, & i = 1, 2, \\ 0, & i = 3, 4, 5, \end{cases} \qquad v_{\Gamma_i} = 0, \quad i = 1, 2, 3, 4, 5.$$

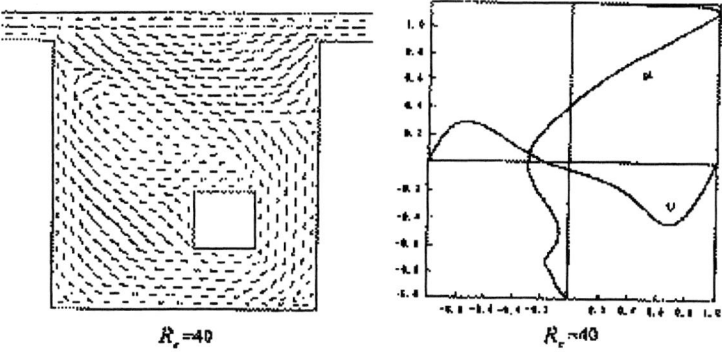

Fig3 flow field distribution   Fig4 velocity distribution

Using a triangular grid, the viscous flow is calculated when $R_e = 30, 40$. The flow field distribution is shown in Fig. 3. In Fig. 3, a big vortex is formed on the top of cavity and inclining to the left, two small vortexes are formed on both angles of the cavity. Our results agree with [3] well. The velocity distribution in the x direction in vertical centerlines and the y direction in horizontal centerlines are shown in Fig. 4. Its changing laws agree with experimental data. For accelerating convergence, we use a operator

$$v_{k\alpha}^{l+1} = v_{k\alpha}^{l} + 0.6(v_{k\alpha}^{l+1} - v_{k\alpha}^{l}).$$

The resulting scheme is effective.

Acknowledgements. This project is in part supported by NSFC.

References

[1] Chung, T. J., *Finite Element Analysis in Fluid Dynamics*, 1978.

[2] Connor, J. J. and Brebbia, C. A., *Finite Element Techniques for Fluid Flow*, New Butterworths, 1976.

[3] Mizukami, A., A stream function-vorticity finite element formulation for Navier-Stokes equations in multi-connected domains, *Inter. J. Numer. Methods Eng.* **19** (1983), 1730-1741.

[4] Yang, Q. and Chen, G., *Progress in Boundary Element Method of 2-D Viscous Flow*, Numerical Methods in Laminar and Turbuler Flow, Vol. VIII, Pau, 2, 1993.

[5] Zhao, Z. and Yang, D., *Boundary Element Methods*, Inner Mongolia University Press, 1994.

A Characteristic Difference Method for 2D Nonlinear Convection-Diffusion Problems

Xi-Jun Yu   Yonghong Wu

Abstract

In this paper we construct a characteristic difference method for two-dimensional nonlinear convection-diffusion problems. These problems arise in the modeling of fluid flow and transport in porous media, for example. The method is analysed mathematically and an error bound is derived. Convergence of the method can be achieved under a milder restriction in the temporal stepsize and spatial stepsize than those required by existing methods.

KEYWORDS: convection, diffusion, characteristic, difference method, convergence, error estimate

1 Introduction

Nonlinear convection-diffusion problems arise from many real world problems such as underground percolation, air-pollution, flows, and chemical diffusion. Various characteristic finite element and characteristic finite difference methods [1, 4, 5] have been developed to solve the convection dominated linear problems. However, the convergence of those methods requires Δt to be in the same order as h, namely $\Delta t = O(h)$. Durán [2] improved the methods to allow Δt to be in the higher order of h. But his analytic way of convergence can not be applied into the nonlinear problems.

In this paper, we construct two characteristic difference methods based on the linear and quadratic interpolations for two-dimensional convection dominated nonlinear convection-diffusion problems. The methods are then analysed mathematically and formulas for the estimation of errors are developed. The problem considered is the two-dimensional nonlinear convection-diffusion problem with the Dirichlet boundary conditions:

$$c(\bar{x})\frac{\partial u}{\partial t} + \bar{b}(\bar{x}, u) \cdot \nabla u - [\frac{\partial}{\partial x}(a_1(\bar{x}, u)\frac{\partial u}{\partial x}) + \frac{\partial}{\partial y}(a_1(\bar{x}, u)\frac{\partial u}{\partial y})]$$
$$= f(t, \bar{x}, u) \quad \text{in } \Omega \times (0, T], \tag{1.1}$$
$$u(t, \bar{x}) = 0 \text{on } \partial\Omega \times (0, T], \quad u(0, \bar{x}) = u_0(\bar{x}) \quad \text{in } \Omega,$$

where Ω is a rectangular domain, $\partial\Omega$ is the boundary of Ω, $\bar{x} = (x, y)$, $\bar{b}(\bar{x}, u) = (b_1(\bar{x}, u), b_2(\bar{x}, u))$, and $a_i(\bar{x}, u)$ $(i = 1, 2)$ are much smaller than $c(\bar{x})$ and $\bar{b}(\bar{x}, u)$.

Assume that the solution of equation (1.1) exists and is unique and the coefficients c, $\bar{b}$, a_i $(i = 1, 2)$ and initial function u_0 satisfy the following conditions:

    i.) $\|u_0\|_{L^\infty(\Omega)} \leq K_1$, $\|u\|_{L^\infty(0,T;L^\infty(\Omega))} \leq K_1$.

    ii.) For $\forall(\bar{x}, p) \in \Omega \times [-2K_1, 2K_1]$, there exist $0 < c_1 \leq c(\bar{x}) \leq c_2$, $|b(\bar{x}, p)| \leq K_2$, $a_i(\bar{x}, p) \geq a_0 > 0$, $i = 1, 2$.

    iii.) The first order partial derivatives of $b_i(\bar{x}, p)$ and $a_i(\bar{x}, p)$ are bounded and $f(t, \bar{x}, p)$ is Lipschitz continuous with respect to $p \in [-2K_1, 2K_1]$.

Here K_1, K_2, c_1, c_2, and a_0 are positive constants, $W^{m,p}(\Omega)$ denotes the Sobolev space of m degree on Ω, and $W^{k,\infty}(0, T; W^{m,p}(\Omega)) = \{v, \frac{\partial^s v}{\partial t^s} : [0, T] \to W^{m,p}(\Omega)$ and $\|v\|_{W^{s,\infty}(0,T;W^{m,p}(\Omega))} < +\infty, s = 1, 2, \ldots, k\}$, where

$$\|v\|_{W^{s,\infty}(0,T;W^{m,p}(\Omega))} = \max_{0 \leq i \leq s} ess \sup_{0 \leq t \leq T} \|\frac{\partial^i v}{\partial t^i}\|_{W^{m,p}(\Omega)}.$$

For simplicity, in the numerical analysis, let $X(0, T; Z(\Omega)) = X(Z(\Omega))$, M_i $(i = 1, 2, \ldots)$ denote bounded positive constants, and $\tilde{M}(s_1, s_2, \ldots, s_r)$ be a positive constant depending on $s_1, s_2, \ldots, s_r$.

2   Characteristic Difference Methods

Let $\psi(\bar{x}, u) = \sqrt{c(\bar{x})^2 + |\bar{b}(\bar{x}, u)|^2} = \sqrt{c(\bar{x})^2 + b_1(\bar{x}, u)^2 + b_2(\bar{x}, u)^2}$ and $\tau(\bar{x}, u)$ denote the characteristic direction of the operator $c(\bar{x})\frac{\partial u}{\partial t} + \bar{b}(\bar{x}, u) \cdot \nabla u$. Then

$$\frac{\partial u}{\partial \tau} = \frac{c(\bar{x})}{\psi(\bar{x}, u)}\frac{\partial u}{\partial t} + \frac{\bar{b}(\bar{x}, u)}{\psi(\bar{x}, u)} \cdot \nabla u.$$

Along the characteristic direction, equation (1.1) becomes

$$\psi(\bar{x}, u)\frac{\partial u}{\partial \tau} - [\frac{\partial}{\partial x}(a_1(\bar{x}, u)\frac{\partial u}{\partial x}) + \frac{\partial}{\partial y}(a_1(\bar{x}, u)\frac{\partial u}{\partial y})] = f(t, \bar{x}, u),$$
$$u(t, \bar{x}) = 0, \quad u(0, \bar{x}) = u_0(\bar{x}). \tag{2.1}$$

Let $\bar{\Omega}_h$ denote a mesh partition of the domain Ω, X_0 and Y_0 be the side lengths of Ω, $h_1 = \frac{X_0}{J_1}$ and $h_2 = \frac{Y_0}{J_2}$ be the spatial stepsizes, $\Delta t = \frac{T}{N}$ be the temporal stepsize, and $t_n = n\Delta t$ $(n = 0, 1, \ldots, N)$. To solve (2.1) numerically, approximate the derivative along τ by the difference quotient at point (x_i, y_j, t_n), namely

$$\psi(\bar{x}_{i,j}, u_{i,j}^n)(\frac{\partial u}{\partial \tau})_{i,j}^n \approx \psi(\bar{x}_{i,j}, u_{i,j}^n)\frac{u(\bar{x}_{i,j}, t_n) - u(\tilde{x}_{i,j}, t_n)}{\sqrt{(\bar{x}_{i,j} - \tilde{x}_{i,j})^2 + \Delta t^2}} = c(\bar{x}_{i,j})\frac{u_{i,j}^n - \tilde{u}_{i,j}^{n-1}}{\Delta t},$$

where $\bar{x}_{i,j} = (x_i, y_j)$, $\tilde{x}_{i,j} = \bar{x}_{i,j} - \bar{b}(\bar{x}_{i,j}, u_{i,j}^n)\Delta t/c(\bar{x}_{i,j})$, $c(\bar{x}_{i,j}) = c(x_i, y_j)$, $b_k(\bar{x}_{i,j}, u_{i,j}^n) = b_k(x_i, y_j, u_{i,j}^n)$ $(k = 1, 2)$, $u_{i,j}^n = u(x_i, y_j, t_n)$, and $\tilde{u}_{i,j}^{n-1} = u(\tilde{x}_{i,j}, t_{n-1})$. Further, by intruducing

$$\delta_{x,+}u_{i,j} = \frac{u_{i+1,j}^n - u_{i,j}^n}{h_1}, \quad \delta_{x,-}u_{i,j} = \frac{u_{i,j}^n - u_{i-1,j}^n}{h_1},$$

the partial derivatives of u with respect to x and y in (1.1) can be expressed by

$$\delta_{x,-}(a_1\delta_{x,+}u)_{i,j}^n = h_1^{-2}[a_{1,i+\frac{1}{2}}(u_{i+1,j}^n - u_{i,j}^n) - a_{1,i-\frac{1}{2}}(u_{i,j}^n - u_{i-1,j}^n)],$$

$$\delta_{y,-}(a_2\delta_{y,+}u)_{i,j}^n = h_2^{-2}[a_{2,i+\frac{1}{2}}(u_{i+1,j}^n - u_{i,j}^n) - a_{2,i-\frac{1}{2}}(u_{i,j}^n - u_{i-1,j}^n)],$$

where $a_{1,i+\frac{1}{2}} = \frac{1}{2}[a_1(\bar{x}_{i,j}, u_{i,j}^n) + a_1(\bar{x}_{i+1,j}\ u_{i+1,j}^n)]$ and $a_{2,i+\frac{1}{2}} = \frac{1}{2}[a_2(\bar{x}_{i,j}, u_{i,j}^n) + a_2(\bar{x}_{i+1,j}, u_{i+1,j}^n)]$.

Assume that $\{U_{i,j}^{n-1}\}$ is known and let

$$\hat{\tilde{x}}_{i,j} = \bar{x}_{i,j} - \bar{b}(\bar{x}_{i,j}, U_{i,j}^{n-1})\Delta t/c(\bar{x}_{i,j}) \tag{2.2}$$

be an approximation value of $\tilde{x}_{i,j}$, where $\hat{\tilde{x}}_{i,j}$ and $\tilde{x}_{i,j}$ refer to quantities at t_n but for simplicity, we omitted the superscript n.

Let $U^{n-1}(\bar{x})$ be the piecewise interpolation function obtained from $\{U_{i,j}^{n-1}\}$ on the spatial mesh $\bar{\Omega}_h$ and $\tilde{U}_{i,j}^{n-1} = U^{n-1}(\hat{\tilde{x}}_{i,j})$; then a characteristic difference scheme for the solution of equation (2.1) is

$$c_{i,j}\frac{U_{i,j}^n - \tilde{U}_{i,j}^{n-1}}{\Delta t} - [\delta_{x,-}(A_1\delta_{x,+}U)_{i,j}^n + \delta_{y,-}(A_2\delta_{y,+}U)_{i,j}^n] = f(t_n, \bar{x}_{i,j}, U_{i,j}^{n-1})$$

$$i = 1, 2, \ldots, J_1 - 1, \ j = 1, 2, \ldots, J_2 - 1, \ n = 1, 2, \ldots, N,$$

$$U_{i,j}^0 = u_0(\bar{x}_{i,j}), \ i = 1, 2, \ldots, J_1 - 1, \ j = 1, 2, \ldots, J_2 - 1,$$

$$U_{i,j}^n = 0, \ \bar{x}_{i,j} \in \partial\Omega$$

$$\tag{2.3}$$

where $A_{1,i+\frac{1}{2}} = \frac{1}{2}[a_1(\bar{x}_{i,j}, U_{i,j}^{n-1}) + a_1(\bar{x}_{i+1,j}, U_{i+1,j}^{n-1})]$ and $A_{2,i+\frac{1}{2}} = \frac{1}{2}[a_2(\bar{x}_{i,j}, U_{i,j}^{n-1}) + a_2(\bar{x}_{i+1,j}, U_{i+1,j}^{n-1})]$.

Remark 1. The error of the characteristic difference scheme (2.3) on the time direction is of order $O(\Delta t)$.

Remark 2. If the interpolating function $U^{n-1}(\bar{x})$ is piecewise bilinear, (2.3) is called a linear characteristic difference scheme. If $U^{n-1}(\bar{x})$ is piecewise biquadratic, (2.3) is known as a quadratic characteristic difference scheme.

To increase the accuracy of time integration, we discretize the first term of (2.1) by

$$\psi(\bar{x}_{i,j}, u_{i,j}^n)\Big(\frac{\partial u}{\partial \tau}\Big)_{i,j}^n \approx \psi(\bar{x}_{i,j}, u_{i,j}^n)\frac{\frac{3}{2}u(\bar{x}_{i,j},t_n)-2u(\tilde{x}_{i,j},t_{n-1})+\frac{1}{2}u(\tilde{\tilde{x}}_{i,j},t_{n-2})}{\sqrt{(\bar{x}_{i,j}-\tilde{x}_{i,j})^2+\Delta t^2}}$$

$$= c(\bar{x}_{i,j})\frac{\frac{3}{2}u_{i,j}^n-2\tilde{u}_{i,j}^{n-1}+\frac{1}{2}\tilde{\tilde{u}}_{i,j}^{n-2}}{\Delta t}$$

where $\tilde{x}_{i,j} = \bar{x}_{i,j} - \bar{b}(\bar{x}_{i,j}, u_{i,j}^n)\Delta t/c(\bar{x}_{i,j})$, $\tilde{\tilde{x}}_{i,j} = \bar{x}_{i,j} - 2\bar{b}(\bar{x}_{i,j}, u_{i,j}^n)\Delta t/c(\bar{x}_{i,j})$, $\tilde{u}_{i,j}^{n-1} = u(\tilde{x}_{i,j}, t_n)$, and $\tilde{\tilde{u}}_{i,j}^{n-2} = u(\tilde{\tilde{x}}_{i,j}, t_n)$. This scheme has the second-order accuracy on the temporal direction.

Assume that $\{U_{i,j}^{n-1}\}$ and $\{U_{i,j}^{n-2}\}$ are known and $U^{n-1}(\bar{x})$ and $U^{n-2}(\bar{x})$ are obtained by the piecewise interpolation bilinear or biquadratic functions of $\{U_{i,j}^{n-1}\}$ and $\{U_{i,j}^{n-2}\}$ on the spatial mesh $\bar{\Omega}_h$, respectively. Then we obtain the multistep characteristic difference scheme for (2.1)

$$c_{i,j}\frac{\frac{3}{2}U_{i,j}^n-2\tilde{U}_{i,j}^{n-1}+\frac{1}{2}\tilde{\tilde{U}}_{i,j}^{n-2}}{\Delta t} - [\delta_{x,-}(A_1\delta_{x,+}U)_{i,j}^n + \delta_{y,-}(A_1\delta_{y,+}U)_{i,j}^n]$$

$$= f(t_n, \bar{x}_{i,j}, \theta_{i,j}^{n-1}), \quad i=1,\ldots,J_1-1, \quad j=1,\ldots,J_2-1, \quad n=2,\ldots,N,$$

$$U_{i,j}^0 = u_0(\bar{x}_{i,j}), \quad i=1,2,\ldots,J_1-1, \quad j=1,2,\ldots,J_2-1,$$

$$U_{i,j}^n = 0, \quad \bar{x}_{i,j} \in \partial\Omega,$$

$$(2.4)$$

where $\tilde{U}_{i,j}^{n-1} = U^{n-1}(\hat{\tilde{x}}_{i,j}, t_{n-1})$, $\tilde{\tilde{U}}_{i,j}^{n-2} = U(\hat{\tilde{\tilde{x}}}_{i,j}, t_{n-2})$, $\hat{\tilde{x}}_{i,j} = \bar{x}_{i,j} - \bar{b}(\bar{x}_{i,j}, \theta_{i,j}^{n-1})\Delta t/c(\bar{x}_{i,j})$, $\hat{\tilde{\tilde{x}}}_{i,j} = \bar{x}_{i,j} - 2\bar{b}(\bar{x}_{i,j}, \theta_{i,j}^{n-1})\Delta t/c(\bar{x}_{i,j})$, $\theta_{i,j} = 2U_{i,j}^{n-1} - U_{i,j}^{n-2}$, $A_{1,i+\frac{1}{2}} = \frac{1}{2}[a_1(\bar{x}_{i,j}, \theta_{i,j}^{n-1}) + a_1(\bar{x}_{i+1,j}, \theta_{i+1,j}^{n-1})]$, and $A_{2,j+\frac{1}{2}} = \frac{1}{2}[a_2(\bar{x}_{i,j}, \theta_{i,j}^{n-1}) + a_2(\bar{x}_{i,j+1}, \theta_{i,j+1}^{n-1})]$.

With (2.4), we obtain $U_{i,j}^n$ for $n = 2, 3, \ldots$ while $\{U_{i,j}^1\}$ can be obtained by the Taylor expansion combining with equation (1.1), namely

$$U_{i,j}^1 = u_0(\bar{x}_{i,j}) + \frac{c(\bar{x}_{i,j})}{\psi(\bar{x}_{i,j}, u_0(\bar{x}_{i,j}))}\Big(\frac{\partial u}{\partial t}\Big)_0\Delta t,$$

where $\|u^1 - U^1\| = O(\Delta t^2)$.

Similar to the characteristic difference scheme (2.3), (2.4) is known as the multistep characteristic difference scheme.

To analyse the convergence property of the characteristic difference methods (2.3) and (2.4), we define the inner products and norms for the mesh functions Y and Z on $\bar{\Omega}_h$ as follows:

$$(Y, Z) = \sum_{i,j=1}^{J_1-1,J_2-1} y_{i,j} z_{i,j} h_1 h_2, \quad ||Y||^2 = (Y, Y),$$

$$[Y, Z) = \sum_{i=0,j=1}^{J_1-1,J_2-1} y_{i,j} z_{i,j} h_1 h_2, \quad |[Y||^2 = (Y, Y),$$

$$(Y, Z] = \sum_{I=1,j=0}^{J_1-1,J_2-1} y_{i,j} z_{i,j} h_1 h_2, \quad ||Y]|^2 = (Y, Y),$$

$$[Y, Z] = \sum_{i,j=0}^{J_1,J_2} y_{i,j} z_{i,j} h_1 h_2, \quad |[Y]|^2 = [Y, Y].$$

3   Convergence
of the Characteristic Difference Scheme

In this section, we give the convergence analysis of the singlestep characteristic difference scheme (2.3). The convergence of multistep characteristic difference scheme (2.4) can be proved similarly.

Let $u(\bar{x}, t)$ be the exact solution of (2.1) and $U_{i,j}^n$ be the approximate solution obtained from the characteristic difference scheme (2.3). (2.1) is discretized at $(\bar{x}_{i,j}, t_n)$. We have

$$c_{i,j} \frac{u_{i,j}^n - \tilde{u}_{i,j}^{n-1}}{\Delta t} - [\delta_{x,-}(a_1 \delta_{x,+} u)_{i,j}^n + \delta_{y,-}(a_2 \delta_{y,+} u)_{i,j}^n] = f(t_n, \bar{x}_{i,j}, u_{i,j}^n) + e_{i,j}^n \tag{3.1}$$

where $\tilde{u}_{i,j}^{n-1} = u^{n-1}(\tilde{x}_{i,j},) = u(\tilde{x}_{i,j,}, t_{n-1})$ and $e_{i,j}^n$ is a local truncation error.

Let $\xi^n = u^n - U^n$. Then from (2.3) and (3.1), we have

$$c_{i,j} \frac{\xi_{i,j}^n - \xi^{n-1}(\tilde{x}_{i,j})}{\Delta t} - [\delta_{x,-}(A_1 \delta_{x,+} \xi)_{i,j}^n + \delta_{y,-}(A_2 \delta_{y,+} \xi)_{i,j}^n] = (f(t_n, \bar{x}_{i,j}, u_{i,j}^n)$$

$$-f(t_n, \bar{x}_{i,j}, U_{i,j}^{n-1})) + [\delta_{x,-}((a_1 - A_1)\delta_{x,+} u)_{i,j}^n + \delta_{y,-}((a_2 - A_2)\delta_{y,+} u)_{i,j}^n]$$

$$+c_{i,j} \frac{u^{n-1}(\tilde{x}_{i,j}) - u^{n-1}(\hat{x}_{i,j})}{\Delta t} + e_{i,j}^n.$$

Multipying equation this equation by $\xi_{i,j}^n h_1 h_2$, summing for $i = 1$ to $J_1 - 1$ and $j = 1$ to $J_2 - 1$, noting that $\xi_{0,j}^n = \xi_{J_1,j} = \xi_{i,0} = \xi_{i,J_2} = 0$, and using the discrete Green formula, we have

$$(\frac{\xi^n - \tilde{\xi}^{n-1}}{\Delta t}, \xi^n) + [A_1 \delta_{x,+} \xi^n, \delta_{x,+} \xi^n) + (A_2 \delta_{y,+} \xi^n, \delta_{y,+} \xi^n] = (f(t_n, \bar{x}, u^n)$$

$$-f(t_n, \bar{x}, U^{n-1}), \xi^n) + (\delta_{x,-}(a_1 - A_1)\delta_{x,+} u)^n, \xi^n) + (\delta_{y,-}((a_2 - A_2)\delta_{y,+} u)^n,$$

$$\xi^n) + (c\frac{u^{n-1}(\tilde{x}_{i,j}) - u^{n-1}(\hat{x}_{i,j})}{\Delta t}, \xi^n) + (e^n, \xi^n) = I_1 + I_2 + I_3 + I_4, \tag{3.2}$$

where c, ξ, a_k, A_k ($k = 1, 2$), and $c\xi$ are the mesh functions on $\bar{\Omega}_h$ such that $c = \{c(\bar{x}_{1,1}), c(\bar{x}_{1,2}), \ldots, c(\bar{x}_{J_1-1,J_2-1})\}$, etc. As

$$||[u^n - u^{n-1}]||^2 = \Delta t^2 \sum_{i,j=0}^{J_1,J_2} (\frac{u_{i,j}^n - u_{i,j}^{n-1}}{\Delta t})^2 h_1 h_2 \leq \tilde{M}(||u||_{W^{1,\infty}(L^\infty)})\Delta t^2, \quad (3.3)$$

by the assumption iii.), we have an estimate for the first term on the right hand side of (3.2)

$$|I_1| \leq K_3 ||u^n - U^{n-1}|| ||\xi^n|| \leq K_3 (||u^n - u^{n-1}|| + ||\xi^{n-1}||)||\xi^n|| \qquad (3.4)$$
$$\leq \tilde{M}(||u||_{W^{1,\infty}(L^\infty)})(||\xi^n||^2 + ||\xi^{n-1}||^2 + \Delta t^2),$$

where K_3 is a positive constant. For the second term on the right hand side of (3.2), we have

$$|I_2| = |((a_1 - A_1)^n \delta_{x,+} u^n, \delta_{x,+} \xi^n)| + |((a_2 - A_2)^n \delta_{y,+} u^n, \delta_{y,+} \xi^n)|$$
$$\leq \varepsilon[||\delta_{x,+}\xi^n||^2 + ||\delta_{y,+}\xi^n||^2] + M_1[||(a_1 - A_1)^n \delta_{x,+} u^n||^2$$
$$+ ||(a_2 - A_2)^n \delta_{y,+} u^n||^2],$$

where ε is a positive constant. Since

$$||[(a_1 - A_1)^n \delta_{x,+} u^n||^2 \leq ||u^n||_{W^{1,\infty}}^2 \sum_{i=0,j=1}^{J_1-1,J_2-1} ((a_1 - A_1)_{i+\frac{1}{2},j}^n)^2 h_1 h_2,$$

from assumption iii.), we have

$$|(a_1 - A_1)_{i+\frac{1}{2},j}^n| \leq \tfrac{1}{2} K_3 \{|u_{i,j}^n - U_{i,j}^{n-1}| + |u_{i+1,j}^n - U_{i+1,j}^{n-1}|\}$$
$$\leq \tfrac{1}{2} K_3 \{|u_{i,j}^n - u_{i,j}^{n-1}| + |u_{i+1,j}^n - u_{i+1,j}^{n-1}| + |\xi_{i,j}^{n-1}| + |\xi_{i+1,j}^{n-1}|\},$$

which, together with (3.3), yields

$$||(a_1 - A_1)^n \delta_{x,+} u^n||^2 \leq \tilde{M}(||u^n||_{W^{1,\infty}}, ||u||_{W^{1,\infty}(L^\infty)})(||\xi^{n-1}||^2 + \Delta t^2).$$

Smiliarly, we can obtain

$$||(a_2 - A_2)^n \delta_{y,+} u^n||^2 \leq \tilde{M}(||u^n||_{W^{1,\infty}}, ||u||_{W^{1,\infty}(L^\infty)})(||\xi^{n-1}||^2 + \Delta t^2).$$

Therefore, we have an estimate for the second term of the right hand side of (3.2), namely

$$|I_2| \leq \varepsilon[||\delta_{x,+}\xi^n||^2 + ||\delta_{x,+}\xi^n||^2] + \tilde{M}(||u^n||_{W^{1,\infty}}, ||u||_{W^{1,\infty}(L^\infty)})(||\xi^{n-1}||^2 + \Delta t^2).$$

We now estimate the third term of the right hand side of (3.2). By assumption iii.), we have

$$|\tfrac{c_{i,j}}{\Delta t}(\tilde{x}_{i,j} - \hat{\tilde{x}}_{i,j})| = |b_1(x_{i,j}, u_{i,j}^n) - b_1(x_{i,j}, U_{i,j}^{n-1})|$$
$$+|b_2(x_{i,j}, u_{i,j}^n) - b_2(x_{i,j}, U_{i,j}^{n-1})| \leq K_3(|u_{i,j}^n - u_{i,j}^{n-1}| + |\xi_{i,j}^{n-1}|),$$

which yields

$$|I_3| \leq \tilde{M}(||u^{n-1}||_{W^{1,\infty}})(||u^n - u^{n-1}|| + ||\xi^{n-1}||)||\xi^n||$$
$$\leq \tilde{M}(||u^{n-1}||_{W^{1,\infty}}, ||u||_{W^{1,\infty}(L^\infty)})(||\xi^n||^2 + ||\xi^{n-1}||^2 + \Delta t^2).$$

If $|\nabla u| \in L^\infty(L^\infty)$, from assumptions i.)–iii.), there exists a constant K^* such that

$$|\frac{\bar{b}(\bar{x}, u(\bar{x}, t))}{c(\bar{x})}| + |\frac{d}{d\bar{x}} \frac{\bar{b}(\bar{x}, u(\bar{x}, t))}{c(\bar{x})}| \leq K^*.$$

Similar to [1], we define the norm for $\alpha(\bar{x}) \in L^\infty(\bar{\Omega})$ by

$$||\alpha||_{\tilde{l}^2}^2 = \sum_{i,j=1}^{J_1-1, J_2-1} \max\{|\alpha(\bar{x})|^2, \ |\bar{x} - \bar{x}_{i,j}| \leq K^*\Delta t\}h_1 h_2.$$

It is also not difficult to obtain

$$||e^n||^2 \leq M_1(||u^n||_{H^4}^2 (h_1^4 + h_2^4) + ||\frac{\partial^2 u}{\partial \tau^2}||_{L^2(t_{n-1}, t_n; \tilde{l}^2)}^2 \Delta t^2).$$

Therefore, we have

$$|I_4| \leq \frac{1}{2}||\xi^n||^2 + \frac{1}{2}M_1(||u^n||_{H^4}^2 (h_1^4 + h_2^4) + ||\frac{\partial^2 u}{\partial \tau^2}||_{L^2(t_{n-1}, t_n; \tilde{l}^2)}^2 \Delta t^2).$$

We now consider the estimate of left hand side of (3.2). By the triangle inequality and assumption ii.), we have

$$I_1 + I_2 + I_3 + I_4 \geq \frac{1}{2\Delta t}[(c\xi^n, \xi^n) - (c\tilde{\xi}^{n-1}, \tilde{\xi}^{n-1})] + a_0[||\delta_{x,+}\xi^n||^2 + ||\delta_{y,+}\xi^n||^2].$$

where $\tilde{\xi}^{n-1} = \xi^{n-1}(\hat{\tilde{x}})$. To simplify this inequality, we need to find a relation between $(c\xi^{n-1}, \xi^{n-1})$ and $(c\tilde{\xi}^{n-1}, \tilde{\xi}^{n-1})$. Set

$$\alpha_{i,j}^n = \frac{b_1(\bar{x}_{i,j}, U_{i,j}^{n-1})\Delta t}{c_{i,j}h_1}, \quad \beta_{i,j}^n = \frac{b_1(\bar{x}_{i,j}, U_{i,j}^{n-1})\Delta t}{c_{i,j}h_2}.$$

Then $\hat{\tilde{x}}_{i,j} = (\hat{\tilde{x}}_i, \hat{\tilde{y}}_j)$, and $\hat{\tilde{x}}_i = x_i - \alpha_{i,j}^n h_1$, $\hat{\tilde{y}}_j = y_j - \beta_{i,j}^n h_2$. By assumptions i.) and ii.), we can choose Δt and h_1, h_2 such that

$$|\alpha_{i,j}^n| \leq 1, \quad |\beta_{i,j}^n| \leq 1.$$

Hence $\hat{\bar{x}}_{i,j} = (\hat{\bar{x}}_i, \hat{\bar{y}}_j) \in [x_{i-1}, x_{i+1}] \times [y_{j-1}, y_{j+1}]$ and the point $\hat{\bar{x}}_{i,j}$ must be in one of the four domains I, II, III and IV as shown in Fig. 1.

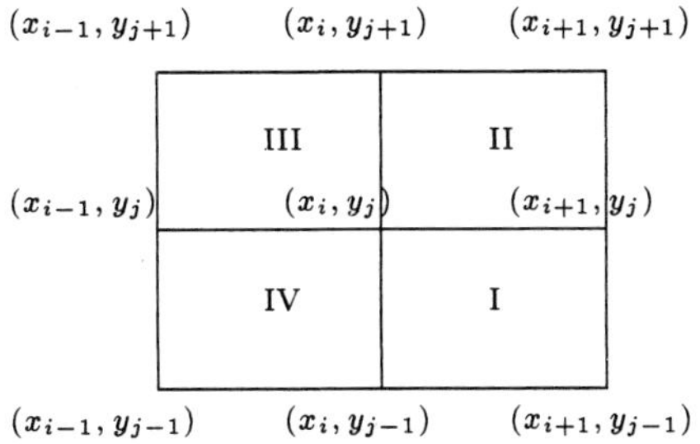

(x_{i-1}, y_{j+1}) (x_i, y_{j+1}) (x_{i+1}, y_{j+1}) $4(x_i, y_{j+1})$ $3(x_{i+1}, y_{j+1})$

III	II
IV	I

(x_{i-1}, y_j) (x_i, y_j) (x_{i+1}, y_j)

II

(x_{i-1}, y_{j-1}) (x_i, y_{j-1}) (x_{i+1}, y_{j-1}) $1(x_i, y_j)$ $2(x_{i+1}, y_j)$

Figure 1 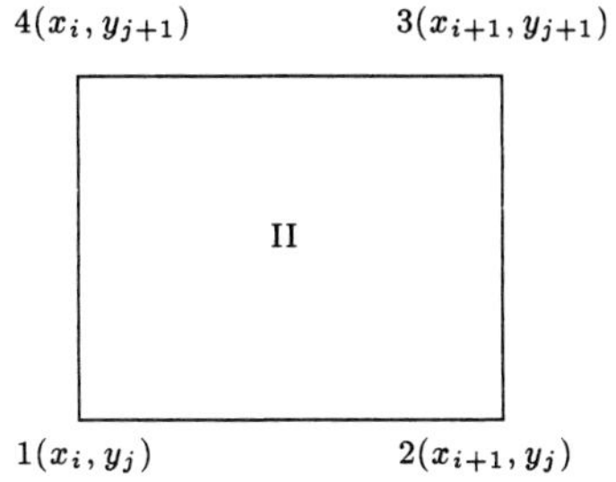   Figure 2

For simplicity, we have omitted the superscripts n and $n-1$ for $\xi_{i,j}$. Assume that the point $\hat{\bar{x}}_{i,j}$ is in domain II and the value of $U^{n-1}(\bar{x})$ at $\hat{\bar{x}}_{i,j}$ is obtained by the piecewise bilinear interpolation of $\{U_{i,j}^{n-1}\}$. The bilinear interpolation function is

$$L_1(\xi)(x,y) = N_1(x,y)\xi_{i,j} + N_2(x,y)\xi_{i+1,j} + N_3(x,y)\xi_{i+1,j+1} + N_4(x,y)\xi_{i,j+1},$$

where $\xi = u - U$ and $N_i(x,y)$ denotes the interpolating function corresponding to (x_i, y_j) as shown in Fig. 2. Thus we have

$$L_1(\xi)(\hat{\bar{x}}_i, \hat{\bar{y}}_j) = \xi_{i,j} - \alpha_{i,j}\delta_{x,+}\xi_{i,j}h_1 - \beta_{i,j}\delta_{y,+}\xi_{i,j}h_2 + \alpha_{i,j}\beta_{i,j}\delta_{x,+}(\xi_{i,j+1} - \xi_{i,j})h_1.$$

Hence we see that

$$\sum_{\substack{i,j=1 \\ (\hat{\bar{x}}_i,\hat{\bar{y}}_j)\in II}}^{J_1-1,J_2-1} c_{i,j} L_1^2(\xi)(\hat{\bar{x}}_i, \hat{\bar{y}}_j) h_1 h_2$$
$$\leq \sum_{\substack{i,j=1 \\ (\hat{\bar{x}}_i,\hat{\bar{y}}_j)\in II}}^{J_1-1,J_2-1} c_{i,j}\xi_{i,j}^2 h_1 h_2 + \tfrac{\varepsilon}{4}\Delta t \sum_{\substack{i,j=1 \\ (\hat{\bar{x}}_i,\hat{\bar{y}}_j)\in II}}^{J_1,J_2} [|\delta_{x,+}\xi_{i,j}|^2$$
$$+ |\delta_{y,+}\xi_{i,j+1}|^2] h_1 h_2 + M_2 \Delta t \sum_{\substack{i,j=1 \\ (\hat{\bar{x}}_i,\hat{\bar{y}}_j)\in II}}^{J_1-1,J_2-1} |\xi_{i,j}|^2 h_1 h_2.$$

Similarly, for $\hat{\bar{x}}_{i,j} \in II$, we can prove that the above inequality holds for $\hat{\bar{x}}_{i,j} \in$ I, III, and IV. Therefore, we have

$$\sum_{i,j=1}^{J_1-1,J_2-1} c_{i,j} L_1^2(\xi^{n-1})(\hat{\bar{x}}_i, \hat{\bar{y}}_j) h_1 h_2 \leq (c\xi^{n-1}, \xi^{n-1})$$
$$+ \varepsilon [[|\delta_{x,+}\xi^{n-1}||^2 + ||\delta_{y,+}\xi^{n-1}||^2]\Delta t + M_2 ||\xi^{n-1}||^2 \Delta t.$$

Noting that

$$\xi(\hat{\bar{x}}_i, \hat{\bar{y}}_j) = u(\hat{\bar{x}}_i, \hat{\bar{y}}_j) - U(\hat{\bar{x}}_i, \hat{\bar{y}}_j) = u(\hat{\bar{x}}_i, \hat{\bar{y}}_j) - L_1(u)(\hat{\bar{x}}_i, \hat{\bar{y}}_j) + L_1(\xi)(\hat{\bar{x}}_i, \hat{\bar{y}}_j),$$

where $L_1(u)(x,y)$ denotes the bilinear interpolation function of $\{u_{i,j}\}$ on $\bar{\Omega}_h$, we can estimate $u(\hat{\bar{x}}_i, \hat{\bar{y}}_j) - L_1(u)(\hat{\bar{x}}_i, \hat{\bar{y}}_j)$ as follows. Let $\hat{\bar{x}}_{i,j} \in II$ and assume that II is divided into four equal rectangular elements Ω_1, Ω_2, Ω_3, Ω_4, and $\hat{\tilde{x}}_{i,j} \in \Omega_1$. Then we have

$$|\rho_1| = |\hat{\tilde{x}}_i - x_i| \leq \min(\tfrac{h_1}{2}, k\Delta t),$$

$$|\rho_2| = |\hat{\tilde{y}}_j - x_j| \leq \min(\tfrac{h_2}{2}, k\Delta t).$$

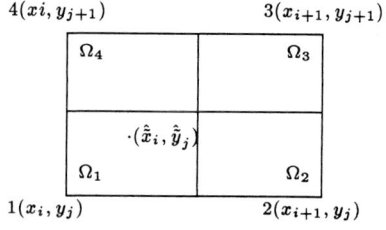

Figure 3.

Further, we expand $u(x_{i+1}, y_j)$, $u(x_{i+1}, y_{j+1})$, $u(x_i, y_{j+1})$, and $u(\hat{\tilde{x}}_i, \hat{\tilde{y}}_j)$ as the Taylor series at the point (x_i, y_j) by

$$u(x_{i+1}, y_j) = u_{i,j} + h_1(\tfrac{\partial u}{\partial x})_{i,j} + \tfrac{h_1^2}{2}(\tfrac{\partial^2 u}{\partial x^2})(x^{\theta_1}, y_j),$$

$$u(x_{i+1}, y_{j+1}) = u_{i,j} + h_1(\tfrac{\partial u}{\partial x})_{i,j} + h_2(\tfrac{\partial u}{\partial y})_{i,j} + \tfrac{h_1^2}{2}\tfrac{\partial^2 u}{\partial x^2}(x^{\theta_2}, y^{\theta_2})$$

$$+ h_1 h_2 \tfrac{\partial^2 u}{\partial x \partial y}(x^{\theta_2}, y^{\theta_2}) + \tfrac{h_2^2}{2}\tfrac{\partial^2 u}{\partial y^2}(x^{\theta_2}, y^{\theta_2}),$$

$$u(x_i, y_{j+1}) = u_{i,j} + h_2(\tfrac{\partial u}{\partial y})_{i,j} + \tfrac{h_2^2}{2}(\tfrac{\partial^2 u}{\partial y^2})(x_i, y^{\theta_3}),$$

$$u(\hat{\tilde{x}}_i, \hat{\tilde{y}}_j) = u_{i,j} + \rho_1(\tfrac{\partial u}{\partial x})_{i,j} + \rho_2(\tfrac{\partial u}{\partial y})_{i,j} + \tfrac{\rho_1^2}{2}\tfrac{\partial^2 u}{\partial x^2}(x', y') + \rho_1\rho_2\tfrac{\partial^2 u}{\partial x \partial y}(x', y')$$

$$+ \tfrac{\rho_2^2}{2}\tfrac{\partial^2 u}{\partial y^2}(x', y'),$$

where $x^{\theta_k} = (1-\theta_k)x_{i+1} + \theta_k x_i$, $y^{\theta_k} = (1-\theta_k)y_{j+1} + \theta_k y_j$, $x' = (1-\theta)\hat{\tilde{x}}_i + \theta x_i$, $y' = (1-\theta)\hat{\tilde{y}}_j + \theta y_j$, $0 \leq \theta_k \leq 1$ ($k = 1, 2, 3$), and $0 \leq \theta \leq 1$. Thus we have

$$|L_1(u^{n-1})(\hat{\tilde{x}}_i, \hat{\tilde{y}}_j) - u^{n-1}(\hat{\tilde{x}}_i, \hat{\tilde{y}}_j)|$$

$$\leq \tilde{M}(\|u^{n-1}\|_{H^2})[h_1 \min(\tfrac{h_1}{2}, K\Delta t) + h_2 \min(\tfrac{h_2}{2}, K\Delta t)].$$

Similarly, for $\hat{\tilde{x}}_{i,j} \in \Omega_1$, we can prove that this inequality holds for $\hat{\tilde{x}}_{i,j} \in \Omega_2$, Ω_3, Ω_4. Therefore we have the following lemma.

Lemma 3.1 *If $U^{n-1}(\tilde{x})$ is obtained by the bilinear interpolation, we have*

$$(c\tilde{\xi}^{n-1}, \tilde{\xi}^{n-1}) = \sum_{i,j=1}^{J_1-1, J_2-1} c_{i,j}(\tilde{\xi}_{i,j}^{n-1})^2 h_1 h_2 \leq \sum_{i,j=1}^{J_1-1, J_2-1} c_{i,j} L_1^2(\xi^{n-1}) h_1 h_2$$

$$+ M \sum_{i,j=1}^{J_1-1, J_2-1} c_{i,j}|u^{n-1}(\hat{\tilde{x}}_{i,j}, \hat{\tilde{y}}_{i,j}) - L_1(u^{n-1})(\hat{\tilde{x}}_i, \hat{\tilde{y}}_j)|^2 h_1 h_2$$

$$\leq (c\xi^{n-1}, \xi^{n-1}) + \varepsilon[\|\delta_{x,+}\xi^{n-1}\|^2 + \|\delta_{y,+}\xi^{n-1}\|^2]\Delta t + M_2\|\xi^{n-1}\|^2\Delta t$$

$$+ \tilde{M}(\|u^{n-1}\|_{H^2})[h_1 \min(\tfrac{h_1}{2}, K\Delta t) + h_2 \min(\tfrac{h_2}{2}, K\Delta t)]^2.$$

If the value of $U^{n-1}(\bar{x})$ at point $\hat{\bar{x}}_{i,j} \in \text{II}$ is obtained by the piecewise biquadratic interpolation of $\{U_{i,j}^{n-1}\}$, the biquadratic interpolation function is

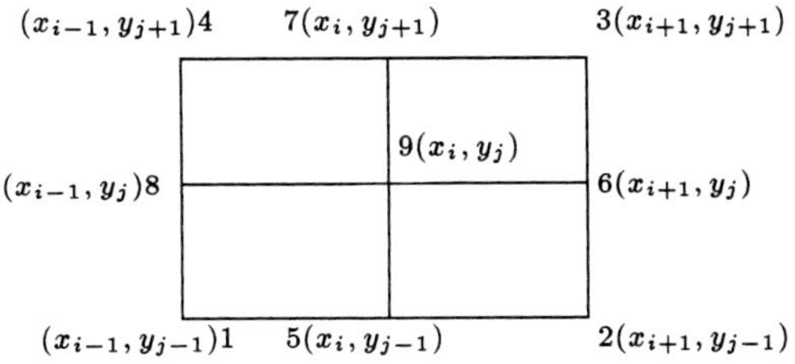

Figure 4.

$$\begin{aligned}L_2(\xi)(x,y) &= N_1(x,y)\xi_{i-1,j-1} + N_2(x,y)\xi_{i+1,j-1} + N_3(x,y)\xi_{i+1,j+1}\\ &+N_4(x,y)\xi_{i-1,j+1} + N_5(x,y)\xi_{i,j-1} + N_6(x,y)\xi_{i+1,j} + N_7(x,y)\xi_{i,j+1}\\ &+N_8(x,y)\xi_{i-1,j} + N_9(x,y)\xi_{i,j},\end{aligned}$$

where $N_i(x,y)$ $(i = 1, \ldots, 9)$ denotes the interpolating function corresponding to node i as shown in Fig. 4. Then we have

$$\begin{aligned}L_2(\xi)(\hat{\bar{x}}_i, \hat{\bar{y}}_j) &= \xi_{i,j} - \tfrac{1}{2}\alpha_{i,j}(\delta_{x,+}\xi_{i,j} + \delta_{x,-}\xi_{i,j})h_1 - \tfrac{1}{2}\beta_{i,j}(\delta_{y,+}\xi_{i,j}\\ &+\delta_{y,-}\xi_{i,j})h_2 + \tfrac{1}{2}\alpha_{i,j}^2\delta_{x,+}(\xi_{i,j} - \xi_{i-1,j})h_1 + \tfrac{1}{2}\beta_{i,j}^2\delta_{y,+}(\xi_{i,j} - \xi_{i,j-1})h_2\\ &+\tfrac{1}{4}\alpha_{i,j}\beta_{i,j}(\delta_{x,+}\xi_{i,j+1} + \delta_{x,-}\xi_{i,j+1} - \delta_{x,+}\xi_{i,j-1} - \delta_{x,-}\xi_{i,j-1})h_1\\ &+\tfrac{1}{4}\alpha_{i,j}^2\beta_{i,j}[\delta_{x,+}(\xi_{i,j-1} - \xi_{i-1,j-1}) - \delta_{x,+}(\xi_{i,j+1} - \xi_{i-1,j+1})]h_1\\ &+\tfrac{1}{4}\alpha_{i,j}\beta_{i,j}^2[\delta_{y,+}(\xi_{i-1,j} - \xi_{i-1,j-1}) - \delta_{y,+}\xi_{i+1,j} - \xi_{i+1,j-1})]h_2\\ &+\tfrac{1}{4}\alpha_{i,j}^2\beta_{i,j}^2[\delta_{y,+}(\xi_{i-1,j} - \xi_{i-1,j-1}) - 2\delta_{y,+}(\xi_{i,j} - \xi_{i,j-1})\\ &+\delta_{y,+}(\xi_{i+1,j} - \xi_{i+1,j-1})]h_2^2.\end{aligned}$$

Similar to the bilinear interpolation, we also obtain an inequality for the biquadratic interpolation through a tediously long computation, namely

$$\begin{aligned}&\textstyle\sum_{i,j=1}^{J_1-1,J_2-1} c_{i,j}L_2^2(\xi)(\hat{\bar{x}}_i, \hat{\bar{y}}_j)h_1h_2\\ &\qquad\le (c\xi^{n-1}, \xi^{n-1}) + M_3\|\xi^{n-1}\|^2\Delta t + \varepsilon[\|[\delta_{x,+}\xi^{n-1}\|^2 + \|\delta_{y,+}\xi^{n-1}]\|^2]\Delta t\\ &|L_2(u^{n-1})(\hat{\bar{x}}_i, \hat{\bar{y}}_j) - u^{n-1}(\hat{\bar{x}}_i, \hat{\bar{y}}_j)|\\ &\qquad\le \tilde{M}(\|u^{n-1}\|_{H^3})[h_1^2\min(\tfrac{h_1}{2}, K\Delta t) + h_2^2\min(\tfrac{h_2}{2}, K\Delta t)].\end{aligned}$$

Therefore, we obtain

Lemma 3.2 *if $U^{n-1}(\bar{x})$ is obtained by the biquadratic interpolation, we have*

$$\begin{aligned}(c\tilde{\xi}^{n-1}, \tilde{\xi}^{n-1}) &\le (c\xi^{n-1}, \xi^{n-1}) + M_3\|\xi^{n-1}\|^2\Delta t + \varepsilon[\|[\delta_{x,+}\xi^{n-1}\|^2\\ &+\|\delta_{y,+}\xi^{n-1}]\|^2]\Delta t + \tilde{M}(\|u^{n-1}\|_{H^3})[h_1^2\min(\tfrac{h_1}{2}, K\Delta t) + h_2^2\min(\tfrac{h_2}{2}, K\Delta t)]^2.\end{aligned}$$

From (3.4), the inequalities for I_i $(i = 1, 2, 3, 4)$, and $U^{n-1}(\bar{x})$ obtained from the bilinear interpolation, we have

$$\frac{1}{2}[(c\xi^n, \xi^n) - (c\xi^{n-1}, \xi^{n-1})] + a_0[||\delta_{x,+}\xi^{n-1}||^2 + ||\delta_{y,+}\xi^{n-1}||^2]\Delta t$$

$$\leq \varepsilon[||\delta_{x,+}\xi^n||^2 + ||\delta_{y,+}\xi^n||^2]\Delta t + \varepsilon[||\delta_{x,+}\xi^{n-1}||^2 + ||\delta_{y,+}\xi^{n-1}||^2]\Delta t$$

$$+\tilde{M}(||u||_{W^{1,\infty}(L^\infty)}, ||u||_{L^\infty(W^{1,\infty})})(||\xi^n||^2 + ||\xi^{n-1}||^2 + \Delta t^2)\Delta t$$

$$+\frac{1}{2}M_1(||u||_{L^\infty(H^4)})(h_1^4 + h_2^4) + ||\frac{\partial^2 u}{\partial \tau^2}||_{L^2(t_{n-1}, t_n; \tilde{l}^2)}\Delta t^2)\Delta t$$

$$+\tilde{M}(||u^{n-1}||_{H^2})[h_1 \min(\frac{h_1}{2}, K\Delta t) + h_2 \min(\frac{h_2}{2}, K\Delta t)]^2.$$

Choosing $\varepsilon = \frac{a_0}{4}$ and summing up the inequality in Lemma 3.2 for n, we obtain

$$||\xi^n||^2 + a_0 \sum_{m=0}^{n}[||\delta_{x,+}\xi^m||^2 + ||\delta_{y,+}\xi^m||^2]\Delta t$$

$$\leq \tilde{M}(||u||_{W^{1,\infty}(L^\infty)}, ||u||_{L^\infty(H^4)}, ||\frac{\partial^2 u}{\partial \tau^2}||_{L^2(\tilde{l}^2)})(\sum_{m=0}^{n} ||\xi^m||^2 \Delta t + h_1^4 + h_2^4$$

$$+\Delta t^2) + \tilde{M}(||u^{n-1}||_{H^2})[h_1 \min(\frac{h_1}{2\sqrt{\Delta t}}, K\Delta t) + h_2 \min(\frac{h_2}{2\sqrt{\Delta t}}, K\Delta t)]^2.$$

Applying the Gronwell inequality, we obtain

$$||\xi^n|| + (\sum_{m=0}^{n}[||\delta_{x,+}\xi^m||^2 + ||\delta_{y,+}\xi^m||^2]\Delta t)^{\frac{1}{2}} \leq \tilde{M}(||u||_{W^{1,\infty}(L^\infty)}, ||u||_{L^\infty(H^4)},$$

$$||\frac{\partial^2 u}{\partial \tau^2}||_{L^2(\tilde{l}^2)})([h_1 \min(\frac{h_1}{2\sqrt{\Delta t}}, K\Delta t) + h_2 \min(\frac{h_2}{2\sqrt{\Delta t}}, K\Delta t)] + \Delta t).$$

$$(3.5)$$

Therefore, we have the theorems.

Theorem 3.3 *Let* $u \in W^{1,\infty}(L^\infty) \cap L^\infty(H^4)$ *be the solution of equation* (1.1) *and* $\frac{\partial^2 u}{\partial \tau^2} \in L^2(\tilde{l}^2)$, *and let* $\{U_{i,j}^n\}$ *be the solution of equation* (2.3) *based on the bilinear interpolation. Then the error estimation* (3.5) *holds.*

Theorem 3.4 *Under the assumptions of Theorem 3.3 about u and the solutions* $\{U_{i,j}^n\}$ *of equation* (2.3) *based on the biquadratic interpolations, the error* $\xi^n = u^n - U^n$ *satisfies the following inequality*

$$||\xi^n|| + (\sum_{0}^{N}[||\delta_{x,+}\xi^n||^2 + ||\delta_{y,+}\xi^n||^2]\Delta t)^{\frac{1}{2}}$$

$$\leq \tilde{M}(||u||_{W^{1,\infty}(L^\infty)}, ||u||_{L^\infty(H^4)}, ||\frac{\partial^2 u}{\partial \tau^2}||_{L^2(\tilde{l}^2)})(h_1^2 + h_2^2 + \Delta t).$$

Acknowledgments. The first author is greatly indebted to Prof. Sun Che in the Department of Mathematics, Nankai University for many helpful discussions. This work is supported by the National Natural Science Foundation of China (19771012) and the Foundation of China Academy of Engineering Physics (970683).

References

[1] Douglas, J., Jr. and Russell, T. F., Numerical methods for convection-dominated diffusion probles based on combing the method of characteristics with finite or finite difference procedures, *SIAM J. Numer. Anal.* **19** (1982), 871–885.

[2] Durán, R., On the approximation of miscible displacement in porous media by a method of characteristics combined with a mixed method, *SIAM J. Numer. Anal.* **25** (1988), 989–1001.

[3] Ewing, R. E., Multistep Galerkin methods along characteristics for convection-diffusion problems, in Advances in Computer Methods for Partial Differential Equations-IV, IMACS, Rutgers Univ. New Brunwith, 1981, 28–36.

[4] Ewing, R. E., Russell, T. F. and Wheeler, M. F., Convergence analysis of an approximation of miscible displacement in porous media by mixed finite elements and a modified method of characteristics, *Comput. Meth. Appl. Mech. Engrg.* **47** (1984), 73–92.

[5] Russell, T. F., Time stepping along characteristics with incomplete interation for a Galerkin approximation of miscible displacement in porous media, *SIAM J. Numer. Anal.* **22** (1985), 970–1013.

Fractional Step Methods for Compressible Multicomponent Flow in Porous Media

YIRANG YUAN

Abstract

This paper discusses characteristic finite difference and finite element fractional step methods for three-dimensional multicomponent flow in porous media. Optimal order estimates in the L^2-norm are derived for the errors in the approximate solution.

KEYWORDS: compressible flow, characteristic finite difference and finite element, fractional steps, L^2-error estimates

1   Introduction

In modern numerical simulation of prospecting and exploiting oil-gas resources, the problems met are often three-dimensional, large-scale, large-scope and extralong-term ones. The node number is as large as tens of thousands or even hundreds of millions, which calls for the new technology of fractional steps to solve the problems. First, a kind of characteristic finite difference fractional step methods is developed. Thick and thin grids are used to form a complete set of techniques, such as the piecewise product threefold-quadratic interpolation, the calculus of variations, the multiplicative commutation rule of difference operators, and the decomposition of high order difference operators. The prior estimates and techniques are adopted. Optimal order estimates in the L^2-norm are derived to determine the errors in the approximate solution. Next, we establish a kind of characteristic finite element operator-splitting methods and use the operator-splitting, characteristic method, the calculus of variations, the energy method, negative norm estimates, and the theory of prior estimates and techniques. Optimal order estimates in the L^2 norm are derived for the errors in the approximate solution. These methods are successfully used in oil-gas resource estimation, enhanced oil recovery simulation, and seawater intrusion numerical simulation.

The mathematical model for the three-dimensional compressible multi-component displacement problems is the nonlinear partial equations with initial-boundary conditions [2, 8, 9, 4, 6]

$$\text{(a)} \quad d(c)\frac{\partial p}{\partial t} + \nabla \cdot u = q(x,t), \quad x \in \Omega, \quad t \in J = (0,T],$$
$$\text{(b)} \quad u = -a(c)\nabla p, \quad\quad\quad\quad x \in \Omega, \quad t \in J, \tag{1.1}$$

and

$$\Phi(x)\frac{\partial c_\alpha}{\partial t} + b_\alpha(c)\frac{\partial p}{\partial t} + u \cdot \nabla c_\alpha - \nabla \cdot (D\nabla c_\alpha) = g(x,t,c_\alpha),$$
$$x \in \Omega, \quad t \in J, \quad \alpha = 1,2,\ldots,n_c - 1, \tag{1.2}$$

where $p(x,t)$ is the pressure function, $c_\alpha(x,t)$ is the concentration of αth component, $\alpha = 1,2,\ldots,n_c$, n_c is the number of components, $\sum_{\alpha=1}^{n_c} c_\alpha(x,t) = 1$, $d(c) = \Phi(x)\sum_{\alpha=1}^{n_c} z_\alpha c_\alpha$, $\Phi(x)$ is the porosity, z_α is the "constant compressibility" factor for the αth component, u is the Darcy velocity, $a(c) = k(x)\mu(c)^{-1}$, $k(x)$ is the permeability of the rock, $\mu(c)$ is the viscosity of the fluid, $b_\alpha(c) = \Phi(x)c_\alpha\{z_\alpha - \sum_{j=1}^{n_c} z_j c_j\}$, and $D(x)$ is the diffusion coefficient. Let $c(x,t) = (c_1(x,t), c_2(x,t),\ldots,c_{n_c-1}(x,t))^T$. Then the pressure function $p(x,t)$ and the concentration functions $c(x,t)$ are obtained.

Here we assume that no flow occurs across the boundary:

$$\text{(a)} \quad u \cdot \gamma = 0, \quad x \in \partial\Omega,$$
$$\text{(b)} \quad (D\nabla c_\alpha - c_\alpha u) \cdot \gamma = 0, \quad x \in \partial\Omega, \quad \alpha = 1,2,\ldots,n_c - 1, \tag{1.3}$$

where γ is the outer normal to $\partial\Omega$. In addition, the initial conditions

$$\text{(a)} \quad p(x,0) = p_0(x), \quad x \in \Omega,$$
$$\text{(b)} \quad c_\alpha(x,0) = c_{\alpha,0}(x), \quad x \in \Omega, \quad \alpha = 1,2,\ldots,n_c - 1, \tag{1.4}$$

must be given.

For planar two-phase immiscible flow, Douglas, Ewing, Russell, and many others have published a series of fundamental papers on the characteristic finite difference and finite element methods [1, 7, 5, 3]. For compressible two-phase displacement problems, Douglas and others have published their papers [2, 8, 9, 4]. On the basis of the preceding works this article makes a further study of the characteristic finite difference and finite element fractional step methods and L^2-error estimates for a three-dimensional compressible multicomponent displacement problem. The three-dimensional problem is decomposed to solve three one-dimensional problems continuously, thus greatly reducing the amount of computation work and making the actual computation of the project possible. This method is used successfully in oil-gas resources estimation [15, 12], enhanced oil recovery simulation [10, 11], and seawater intrusion numerical simulation [13, 14].

Generally, this is a positive definite problem

$$(a) \quad 0 < a_* \le a(c) \le a^*, 0 < d_* \le d(c) \le d^*, 0 < D_* \le D(x) \le D^*,$$

$$(b) \quad \left| \frac{\partial a}{\partial c}(x,c) \right| + \left| \frac{\partial d}{\partial c}(x,c) \right| \le K^*, \tag{1.5}$$

where a_*, a^*, d_*, d^*, D_*, D^*, and K^* are constants. Our assumptions on the regularity of the solution of (1.1)–(1.5) are given collectively by

$$p, c_\alpha \in L^\infty(W^{4,\infty}) \bigcap W^{1,\infty}(W^{1,\infty}), \quad \frac{\partial^2 p}{\partial t^2}, \frac{\partial^2 c_\alpha}{\partial \tau^2} \in L^\infty(L^\infty),$$

$$\alpha = 1, 2, \ldots, n_c - 1.$$

Assume that $\Omega = \{[0,1]\}^3$ and problem (1.1)–(1.5) is Ω-periodic. Then the boundary condition (1.3) can be dropped [2, 7, 5].

In this paper M and ε express a general positive constant and a general positive small constant, respectively, and they may have different meanings in different places.

2 Finite Difference Fractional Steps

Let $\Omega = \{[0,1]\}^3$, $h = 1/N$, $X_{ijh} = (ih, jh, kh)^T$, $t^n = n\Delta t$ $W(X_{ijk}, t^n) = W_{ijk}^n$,

$$(a) \quad A_{i+1/2,jk}^n = \left[a(X_{ijk}, C_{ijk}^n) + a(X_{i+1,jk}, C_{i+1,jk}^n) \right]/2,$$

$$(b) \quad a_{i+1/2,jk}^n = \left[a(X_{ijk}, c_{ijk}^n) + a(X_{i+1,jk}, c_{i+1,jk}^n) \right]/2, \tag{2.1}$$

and $A_{i,j+1/2,k}^n$, $A_{ij,k+1/2}^n$, $a_{i,j+1/2,k}^n$, and $a_{ij,k+1/2}^n$ be defined analogously. Let

$$(a) \quad \delta_{\bar{x}_1}(A^n \delta_{x_1} P^{n+1})_{ijk} = h^{-2} \left[A_{i+1/2,jk}^n (P_{i+1,jk}^{n+1} - P_{ijk}^{n+1}) \right.$$
$$\left. - A_{i-1/2,jk}^n (P_{ijk}^{n+1} - P_{i-1,jk}^{n+1}) \right],$$

$$(b) \quad \delta_{\bar{x}_2}(A^n \delta_{x_2} P^{n+1})_{ijk} = h^{-2} \left[A_{i,j+1/2,k}^n (P_{i,j+1,k}^{n+1} - P_{ijk}^{n+1}) \right. \tag{2.2}$$
$$\left. - A_{i,j-1/2,k}^n (P_{ijk}^{n+1} - P_{i,j-1,k}^{n+1}) \right],$$

$$(c) \quad \delta_{\bar{x}_3}(A^n \delta_{x_3} P^{n+1})_{ijk} = h^{-2} \left[A_{ij,k+1/2}^n (P_{ij,k+1}^{n+1} - P_{ijk}^{n+1}) \right.$$
$$\left. - A_{ij,k-1/2}^n (P_{ijk}^{n+1} - P_{ij,k-1}^{n+1}) \right],$$

$$(d) \quad \nabla_h(A^n \nabla P^{n+1})_{ijk} = \delta_{\bar{x}_1}(A^n \delta_{x_1} P^{n+1})_{ijk}$$
$$+ \delta_{\bar{x}_2}(A^n \delta_{x_2} P^{n+1})_{ijk} + \delta_{\bar{x}_3}(A^n \delta_{x_3} P^{n+1})_{ijk}.$$

For the fluid equation (1.1), the finite difference fractional step scheme is given by

$$(a) \quad d(C_{ijk}^n)\frac{P_{ijk}^{n+1/3} - P_{ijk}^n}{\Delta t} = \delta_{\bar{x}_1}(A^n\delta_{x_1}P^{n+1/3})_{ijk} + \delta_{\bar{x}_2}(A^n\delta_{x_2}P^n)_{ijk}$$
$$+\delta_{\bar{x}_3}(A^n\delta_{x_3}P^n)_{ijk} + q(X_{ijk}, t^{n+1}), \quad 1 \le i \le N,$$

$$(b) \quad d(C_{ijk}^n)\frac{P_{ijk}^{n+2/3} - P_{ijk}^{n+1/3}}{\Delta t} = \delta_{\bar{x}_2}\left(A^n\delta_{x_2}(P^{n+2/3} - P^n)\right)_{ijk}, 1 \le j \le N,$$

$$(c) \quad d(C_{ijk}^n)\frac{P_{ijk}^{n+1} - P_{ijk}^{n+2/3}}{\Delta t} = \delta_{\bar{x}_3}\left(A^n\delta_{x_3}(P^{n+1} - P^n)\right)_{ijk}, \quad 1 \le k \le N.$$

$$(2.3)$$

Compute the approximate Darcy velocity $U = (U_1, U_2, U_3)^T$ as follows:

$$U_{1,ijk}^n = -\frac{1}{2}\left[A_{i+1/2,jk}^n\frac{P_{i+1,jk}^n - P_{ijk}^n}{h} + A_{i-1/2,jk}^n\frac{P_{ijk}^n - P_{i-1,jk}^n}{h}\right], \quad (2.4)$$

with $U_{2,ijk}^n$ and $U_{3,ijk}^n$ being the corresponding average in the other direction.

As the flow is essentially in the characteristic direction, we apply the modified method of characteristic to the first-order parts of (1.2), thus ensuring a high accuracy of the numerical results [1, 7, 5, 3]. Let $\Psi(x, u) = [\Phi^2(x) + |u|^2]^{1/2}$ and $\partial/\partial\tau = \frac{1}{\Psi}\{\Phi\partial/\partial t + u \cdot \nabla\}$. Equation (1.2) can be rewritten in the form

$$\psi\frac{\partial c_\alpha}{\partial\tau} - \nabla \cdot (D\nabla c_\alpha) + b_\alpha(c)\frac{\partial p}{\partial t} = g(x, t, c_\alpha), \quad x \in \Omega, \quad t \in J.$$

Approximate $\frac{\partial c_\alpha^{n+1}}{\partial\tau} = \frac{\partial c_\alpha}{\partial\tau}(x, t^{n+1})$ by a backward difference quotient in τ-direction, $\frac{\partial c_\alpha^{n+1}}{\partial\tau} \approx \frac{c_\alpha^{n+1} - c_\alpha^n\left(x - \Phi^{-1}(x)u^{n+1}(x)\Delta t\right)}{\Delta t\sqrt{1 + \Phi^{-2}(x)|u^{n+1}(x)|^2}}.$

For the system of concentration equations, the characteristic difference fractional step schemes are given by

$$(a) \quad \Phi_{ijk}\frac{C_{\alpha,ijk}^{n+1/3} - \hat{C}_{\alpha,ijk}^n}{\Delta t} = \delta_{\bar{x}_1}(D\delta_{x_1}C_\alpha^{n+1/3})_{ijk}$$
$$+\delta_{\bar{x}_2}(D\delta_{x_2}C_\alpha^n)_{ijk} + \delta_{\bar{x}_3}(D\delta_{x_3}C_\alpha^n)_{ijk}$$
$$-b_\alpha(C_{ijk}^n)\frac{P_{ijk}^{n+1} - P_{ijk}^n}{\Delta t} + g(X_{ijk}, t^n, \hat{C}_{\alpha,ijk}^n),$$
$$1 \le i \le N, \quad \alpha = 1, 2, \ldots, n_c - 1,$$

(b) $\Phi_{ijk} \dfrac{C^{n+2/3}_{\alpha,ijk} - C^{n+1/3}_{\alpha,ijk}}{\Delta t} = \delta_{\bar{x}_2}\big(D\delta_{x_2}(C^{n+2/3}_\alpha - C^n_\alpha)\big)_{ijk},$

$1 \le j \le N, \quad \alpha = 1,2,\ldots,n_c - 1,$

(c) $\Phi_{ijk} \dfrac{C^{n+1}_{\alpha,ijk} - C^{n+2/3}_{\alpha,ijk}}{\Delta t} = \delta_{\bar{x}_3}\big(D\delta_{x_3}(C^{n+1}_\alpha - C^n_\alpha)\big)_{ijk},$

$1 \le k \le N, \quad \alpha = 1,2,\ldots,n_c - 1,$

$$(2.5)$$

where we interpret $C^n_\alpha(x)$ as the piecewise threefold-quadratic interpolation [15], $\hat{C}^n_{\alpha,ijk} = C^n_\alpha(\hat{X}^n_{ijk})$, and $\hat{X}^n_{ijk} = X_{ijk} - \Phi^{-1}_{ijk}U^n_{ijk}\Delta t$.

The initial approximation is given by

$$P^0_{ijk} = p_0(X_{ijk}), \quad C^0_{\alpha,ijk} = c_{\alpha,0,ijk},$$
$$1 \le i,j,k \le N, \quad \alpha = 1,2,\ldots,n_c - 1. \tag{2.6}$$

The algorithm for a time step is as follows: Assume the approximate solution $\{P^n_{ijk}, C^n_{\alpha,ijk}(\alpha = 1,2,\ldots,n_c - 1)\}$ at time t^n is known. First, from scheme (2.3a), the method of speed up is used to get the solution of transition sheaf $\{P^{n+1/3}_{ijk}\}$ along the x_1 direction. From (2.3b), we obtain $\{P^{n+2/3}_{ijk}\}$ and from (2.3c), $\{P^{n+1}_{ijk}\}$. Next, from scheme (2.5a), by using the method of speed up, we get the solution of transition sheaf $\{C^{n+1/3}_{\alpha,ijk}\}$ along the x_1 direction. From (2.5b), we obtain $\{C^{n+2/3}_{\alpha,ijk}\}$ and from (2.5c), $\{C^{n+1}_{\alpha,ijk}\}$. So a complete time step can be taken. Finally, because of the positive definite condition, only one solution of this problem can be obtained.

3 Convergence Analysis of Finite Difference

Let $\pi = p - P$ and $\xi_\alpha = c_\alpha - C_\alpha$, where p and $c_\alpha(\alpha = 1,2,\ldots,n_c - 1)$ are the exact solutions of this problem and P and $C_\alpha(\alpha = 1,2,\ldots,n_c - 1)$ are the difference solutions.

First, consider the fluid equation for equation (2.3a), by using (2.3b) and (2.3c), we get the equivalent form

$$d(C^n_{ijk})\dfrac{P^{n+1}_{ijk} - P^n_{ijk}}{\Delta t} - \nabla_h(A^n\nabla_h P^{n+1})_{ijk} = q(X_{ijk}, t^{n+1})$$
$$-(\Delta t)^2\big\{\delta_{\bar{x}_1}(A^n\delta_{x_1}(d^{-1}(C^n)\delta_{\bar{x}_2}(A^n\delta_{x_2}))) + \delta_{\bar{x}_1}(A^n\delta_{x_1}(d^{-1}(C^n)$$
$$\delta_{\bar{x}_3}(A^n\delta_{x_3}))) + \delta_{\bar{x}_2}(A^n\delta_{x_2}(d^{-1}(C^n)\delta_{\bar{x}_3}(A^n\delta_{x_3})))\big\}d_t P^n_{ijk} \tag{3.1}$$
$$+(\Delta t)^3\delta_{\bar{x}_1}(A^n\delta_{x_1}(d^{-1}(C^n)\delta_{\bar{x}_2}(A^n\delta_{x_2}(d^{-1}(C^n)$$
$$\delta_{\bar{x}_3}(A^n\delta_{x_3}d_t P^n)\ldots))_{ijk}, \quad 1 \le i,j,k \le N,$$

where $d_t P_{ijk}^n = \{P_{ijk}^{n+1} - P_{ijk}^n\}/\Delta t$. By (1.1) $(t = t^{n+1})$ and (3.1), we have the pressure error equations

$$
\begin{aligned}
d(C_{ijk}^n)\frac{\pi_{ijk}^{n+1} - \pi_{ijk}^n}{\Delta t} &- \nabla_h(A^n \nabla_h \pi^{n+1})_{ijk} = -(\Delta t)^2\{\delta_{\bar{x}_1}(A^n \delta_{x_1} \\
(d^{-1}(C^n)\delta_{\bar{x}_2}(A^n \delta_{x_2}))) &+ \delta_{\bar{x}_1}(A^n \delta_{x_1}(d^{-1}(C^n)\delta_{\bar{x}_3}(A^n \delta_{x_3}))) \\
+\delta_{\bar{x}_2}(A^n \delta_{x_2}(d^{-1}(C^n)\delta_{\bar{x}_3}(A^n \delta_{x_3})))\}&d_t \pi_{ijk}^n + (\Delta t)^3 \delta_{\bar{x}_1}(A^n \delta_{x_1} \\
(d^{-1}(C^n)\delta_{\bar{x}_2}(A^n \delta_{x_2}(d^{-1}(C^n)\delta_{\bar{x}_3}&(A^n \delta_{x_3} d_t \pi^n)\ldots)_{ijk} \\
+(\Delta t)^2\{\delta_{\bar{x}_1}(A^n \delta_{x_1}(d^{-1}(C^n)\delta_{\bar{x}_2}(A^n \delta_{x_2}))) &+ \delta_{\bar{x}_1}(A^n \delta_{x_1}(d^{-1}(C^n) \\
\delta_{\bar{x}_3}(A^n \delta_{x_3}))) + \delta_{\bar{x}_2}(A^n \delta_{x_2}(d^{-1}(C^n)&\delta_{\bar{x}_3}(A^n \delta_{x_3})))\}d_t p_{ijk}^n \\
-(\Delta t)^3 \delta_{\bar{x}_1}(A^n \delta_{x_1}(d^{-1}(C^n)&\delta_{\bar{x}_2}(A^n \delta_{x_2}(d^{-1}(C^n) \\
\delta_{\bar{x}_3}(A^n \delta_{x_3} d_t p^n)\ldots)_{ijk} &+ \sigma_{ijk}^{n+1}, \quad 1 \le i,j,k \le N;
\end{aligned}
\tag{3.2}
$$

here $d_t \pi^n = \frac{1}{\Delta t}(\pi^{n+1} - \pi^n)$ and $|\sigma_{ijk}^{n+1}| \le M\Big\{\big\|\frac{\partial^2 p}{\partial t^2}\big\|_{L^\infty(L^\infty)}, \big\|\frac{\partial p}{\partial t}\big\|_{L^\infty(W^{4,\infty})},$
$\|p\|_{L^\infty(W^{4,\infty})}, \|c_\alpha\|_{L^\infty(W^{3,\infty})}$ $(\alpha = 1,2,\ldots,n_c - 1)\Big\}(h^2 + \Delta t)$. Suppose that the space and time steps satisfy

$$\Delta t = O(h^2). \tag{3.3}$$

By testing (3.2) against $\delta_t \pi_{ijk}^n = d_t \pi_{ijk}^n \Delta t = \pi_{ijk}^{n+1} - \pi_{ijk}^n$ and summing by parts, we have

$$
\begin{aligned}
< d(C^n)d_t \pi^n, d_t \pi^n > &\Delta t + \frac{1}{2}\{< A^n \nabla_h \pi^{n+1}, \nabla_h \pi^{n+1} > \\
- < A^n \nabla_h \pi^n, \nabla_h \pi^n >\} &\le M\{h^4 + (\Delta t)^2\}\Delta t + \varepsilon|d_t \pi^n|_0^2 \Delta t \\
-(\Delta t)^3\{ < \delta_{\bar{x}_1}(A^n \delta_{x_1}(d^{-1}(C^n)&\delta_{\bar{x}_2}(A^n \delta_{x_2} d_t \pi^n))), d_t \pi^n > \\
+ < \delta_{\bar{x}_1}(A^n \delta_{x_1}(d^{-1}(C^n)\delta_{\bar{x}_3}(A^n \delta_{x_3} d_t \pi^n))), d_t \pi^n > &+ < \delta_{\bar{x}_2}(A^n \delta_{x_2} \\
(d^{-1}(C^n)\delta_{\bar{x}_3}(A^n \delta_{x_3} d_t \pi^n))), d_t \pi^n > \} &+ (\Delta t)^4 < \delta_{\bar{x}_1}(A^n \delta_{x_1} \\
(d^{-1}(C^n)\delta_{\bar{x}_2}(A^n \delta_{x_2}(d^{-1}(C^n)&\delta_{\bar{x}_3}(A^n \delta_{x_3} d_t \pi^n)\ldots)_{ijk}, d_t \pi^n > .
\end{aligned}
\tag{3.4}
$$

By the multiplicative commutation rule of difference operators and the decomposition of high order difference operators, we obtain

$$
\begin{aligned}
|d_t \pi^n|_0^2 \Delta t + \frac{1}{2}\{< A^n \nabla_h \pi^{n+1}, \nabla_h \pi^{n+1} > &- < A^n \nabla_h \pi^n, \nabla_h \pi^n >\} \\
\le M\{|\pi^{n+1}|_1^2 + |\pi^n|_1^2 + h^4 + (\Delta t)^2\}(\Delta t).
\end{aligned}
\tag{3.5}
$$

Next, consider the concentration equations. For equations (2.5a), (2.5b), and (2.5c), we get the equivalent from

$$
\Phi_{ijk}\frac{C^{n+1}_{\alpha,ijk} - \hat{C}^n_{\alpha,ijk}}{\Delta t} - \nabla_h(D\nabla_h C^{n+1})_{ijk} = -b_\alpha(C^n_{ijk})\frac{P^{n+1}_{ijk} - P^n_{ijk}}{\Delta t}
$$

$$
\begin{aligned}
&+g(X_{ijk}, t^n, \hat{C}^n_{\alpha,ijk}) - (\Delta t)^2\{\delta_{\bar{x}_1}(D\delta_{x_1}(\Phi^{-1}\delta_{\bar{x}_2}(D\delta_{x_2}))) \\
&+\delta_{\bar{x}_1}(D\delta_{x_1}(\Phi^{-1}\delta_{\bar{x}_3}(D\delta_{x_3}))) + \delta_{\bar{x}_2}(D\delta_{x_2}(\Phi^{-1}\delta_{\bar{x}_3}(D\delta_{x_3})))\}d_t C^n_{\alpha,ijk} \\
&+(\Delta t)^3 \delta_{\bar{x}_1}(D\delta_{x_1}(\Phi^{-1}\delta_{\bar{x}_2}(D\delta_{x_2}(\Phi^{-1}\delta_{\bar{x}_3}(D\delta_{x_3}d_t C^n_\alpha)\ldots)_{ijk},
\end{aligned} \tag{3.6}
$$

$$
1 \le i, j, k \le N, \quad \alpha = 1, 2, \ldots, n_c - 1.
$$

Similarly, we can obtain

$$
\begin{aligned}
&|d_t\xi^n_\alpha|^2_0\Delta t + <D\nabla_h\xi^{n+1}_\alpha, \nabla_h\xi^{n+1}_\alpha> - <D\nabla_h\xi^n_\alpha, \nabla_h\xi^n_\alpha> \\
&\le M\{|\xi^n|^2_1 + |\xi^{n+1}|^2_1 + |\nabla_h\pi^n|^2_0 + h^4 + (\Delta t)^2\},
\end{aligned} \tag{3.7}
$$

where $|\xi|^2_1 = |\xi|^2_0 + |\nabla_h\xi|^2_0$.

For (3.5), summing on $0 \le n \le L$ and for (3.7), summing on $0 \le n \le L$, $1 \le \alpha \le n_c - 1$, we have

$$
\begin{aligned}
&\sum_{n=0}^{L} |d_t\pi^n|^2_0\Delta t + |\pi^{L+1}|^2_1 \\
&\le \varepsilon\sum_{n=0}^{L-1}|d_t\xi^n|^2_0\Delta t + M\{h^4 + (\Delta t)^2 + \sum_{n=1}^{L}|\pi^{n+1}|^2_1\Delta t\},
\end{aligned} \tag{3.8}
$$

and

$$
\sum_{n=0}^{L}|d_t\xi^n|^2_0\Delta t + |\xi^{L+1}|^2_1 \le M\{h^4 + (\Delta t)^2 + \sum_{n=1}^{L}\left[|\xi^{n+1}|^2_1 + |\pi^n|^2_1\right]\Delta t\}. \tag{3.9}
$$

Combining (3.8) and (3.9) and applying the discrete Gronwall inequality, we have

$$
\sum_{n=0}^{L}\left[|d_t\pi^n|^2_0 + |d_t\xi^n|^2_0\right]\Delta t + |\pi^{L+1}|^2_1 + |\xi^{L+1}|^2_1 \le M\{h^4 + (\Delta t)^2\}. \tag{3.10}
$$

Theorem I *Suppose that the exact solution of problems (1.1)-(1.5) satisfies condition: p, $c_\alpha(\alpha = 1, 2, \ldots, n_c - 1) \in W^{1,\infty}(W^{1,\infty})\bigcap L^\infty(L^\infty)$, $\partial p/\partial t$, $\partial c_\alpha/\partial t(\alpha = 1, 2, \ldots, n_c - 1) \in L^\infty(W^{4,\infty})$, $\partial^2 p/\partial t^2$, $\partial^2 c_\alpha/\partial\tau^2(\alpha = 1, 2, \ldots, n_c - 1) \in L^\infty(L^\infty)$. Adopt the characteristic finite difference fractional steps schemes (2.3) and (2.5). Let the discretization parameter satisfy relation (3.3). The error estimates hold*

$$
\begin{aligned}
&\|p - P\|_{\bar{L}^\infty(J;h^1)} + \sum_{\alpha=1}^{n_c-1}\|c_\alpha - C_\alpha\|_{\bar{L}^\infty(J;h^1)} + \|d_t(p - P)\|_{\bar{L}^2(J;l^2)} \\
&+ \sum_{\alpha=1}^{n_c-1}\|d_t(c_\alpha - C_\alpha)\|_{\bar{L}^\infty(J;l^1)} \le M^*\{h^2 + (\Delta t)\},
\end{aligned} \tag{3.11}
$$

where $\|f\|_{\bar{L}^{\infty}(J;X)} = \sup_{n\Delta t \leq T} \|f^n\|_X$, $\|g\|_{\bar{L}^2(J;X)} = \sup_{n\Delta t \leq T} \{ \sum_{n=0}^{N} \|g^n\|_X^2 \Delta t \}^{1/2}$,

$$M^* = M^* \Big\{ \|p\|_{W^{1,\infty}(W^{1,\infty})}, \ \|p\|_{L^{\infty}(W^{4,\infty})}, \ \Big\| \frac{\partial p}{\partial t} \Big\|_{L^{\infty}(W^{4,\infty})}, \ \Big\| \frac{\partial^2 p}{\partial t^2} \Big\|_{L^{\infty}(L^{\infty})},$$

$$\|c_{\alpha}\|_{W^{1,\infty}(W^{1,\infty})}, \ \|c_{\alpha}\|_{L^{\infty}(W^{4,\infty})}, \ \Big\| \frac{\partial c_{\alpha}}{\partial t} \Big\|_{L^{\infty}(W^{4,\infty})}, \ \Big\| \frac{\partial^2 c_{\alpha}}{\partial \tau^2} \Big\|_{L^{\infty}(L^{\infty})}, \quad (\alpha = $$

$$1, 2, \ldots, n_c - 1) \Big\}.$$

4   Finite Element Operator-Splitting

The finite element method for problem (1.1)–(1.4) based on the weak form is given by

$$\begin{array}{ll}
(a) & \Big(d\frac{\partial p}{\partial t}, v \Big) + \big(a(c) \nabla p, \nabla v \big) = \big(q(c), v \big), \\
& v \in H^1(\Omega), \quad t \in J = (0, T], \\
(b) & \Big(\Phi \frac{\partial c_{\alpha}}{\partial t}, z \Big) + \big(u \cdot \nabla c_{\alpha}, z \big) + \big(D\nabla c_{\alpha}, \nabla z \big) + \Big(b_{\alpha}(c)\frac{\partial p}{\partial t}, z \Big) \\
& = \big(g(c_{\alpha}), z \big), \quad z \in H^1(\Omega), \quad t \in J, \quad \alpha = 1, 2, \ldots, n_c - 1.
\end{array}$$

(4.1)

Assume that $d = d_1(x_1)d_2(x_2)d_3(x_3)$ and $\Phi = \Phi_1(x_1)\Phi_2(x_2)\Phi_3(x_3)$. We discuss a finite element operator-splitting method approximation of the fluid equation (4.1a). We equidistantly subdivide the region Ω. The coding of nodes: $\{x_{1,\alpha}|0 \leq \alpha \leq N_{x_1}\}$, $\{x_{2,\beta}|0 \leq \beta \leq N_{x_2}\}$, and $\{x_{3,\gamma}|0 \leq \gamma \leq N_{x_3}\}$ is used. The global coding of three-dimensional mesh region i ($i = 1, 2, \ldots, N$), $N = (N_{x_1} + 1)(N_{x_2} + 1)(N_{x_3} + 1)$ is obtained. The tensor product index of node i is $(\alpha(i), \beta(i), \gamma(i))$, where $\alpha(i)$ is the number of the x_1-axis, $\beta(i)$ is the number of the x_2-axis, and $\gamma(i)$ is the number of the x_3-axis. The tensor product basis can be rewritten as products of one-dimensional basis functions in the manner

$$\begin{aligned}
N_i(x_1, x_2, x_3) &= \varphi_{\alpha(i)}(x_1)\psi_{\beta(i)}(x_2)\omega_{\gamma(i)}(x_3) \\
&= \varphi_{\alpha}(x_1)\psi_{\beta}(x_2)\omega_{\gamma}(x_3), \quad 1 \leq i \leq N.
\end{aligned}$$

(4.2)

If $N_{h_p} = \varphi \otimes \psi \otimes \omega$ is a finite element space, let

$$W = \Big\{ w \Big| w, \ \frac{\partial w}{\partial x_i}, \ \frac{\partial^2 w}{\partial x_i \partial x_j}(i \neq j), \ \frac{\partial^3 w}{\partial x_1 \partial x_2 \partial x_3} \in L_2(\Omega) \Big\}.$$

Note that $N_{h_p} \subset W$ [22,23]. The approximation properties are given by the inequalities

$$\inf_{\chi \in N_h} \left\{ \sum_{m=0}^{3} h_p^m \sum_{\substack{i,j,k=0,1 \\ i+j+k=m}} \left\| \frac{\partial^m (u-\chi)}{\partial x_1^i \partial x_2^j \partial x_3^k} \right\|_0 \right\} \leq M h_p^{k+1} \|u\|_{k+1}, \quad (4.3)$$

where $h_p = \max\{N_{x_1}^{-1}, N_{x_2}^{-1}, N_{x_3}^{-1}\}$ is the subdivision step.

We discuss a characteristic finite element operator-splitting method approximation of the concentration equations (4.1b); similarly, we equidistantly subdivide the region Ω and use the coding of nodes: $\{x_{1,\lambda}|0 \leq \lambda \leq N_{x_1}\}$, $\{x_{2,\mu}|0 \leq \mu \leq N_{x_2}\}$, and $\{x_{3,\chi}|0 \leq \chi \leq N_{x_3}\}$. The global coding j $(j = 1, 2, \ldots, M)$, $M = (M_{x_1}+1)(M_{x_2}+1)(M_{x_3}+1)$ is obtained. The tensor product index of node j is $(\lambda(j), \mu(j), \chi(j))$. The tensor product basis can be rewritten as products of one-dimensional basis functions in the manner

$$\begin{aligned} M_j(x_1, x_2, x_3) &= \Phi_{\lambda(j)}(x_1) \Psi_{\mu(j)}(x_2) \Omega_{\chi(j)}(x_3) \\ &= \Phi_\lambda(x_1) \Psi_\mu(x_2) \Omega_\chi(x_3), \quad 1 \leq j \leq M. \end{aligned} \qquad (4.4)$$

Let the finite element space $M_{h_c} = M_h = \Phi \otimes \Psi \otimes \Omega$. Note that $M_h \subset W$. The approximation properties are given by the inequalities

$$\inf_{\varphi \in M_h} \left\{ \sum_{m=0}^{3} h_c^m \sum_{\substack{i,j,k=0,1 \\ i+j+k=m}} \left\| \frac{\partial^m (u-\varphi)}{\partial x_1^i \partial x_2^j \partial x_3^k} \right\|_0 \right\} \leq M h_c^{l+1} \|u\|_{l+1}, \quad (4.5)$$

where $h_c = \max\{M_{x_1}^{-1}, M_{x_2}^{-1}, M_{x_3}^{-1}\}$ is the subdivision step.

The characteristic finite element two-level operator-splitting scheme of problem (4.1): When $t = t^n$, if $\{P_h^n, C_h^n\} \in N_h \times M_h^{n_c-1}$ are known, we find the finite element solution $\{P_h^{n+1}, C_h^{n+1}\} \in N_h \times M_h^{n_c-1}$, $t = t^{n+1}$. First, the finite element scheme of the fluid equation (4.1a):

$$\begin{aligned} &(d\, d_t p_h^n, v_h) + \left(a(c_h^n) \nabla P_h^n, \nabla v_h \right) + \lambda_p \Delta t (d \nabla d_t P_h^n, \nabla v_h) \\ &+ (\lambda_p \Delta t)^2 \sum_{\substack{i \neq j, i,j=1,2,3}} \left(d \frac{\partial^2 d_t P_h^n}{\partial x_i \partial x_j}, \frac{\partial^2 v_h}{\partial x_i \partial x_j} \right) + (\lambda_p \Delta t)^3 \\ &+ \left(d \frac{\partial^3 d_t P_h^n}{\partial x_1 \partial x_2 \partial x_3}, \frac{\partial^3 v_h}{\partial x_1 \partial x_2 \partial x_3} \right) = (q(x, t^n, C_h^n), v_h), \quad \forall v_h \in N_h, \end{aligned} \qquad (4.6)$$

$$U_h^n = -a(C_h^n) \nabla P_h^n, \qquad (4.7)$$

where $d_t P_h^n = (P_h^{n+1} - P_h^n)/\Delta t$,

$$\sum_{\substack{i \neq j, i,j=1,2,3}}^{3} \left(d\frac{\partial^2 d_t P_h^n}{\partial x_i \partial x_j}, \frac{\partial^2 v_h}{\partial x_i \partial x_j}\right) = \left(d\frac{\partial^2 d_t P_h^n}{\partial x_1 \partial x_2}, \frac{\partial^2 v_h}{\partial x_1 \partial x_2}\right)$$
$$+\left(d\frac{\partial^2 d_t P_h^n}{\partial x_2 \partial x_3}, \frac{\partial^2 v_h}{\partial x_2 \partial x_3}\right) + \left(d\frac{\partial^2 d_t P_h^n}{\partial x_3 \partial x_1}, \frac{\partial^2 v_h}{\partial x_3 \partial x_1}\right),$$

and λ_p is a chosen constant.

As the flow is essentially in the characteristic direction, we apply the modified method of characteristic to the first-order parts of (4.1b), thus ensuring a high accuracy of the numerical results [1, 7, 5, 3]. Let $\psi(x, u) = [\Phi^2(x) + |u|^2]^{1/2}$ and $\partial/\partial\tau = \psi^{-1}\{\Phi\partial/\partial t + u \cdot \nabla\}$. We write (4.1b) in the form

$$\left(\psi\frac{\partial c_\alpha}{\partial\tau}, z\right) + (D\nabla c_\alpha, \nabla z) + \left(b_\alpha(c)\frac{\partial p}{\partial t}, z\right) = (g(c_\alpha, z)), \qquad (4.8)$$
$$z \in H^1(\Omega), \quad t \in J, \quad \alpha = 1, 2, \ldots, n_c - 1.$$

Approximate $\dfrac{\partial c_\alpha^{n+1}}{\partial\tau} = \dfrac{\partial c_\alpha}{\partial\tau}(x, t^{n+1})$ by a backward difference quotient in the τ-direction, $\dfrac{\partial c_\alpha^{n+1}}{\partial\tau} \approx \dfrac{c_\alpha^{n+1}(x) - c_\alpha^n(x - u^{n+1}\Delta t/\Phi(x))}{\Delta t(1 + \Phi^{-2}|u^{n+1}|^2)^{1/2}}$.

For the concentration equations (4.8), the characteristic finite element operator-splitting procedure is

$$\left(\Phi\frac{C_{\alpha,h}^{n+1} - \hat{C}_{\alpha,h}^n}{\Delta t}, z_h\right) + (D\nabla C_{\alpha,h}^n, \nabla z_h) + \lambda_c \Delta t(\Phi\nabla d_t C_{\alpha,h}^n, \nabla z_h)$$
$$+(\lambda_c \Delta t)^2 \sum_{\substack{i \neq j, i,j=1,2,3}}^{3} \left(\Phi\frac{\partial^2 d_t C_{\alpha,h}^n}{\partial x_i \partial x_j}, \frac{\partial^2 z_h}{\partial x_i \partial x_j}\right) + (\lambda_c \Delta t)^3$$
$$+\left(\Phi\frac{\partial^3 d_t C_{\alpha,h}^n}{\partial x_1 \partial x_2 \partial x_3}, \frac{\partial^3 z_h}{\partial x_1 \partial x_2 \partial x_3}\right) + (b_\alpha(C_h^n)d_t P_h^n, z_h) \qquad (4.9)$$
$$= (g(\hat{C}_{\alpha,h}^n), z_h), \quad \forall z_h \in M_h, \quad \alpha = 1, 2, \ldots, n_c - 1,$$

where $\hat{C}_{\alpha,h}^n = C_{\alpha,h}^n(\hat{x})$, $\hat{x} = x - U_h^n \Delta t/\Phi(x)$, and λ_c is a chosen constant.

For the fluid equation (4.6), if $P_h^{n+1} = \sum_{\alpha,\beta,\gamma} \xi_{\alpha\beta\gamma}^{n+1}\varphi_\alpha\psi_\beta\omega_\gamma$, then (4.6) can be written in the form

$$\sum_{\alpha,\beta,\gamma} (\xi_{\alpha\beta\gamma}^{n+1} - \xi_{\alpha\beta\gamma}^n)(d\varphi_\alpha \otimes \psi_\beta \otimes \omega_\gamma, \varphi_\alpha \otimes \psi_\beta \otimes \omega_\gamma)$$
$$+\lambda_p\Delta t \sum_{\alpha,\beta,\gamma} (\xi_{\alpha\beta\gamma}^{n+1} - \xi_{\alpha\beta\gamma}^n)\{(d\varphi_\alpha' \otimes \psi_\beta \otimes \omega_\gamma, \varphi_\alpha' \otimes \psi_\beta \otimes \omega_\gamma)$$
$$+(d\varphi_\alpha \otimes \psi_\beta' \otimes \omega_\gamma, \varphi_\alpha \otimes \psi_\beta' \otimes \omega_\gamma) + (d\varphi_\alpha \otimes \psi_\beta \otimes \omega_\gamma', \varphi_\alpha \otimes \psi_\beta \otimes \omega_\gamma')\}$$
$$+(\lambda_p\Delta t)^2 \sum_{\alpha,\beta,\gamma} (\xi_{\alpha\beta\gamma}^{n+1} - \xi_{\alpha\beta\gamma}^n)\{(d\varphi_\alpha' \otimes \psi_\beta' \otimes \omega_\gamma, \varphi_\alpha \otimes \psi_\beta' \otimes \omega_\gamma)$$

$$(4.10)$$

$$+(d\varphi_\alpha \otimes \psi'_\beta \otimes \omega'_\gamma, \varphi_\alpha \otimes \psi'_\beta \otimes \omega'_\gamma) + (d\varphi'_\alpha \otimes \psi_\beta \otimes \omega'_\gamma, \varphi'_\alpha \otimes \psi_\beta \otimes \omega'_\gamma)\}$$
$$+(\lambda_p\Delta t)^3 \sum_{\alpha,\beta,\gamma}(\xi^{n+1}_{\alpha\beta\gamma} - \xi^n_{\alpha\beta\gamma})(d\varphi'_\alpha \otimes \psi'_\beta \otimes \omega'_\gamma, \varphi'_\alpha \otimes \psi'_\beta \otimes \omega'_\gamma) = \Delta t F^n.$$

Let

$$C_{x_1} = \left(\int_0^1 d_1\varphi_{\alpha_1}\varphi_{\alpha_2}dx_1\right), \quad A_{x_1} = \left(\int_0^1 d_1\varphi'_{\alpha_1}\varphi'_{\alpha_2}dx_1\right),$$
$$C_{x_2} = \left(\int_0^1 d_2\psi_{\beta_1}\psi_{\beta_2}dx_2\right), \quad A_{x_2} = \left(\int_0^1 d_2\psi'_{\beta_1}\psi'_{\beta_2}dx_2\right),$$
$$C_{x_3} = \left(\int_0^1 d_3\omega_{\gamma_1}\omega_{\gamma_2}dx_3\right), \quad A_{x_3} = \left(\int_0^1 d_3\omega'_{\gamma_1}\omega'_{\gamma_2}dx_3\right).$$

Then we have

$$(C_{x_1} + \lambda_p\Delta t A_{x_1}) \otimes (C_{x_2} + \lambda_p\Delta t A_{x_2}) \otimes (C_{x_3} + \lambda_p\Delta t A_{x_3})$$
$$(\xi^{n+1} - \xi^n) = \Delta t F^n, \tag{4.11}$$

where

$$F^n_{\alpha\beta\gamma} = -(a(C^n_h)\nabla P^n_h, \nabla(\varphi_\alpha \otimes \psi_\beta \otimes \omega_\gamma)) + (q(C^n_h), \varphi_\alpha \otimes \psi_\beta \otimes \omega_\gamma). \tag{4.12}$$

We point out that (4.11) can be solved by an alternating-direction.

For the concentration equation (4.9), if $C^{n+1}_{\alpha,h} = \sum_{\lambda,\mu,\chi} \zeta^{n+1}_{\alpha,\lambda\mu\chi}\Phi_\lambda\Psi_\mu\Omega_\chi$, it can be written in the form

$$\sum_{\lambda,\mu,\chi}(\zeta^{n+1}_{\alpha,\lambda\mu\chi} - \zeta^n_{\alpha,\lambda\mu\chi})(\Phi\Phi_\lambda \otimes \Psi_\mu \otimes \Omega_\chi, \Phi_\lambda \otimes \Psi_\mu \otimes \Omega_\chi)$$
$$+\lambda_c\Delta t \sum_{\lambda,\mu,\chi}(\zeta^{n+1}_{\alpha,\lambda\mu\chi} - \zeta^n_{\alpha,\lambda\mu\chi})\{(\Phi\Phi'_\lambda \otimes \Psi_\mu \otimes \Omega_\chi, \Phi'_\lambda \otimes \Psi_\mu \otimes \Omega_\chi)$$
$$+(\Phi\Phi_\lambda \otimes \Psi'_\mu \otimes \Omega_\chi, \Phi_\lambda \otimes \Psi'_\mu \otimes \Omega_\chi) + (\Phi\Phi_\lambda \otimes \Psi_\mu \otimes \Omega'_\chi, \Phi_\lambda \otimes \Psi_\mu \otimes \Omega'_\chi)\}$$
$$+(\lambda_c\Delta t)^2 \sum_{\lambda,\mu,\chi}(\zeta^{n+1}_{\alpha,\lambda\mu\chi} - \zeta^n_{\alpha,\lambda\mu\chi})\{(\Phi\Phi'_\lambda \otimes \Psi'_\mu \otimes \Omega_\chi, \Phi'_\lambda \otimes \Psi'_\mu \otimes \Omega_\chi)$$
$$+(\Phi\Phi_\lambda \otimes \Psi'_\mu \otimes \Omega'_\chi, \Phi_\lambda \otimes \Psi'_\mu \otimes \Omega'_\chi) + (\Phi\Phi'_\lambda \otimes \Psi_\mu \otimes \Omega'_\chi,$$
$$\Phi'_\lambda \otimes \Psi_\mu \otimes \Omega'_\chi)\} + (\lambda_c\Delta t)^3 \sum_{\lambda,\mu,\chi}(\zeta^{n+1}_{\alpha,\lambda\mu\chi} - \zeta^n_{\alpha,\lambda\mu\chi})(\Phi\Phi'_\lambda \otimes \Psi'_\mu \otimes \Omega'_\chi,$$
$$\Phi'_\lambda \otimes \Psi'_\mu \otimes \Omega'_\chi) = \Delta t G^n_\alpha, \quad \alpha = 1, 2, \ldots, n_c - 1, \tag{4.13}$$

where

$$D_{x_1} = \left(\int_0^1 \Phi_1\Phi_{\lambda_1}(x_1)\Phi_{\lambda_2}(x_1)dx_1\right), \quad B_{x_1} = \left(\int_0^1 \Phi_1\Phi'_{\lambda_1}\Phi'_{\lambda_2}dx_1\right),$$
$$D_{x_2} = \left(\int_0^1 \Phi_2\Psi_{\mu_1}\Psi_{\mu_2}dx_2\right), \quad B_{x_2} = \left(\int_0^1 \Phi_2\Psi'_{\mu_1}\Psi'_{\mu_2}dx_2\right),$$
$$D_{x_3} = \left(\int_0^1 \Phi_3\Omega_{\chi_1}\Omega_{\chi_2}dx_3\right), \quad B_{x_3} = \left(\int_0^1 \Phi_3\Omega'_{\chi_1}\Omega'_{\chi_2}dx_3\right),$$

$$(D_{x_1} + \lambda_c\Delta t B_{x_1}) \otimes (D_{x_2} + \lambda_c\Delta t B_{x_2}) \otimes (D_{x_3} + \lambda_c\Delta t B_{x_3})$$
$$(\zeta^{n+1}_\alpha - \zeta^n_\alpha) = \Delta t G^n_\alpha, \quad \alpha = 1, 2, \ldots, n_c - 1, \tag{4.14}$$

and

$$G_{\alpha,\lambda\mu\chi}^n = \frac{1}{\Delta t}\left(\Phi(\hat{C}_{\alpha,h}^n - C_{\alpha,h}^n), \Phi_\lambda \otimes \Psi_\mu \otimes \Omega_\chi\right)$$
$$- \left(D\nabla C_{\alpha,h}^n, \nabla(\Phi_\lambda \otimes \Psi_\mu \otimes \Omega_\chi)\right) - \left(b_\alpha(C_h^n)d_t P_h^n, \Phi_\lambda \otimes \Psi_\mu \otimes \Omega_\chi\right)$$
$$+ \left(g(\hat{C}_{\alpha,h}^n), \Phi_\lambda \otimes \Psi_\mu \otimes \Omega_\chi\right).$$

$$(4.15)$$

Similarly, we point out that (4.13) can be solved by an alternating-direction.

Theorem II *Suppose that the exact solution of problem (1.1)–(1.5) is smooth. Adopt the characteristic finite element operator-splitting scheme (4.6), (4.7), and (4.9). Suppose that $k \geq 1$, $l \geq 1$, and the spatial and time discretizations satisfy the relations $\Delta t = O(h_p^2) = O(h_c^2)$, $h_p^{k+1} = o(h_c^{3/2})$, and $h_c^{l+1} = o(h_p^{3/2})$. Then the error estimate holds*

$$\|p - P_h\|_{\bar{L}_\infty(J;L^2(\Omega))} + \sum_{\alpha=1}^{n_c-1} \|c_\alpha - C_{\alpha,h}\|_{\bar{L}_\infty(J;L^2(\Omega))}$$
$$+ \|d_t(p - P_h)\|_{\bar{L}_2(J;L^2(\Omega))} + h_c \sum_{\alpha=1}^{n_c-1} \|c_\alpha - C_{\alpha,h}\|_{\bar{L}_2(J;H^1(\Omega))}$$
$$\leq M^*\{\Delta t + h_p^{k+1} + h_c^{l+1}\},$$

$$(4.16)$$

where $\|g\|_{\bar{L}_\infty(J;X)} = \sup_{n\Delta t \leq T} \|g^n\|$, $\|g\|_{\bar{L}_2(J;X)} = \sup_{N\Delta t \leq T} (\sum_{n=0}^{N} \|g^n\|_X^2)^{1/2}$, and the constant M^* depends on p, c, and its derivatives.

5 Applications

The numerical method has already been used in the numerical simulation for evolutionary history [15, 12], their mathematical model:

$$(a) \quad \nabla \cdot \left(\frac{K}{\mu}\nabla p\right) = (\alpha(1 - \Phi) + \beta\Phi)\frac{\partial p}{\partial t} - \alpha(1 - \Phi)\frac{\partial s}{\partial t}$$
$$+ (\alpha(1 - \Phi) + \beta\Phi)\frac{\partial p_h}{\partial t}, \quad X = (x_1, x_2, x_3)^T \in \Omega, \quad t \in J,$$

$$(b) \quad \nabla \cdot [K_s\nabla T] - c_\omega \rho_\omega \nabla \cdot (VT) + Q = c_s \rho_s \frac{\partial T}{\partial t}, \quad x \in \Omega, \quad t \in J,$$

$$(c) \quad \frac{\partial \Phi}{\partial t} = -\alpha(1 - \Phi)\left(\frac{\partial s}{\partial t} - \frac{\partial p}{\partial t} + \frac{\partial p_h}{\partial t}\right), \quad x \in \Omega, \quad t \in J.$$

$$(5.1)$$

The numerical method has also been used in the numerical simulation for enhanced oil recovery simulation [10, 11]. Their mathematical model is

$$\Phi\frac{\partial c_i}{\partial t} + \nabla \cdot \left(\sum_{j=1}^{n_p} c_{ij}u_j\right) - \nabla \cdot \left(\sum_{j=1}^{n_p} \Phi s_j D_j(u_j)\nabla c_{ij}\right) = Q_i(c_i),$$
$$x \in \Omega, \quad t \in J, \quad i = 1, 2, \ldots, n_c.$$

$$(5.2)$$

Moreover, this method has been used in the numerical simulation for seawater intrusion [14,15]. Their mathematical model is

$$(a) \quad \nabla \cdot (a\nabla H) = S_s \frac{\partial H}{\partial t} - q, \quad x \in \Omega, \quad t \in J,$$

$$(b) \quad \nabla \cdot (\Psi D \nabla c) - V \cdot \nabla c = \Psi \frac{\partial c}{\partial t} + \Psi S_s c \frac{\partial H}{\partial t} \qquad (5.3)$$
$$-q(c - c^*), \quad x \in \Omega, \quad t \in J.$$

Acknowledgments. This research is supported by the National Natural Science Foundation of China, the National Scaling Program, the National Tackling Key Problems Program, and the Doctoral Foundation of the State Education Department of China.

References

[1] Douglas, J. Jr., Finite difference methods for two-phase incompressible flow in porous media, *SIAM. J. Numer. Anal.* **4** (1983), 681–689.

[2] Douglas, J., Jr., and Roberts , J. E., Numerical method for a model for compressible miscible displacement in porous media, *Math. Comp.* **41**(1983), 441–459.

[3] Douglas, J,Jr. and Yuan, Y., Numerical simulation of immiscible flow in porous media based on combining the method of characteristics with mixed finite element procedure, *IMA Vol. in Math. and its Appl.* **11** (1986), 119–131.

[4] Ewing, R. E. (ed.), *The Mathematics of Reservoir Simulation*, SIAM, Philadelphia, 1983.

[5] Ewing, R. E., Russell, T. F., and Wheeler, M. F., Convergence analysis of an approximation of miscible displacement in porous media by mixed finite elements and a modified method of characteristics, *Comput. Meth. Appl. Mech. Engrg.*, **47** (1984), 73–92.

[6] Marchuk, G. I., *Splitting and alternating direction methods*, Handbook of Numerical Analysis, Ciarlet P. G., Lions J. L., Elsevic, eds., Science Publishers B.V., 1996, 197–460.

[7] Russell, T. F., Time stepping along characteristics with incomplete iteration for a Galerkin approximation of miscible displacement in porous media, *SIAM. J. Numer. Anal.* **5** (1985), 970–1013.

[8] Yuan, Y., Time stepping along characteristics for the finite element approximation of compressible miscible displacement in porous media, *Math. Numer. Sinica* **4** (1992), 385–400.

[9] Yuan, Y., Finite difference methods for a compressible miscible displacement problem in porous media, *Math. Numer. Sinica* **1** (1993), 16–28.

[10] Yuan, Y., The characteristics-mixed finite element method for enhanced oil recovery simulation and optimal order L^2 error estimate, *Chinese Science Bulletin* **21** (1993), 1761–1766.

[11] Yuan, Y., FDM for enhanced oil recovery simulation, *Science in China, Ser.A.* **1** (1993), 1296–1307.

[12] Yuan, Y., CFD method for moving boundary value problem, *Science in China, Ser. A* **12** (1994), 1442–1453.

[13] Yuan, Y., Liang, D., and Rui, H., Characteristics finite element methods for seawater intrusion numerical simulation and theoretical analysis, *Acta Math. Appl. Sinica* **1** (1998), 11–23.

[14] Yuan, Y., Liang, D., Rui, H., and Wang, G., The characteristics finite difference method for seawater intrusion numerical simulation and optimal order L^2 error estimates, *Acta Mathematical Appl. Sinica* **3** (1996), 395–404.

[15] Yuan, Y., Wang, W., Zhao, W., and Han, Y., Numerical method and software for numerical simulation of oil-gas resources, in *CSIAM'96*, Collected Works on Industry and Applied Mathematics (Cang, Q. and Li,D., eds.), Shanghai, Fudan University, 1996, 576–580.

A Model and its Solution Method for a Generalized Unsteady Seepage Flow Problem

Guoyou Zhang   Tigui Fan   Zhongsheng Zhao
Dequan Yang

Abstract

A model and its solution method for a generalized unsteady seepage flow problem are given. Using this method, many generalized unsteady seepage flow control equations are described. Using the solutions to these equations, the pressure distribution of vertical crevice and level wells is obtained. We can analyze the unsteady testing material of oil wells.

KEYWORDS: generalized unsteady seepage flow, control equations, oil well

1 Introduction

The study of seepage flow is important in engineering design. The new technology of oil field development is hard to be treated. They are related to the study of seepage flow. Some models and their solution methods have been developed for seepage flow. Using Green's functions and minor images, Ozcan and Raghavan [1, 2] gave some important results; also, see [3]. A model and its solution method for a generalized advective seepage flow problem is given. Using this method, many generalized unsteady seepage flow control equations are described. The pressure distribution of the seepage flow is obtained. Three formulas are given with different α and β. The formulas become approximately linear or steady state with the dimensionless time increasing. The source solutions are obtained when $\alpha - \beta + 2 > 0$. The pressure distribution of vertical crevice and level wells is given from the solutions. We can analyze the unsteady testing material of oil wells with these results. All these results show that the method is effective.

2   A Mathematical Model

If α and β are control parameters, we can change many seepage flow problems into generalized control equations for the unsteady Darcy flow of a weak compressible fluid in a porous medium:

$$\frac{\partial}{\partial r_D}\left(r_D^{\beta}\frac{\partial p_D}{\partial r_D}\right) = r_D^{\alpha}\frac{\partial p_D}{\partial t_D}, \qquad 0 \le \alpha \le 3, \quad 0 \le \beta \le 5, \qquad (2.1)$$

where $\alpha = \theta_\varphi + \gamma - 1$, $\beta = \theta_\mu + \gamma - 1$, $p_D = k_0 h \Delta p/(c_p Q \mu_0 b)$, $\Delta p = p - p_i$ for injection wells, $\Delta p = p_i - p$ for production wells, $t_D = c_t k_0 t/(\phi \mu_0 C_t r_w^2)$, and $r_D = r/r_w$. Here h is the medium thickness. ϕ the porosity, μ_0 the viscosity, c_t the compressible coefficient, r_w the well radius, c_p and C_t unit conversion coefficients, p_i the initial pressure, θ_ϕ the interval changing coefficient, θ_k the osmosis changing coefficient, θ_μ the fluid viscosity changing coefficient, and γ the fractal dimension. Form (2.1) owns an extensive meaning. It can be changed to various generalized unsteady seepage flow control equations with different α and β. If Q is out of a vertical well, the unsteady control equation can be obtained for an even equal thickness

$$\frac{\partial^2 p_D}{\partial r_D^2} + \frac{\beta}{r_D}\frac{\partial p_D}{\partial r_D} = r_D^{\alpha-\beta}\frac{\partial p_D}{\partial t_D}, \qquad (2.2)$$

with the initial condition

$$p_D(r_D, 0) = 0, \qquad (2.3)$$

the inner condition of the well

$$\left(r_D^{\beta}\frac{\partial p_D}{\partial r_D}\right)_{r_D=1} = -1, \qquad (2.4)$$

the out boundary condition

$$p_D(\infty, t_D) = 0, \qquad (2.5)$$

and the pressure at the well wall

$$\tilde{p}_{SD}(s) = \tilde{p}_{SD}(1, t_D). \qquad (2.6)$$

Eqs. (2.2)–(2.5) are unsteady seepage flow formulas of an even equal thickness vertical production well.

3   Solution

Making the Laplace transform for form (2.2), if s is the Laplace transform variable and

$$\tilde{p}(r_D, s) = \int_0^\infty p_D(r_D, t_D) e^{-s t_D} dt_D,$$

we can obtain

$$\frac{\partial^2 \tilde{p}_D}{\partial r_D^2} + \frac{\beta}{r_D} \frac{\partial \tilde{p}_D}{\partial r_D} = s r_D^{\alpha - \beta} \tilde{p}_D,$$

whose complete solution is

$$\tilde{p}_D(r_D, s) = \begin{cases} \sqrt{r_D^{1-\beta}} \left(c_1 I_\gamma \left(Z_1 \sqrt{r_D^{(\alpha-\beta+2)}} \right) + c_2 K_\gamma \left(Z_1 \sqrt{r_D^{(\alpha-\beta+2)}} \right) \right), \\ \qquad\qquad \alpha - \beta + 2 > 0, \\[2mm] \sqrt{r_D^{1-\beta}} \left(c_1 \sqrt{r_D^{\sqrt{(1-\beta)^2+4s}}} + c_2 \sqrt{r_D^{-\sqrt{(1-\beta)^2+4s}}} \right), \\ \qquad\qquad \alpha - \beta + 2 = 0, \\[2mm] \sqrt{r_D^{1-\beta}} \left(c_1 I_\gamma \left(Z_2 \sqrt{r_D^{(\alpha-\beta+2)}} \right) c_2 K_\gamma \left(Z_2 \sqrt{r_D^{\alpha-\beta+2}} \right) \right), \\ \qquad\qquad \alpha - \beta + 2 < 0, \end{cases}$$

where $I_\gamma(\cdot)$ and $K_\gamma(\cdot)$ are the γ step Bessel functions of second type:

$$\gamma = \frac{1-\beta}{\alpha-\beta+2}, \quad Z_1 = \frac{2\sqrt{s}}{\alpha-\beta+2}, \quad Z_2 = \frac{2\sqrt{s}}{\beta-\alpha-2}.$$

We can also obtain the pressure distribution in the medium from (2.4) and (2.5)

$$\tilde{p}_D(r_D, t_D) = \begin{cases} \dfrac{\sqrt{r_D^{(1+\beta)}} I_r(Z_1 \sqrt{r_D^{(\alpha-\beta+2)}})}{S\sqrt{S} I_{r-1}(Z_1)}, & \alpha - \beta + 2 > 0, \\[4mm] \dfrac{2\sqrt{r_D^{(1-\beta)}} \sqrt{r_D^{-\sqrt{(r-\beta)^2+4s}}}}{s(\beta + \sqrt{(1-\beta)^2 + 4s} - 1)}, & \alpha - \beta + 2 = 0, \\[4mm] \dfrac{\sqrt{r_D^{(1-\beta)}} I_r Z_2 \sqrt{r_D^{(\alpha-\beta+2)}}}{s\sqrt{s} I_{r-1}(Z_2)}, & \alpha - \beta + 2 < 0. \end{cases} \qquad (3.1)$$

When the dimensionless time become longer, the gradual results of form (3.1) are

$$p_{SD}(t_D) \approx \frac{1}{\beta - 1} \quad if\, \alpha - \beta + 2 < 0,$$

and if $\alpha - \beta + 2 > 1$,

$$
p_{SD}(t_D) \approx \begin{cases} \dfrac{1}{\beta - 1}, & \beta > 1, \\[2mm] \dfrac{1}{\alpha + 1}(L_n t_D + 2L_n(\alpha + 1) + \gamma), & \beta = 1, \\[2mm] \dfrac{1}{\gamma}\dfrac{1}{\Gamma_{1-\gamma}}(\alpha - \beta + 2)^{2\gamma - 1} t_D^\gamma, & \beta < 1, \end{cases} \quad .
$$

We can see the change trend of the pressure and pressure derivative from this equation.

There is a real space solution under the point source inner boundary condition for the first case of the complete solution. Other solution conditions do not change. The point source inner boundary condition is

$$
\left(\frac{r_D^\beta \partial p_D}{\partial r_D} \right)_{r_D \to 0} = -1.
$$

When making the Laplace transform, the pressure distribution is

$$
\tilde{p}_D(r_D, s) = \frac{1}{s\sqrt{s}} \frac{2}{\Gamma(1-\gamma)} \left(\frac{\sqrt{s}}{\alpha - \beta + 2} \right)^{1-\gamma} \sqrt{r_D^{(1-\beta)}} K_\gamma \left(Z_1 \sqrt{r_D^{(\alpha-\beta+2)}} \right),
$$

which can be inverted as follows:

$$
p_D(r_D, t_D) = \frac{1}{\Gamma(1-\gamma)(\alpha - \beta + 2)^{1-2\gamma}} \int_0^{t_D} \tau^{\gamma-1} \exp\left(-\frac{r_D^{\alpha-\beta+2}}{(\alpha - \beta + 2)^2 \tau} \right) d\tau.
$$

Beier's [4] point source solution can be obtained from this equation.

Using numerical methods, the dimensionless well wall pressure and pressure derivative can be obtained from (3.1) with different α and β. The unsteady testing material is analyzed by using these results.

4 Conclusions

(1) The generalized unsteady seepage flow model which we obtained can be changed into unsteady seepage flow control equations.
(2) There are three forms for the pressure distribution of seepage flow which we obtained with different α and β. The analysis shows that these three forms are steady state or approximately linear with longer dimensionless time.
(3) When $\alpha - \beta + 2 > 0$, the solutions of the point source in real space are obtained by using numerical methods. The pressure distribution of vertical

crevice and level wells is obtained from these solutions. The unsteady testing material of the oil well pressure is analyzed using these results.

Acknowledgements. This work is partly supported by NSFC.

References

[1] Ozcan, E. and Raghavan, R., New solutions for well-test-analysis problems: Part 4 analytical considerations, Paper SPEFE, September, 1999.

[2] Ozcan, E. and Raghavan, R., New solutions for well-test-analysis problems: Part 3 additional algorithms, Paper SPEFE, September, 1994.

[3] Liu, C. and Li, F., Unsteady flow of non-Newtonian fluid between double layer medium, Collection of 5th Conference on Fluid Mechanics, 1996.

[4] Beier, R. A., Pressure transient model of a vertically fractured well in a fractal reservoir, Paper SPE 20582, 1990.

Domain Decomposition Preconditioners for Non-Selfconjugate Second Order Elliptic Problems

HUAIYU ZHANG   JIACHANG SUN

Abstract

A non-symmetric interface Schur complement arises from non- self-conjugate second order elliptic problems with domain decomposition methods. The usual numerical methods for solving it are GMRES, ORTHOMIN, and BICGSTAB, but they take a large amount of computer time and memory. The authors find in this paper that the non-symmetric Schur complement can in fact be changed into a symmetric one by scaling. Then an efficient preconditioner can be provided by which the preconditioned system can be solved iteratively by a modified PCG method. When the problem is imposed on a rectangular region, the condition number is estimated and is nearly one. Numerical experiments are also presented. Non-selfconjugate problems arise in mathematical modeling and numerical simulation of fluid flows and transport in porous media.

KEYWORDS: non-selfconjugate, elliptic equation, domain decomposition, Schur complement, preconditioner

1 Introduction

Consider the non-selfconjugate second order elliptic equation

$$- \triangle u + \omega \cdot \bigtriangledown u = f \quad \text{in } \Omega, \tag{1.1}$$

where Ω is an L-shaped region in R^2; i.e., $\Omega = \Omega^{(1)} \cup \Omega^{(2)}$, with (see Fig. 1)

$$\Omega^{(1)} = (0, a_1) \times (0, b_1), \quad \Omega^{(2)} = (a_1, a_2) \times (0, b_2).$$

The Dirichlet boundary condition is imposed

$$u = 0 \quad \text{on } \partial\Omega. \tag{1.2}$$

We assume that we can construct the rectangular grids $\bar{\Omega}_h^{(i)}$ on the rectangles $\bar{\Omega}^{(i)}$, $i = 1, 2$. With the common step h, we have

$$\bar{\Omega}_h^{(1)} = \{(nh, mh) : n = 0, \dots, N_1, m = 0, \dots, M_1, N_1 h = a_1, M_1 h = b_1\},$$
$$\bar{\Omega}_h^{(2)} = \{(nh, mh) : n = N_1, \dots, N_1 + N_2, m = 0, \dots, M_2,$$
$$(N_1 + N_2)h = a_2, M_2 h = b_2\}.$$

Let $\bar{\Omega}_h$ denote $\bar{\Omega}_h = \bar{\Omega}_h^{(1)} \cup \bar{\Omega}_h^{(2)}$. We decompose $\bar{\Omega}_h$ into two parts: $\bar{\Omega}_h =$

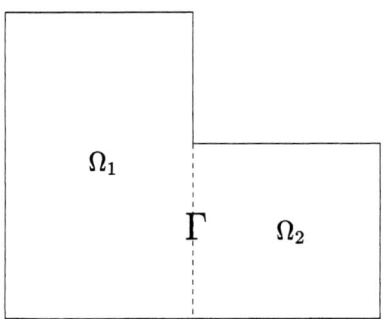

Figure 1: An L-shaped region.

$\Omega_h \cup \partial\Omega_h$, where $\partial\Omega_h$ is the set of the grid points belonging to the boundary $\partial\Omega$ of the region Ω. For the discretization of (1.1) and (1.2), we use either finite difference methods or finite element methods. The stiffness matrix associated with the individual subdomains may be written as

$$A^{(i)} = \begin{pmatrix} A_{II}^{(i)} & A_{IB}^{(i)} \\ A_{BI}^{(i)} & A_{BB}^{(i)} \end{pmatrix}, \quad i = 1, 2. \tag{1.3}$$

Combining them into one we obtain the stiffness matrix

$$A = \begin{pmatrix} A_{II}^{(1)} & 0 & A_{IB}^{(1)} \\ 0 & A_{II}^{(2)} & A_{IB}^{(2)} \\ A_{BI}^{(1)} & A_{BI}^{(2)} & A_{BB}^{(1)} + A_{BB}^{(2)} \end{pmatrix}. \tag{1.4}$$

We now partition the vector of unknown coefficients in the same way as the matrix: $u = (u_I^{(1)}, u_I^{(2)}, u_B)^T$. The linear system can then be written as

$$
\begin{pmatrix}
A_{II}^{(1)} & 0 & 0 \\
0 & A_{II}^{(2)} & 0 \\
A_{BI}^{(1)} & A_{BI}^{(2)} & I
\end{pmatrix}
\begin{pmatrix}
I & 0 & A_{II}^{(1)-1} A_{IB}^{(1)} \\
0 & I & A_{II}^{(2)-1} A_{IB}^{(2)} \\
0 & 0 & S^{(1)} + S^{(2)}
\end{pmatrix}
\begin{pmatrix}
u_I^{(1)} \\
u_I^{(2)} \\
u_B
\end{pmatrix} = f, \qquad (1.5)
$$

where $S^{(i)} = A_{BB}^{(i)} - A_{BI}^{(i)} A_{II}^{(i)-1} A_{IB}^{(i)}$ is called the Schur completement and each can be calculated independently. Perform a forward solver to obtain

$$
\begin{pmatrix}
I & 0 & A_{II}^{(1)-1} A_{IB}^{(1)} \\
0 & I & A_{II}^{(2)-1} A_{IB}^{(2)} \\
0 & 0 & S^{(1)} + S^{(2)}
\end{pmatrix}
\begin{pmatrix}
u_I^{(1)} \\
u_I^{(2)} \\
u_B
\end{pmatrix}
$$

$$
=
\begin{pmatrix}
I & 0 & 0 \\
0 & I & 0 \\
-A_{BI}^{(1)} & -A_{BI}^{(2)} & I
\end{pmatrix}
\begin{pmatrix}
A_{II}^{(1)-1} & 0 & 0 \\
0 & A_{II}^{(2)-1} & 0 \\
0 & 0 & I
\end{pmatrix}
\begin{pmatrix}
f_I^{(1)} \\
f_I^{(2)} \\
f_B
\end{pmatrix}.
$$

(1.6)

Next, we must solve the reduced Schur complement problem

$$
(S^{(1)} + S^{(2)}) u_B = f_B - A_{BI}^{(1)} A_{II}^{(1)-1} f_I^{(1)} - A_{BI}^2 A_{II}^{(2)-1} f_I^{(2)} \equiv g. \qquad (1.7)
$$

The solution of this equation gives us the values along the interior boundary and the problem then splits into two independent subproblems. We can thus back solve in parallel for the interior unknowns in each substructure:

$$
u_I^{(i)} = A_{II}^{(i)-1} (f_I^{(i)} - A_{IB}^{(i)} u_B^{(i)}). \qquad (1.8)
$$

This procedure can be generalized to the case of N subdomains. The Schur complement arising from non-selfconjugate second order elliptic equations is non-symmetric. Its condition number grows like $O(1/h)$. Since it is denser than the original stiffness matrix, in general, the linear system (1.7) is solved iteratively. The usual iterative methods for non-symmetric problems are GMRES, ORTHOMIN, and BICGSTAB.

Several papers of Chan, et al. have been devoted to the construction of preconditioners for this kind of non-symmetric Schur complement; see

[2, 3]. These preconditioners are referred to as interface solvers or interface preconditioners. The authors in this paper show that the non-symmetric Schur complement can be changed into a symmetric one only through the method of diagonal scaling. An efficient preconditioner is also provided, by which the preconditioned system can then be solved iteratively by a modified PCG method instead of those mentioned above.

2   Diagonal Scaling and Preconditioners

Suppose that the standard five-point central difference scheme is applied, which has the form

$$-1 + \tfrac{1}{2}\omega_2 h$$

$$-1 - \tfrac{1}{2}\omega_1 h \qquad 4 \qquad -1 + \tfrac{1}{2}\omega_1 h$$

$$-1 - \tfrac{1}{2}\omega_2 h$$

Let T_i, $i = 1, 2$, be an $M_i \times M_i$ matrix having the form

$$
T^{[i]} = \begin{pmatrix}
4 & -1 + \tfrac{1}{2}\omega_2 h & & & \\
-1 - \tfrac{1}{2}\omega_2 h & 4 & & \ddots & \\
& \ddots & & \ddots & -1 + \tfrac{1}{2}\omega_2 h \\
& & & -1 - \tfrac{1}{2}\omega_2 h & 4
\end{pmatrix}.
$$

The Schur complement then becomes

$$S = T^{[2]} - (1 - \tfrac{1}{4}\omega_1^2 h^2)[I, 0]B_{N_1}^{[1]}\begin{bmatrix} I \\ 0 \end{bmatrix} - (1 - \tfrac{1}{4}\omega_1^2 h^2)B_{N_2}^{[2]}, \qquad (2.1)$$

where $B_n^{[i]}$ has the recursive relations

$$B_n^{[i]} = (T^{[i]} - (1 - \tfrac{1}{4}\omega_1^2 h^2)B_{n-1}^{[i]})^{-1}, \ B_1^{[i]} = T^{[i]-1}, \ i = 1, 2. \qquad (2.2)$$

Let D be a uniformly bounded diagonal matrix with its elements given by

$$d_i = \left(\sqrt{\frac{1 + \tfrac{1}{2}\omega_2 h}{1 - \tfrac{1}{2}\omega_2 h}}\right)^i.$$

Let $\hat{S} = DSD^{-1}$. It can be easily seen that $\hat{S}$ becomes a symmetric matrix so that the CG method can be used. Numerical experiments show that the times of iteration by the CG method reduce largely.

Let

$$C^2 = K_{\omega_2} + \frac{1}{4}K_{\omega_2}^2 + \frac{1}{4}\omega_1^2 h^2 I, \qquad (2.3)$$

where

$$K_{\omega_2} = \begin{pmatrix} 2 & -1+\frac{1}{2}\omega_2 h & & & \\ -1-\frac{1}{2}\omega_2 h & 2 & \ddots & & \\ & \ddots & \ddots & -1+\frac{1}{2}\omega_2 h \\ & & -1-\frac{1}{2}\omega_2 h & 2 \end{pmatrix}.$$

Also, let $\hat{C} = DCD^{-1}$. Now, instead of solving (1.7), we can solve any of the two preconditioned systems by the PCG method

$$\hat{C}^{-2}\hat{S}^2 v_B = DC^{-2}Sg, \quad u_B = D^{-1}v_B, \qquad (2.4a)$$

or

$$\hat{S}^2\hat{C}^{-2}v_B = DSg, \quad u_B = C^{-2}D^{-1}v_B. \qquad (2.4b)$$

For the PCG method, see Chapter 9 in [5]. Or, equivalently, we solve any of the two by a modified PCG method

$$DC^{-2}S^2 u_B = DC^{-2}Sg, \qquad (2.5a)$$

or

$$DS^2C^{-2}v_B = DSg, \quad u_B = C^{-2}v_B. \qquad (2.5b)$$

MODIFIED PCG ALGORITHM:

1. Compute $r_0 := g - Su_B^0$, $z_0 := C^{-2}Sr_0$, and $p_0 := z_0$
2. For $j = 0, 1, \ldots$, until convergence Do:
3. $\alpha_j := (Dr_j, z_j)/(D^2 p_j, S^2 p_j)$
4. $u_B^{j+1} := u_B^j + \alpha_j p_j$
5. $r_{j+1} := r_j - \alpha_j Sp_j$
6. $z_{j+1} := C^{-2}Sr_{j+1}$
7. $\beta_j := (Dr_{j+1}, z_{j+1})/(Dr_j, z_j)$
8. $p_{j+1} := z_{j+1} + \beta_j p_j$
9. EndDo

In the above algorithm, C^2 is a tri-diagnoal matrix, so it is easy for C^{-2} to do the matrix-vector operations.

3   The Condition Number

Define the $\hat{S}^2$-inner product by

$$(x,y)_{\hat{S}^2} = (\hat{S}^2 x, y) = (\hat{S}x, \hat{S}y) = (x, \hat{S}^2 y). \tag{3.1}$$

For a linear operator L, which is self-adjoint with respect to the $\hat{S}^2$-inner product, we use the Rayleigh quotient characterization of the extreme weighted eigenvalues

$$\lambda_{\min}^{\hat{S}^2}(L) = \min_{x \neq 0} \frac{(Lx, x)_{\hat{S}^2}}{(x, x)_{\hat{S}^2}}, \quad \lambda_{\max}^{\hat{S}^2}(L) = \max_{x \neq 0} \frac{(Lx, x)_{\hat{S}^2}}{(x, x)_{\hat{S}^2}}. \tag{3.2}$$

Then the condition number of L is given by

$$\kappa(L) = \lambda_{\max}^{\hat{S}^2}(L)/\lambda_{\min}^{\hat{S}^2}(L). \tag{3.3}$$

Lemma 3.1 *It holds that* $\kappa(\hat{C}^{-2}\hat{S}^2) = \max_{x \neq 0} \dfrac{(\hat{S}^2 x, x)}{(\hat{C}^2 x, x)} / \min_{x \neq 0} \dfrac{(\hat{S}^2 x, x)}{(\hat{C}^2 x, x)}.$

When the problem is imposed on a rectangular region, i.e., $b_1 = b_2$, we show that the condition number of the preconditioned Schur complement is nearly one. From the numerical experiments, it seems still true for the L-shaped regions.

Theorem 3.1 *Assume that* $a = \min(a_1, a_2 - a_1) \geq 1$ *and* $\pi h \leq 1/3$. *Then the condition number of the preconditioned Schur complement is bounded by*

$$\kappa(\hat{C}^{-2}\hat{S}^2) \leq (1 + e^{-2\pi a})^2.$$

Proof: When a rectangular region is imposed, let B_n have the recursive relation

$$B_n = (T - (1 - \tfrac{1}{4}\omega_1^2 h^2)B_{n-1})^{-1}, \quad B_1 = T^{-1},$$

where $T = T^{[1]} = T^{[2]}$. Then the Schur complement S has the form

$$S = T - (1 - \tfrac{1}{4}\omega_1^2 h^2)(B_{N_1} + B_{N_2}).$$

Under this case, $\hat{S}^2$ and $\hat{C}^2$ have the same eigenvectors

$$\hat{S}^2 = (D\,Q)\Lambda_S^2(D\,Q)^{-1}, \quad \Lambda_S = \mathrm{Diag}\{\lambda_j\},$$
$$\hat{C}^2 = (D\,Q)\Lambda_C^2(D\,Q)^{-1}, \quad \Lambda_C = \mathrm{Diag}\{\mu_j\},$$

where λ_i and μ_i are the eigenvalues of S and C, respectively. It yields that

$$\kappa(\hat{C}^{-2}\hat{S}^2) = \max(\frac{\lambda_j}{\mu_j})^2 / \min(\frac{\lambda_j}{\mu_j})^2.$$

Let

$$\alpha_j = ch^{-1}\left(\frac{t_j}{2\sqrt{1-\frac{1}{4}\omega_1^2 h^2}}\right) > 0, \quad t_j = 4 - 2\sqrt{1 - \frac{1}{4}\omega_2^2 h^2}\cos j\pi h.$$

We see from the corresponding recursive relations of eigenvalues that

$$\lambda_j = \sqrt{1 - \frac{1}{4}\omega_1^2 h^2}\left(\frac{ch(N_1+1)\alpha_j}{sh(N_1+1)\alpha_j} + \frac{ch(N_2+1)\alpha_j}{sh(N_2+1)\alpha_j}\right)sh\alpha_j,$$
$$\mu_j = 2\sqrt{1 - \frac{1}{4}\omega_1^2 h^2}\, sh\alpha_j.$$

Therefore, we have

$$\frac{\lambda_j}{\mu_j} = \frac{1}{2}\left(\frac{ch(N_1+1)\alpha_j}{sh(N_1+1)\alpha_j} + \frac{ch(N_2+1)\alpha_j}{sh(N_2+1)\alpha_j}\right)$$
$$= 1 + \frac{e^{-2(N_1+1)\alpha_j}}{1 - e^{-2(N_1+1)\alpha_j}} + \frac{e^{-2(N_2+1)\alpha_j}}{1 - e^{-2(N_2+1)\alpha_j}} > 1.$$

When $\pi h \le 1/3$, it follows that

$$e^{-\alpha_j} \le \frac{t_1}{2\sqrt{1-\frac{1}{4}\omega_1^2 h^2}} - \sqrt{\left(\frac{t_1}{2\sqrt{1-\frac{1}{4}\omega_1^2 h^2}}\right)^2 - 1}$$
$$\le 2 - \cos\pi h - \sqrt{(2-\cos\pi h)^2 - 1} \le 1 - 2\pi h.$$

Therefore, we obtain

$$\frac{\lambda_j}{\mu_j} \le 1 + e^{-2\pi a}.$$

This completes the proof. □

4 Numerical Experiments

Let

$$K = \begin{pmatrix} 2 & -1 & & \\ -1 & 2 & \ddots & \\ & \ddots & \ddots & -1 \\ & & -1 & 2 \end{pmatrix}.$$

and

$$\hat{C}_1^2 = K + \frac{1}{4}K^2, \quad \hat{C}_2^2 = D(K_{\omega_2} + \frac{1}{4}K_{\omega_2}^2)D^{-1},$$
$$\hat{C}_3^2 = K + \frac{1}{4}K^2 + \frac{1}{4}\omega_1^2 h^2, \quad \hat{C}_4^2 = K_{\omega_2} + \frac{1}{4}K_{\omega_2}^2 + \frac{1}{4}\omega_1^2 h^2.$$

The following numerical experiments are done using Matlab, where the parameter ω is given in the table and $h=1/N_1$.

ω	$\kappa(S)$	$\kappa(\hat{C}_1^{-2}\hat{S}^2)$	$\kappa(\hat{C}_2^{-2}\hat{S}^2)$	$\kappa(\hat{C}_3^{-2}\hat{S}^2)$	$\kappa(\hat{C}_4^{-2}\hat{S}^2)$
(5,5)	19.87208	1.50997	1.20197	1.20464	1.00002
(10,5)	16.04520	2.26958	1.80664	1.12779	1.000001
(5,10)	17.97331	2.28940	1.12501	1.82647	1.000001

Table 1. $N_1 = N_2 = 40, M_1 = M_2 = 25$.

ω	$\kappa(S)$	$\kappa(\hat{C}_1^{-2}\hat{S}^2)$	$\kappa(\hat{C}_2^{-2}\hat{S}^2)$	$\kappa(\hat{C}_3^{-2}\hat{S}^2)$	$\kappa(\hat{C}_4^{-2}\hat{S}^2)$
(30,30)	7.04838	2.06521	1.81486	2.06983	1.000000
(40,20)	5.64662	21.28800	4.01108	1.27035	1.000000
(20,40)	8.66044	25.70777	1.19912	5.11383	1.000000

Table 2. $N_1 = N_2 = 40, M_1 = M_2 = 25$.

ω	$\kappa(S)$	$\kappa(\hat{C}_1^{-2}\hat{S}^2)$	$\kappa(\hat{C}_2^{-2}\hat{S}^2)$	$\kappa(\hat{C}_3^{-2}\hat{S}^2)$	$\kappa(\hat{C}_4^{-2}\hat{S}^2)$
(5,5)	17.25141	1.44270	1.27717	1.27788	1.18549
(10,5)	14.50527	1.90157	1.64653	1.21589	1.14762
(5,10)	15.79645	1.91053	1.21513	1.65534	1.14760

Table 3. $N_1 = N_2 = 40, M_1 = 25, M_2 = 20$.

ω	$\kappa(S)$	$\kappa(\hat{C}_1^{-2}\hat{S}^2)$	$\kappa(\hat{C}_2^{-2}\hat{S}^2)$	$\kappa(\hat{C}_3^{-2}\hat{S}^2)$	$\kappa(\hat{C}_4^{-2}\hat{S}^2)$
(30,30)	6.79150	13.80025	1.77435	2.01674	1.03194
(40,20)	5.54002	14.20308	3.72434	1.26206	1.02831
(20,40)	8.07529	17.10853	1.19631	4.72469	1.02811

Table 4. $N_1 = N_2 = 40, M_1 = 25, M_2 = 20$.

ω	$\kappa(S)$	$\kappa(\hat{C}_1^{-2}\hat{S}^2)$	$\kappa(\hat{C}_2^{-2}\hat{S}^2)$	$\kappa(\hat{C}_3^{-2}\hat{S}^2)$	$\kappa(\hat{C}_4^{-2}\hat{S}^2)$
(5,5)	9.73994	1.18636	1.16951	1.16972	1.16500
(10,5)	9.10944	1.25928	1.21972	1.14636	1.13931
(5,10)	9.37308	1.26280	1.14621	1.22346	1.13936

Table 5. $N_1 = N_2 = 40, M_1 = 30, M_2 = 10$.

ω	$\kappa(S)$	$\kappa(\hat{C}_1^{-2}\hat{S}^2)$	$\kappa(\hat{C}_2^{-2}\hat{S}^2)$	$\kappa(\hat{C}_3^{-2}\hat{S}^2)$	$\kappa(\hat{C}_4^{-2}\hat{S}^2)$
(30,30)	5.60318	4.47593	1.55878	1.73865	1.03195
(40,20)	4.90427	4.56155	2.56449	1.21294	1.02832
(20,40)	6.08107	5.40245	1.15975	3.16567	1.02812

Table 6. $N_1 = N_2 = 40, M_1 = 30, M_2 = 10$.

References

[1] Bramble, J. H., *Multigrid Methods*, Cornell Mathematics Department Lecture Notes, 1992.

[2] Chan, T. F. and Keyes, D. E., Interface preconditioning for domain-decomposed convection-diffusion operators, *Third International Symposium on Domain Decomposition Methods for PDEs*, SIAM, Philadelphia, 1990, 245–262.

[3] Chan, T. F. and Mathew, T. P., Domain decomposition preconditioners for convection diffusion problems, *Sixth International Conference on Domain Decomposition*, SIAM, Como, 1992, 157–175.

[4] Dryja, M., A capacitance matrix method for Dirichlet problem on polygon region, *Numer. Math.* **58** (1982), 51–64.

[5] Saad, Y., *Iterative Methods for Sparse Linear Systems*, PWS Publishing Company, 1982.

[6] Smith, B. F., Bjorstad, P. E., and Gropp, W. D., *Domain Decomposition: Parallel Multilevel Methods for Elliptic Partial Differential Equations*, Cambridge University Press, 1996.

Performance of MOL for Surface Motion Driven by a Laplacian of Curvature

WEN ZHANG   IAN GLADWELL

Abstract

We analyze the performance of the method of lines when solving a partial differential equation system describing microstructural evolution in a sintering process. The system involves a fourth order nonlinear partial differential equation with a moving boundary. Both sequential and parallel ordinary differential equation solvers are applied.

KEYWORDS: method of lines, parallel ODE solver, partial differential equations, sintering, surface diffusion, motion by curvature, moving boundary

1   Introduction

We discuss and analyze the performance of the method of lines (MOL) when solving a system of partial differential equations (PDEs) describing microstructural evolution in a sintering process. Sintering is a material manufacture process where powdered material is densified to form a solid body. From a microstructural perspective, the sintering process can be viewed as the combination of two diffusion processes: surface and grain boundary diffusion. The mathematical model for surface diffusion is a fourth order nonlinear PDE and for grain boundary diffusion is a second order linear PDE coupled with an integral equation. The two models join at the moving boundary and constitute a closed mathematical system governing the movement of the particle surfaces during sintering; detailed descriptions can be found in [17, 18, 19].

Here, we focus our attention on the surface diffusion model. The characteristic feature of surface diffusion is that the mass flow is restricted along the surface. Following the theory of Herring [7, 11, 13], mass flow is generated by the surface gradient of the chemical potential, which is proportional to the surface curvature. Hence, locally the motion is driven by a surface Laplacian of the curvature. This motion is nonlinear and complex. There are limited analytical studies [9, 10]. Particularly, Elliott and Garche [10] prove existence of a solution under certain conditions. We have used MOL successfully to solve the system [16, 17, 19] employing the ordinary differential equation (ODE) solvers LSODE [12, 14] and DASSL [15]. Here, we explore the use of a new generation of ODE solvers, both sequential and parallel, and compare them for convergence of the solution of the PDE.

We use four ODE solvers, the sequential solvers VODE [4] and VODPK [3, 5], and the parallel solvers PVODE [6] and the version of PVODE in PETSc [1, 2]. VODE was developed from LSODE. VODPK is identical to VODE except for the linear solver. VODE uses a direct method as in DASSL and LSODE, while VODPK uses an iterative method; i.e., a scaled, preconditioned, incomplete generalized minimum residual method (SPIGMR). PVODE is a parallel solver based on VODPK and its modularized C version CVODE [8]. It is built upon the Message Passing Interface (MPI) library and is intended for a single program multiple data environment on distributed memory computers. PETSc is a parallel computation library that includes linear and nonlinear equation solvers, unconstrained minimization modules and ODE solvers. It is also built upon MPI and includes PVODE (with some modifications) as one of its ODE solvers. For simplicity, we shall use the notation PETSc to represent the PETSc/PVODE solver.

An advantage of using MOL to solve PDEs is the stability of the numerical solution achieved from employing the dynamically chosen time steps and formulas of the integration. Here, stability has two meanings. One is stability of the solution as time proceeds. The other is stability of the solution as the spatial step size decreases. The latter is a virtue of the MOL approach. With a reliable ODE solver, MOL can guarantee error control on the solution for any sufficiently small integration error tolerance and all small spatial step sizes. That is, once we decide on a spatial discretization scheme and choose a sufficiently small integration error tolerance, the error in the solution will not grow as we refine the spatial mesh.

2 The Mathematical Model

To evaluate the performance of MOL, we consider 2D sintering; 3D cases are discussed in [17, 18]. Since we are concerned with only the surface profile, this yields a computational 1D problem. Consider a periodic 2D particle string shown in Fig. 1 with the x-axis as a line of symmetry and vertical grain boundaries.

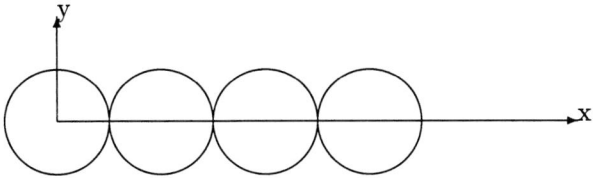

Figure 1.   Periodic 2D particle string.

The surface is represented by function $y(t, x)$ with t being the time variable and x the spatial variable. For this system, the dimensionless surface diffusion equation is

$$\frac{\partial y}{\partial t} = -\frac{\partial J}{\partial x},$$

with surface flux

$$J = -\left[1 + \left(\frac{\partial y}{\partial x}\right)^2\right]^{-1/2}\frac{\partial K}{\partial x}$$

and surface curvature

$$K = -\frac{\partial^2 y}{\partial x^2}\left[1 + \left(\frac{\partial y}{\partial x}\right)^2\right]^{-3/2}.$$

Exploiting symmetry, we need compute only that part of the surface profile between a grain boundary and the central vertical line of a particle. By placing the origin at the center of a particle, we have symmetric boundary conditions (BCs) at $x = 0$:

$$y_x = 0, \qquad y_{xxx} = 0$$

At the other boundary, $x = x_{GB}(t)$, we impose the BCs of a fixed dihedral angle, grain boundary flux and material plating rate:

$$y_x = -\cot(\frac{A}{2}),$$

$$J = -\frac{J_{GB}}{2} = -\frac{3\Gamma}{2y^2(t, x_{GB})}\left(\sin\frac{A}{2} - K(t, x_{GB})y(t, x_{GB})\right),$$

$$x'_{GB}(t) = -\frac{J_{GB}}{2y(t, x_{GB})},$$

where A is the dihedral angle, J_{GB} is the grain boundary flux at $x = x_{GB}(t)$ and Γ is the ratio of grain boundary to surface diffusion. All variables are dimensionless. Their relation to the physical units is shown in [19]. When $J_{GB} \neq 0$, we have a moving boundary problem.

3   The Computational Model

We select a mass conserving finite volume scheme for spatial discretization of Eqn. (1). In this scheme we use second order central differences for all derivatives except near the triple point, $x = x_{GB}$, where we use one-sided second order differences. It can be shown that the spatial discretization error is of second order for solutions in C^6.

Since the moving boundary is unidirectional in our model, we select a front tracking method which maps our coordinates to the moving boundary. Hence our computational grids are fixed but our physical grids move with the moving boundary. In 1D this simply implies scaling. Throughout, we use uniform meshes in space with N equally spaced mesh points and a step size $h = 1/(N-1)$. We let $y_n = y(t, s_n)$ where s_n is a mesh point of the scaled spatial variable $s = x/x_{GB}$. After discretization, the PDE system becomes the ODE system

$$y_{n_t} = -\frac{J_{n+1/2} - J_{n-1/2}}{hx_{GB}} + \frac{y_{n+1} - y_{n-1}}{2hx_{GB}} s_n x'_{GB},$$

$$J_{n+1/2} = -\frac{1}{2}\left\{\left[1 + \left(\frac{y_{n+1} - y_{n-1}}{2hx_{GB}}\right)^2\right]^{-1/2}\right.$$
$$\left. + \left[1 + \left(\frac{y_{n+2} - y_n}{2hx_{GB}}\right)^2\right]^{-1/2}\right\} \frac{K_{n+1} - K_n}{hx_{GB}},$$

$$K_n = -\frac{y_{n+1} - 2y_n + y_{n-1}}{h^2 x_{GB}^2}\left[1 + \left(\frac{y_{n+1} - y_{n-1}}{2hx_{GB}}\right)^2\right]^{-3/2},$$

$$x'_{GB}(t) = -\frac{3\Gamma}{2y_N^3}\left(\sin\frac{A}{2} - K_N y_N\right),$$

$$y_{N_t} = -\frac{3J_N - 4J_{N-1/2} + (J_{N-1/2} + J_{N-3/2})/2}{hx_{GB}} + y_x x'_{GB},$$

$$J_{N-1/2} = -\frac{1}{2}\left\{\left[1 + \left(\frac{y_{n+1} - y_{n-1}}{2hx_{GB}}\right)^2\right]^{-1/2} + (1 + y_x^2)^{-1/2}\right\}$$
$$\times \left\{\frac{K_{n+1} - K_n}{hx_{GB}}\right\},$$

for $n = 1, 2, \ldots, N-1$ with $y_0 = y_2$, $y_{-1} = y_3$. The Jacobian of this ODE system typically has the structure shown in Fig. 2. It has band width 5

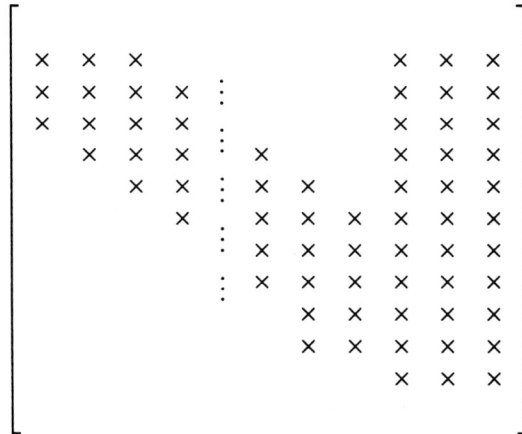

Figure 2. The two band structure of the Jacobian.

everywhere except near the right end where it has three full columns. Also, the lower band width increased by 1 due to using one-sided differences. This structure often occurs in moving boundary problems and can be exploited computationally by a combination of algorithms for banded and full matrices.

4 Computational Performance and Analysis

To be thorough, we have selected a wide range of integration error tolerances, *tol*, and numbers of spatial mesh points, N. The range of tolerances used is $tol = 10^{-4}$ to $tol = 10^{-12}$ and N ranges from 33 to 1025. As an 'analytical' solution, we use the numerically proved 'exact' solution obtained with the finest spatial mesh, $N = 1025$, and the most stringent tolerance, $tol = 10^{-12}$. We fix the physical parameters in the sintering model at physically realistic values: the dihedral angle $A = 160°$ and the ratio of grain boundary to surface diffusion $\Gamma = 0.1$. To better illustrate the error behavior, we use an initial sinusoidal surface with maximum height 1 and initial grain boundary of height 0.156. Our computation reaches the equilibrium state stably. Fig. 3 shows the initial and final surface profile. To compare errors, we stop the integration at a fixed time, $t = 1$, before equilibrium is reached. Then, the height of the entire surface lies between 0.85 and 1 so the absolute and relative errors in the solution are almost equivalent. For simplicity, we measure absolute error and use the scaled variable s instead of x.

We use the backward differentiation formula (BDF) option in the ODE solvers. Fig. 4 shows the errors as the spatial step size decreased by halving. The errors behave almost uniformly on the surface and converge to zero as the spatial step size approaches zero. The calculated orders of convergence at the midpoint in s in Fig. 4 are shown in Table 1. We observe the anticipated second order convergence for a sufficiently large number of mesh points. In VODE, we implement the full exact Jacobian matrix. In VODPK, we use the full dimension of Krylov subspace and no preconditioning. In PVODE, we use the built-in block diagonal with banded block preconditioner (PVBBDPRE). In PETSc, we use the full dimension of Krylov subspace and 5×10^{-8} for the ratio of tolerances of the linear to the nonlinear solver. The four ODE solvers, VODE, VODPK, PVODE, PETSc produce essentially the same results. For a less stringent integration error tolerance, $tol = 10^{-4}$, we no longer observe second order convergence as the number of spatial mesh points increases, see Figs. 5(a)-(c). This is due to the integration error dominating the spatial discretization error. Note, from Figs. 5, the errors are bounded as the number of mesh points increases, demonstrating that stability is preserved by the MOL approach, due to the high quality of the ODE solvers.

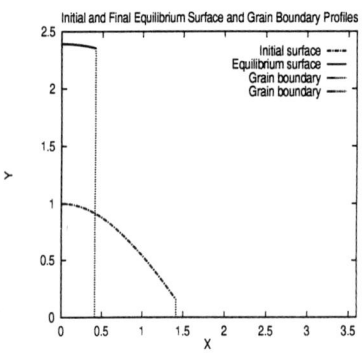
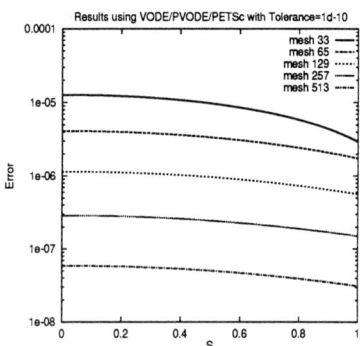

Figure 3: Profiles.   Figure 4: Errors.

We also observed that error growth can be induced by poor approximations to the Jacobian, and by errors from the linear and nonlinear solvers. The iterative linear system solvers in PVODE and PETSc can perform as accurately as the direct linear solver in VODE with an appropriate tolerance in the linear solver, Krylov subspace dimension and preconditioner. Without properly adjusted parameters, the iterative solver can produce poor results and MOL can loose stability as the spatial mesh size decreases. Fig. 6 shows results when using the default parameters in the linear solver in PETSc; i.e., the linear to nonlinear iteration convergence factor is 0.05 and the Krylov subspace dimension is 5. The errors grow significantly as N increases and eventually stability is lost. Fig. 7 shows corresponding results when we use the banded Jacobian approximation with band width 5 in VODE; hence cutting out the moving boundary information in the Jacobian. Again, observe the growth of the errors and the instability induced as the number of mesh points increases. In these cases, MOL behaves like a finite difference method where the constraint for stability is a simple relation between the time and spatial step sizes. The instability can be suppressed by decreasing *tol*. However, our tests show that this is less efficient than using properly adjusted parameters and an appropriate Jacobian approximation.

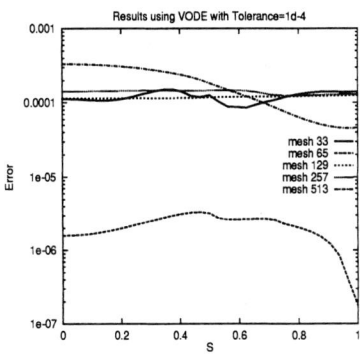

Figure 5a: Errors with VODE.

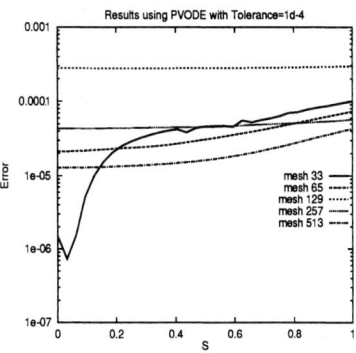

Figure 5b: Errors with PVODE.

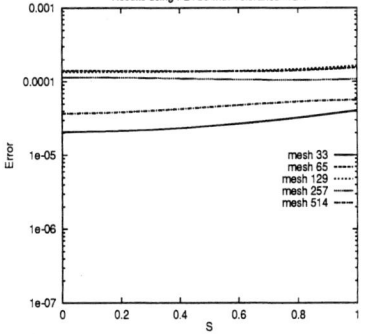

Figure 5c: Errors with PETSc.

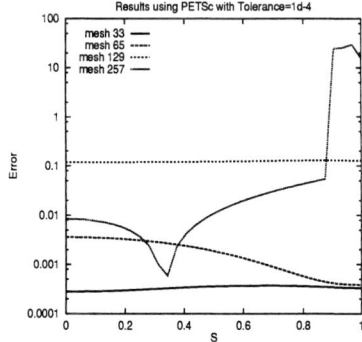

Figure 6: Errors with PETSc-default.

Mesh	33	65	129
Order	1.53	1.82	1.98

Table 1: Convergence rate with spatial mesh size

Band Width	5	7	9	11	13	65	full
Computing Time	1302	9.261	9.578	12.80	12.94	22.18	29.16

Table 2: Computing time in seconds for varying band width; 257 spatial mesh points; integration error tolerance 10^{-12}.

Although the Jacobian is not banded, it can be well-approximated by a banded matrix with an appropriate band width. We use a Jacobian approximation with band width 7 instead of the band width 5 which actually applies in most parts of the Jacobian. The numerical results from VODE, see Fig. 8, show that the errors and instability are completely under the control. The results here are as good as in Fig. 5a where the exact full Jacobian is used. Using this banded approximation significantly reduces both the computing time and the coding effort for the Jacobian. (In our tests, we modified VODE slightly in the way it forms the approximated Jacobian to achieve better results.) Further increases in the band width gave no significant improvement, whereas reducing the band width led to growth in the error and to instability. We also tested using different band widths in the preconditioner PVBBDPRE employed in PVODE, see Table 2. The results are similar to those when using VODE. Table 2 presents results for equal upper and lower half band width.

The excessive computing time for band width 5 indicates that the approximation for the Jacobian is too poor, so instability occurs, and time is wasted integrating through the instability. For band widths greater than 5, the computing time dramatically decreases and there is no instability. The most efficient band widths are 7 and 9, just above the critical band width 5. The critical band width case can be analyzed by studying the eigenvalues of the Jacobian. A good approximation to the Jacobian should have eigenvalues close to those of the exact Jacobian. Since our problem is nonlinear we computed the eigenvalues of the Jacobian and its banded approximations at a fixed time in the integration. Table 3 shows the largest 4 eigenvalues

Band Width	1	3	5	7	9	exact
Eigenvalue 1	1.8	1.7+6	-5.7-2	1.1	1.0	-0.21+i0.19
Eigenvalue 2	-4.9+6	1.6+6	-2.6+1	-1.7-1	-2.0-1	-0.21-i0.19
Eigenvalue 3	-4.9+6	1.5+6	-7.8+2	-8.0+1	-8.1+1	-8.3+1
Eigenvalue 4	-5.0+6	1.3+6	-4.8+3	-1.2+3	-1.2+3	-1.2+3

Table 3: Largest 4 eigenvalues for the approximation to the Jacobian, band widths $1, 3, 5, 7, 9$ and the exact Jacobian.

computed at the same time as when we compare the errors in the solution. When the band width is less than 7, these eigenvalues are far from those of the exact Jacobian. They change significantly with increasing band width. For band width 7, there is a significant improvement in the accuracy of the eigenvalues. The largest two eigenvalues correspond to a split of the only pair of complex eigenvalues of the exact Jacobian. The remaining eigenvalues are within three correct digits of those of the exact Jacobian. As the band width increases further, the eigenvalues do not change significantly. Hence, by choosing the smallest band width above the critical value we are able to achieve most efficiency when using the ODE solvers.

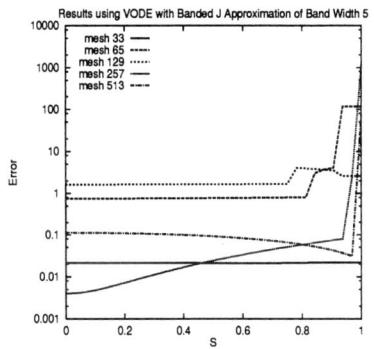

Figure 7: Errors with VODE-band 5.

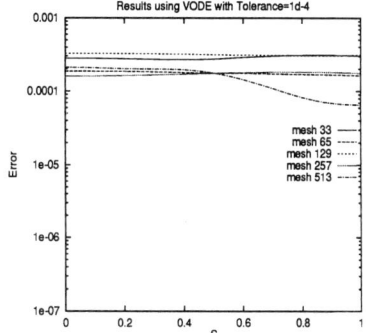

Figure 8: Errors with VODE-band 7.

5   Conclusions

We have shown that when solving surface diffusion problems with a moving boundary using MOL and quality ODE solvers, for stability and accuracy it is essential to use sufficiently stringent integration tolerances and sufficiently accurate approximations to the Jacobian. The Jacobian resulting from MOL typically has a structure of a diagonal band and a vertical band for our problem. It can be well approximated by a banded matrix with a sufficiently wide band width. Also, the block diagonal preconditioner performs well with the iterative linear solver.

Acknowledgments. This material is based upon work supported by the National Science Foundation under Grant No. DMR-9996087. The authors thank Drs. A.C. Hindmarsh and G.D. Byrne for the helpful discussions, Drs. L.C. McInnes and B.F. Smith for help on programming and debugging with PETSc, and Dr. S. Balay for help in using and installation of PETSc.

References

[1] Gropp, W. D. and Smith, B. F., Scalable, extensible, and portable numerical libraries, *Proceedings of the Scalable Parallel Libraries Conference, IEEE* (1994), 87–93.

[2] Balay, S., McInnes, L. C., and Smith, B. F., PETSc 2.0 Users Manual, *ANL-95/11 - Revision 2.0.24*, 1999.

[3] Brown, P. N. and Hindmarsh, A. C., Reduced storage matrix methods in stiff ODE systems, *J. Appl. Math. Comp.* **31** (1989), pp. 40–91.

[4] Brown, P. N., Byrne, G. D., and Hindmarsh, A. C., VODE: a variable coefficient ODE solver, *SIAM J. Sci. Stat. Comput.* **10** (1989), pp. 1038–1051.

[5] Byrne, G. D., Pragmatic experiments with Krylov methods in the stiff ODE setting, *Computational Ordinary Differential Equations* J. Cash and I. Gladwell, eds., Oxford Univ. Press, Oxford, 1992, 323–356.

[6] Byrne, G. D. and Hindmarsh, A. C., User documentation for PVODE, an ODE solver for parallel computers, *LLNL Report* UCRL-ID-130884, 1989.

[7] Cahn, J. W. and Taylor, J. E., Surface motion by surface diffusion, *Acta Metall. Mater.* **42** (1994) 1045–1063.

[8] Cohen, S. D. and Hindmarsh, A. C., CVODE, a stiff/nonstiff ODE solver in C, *Comput. Phys.* **10** (1996), 138–143.

[9] Elliott, C. M. and Garche, H., On the Cahn-Hilliard equation with degenerate mobility, *SIAM J. Math. Anal.* **27** (1996), 404–423.

[10] Elliott, C. M. and Garche, H., Existence results for diffusive surface motion laws, *Adv. Math. Sci. Appl.* **7** (1997), 467-481.

[11] Herring, C., Surface tension as a motivation for sintering, *The Physics of Powder Metallurgy*, W. E. Kingston, ed., McGraw-Hill, New York, 1951, pp.143–152.

[12] Hindmarsh, A. C., ODEPACK, a systematized collection of ODE solvers, *Scientific Computing*, S. Stepleman et al. eds., North-Holland, Amsterdam, 1983.

[13] Mullins, W. W., Theory of Thermal Grooving, *J. Appl. Phys.* **28** (1957) 333–339.

[14] Radhakrishnan, K. and Hindmarsh, A. C., Description and use of LSODE, the Livermore solver for ordinary differential equations, *NASA reference publication* **1327**, and *Lawrence Livermore National Laboratory technical report* **UCRL-ID-113855** (1994).

[15] Petzold, L. R., A description of DASSL: A differential/algebraic system solver, *Scientific Computing*, R. S. Stepleman et. al., eds., North-Holland, Amsterdam, 1983, 65–68.

[16] Zhang, W., Using MOL to solve a high order nonlinear PDE with a moving boundary in the simulation of a sintering process, *Appl. Numer. Math.* **20** (1996), 235–244.

[17] Zhang, W. and Gladwell, I., The sintering of two particles by surface and grain boundary diffusion - a three-dimensional numerical study, *Comp. Mater. Sci.* **12** (1998), 84–104.

[18] Zhang, W. and Gladwell, I., A mathematical and computational model for stress on a planar grain boundary under diffusion, submitted.

[19] Zhang, W. and Schneibel, J. H., The sintering of two particles by surface and grain boundary diffusion - a two-dimensional numerical study, *Acta Metall. Mater.* **43** (1995), 4377–4386.

[20] Zhang, W. and Schneibel, J. H., Calculations of internal stresses during sintering, *J. Am. Ceramic Soc.* **79** (1996), 2141–2144.

A High-Order Upwind Method for Convection-Diffusion Equations with the Newmann Boundary Condition

WEIDONG ZHAO

Abstract

In this paper a high-order upwind finite difference method is studied for steady convection-diffusion problems with the Newmann boundary condition. Based on these equations in the conservation form, a conservative high-order upwind finite difference scheme on nonuniform rectangular partition is proposed. The scheme satisfies the maximum value principle and has a second-order error estimate in the discrete H^1 norm. The method and its analysis apply to groundwater pollution and reservoir simulation problems.

KEYWORDS: convection-diffusion, high-order upwind method, error estimate

1 Introduction

Consider the 2D, steady, linear convection-diffusion problem

$$\begin{aligned}
-\nabla \cdot (A\nabla u) + \nabla \cdot (\vec{b}u) + cu &= f, & x \in \Omega, \\
(A\nabla u) \cdot \vec{n} &= 0 & x \in \Gamma = \partial\Omega,
\end{aligned} \tag{1.1}$$

where $u = u(x,y)$ is the transport quantity, $A = \mathrm{diag}(a_1(x,y), a_2(x,y))$ is the diffusion coefficient matrix, $\vec{b} = (b_1(x,y), b_2(x,y))^T$ is the transport velocity, $c = c(x,y)$ and $f = f(x,y)$ are known functions, Ω is a bounded region in R^2, $\partial\Omega$ is the boundary of Ω, and $\vec{n}$ is the normal exterior unit vector to $\partial\Omega$.

Many realistic fluid flow procedures, such as groundwater pollution and reservoir simulation problems, can be described in the form of convection-diffusion equations [1, 2] and these equations are often convection-dominated. It is very difficult to simulate this kind of problems numerically. The ordinary finite difference and finite element methods, which have been widely used in numerical simulation of the nonconvection-dominated problems, introduce nonphysical oscillations. To overcome nonphysical oscillations, over the past three decades, the upwind schemes have been extensively studied and widely used in the numerical simulation of this kind of problems (e.g., see [3–10]). The standard upwind methods avoid the numerical oscillation, but they only have first-order accuracy. To improve the accuracy of upwind methods, many upwind schemes have been proposed by modifying the second-order differential terms to obtain the second-order accuracy. But some of them do not preserve the maximum value principle and others do not preserve the conservation property. To establish a high-order method which is in the conservation form and preserves the maximum value principle, for 1D problems it has been extensively studied, but many such methods are not as efficient for high dimensional problems as for 1D problems. Thus it is very important and interesting to consider the multi-dimensional high-order upwind methods, which are in the conservation form, satisfy the maximum value principle, and have a high efficiency of applications.

In this paper we study a high-order upwind method for (1.1). In [7], a modified high-order upwind method on uniform rectangular grids has been proposed and analyzed and its theoretical results can only be obtained for uniform partitions. With a new technique of modifying the convective and diffusion terms and the domain partition, a new kind modified upwind finite difference method on nonuniform rectangular partitions is proposed and analyzed in this paper. The proposed scheme is in the conservation form, satisfies the maximum value principle, and has a second-order error estimate in the discrete H^1 norm.

The paper is organized as follows. The modified high-order upwind finite difference scheme for (1.1) is given in §2 and its theoretical analysis is expanded in §3. Some conclusions are given in §4.

2   High-Order Upwind Scheme

In this section, we consider a high-order upwind finite difference method for multi-dimensional problems. For simplicity, only 2D problem is considered. The results obtained are also true for multidimensional problems.

2.1   Discretization and notation

Let $0 = x_0 < x_1 < \ldots < x_N < x_{N+1} = 1$ and $0 = y_0 < y_1 < \ldots < y_M < y_{M+1} = 1$ be a partition of $\Omega = [0,1] \times [0,1]$ in x and y directions, respectively. The partition is denoted as I_h with grid points (i,j), $0 \leq i \leq N$, and $0 \leq j \leq M$, where (i,j) represents (x_i, y_j). For a continuous function $f(x,y)$, f_{ij} means $f(x_i, y_j)$. Let $x_{i+1/2} = x_i + x_{i+1}/2$, $y_{j+1/2} = y_j + y_{j+1}/2$ for $i = 1, 2, \ldots, N-1$, $j = 1, 2, \ldots, M-1$, with $x_{1/2} = y_{1/2} = 0$, $x_{N+1/2} = y_{M+1/2} = 1$, $h_i^x = x_{i+1/2} - x_{i-1/2}$ for $i = 1, 2, \ldots, N$, $h_j^y = y_{j+1/2} - y_{j-1/2}$ for $j = 1, 2, \ldots, M$, $h_{i+1/2}^x = x_{i+1} - x_i$ for $i = 1, 2, \ldots, N-1$, $h_{j+1/2}^y = y_{j+1} - y_j$ for $j = 1, 2, \ldots, M-1$, $h^x = \max\limits_{i=1,N+1} h_i^x$, $h^y = \max\limits_{j=1,M+1} h_j^y$, and $I_{ij}^* = [x_{i-1/2}, x_{i+1/2}] \times [y_{j-1/2}, y_{j+1/2}]$. Let I_h^* denote the dual partition of Ω with I_h. The nodes of I_h^* are $(x_{i+1/2}, y_{j+1/2})$ for $i = 0, 1, \ldots, N$, $j = 0, 1, \ldots, M$.

Now, define V_h as a piecewise constant function space on I_h^*; i.e., $V_h = \{v, v|_{I_{ij}^*} = v_{ij}, i = 0, 1, \ldots, N, j = 0, 1, \ldots, M\}$. Define the discrete H^1 and L^2 norms for V_h as

$$|u|_{a*}^2 = \sum_{i=1}^N \sum_{j=1}^M \left[a_{1i-1/2,j}^* \frac{(u_{ij} - u_{i-1,j})^2}{h_{i-1/2}^x} h_j^y + a_{2i,j-1/2}^* \left(\frac{(u_{ij} - u_{i,j-1})^2}{h_{j-1/2}^y} \right) h_i^x \right],$$

$$|u|_{bh}^2 = \sum_{i=1}^N \sum_{j=1}^M \left[b_{1i-1/2,j} (u_{ij} - u_{i-1,j})^2 h_j^y + b_{2i,j-1/2} (u_{ij} - u_{i,j-1})^2 h_i^x \right],$$

$$\|u\|^2 = \sum_{i=1}^N \sum_{j=1}^M u_{ij}^2 h_i^x h_j^y, \quad \|u\|_{c*}^2 = \sum_{i=1}^N \sum_{j=1}^M c_{ij}^* u_{ij}^2 h_i^x h_j^y.$$

2.2 A high-order upwind scheme

To construct a high-order upwind scheme for (1.1), we first outline the method for a 1D convection-diffusion equation. The 1D convection-diffusion equation is

$$-\frac{d}{dx}\left(a\frac{du}{dx}\right) + \frac{d}{dx}(bu) + cu = f, \quad x \in \Omega,$$
$$\frac{du}{dx}\Big|_{x=0} = \frac{du}{dx}\Big|_{x=1} = 0, \tag{2.1}$$

where $u = u(x)$ is the transport quantity, $a = a(x) > 0$ is the diffusion coefficient, $b = b(x)$ is the transport velocity satisfying $b(0) = b(1) = 0$, and $c = c(x)$ and $f = f(x)$ are known functions.

Let $0 = x_0 < x_1 < \ldots < x_N < x_{N+1} = 1$ be the partition of $\Omega = [0, 1]$. The partition is denoted as I_h. Let $x_{i+1/2} = x_i + x_{i+1}/2$ for $i = 1, 2, \ldots, N-1$, with $x_{1/2} = 0$, $x_{N+1/2} = 1$, $h_i = x_{i+1/2} - x_{i-1/2}$ for $i = 1, 2, \ldots, N$, and $h_{i+1/2} = x_{i+1} - x_i$ for $i = 1, 2, \ldots, N-1$.

Integrate (2.1) over $[x_{i-1/2}, x_{i+1/2}]$ and use Green's formula to get

$$-[a_{i+1/2} u_{i+1/2}' - a_{i-1/2} u_{i-1/2}'] + [b_{i+1/2} u_{i+1/2} - b_{i-1/2} u_{i-1/2}]$$
$$+ \int_{I_i^*} cu \, dx dy = \int_{I_i^*} f \, dx dy, \tag{2.2}$$

for $i = 1, 2, \ldots, N$. In (2.2), the central difference method may be used to approximate the term $u'_{i+1/2}$, but $u_{i+1/2}$ in $b_{i+1/2}u_{i+1/2}$ must be handled carefully in order that a nonphysical numerical oscillation does not occur. To do this, a special technique should be used. Here we approximate the convective term $b_{i+1/2}u_{i+1/2}$ as follows. Define $H(x) = \left\{ \begin{array}{ll} 1, & x \geq 0 \\ 0, & x < 0 \end{array} \right.$. From the equality,

$$b_{i+1/2}u_{i+1/2} = b_{i+1/2}[H(b_{i+1/2})u_{i+1/2} + (1 - H(b_{i+1/2}))u_{i+1/2}], \quad (2.3)$$

we know that $H(b_{i+1/2})u_{i+1/2}$ and $(1 - H(b_{i+1/2}))u_{i+1/2}$ are related to the positive and negative transport velocities, respectively. Thus $H(b_{i+1/2})u_{i+1/2}$ and $(1 - H(b_{i+1/2}))u_{i+1/2}$ must be approximated differently. From Taylor's formula, $u_{i+1/2}$ has the two expressions

$$u_{i+1/2} = u_i + \frac{h_{i+1/2}}{2}u'_{i+1/2} - \frac{h^2_{i+1/2}}{8}u''_{i+1/2} + O(h^3_{i+1/2}), \quad (2.4)$$

$$u_{i+1/2} = u_{i+1} - \frac{h_{i+1/2}}{2}u'_{i+1/2} - \frac{h^2_{i+1/2}}{8}u''_{i+1/2} + O(h^3_{i+1/2}). \quad (2.5)$$

From the nature of transport and the upwind idea, (2.4) and (2.5) should be used in $H(b_{i+1/2})u_{i+1/2}$ and $(1 - H(b_{i+1/2}))u_{i+1/2}$, respectively. The term $u_{i-1/2}$ can be handled similarly. Then (2.2) can be written as

$$-[(a_{i+1/2} - \frac{|b_{i+1/2}|h_{i+1/2}}{2})u'_{i+1/2} - (a_{i-1/2} - \frac{|b_{i-1/2}|h_{i-1/2}}{2})u'_{i-1/2}]$$
$$+b_{i+1/2}[H(b_{i+1/2})u_i + (1 - H(b_{i+1/2}))u_{i+1}] - b_{i-1/2}[H(b_{i-1/2})u_{i-1}$$
$$+(1 - H(b_{i-1/2}))u_i] - [\frac{b_{i+1/2}h^2_{i+1/2}}{8}u''_{i+1/2} - \frac{b_{i-1/2}h^2_{i-1/2}}{8}u''_{i-1/2}]$$

$$+ \int_{I^*_i} cudx + O(h^3_{i-1/2} + h^3_{i+1/2}) = \int_{I^*_i} fdx.$$
$$(2.6)$$

Based on (2.6), a modified high-order upwind finite difference scheme for (2.1) can be proposed as

$$-\nabla_{\tilde{x}}(a^*\nabla_{\tilde{x}}U)_i + \nabla_{\tilde{x}}(bU^{ux})_i + c_iU_i = f_i, \quad (2.7)$$

for $i = 1, \ldots, N$, where

$$\nabla_{\bar{x}}(a^* \nabla_{\bar{x}} U)_i = \frac{1}{h_i}(a^*_{i+1/2}\frac{U_{i+1} - U_i}{h_{i+1/2}} - a^*_{i-1/2}\frac{U_i - U_{i-1}}{h_{i-1/2}}),$$

$$a^*_{i+1/2} = \frac{2a^2_{i+1/2}}{2a_{i+1/2} + |b_{i+1/2}|h_{i+1/2}},$$

$$\nabla_{\bar{x}}(bU^{ux})_i = \frac{1}{h_i}(b_{i+1/2}U^{ux}_{i+1/2} - b_{i-1/2}U^{ux}_{i-1/2}),$$

$$U^{ux}_{i+1/2} = H(b_{i+1/2})U_i + (1 - H(b_{i+1/2}))U_{i+1},$$

$$c_i = \tfrac{1}{h_i}\int_{I^*_i} c\,dx, \quad f_i = \tfrac{1}{h_i}\int_{I^*_i} f\,dx,$$

with $a_{1/2} = a_{N+1/2} = b_{1/2} = b_{N+1/2} = 0$.

Using the idea of getting the scheme (2.7), we construct the high-order upwind scheme for (1.1) as follows:

$$\begin{aligned}
&-\nabla_{\bar{x}}(a^*_1 \nabla_{\bar{x}} U)_{ij} - \nabla_{\bar{y}}(a^*_2 \nabla_{\bar{y}} U)_{ij} \\
&+\nabla_{\bar{x}}(b_1 U^{ux}) + \nabla_{\bar{y}}(b_2 U^{uy})_{ij} + c_{ij}U_{ij} = f_{ij},
\end{aligned} \tag{2.8}$$

for $i = 1, \ldots, N, j = 1, \ldots, M$, with $a_{1/2,j} = a_{N+1/2,j} = a_{i,1/2} = a_{i,M+1/2} = 0$, $b_{1/2,j} = b_{N+1/2,j} = b_{i,1/2} = b_{i,M+1/2} = 0$, where

$$\nabla_{\bar{x}}(a^*_1 \nabla_{\bar{x}} U)_{ij} = \frac{1}{h_i}(a^*_{1i+1/2,j}\frac{U_{i+1,j} - U_{ij}}{h^x_{i+1/2}} - a^*_{1i-1/2,j}\frac{U_{ij} - U_{i-1,j}}{h^x_{i-1/2}}),$$

$$\nabla_{\bar{y}}(a^*_2 \nabla_{\bar{y}} U)_{ij} = \frac{1}{h^y_j}(a^*_{2i,j+1/2}\frac{U_{i,j+1} - U_{ij}}{h^y_{j+1/2}} - a^*_{1i,j-1/2}\frac{U_{ij} - U_{i,j-1}}{h^y_{j-1/2}}),$$

$$a^*_{1i+1/2,j} = \frac{2a^2_{1i+1/2,j}}{2a_{1i+1/2,j} + |b_{1i+1/2,j}|h^x_{i+1/2}},$$

$$a^*_{2i,j+1/2} = \frac{2a^2_{2i,j+1/2}}{2a_{2i,j+1/2} + |b_{2i,j+1/2}|h^y_{j+1/2}},$$

$$\nabla_{\bar{x}}(b_1 U^{ux})_{ij} = \frac{1}{h^x_i}(b_{1i+1/2,j}U^{ux}_{i+1/2,j} - b_{1i-1/2,j}U^{ux}_{i-1/2,j}),$$

$$U^{ux}_{i+1/2,j} = H(b_{1i+1/2,j})U_{ij} + (1 - H(b_{1i+1/2,j}))U_{i+1,j},$$

$$\nabla_{\bar{y}}(b_2 U^{uy})_{ij} = \frac{1}{h^y_j}(b_{2i,j+1/2}U^{uy}_{i,j+1/2} - b_{2i,j-1/2}U^{uy}_{i,j-1/2}),$$

$$U^{uy}_{i,j+1/2} = H(b_{2i,j+1/2})U_{ij} + (1 - H(b_{2i,j+1/2}))U_{i,j+1}.$$

3 Theoretical Analysis

In this section, we prove that scheme (2.8) satisfies the maximum value principle and obtain its error estimate.

3.1   The maximum value principle

It is well known that (1.1) satisfies the maximum value principle if

$$div(\vec{b}) + c > 0, \quad (x,y) \in \Omega. \tag{3.1}$$

Thus, for (2.8), we assume that

$$\frac{b_{1i+1/2,j} - b_{1i-1/2,j}}{h_i^x} + \frac{b_{2i,j+1/2} - b_{2i,j-1/2}}{h_j^y} + c_{ij} > 0 \tag{3.2}$$

holds for $i = 1, 2 \ldots, N$ and $j = 1, 2 \ldots, M$.

Theorem 3.1 *Scheme* (2.8) *is conservative and under condition* (3.2), *it satisfies the maximum value principle.*

Proof: From the definition of (2.8), it is easy to verify that it is conservative. To prove the maximum value principle, we write (2.8) in the equivalent form

$$A_{ij}U_{i-1,j} + B_{ij}U_{i+1,j} + C_{ij}U_{ij} + D_{ij}U_{i,j-1} + E_{ij}U_{i,j+1} = F_{ij}$$
$$i = 1, \ldots, N, \; j = 1, \ldots, M, \tag{3.3}$$

where

$$A_{ij} = [-\frac{a^*_{1i-1/2,j}}{h^x_{i-1/2}} - b_{1i-1/2,j}H(b_{1i-1/2,j})]h^y_j,$$

$$B_{ij} = [-\frac{a^*_{1i+1/2,j}}{h^x_{i+1/2}} + b_{1i+1/2,j}(1 - H(b_{1i+1/2,j}))]h^y_j,$$

$$D_{ij} = [-\frac{a^*_{2i,j-1/2}}{h^y_{j-1/2}} - b_{2i,j-1/2}H(b_{2i,j-1/2})]h^x_i,$$

$$E_{ij} = -[\frac{a^*_{2i,j+1/2}}{h^y_{j+1/2}} + b_{2i,j+1/2}(1 - H(b_{2i,j+1/2}))]h^x_i,$$

$$C_{ij} = [\frac{a^*_{1i+1/2,j}}{h^x_{i+1/2}} + \frac{a^*_{1i-1/2,j}}{h^x_{i+1/2}}]h^y_j + [\frac{a^*_{2i,j+1/2}}{h^y_{j+1/2}} + \frac{a^*_{2i,j-1/2}}{h^x_{j-1/2}}]h^x_i$$
$$+ [b_{1i+1/2,j}H(b_{1i+1/2,j}) - b_{1i-1/2,j}(1 - H(b_{1i-1/2,j}))]h^y_j$$
$$+ [b_{2i,j+1/2}H(b_{2i,j+1/2}) - b_{2i,j-1/2}(1 - H(b_{2i,j-1/2}))]h^x_i + c_{ij}h^x_i h^y_j,$$

$$F_{ij} = f_{ij}h^x_i h^y_j.$$

Further,

$$A_{ij} + B_{ij} + C_{ij} + D_{ij} + E_{ij}$$
$$[b_{1i+1/2,j} - b_{1i-1/2,j}]h^y_j + [b_{2i,j+1/2} - b_{2i,j-1/2}]h^x_i + c_{ij}h^x_i h^y_j.$$

From (3.3) and the definition of $H(x)$, it is easy to verify that A_{ij}, B_{ij}, D_{ij} and E_{ij} are nonpositive and C_{ij} is nonnegative. Under condition (3.2), $A_{ij} + B_{ij} + C_{ij} + D_{ij} + E_{ij}$ is positive too. Thus it satisfies the maximum value principle.   ⬜

3.2   Error estimate

As we know, (1.1) is regular and uniquely solvable in H^2 if $0.5\mathrm{div}\vec{b} + c > 0$. Thus, for (2.8), we also assume that

$$c_{ij}^* = c_{ij} + \frac{1}{2}\left(\frac{b_{1i+1/2,j} - b_{1i-1/2,j}}{h_i^x}\right) + \frac{b_{2i,j+1/2} - b_{2i,j-1/2}}{h_j^y}) \geq c_0 > 0 \quad (3.4)$$

holds for $i = 1, 2, \ldots, N$ and $j = 1, 2, \ldots, M$. For the sake of error estimates, we assume that

$$a_1 \geq a_{10}, \quad a_2 \geq a_{20}, \tag{3.5}$$

for two positive constants a_{10} and a_{20}.

Let u_{ij} be the solution of (1.1) at the point (x_i, y_j). Then u_{ij} satisfies the equality

$$[-(\nabla_{\bar{x}}(a_1^* \nabla_{\bar{x}} u)_{ij} - (\nabla_{\bar{y}}(a_2^* \nabla_{\bar{y}} u)_{ij}$$
$$+ \nabla_{\bar{x}}(bu^{ux})_{ij} + \nabla_{\bar{y}}(bu^{uy})_{ij} + c_{ij} u_{ij})]h_i^x h_j^y = \sum_{l=1}^{11} A_{ij}^l, \tag{3.6}$$

where

$$A_{ij}^1 = -[a_{1i+1/2,j}\frac{\partial u}{\partial x}\big|_{i+1/2,j} - a_{1i-1/2,j}\frac{\partial u}{\partial x}\big|_{i-1/2,j}]h_j^y,$$

$$A_{ij}^2 = -[a_{2i,j+1/2}\frac{\partial u}{\partial y}\big|_{i,j+1/2} - a_{2i,j-1/2}\frac{\partial u}{\partial y}\big|_{i,j-1/2}]h_i^x,$$

$$A_{ij}^3 = [(a_{1i+1/2,j} - \frac{|b_{1i+1/2,j}|h_{i+1/2}^x}{2} - a_{1i+1/2,j}^*)\frac{u_{i+1,j} - u_{ij}}{h_{i+1/2}^x}$$
$$- (a_{1i-1/2,j} - \frac{|b_{1i-1/2,j}|h_{i-1/2}^x}{2} - a_{1i-1/2,j}^*)\frac{u_{ij} - u_{i-1,j}}{h_{i-1/2}^x}]h_j^y,$$

$$A_{ij}^4 = [(a_{2i,j+1/2} - \frac{|b_{2i,j+1/2}|h_{j+1/2}^y}{2} - a_{2i,j+1/2}^*)\frac{u_{i,j+1} - u_{ij}}{h_{j+1/2}^y}$$
$$- (a_{2i,j-1/2} - \frac{|b_{2i,j-1/2}|h_{j-1/2}^y}{2} - a_{2i,j-1/2}^*)\frac{u_{ij} - u_{i,j-1}}{h_{j-1/2}^y}]h_i^x,$$

$$A_{ij}^5 = [(a_{1i+1/2,j} - \frac{|b_{1i+1/2,j}|h_{i+1/2}^x}{2})(\frac{\partial u}{\partial x}|_{i+1/2,j} - \frac{u_{i+1,j} - u_{ij}}{h_{i+1/2}^x})$$

$$-(a_{1i-1/2,j} - \frac{|b_{1i-1/2,j}|h_{i-1/2}^x}{2})(\frac{\partial u}{\partial x}|_{i-1/2,j} - \frac{u_{ij} - u_{i-1,j}}{h_{i-1/2}^x})]h_j^y,$$

$$A_{ij}^6 = [(a_{2i,j+1/2} - \frac{|b_{2i,j+1/2}|h_{j+1/2}^y}{2})(\frac{\partial u}{\partial y}|_{i,j+1/2} - \frac{u_{i,j+1} - u_{ij}}{h_{j+1/2}^y})$$

$$-(a_{2i,j-1/2} - \frac{|b_{2i,j-1/2}|h_{j-1/2}^y}{2})(\frac{\partial u}{\partial y}|_{i,j-1/2} - \frac{u_{ij} - u_{i,j-1}}{h_{j-1/2}^y})]h_i^x,$$

$$A_{ij}^7 = [b_{1i+1/2,j}(H(b_{1i+1/2,j})u_{ij} + (1 - H(b_{1i+1/2,j})u_{i+1,j})$$

$$-b_{1i-1/2,j}(H(b_{1i-1/2,j})u_{i-1,j} + (1 - H(b_{1i-1/2,j})u_{ij})]h_j^y,$$

$$A_{ij}^8 = [b_{2i,j+1/2}(H(b_{2i,j+1/2})u_{ij} + (1 - H(b_{2i,j+1/2})u_{i,j+1})$$

$$-b_{2i,j-1/2}(H(b_{2i,j-1/2})u_{i,j-1} + (1 - H(b_{2i,j-1/2})u_{ij})]h_i^x,$$

$$A_{ij}^9 = -1/2[b_{1i+1/2,j}(1 - 2H(b_{1i+1/2,j}))h_{i+1/2}^x \frac{\partial u}{\partial x}|_{i+1/2,j}$$

$$-b_{1i-1/2,j}(1 - 2H(b_{1i-1/2,j}))h_{i-1/2}^x \frac{\partial u}{\partial x}|_{i-1/2,j}]h_j^y,$$

$$A_{ij}^{10} = -\frac{1}{2}[b_{2i,j+1/2}(1 - 2H(b_{2i,j+1/2}))h_{j+1/2}^y \frac{\partial u}{\partial y}|_{i,j+1/2}$$

$$-b_{2i,j-1/2}(1 - 2H(b_{2i,j-1/2}))h_{j-1/2}^y \frac{\partial u}{\partial y}|_{i,j-1/2}]h_i^x,$$

$$A_{ij}^{11} = c_{ij}u_{ij}h_i^x h_j^y.$$

Since

$$H(b_{1i+1/2,j})u_{ij} + (1 - H(b_{1i+1/2,j}))u_{i+1,j} = u_{i+1/2,j}$$

$$+\frac{1}{2}(1 - 2H(b_{1i+1/2,j}))h_{i+1/2}^x \frac{\partial u}{\partial x}|_{i+1/2,j}$$

$$+\frac{(h_{i+1/2}^x)^2}{8} \frac{\partial^2 u}{\partial x^2}|_{i+1/2,j} + O((h_{i+1/2}^x)^3),$$

$$H(b_{2i,j+1/2})u_{ij} + (1 - H(b_{2i,j+1/2}))u_{i,j+1} = u_{i,j+1/2}$$

$$+\frac{1}{2}(1 - 2H(b_{2i,j+1/2}))h_{j+1/2}^y \frac{\partial u}{\partial y}|_{i,j+1/2}$$

$$+\frac{(h_{j+1/2}^y)^2}{8} \frac{\partial^2 u}{\partial y2}|_{i,j+1/2} + O((h_{j+1/2}^y)^3),$$

$$\varphi_{i+1/2} - \varphi_{i-1/2} = h_i\varphi_i' + \frac{1}{8}(h_{i+1/2}^2\varphi"_{i+1} - h_{i-1/2}^2\varphi"_{i-1/2}) + O(h^3),$$

(3.6) leads to

$$[-(\nabla_{\tilde{x}}(a_1^*\nabla_{\tilde{x}}u)_{ij} - (\nabla_{\tilde{y}}(a_2^*\nabla_{\tilde{y}}u)_{ij} + \nabla_{\tilde{x}}(bu^{ux})_{ij} + \nabla_{\tilde{y}}(bu^{uy})_{ij}$$

$$+c_{ij}u_{ij})]h_i^x h_j^y = [-\frac{\partial}{\partial x}(a_1\frac{\partial u}{\partial x})_{ij} - \frac{\partial}{\partial y}(a_2\frac{\partial u}{\partial y})_{ij} + \frac{\partial}{\partial x}(b_1u)_{ij}$$

$$+\frac{\partial}{\partial y}(b_2 u)_{ij} + c_{ij} u_{ij}]h_i^x h_j^y + \sum_3^6 A_{ij}^l + \sum_3^8 B_{ij}^l \qquad (3.7)$$
$$+O((h_{i-1/2}^x)^2 + (h_{i+1/2}^x)^2 + (h_{j+1/2}^y)^2 + (h_{j+1/2}^y)^2)h_i^x h_j^y,$$

where

$$B_{ij}^3 = \frac{1}{8}[b_{1i+1/2,j}(h_{i+1/2}^x)^2\frac{\partial^2 u}{\partial x^2}|_{i+1/2,j} - b_{1i-1/2,j}(h_{i-1/2}^x)^2\frac{\partial^2 u}{\partial x^2}|_{i-1/2,j}]h_j^y,$$

$$B_{ij}^4 = \frac{1}{8}[b_{2i,j+1/2}(h_{j+1/2}^y)^2\frac{\partial^2 u}{\partial y^2}|_{i,j+1/2} - b_{1i,j-1/2}(h_{j-1/2}^y)^2\frac{\partial^2 u}{\partial y^2}|_{i,j-1/2}]h_i^x,$$

$$B_{ij}^5 = \frac{1}{8}[(h_{i+1/2}^x)^2\frac{\partial^2}{\partial x^2}(b_1 u)_{i+1,j} - (h_{i-1/2}^x)^2\frac{\partial^2}{\partial x^2}(b_1 u)_{ij}]h_j^y,$$

$$B_{ij}^6 = \frac{1}{8}[(h_{j+1/2}^y)^2\frac{\partial^2}{\partial y^2}(b_2 u)_{i,j+1} - (h_{j-1/2}^y)^2\frac{\partial^2}{\partial y^2}(b_2 u)_{ij}]h_i^x,$$

$$B_{ij}^7 = \frac{1}{8}[b_{1i+1/2,j}(h_{i+1/2}^x)^2\frac{\partial^2 u}{\partial x^2}|_{i+1/2,j} - b_{1i-1/2,j}(h_{i-1/2}^x)^2\frac{\partial^2 u}{\partial x^2}|_{i-1/2,j}]h_j^y,$$

$$B_{ij}^8 = \frac{1}{8}[b_{2i,j+1/2}(h_{j+1/2}^y)^2\frac{\partial^2 u}{\partial y^2}|_{i,j+1/2} - b_{1i,j-1/2}(h_{j-1/2}^y)^2\frac{\partial^2 u}{\partial y^2}|_{i,j-1/2}]h_i^x.$$

Now, let $\xi_{ij} = u_{ij} - U_{ij}$. Then, use (2.8) and (3.7) to obtain that ξ_i satisfies

$$[-(\nabla_{\bar{x}}(a_1^*\nabla_{\bar{x}}\xi)_{ij} - (\nabla_{\bar{y}}(a_2^*\nabla_{\bar{y}}\xi)_{ij} + \nabla_{\bar{x}}(b\xi^{ux})_{ij}+$$
$$\nabla_{\bar{y}}(b\xi^{uy})_{ij} + c_{ij}\xi_{ij})]h_i^x h_j^y = \sum_3^6 A_{ij}^l + \sum_3^8 B_{ij}^l \qquad (3.8)$$
$$+O((h_{i-1/2}^x)^2 + (h_{i+1/2}^x)^2 + (h_{j+1/2}^y)^2 + (h_{j+1/2}^y)^2)h_i^x h_j^y.$$

Multiply (3.8) by ξ_{ij}, sum it up for $i = 1, 2, \ldots, N$, $j = 1, 2, \ldots, M$, and use the discrete Green formula to obtain

$$\sum_{i=1}^N \sum_{j=1}^M [a_{1i-1/2,j}^*(\frac{\xi_{ij} - \xi_{i-1,j}}{h_{i-1/2}^x})^2 h_{i-1/2}^x h_j^y + a_{2i,j-1/2}^*(\frac{\xi_{ij} - \xi_{i,j-1}}{h_{j-1/2}^y})^2 h_{j-1/2}^y h_i^x+$$
$$\frac{1}{2}|b_{1i-1/2,j}|h_i^x(\frac{\xi_{ij} - \xi_{i-1,j}}{h_{i-1/2}^x})^2 h_{i-1/2}^x h_j^y + \frac{1}{2}|b_{2i,j-1/2}|h_j^y(\frac{\xi_{ij} - \xi_{i,j-1}}{h_{j-1/2}^y})^2 h_{j-1/2}^y h_i^x]$$
$$+\sum_{i=1}^N \sum_{j=1}^M [c_{ij} + \frac{1}{2}(\frac{b_{1i+1/2,j} - b_{1i-1/2,j}}{h_i^x} + \frac{b_{2i,j+1/2} - b_{2i,j-1/2}}{h_j^y})]\xi_{ij}^2 h_i^x h_j^y$$
$$=\sum_{l=3}^6 \sum_{i=1}^N \sum_{j=3}^M A_{ij}^l \xi_{ij} + \sum_{l=3}^8 \sum_{i=1}^N \sum_{j=3}^M B_{ij}^l v_{ij}$$
$$+\sum_{i=1}^N \sum_{j=1}^M O((h_{i-1/2}^y)^2 + (h_{i+1/2}^x)^2 + (h_{j-1/2}^y)^2 + (h_{j+1/2}^y)^2)\xi_{ij}h_i^x h_j^y.$$
$$(3.9)$$

Thus, use the discrete Green formula and the definition of A_{ij}^l, B_{ij}^l, $a_{1i-1/2,j}^*$, and $a^*_{2i,j-1/2}$ to obtain

$$
|\xi|^2_{a^*} + \frac{1}{2}|\xi|^2_{bh} + \sum_{i=1}^{N}\sum_{j=1}^{M}[c_{ij} + \frac{1}{2}(\frac{b_{1i+1/2,j} - b_{1i-1/2,j}}{h^x_i}
$$
$$
+ \frac{b_{2i,j+1/2} - b_{2i,j-1/2}}{h^y_j})]\xi^2_{ij}h^x_i h^y_j \leq K[(h^x)^4 + (h^y)^4] + \epsilon|\xi|^2_{a^*} + \epsilon\|\xi\|^2,
$$

$$(3.10)$$

where ϵ is an arbutary positive number. From the above analysis, we have

Theorem 3.2 *Assume that u and U are the solutions of* (1.1) *and* (2.8), *respectively, u has continuous 4th derivatives, and* (3.4) *and* (3.5) *are satisfied. Then,*

$$
|\xi|_{a^*} + |\xi|_{bh} + \|\xi\|_{c_0} \leq Q((h^x)^2 + (h^y)^2), \tag{3.11}
$$

holds, where Q is a positive constant which does not depend on the partition I_h.

Proof: Choose ϵ in (3.10) properly. Inequality (3.11) can be obtained from (3.4), (3.5), and (3.10) directly. $\qquad\qquad\qquad\qquad\qquad\qquad\qquad\Box$

4   Conclusions

We have proposed a five point upwind finite difference scheme on nonuniform rectangular partitions for convection-diffusion equations by using a new technique of handling the convective term. The proposed scheme is in the conservation form, satisfies the maximum value principle, and has a second-order error estimate. We have done some numerical experiments, which show that the high-order upwind scheme is as accurate as central methods for nonconvection-dominated problems and more accurate than full upwind methods for strong convection-dominated problems, in which the central methods do not work. Thus we can say that the proposed method can be used to simulate realistic fluid flow problems numerically.

References

[1] Aziz, K. and Settari, A., *Petroleum Reservoir Simulation*, Applied Science Publisher LTD, London, 1979.

[2] Bear, J., *Dynamics of Fluids in Porous Media*, American Elsevier Publishing Company, New York, 1972.

[3] Christe, I., Upwind compact finite difference schemes, *J. Comput. Phys.* **59** (1985), 353–368.

[4] Hegarty, A. F., O'Riordan, E., and Stynes, M., A comparison of uniformly comvergent difference schemes for two-dimensional convection-diffusion problems, *J. Comp. Phy.* **105** (1993), 24–32.

[5] Heinrich, J., Huyakorn, P., Zienkiewicz, O., and Mitchell, A., An 'upwind' finite element scheme for two-dimensional convective transport equation, *Inter. J. Numer. Methods Engrg.* **11** (1977), 131–143.

[6] Huyakorn, P. S. and Nikula, K., Solution of transient transport equation using an upstream finite element scheme, *Appl. Math. Model.* **3** (1979), 7–17.

[7] Lazarov, R. D., Mishev, I. D., and Vassilevski, P. S., Finite volume methods for convection-diffusion problems, *SIAM J. Numer. Anal.* **33** (1996), 31-55.

[8] Liang, D. and Zhao, W. D., A high-order upwind method for yhe convection-diffusion problem, *Comput. Methods Apil. Mech. Engrg.* **147** (1997), 105–115.

[9] Spalding, D. B., A novel finite difference formulation for differential equation involving first and second derivatives, *Int. J. Numer. Meth. Engrg.*, **7** (1973), 551–559.

[10] Tamamidis, P., A new upwind scheme on triangular meshes using the finite volume method, *Comput. Methods Appl. Mech. Engrg.* **124** (1995), 15–31.

Author Index

Subject Index

Druck: Strauss Offsetdruck, Mörlenbach
Verarbeitung: Schäffer, Grünstadt